Stellate Cells in
HEALTH AND DISEASE

Stellate Cells in
HEALTH AND DISEASE

Edited by

CHANDRASHEKHAR R. GANDHI

MASSIMO PINZANI

ELSEVIER

AMSTERDAM • BOSTON • HEIDELBERG • LONDON
NEWYORK • OXFORD • PARIS • SANDIEGO
SAN FRANCISCO • SINGAPORE • SYDNEY • TOKYO
Academic Press is an imprint of Elsevier

Academic Press is an imprint of Elsevier
32 Jamestown Road, London NW1 7BY, UK
525 B Street, Suite 1800, San Diego, CA 92101-4495, USA
225 Wyman Street, Waltham, MA 02451, USA
The Boulevard, Langford Lane, Kidlington, Oxford OX5 1GB, UK

Cover image: The image shows communication between hepatic stellate cells (green) and dendritic cells (red) in a liver section from mouse treated with Flt3 ligand to mobilize dendritic cells. The image was generated in the Center for Biologic Imaging, University of Pittsburgh by Donna Stolz in collaboration with Angus Thomson and Osamu Yoshida at TE Starzl Transplantation Instititute, University of Pittsburgh.

ISBN: 978-0-12-800134-9

British Library Cataloguing-in-Publication Data
A catalogue record for this book is available from the British Library.

Library of Congress Cataloging-in-Publication Data
A catalog record for this book is available from the Library of Congress.

For Information on all Academic Press publications
visit our website at http://store.elsevier.com/

Typeset by MPS Limited, Chennai, India
www.adi-mps.com

Back cover image: Created by Donna Stolz, Angus Thomson, and Osamu Yoshida

DEDICATION

This book is dedicated to Professor Albert "Bert" Geerts (1952–2009), one of the early pioneers of the research in hepatic stellate cell biology, who made several key contributions to the field of liver fibrosis.

The book is also dedicated to all the basic science and clinical fellows, students, and technicians whose diligent efforts helped reveal the many secrets of hepatic stellate cell pathophysiology.

CONTENTS

List of Contributors xiii
Foreword xv
Preface xix

1. History and Early Work **1**

Massimo Pinzani *and* Chandrashekhar R. Gandhi

1.1 Discovery of Hepatic Stellate Cells 1
1.2 HSCs and Vitamin A Homeostasis 3
1.3 Morphological Characteristics of HSCs 4
1.4 HSCs and Liver Fibrosis 6
1.5 Isolation and Culture of HSCs 6
1.6 Activation and Transdifferentiation of HSCs 7
1.7 Markers for HSCs 9
1.8 Perspective 11
References 11

2. Hepatic Stellate Cell Culture Models **15**

Krista Rombouts

2.1 Isolation of Hepatic Stellate Cells 15
2.2 Single Cell Culture 17
2.3 *In Vitro*- Versus *In Vivo*-Activated HSCs 18
2.4 Single Cell Culture and 2D: Importance of Adhesion, Arg–Gly–Asp, and Matrix Components 19
2.5 HSC Co-Cultures with Kupffer Cells, Hepatocytes, LSEC, HCC, and CC Cells 20
2.6 *In Vitro* 3D Culture Systems 21
2.7 Conclusions 23
References 23

3. Hepatic Fibrosis: A Global Clinical Problem **29**

Massimo Pinzani

3.1 Introduction 29
3.2 Chronic Viral Hepatitis 30
3.3 NAFLD and NASH 32
3.4 Autoimmune Hepatitis, Primary Biliary Cirrhosis, and Primary Sclerosing Cholangitis 33
3.5 Etiology-Driven Liver Fibrosis 34
References 38

4. Stellate Cells and Hepatic Fibrosis **41**

Daisuke Hasegawa, Michael C. Wallace *and* **Scott L. Friedman**

4.1 Introduction 41
4.2 Pathogenesis of Hepatic Fibrosis 42
4.3 HSCs and Hepatic Fibrosis 45
4.4 New and Emerging Pathways of HSC Activation 51
4.5 Conclusions 55
References 56

5. Cytokine Production and Signaling in Stellate Cells **63**

Fabio Marra *and* **Alessandra Caligiuri**

5.1 Introduction 63
5.2 Platelet-Derived Growth Factor 63
5.3 Angiogenic Cytokines 67
5.4 TGF-β Superfamily 68
5.5 Chemokines 70
5.6 Tumor Necrosis Factor Superfamily 72
5.7 Interleukins and Interferons 74
5.8 Adipokines and Other Cytokines Related to Metabolism 74
5.9 Osteopontin 77
5.10 Perspective 77
References 78

6. Stellate Cells, Portal Myofibroblasts, and Epithelial-to-Mesenchymal Transition **87**

Yukinori Koyama, Ping Wang, David A. Brenner *and* **Tatiana Kisseleva**

6.1 Introduction 87
6.2 Hepatic Stellate Cells 89
6.3 Portal Fibroblasts 97
6.4 Epithelial-to-Mesenchymal Transition 99
6.5 Perspective 100
References 100

7. Matrix Metalloproteinases and Their Inhibitors **107**

Lara Campana *and* **John Iredale**

7.1 Introduction 107
7.2 The Metalloproteases (MMPs) 109
7.3 Relevance of MMP Activity and Matrix Degradation in Chronic Liver Disease 114
7.4 Conclusions 121
References 121

8. Stellate Cells and Portal Hypertension **125**

Don C. Rockey

Abbreviations 125
8.1 Overview 125
8.2 Cell Culture Based Studies 127
8.3 *In Vivo* Studies 129
8.4 Vasoactive Mediators and Stellate Cells 129
8.5 Therapeutic Implications 139
Acknowledgment 139
References 139

9. Hepatic Stellate Cells and Liver Cancer **145**

Dianne H. Dapito *and* **Robert F. Schwabe**

9.1 Fibrosis and HCC 145
9.2 The Premalignant Microenvironment 145
9.3 The Tumor Microenvironment 146
9.4 HSCs as Potential Contributors to Cancer Promotion 147
9.5 Mechanisms by Which HSCs Promote HCC 150
9.6 Concluding Remarks 158
References 159

10. Stellate Cells in Alcoholic Hepatitis **163**

Palash Mandal *and* **Laura E. Nagy**

10.1 Introduction 163
10.2 Pathophysiology of Fibrosis in ALD 163
10.3 HSC and ASH 165
10.4 Direct Effects of Ethanol and its Metabolites on HSC 166
10.5 Interactions of HSCs with Innate Immune Systems in ASH 168
10.6 Interactions of HSCs with Adaptive Immunity in ASH 169
10.7 Summary: Potential Therapeutic Strategies for Reducing HSC Activation in ASH 170
References 170

11. Hepatic Stellate Cells as Target for Reversal of Fibrosis/Cirrhosis **175**

Sophie Lotersztajn *and* **Ariane Mallat**

11.1 Introduction 175
11.2 Fibrosis Reversibility: General Overview 176
11.3 Mechanisms of HSC Clearance During Fibrosis Resolution 177
11.4 Control of HSC Fate and Matrix Degradation by Neighboring Liver Cells During Fibrosis Resolution 179

11.5 Conclusion 182
Acknowledgments 183
References 183

12. Interactions of Stellate Cells with Other Non-Parenchymal Cells **185**
Norifumi Kawada *and* Maurizio Parola

12.1 Hepatic Stellate Cells: Introductory Remarks 185
12.2 The Crosstalk of HSCs with Macrophages 186
12.3 HSC/MFs and Interactions with Other Cells of Innate and Adaptive Immunity 190
12.4 Interactions of HSCs and HSC/MFs with LSECs: From Liver Specific Pericytes to Pro-Angiogenic Cells 196
12.5 Interactions of HSCs and Portal Fibroblasts with Cholangiocytes 199
12.6 Concluding Remarks 202
References 202

13. Hepatic Stellate Cells and Hepatocyte Survival **209**
Chandrashekhar R. Gandhi

13.1 Introduction 209
13.2 HSC Phenotypes and Liver Injury 210
13.3 HSCs and Liver Regeneration 211
13.4 HSCs in Hepatocyte Injury During Endotoxemia 212
13.5 HSC Depletion Model to Study Acute Liver Injury 217
13.6 Perspective 219
References 219

14. Stellate Cells in Hepatic Immunological Tolerance **227**
Chandrashekhar R. Gandhi

14.1 Hepatic Immune System 227
14.2 Liver Tolerogenicity 228
14.3 Importance of HSCs in Transplantation 230
14.4 HSCs Produce Inflammatory and Immunoregulatory Cytokines and Chemokines 232
14.5 HSC–DC Interactions 234
14.6 HSCs and Conventional Effector T Cells 236
14.7 HSCs and Regulatory T Cells 239
14.8 HSCs and NKT Cells 242
14.9 Perspective 242
Acknowledgment 243
References 243

15. Stellate Cell Depletion Models **251**

Fiona Oakley *and* **Derek A. Mann**

15.1 Introduction 251
15.2 Immune Cell Mediated Clearance of HM 252
15.3 HM Intrinsic Signaling Events that Limit Scar Cell Survival 253
15.4 Chemical Systems 254
15.5 Development of Carriers to Selectively Target Therapeutics to HMs 255
15.6 Advantages of HM Targeted Delivery Vehicles for Therapy 256
15.7 Nanoparticles and Viral Delivery Systems 260
15.8 Genetic Systems 262
15.9 The Role of HM Depletion in Liver Regeneration and Cancer 265
15.10 Future Perspectives 266
References 267

16. Pancreatic Stellate Cells **271**

Minoti V. Apte, Romano C. Pirola *and* **Jeremy S. Wilson**

16.1 Introduction 271
16.2 Pancreatic Stellate Cells—Isolation and Characterization 272
16.3 Role of PSCs in Pancreatic Disease 277
16.4 PSCs in Acute Pancreatitis 281
16.5 PSCs in Chronic Pancreatitis 285
16.6 Reversal of Pancreatic Fibrosis in Chronic Pancreatitis 288
16.7 PSCs in Pancreatic Cancer 289
16.8 Summary and Conclusions 298
Acknowledgments 299
References 299

Index 307

LIST OF CONTRIBUTORS

Minoti V. Apte
Pancreatic Research Group, South Western Sydney Clinical School, University of New South Wales, Ingham Institute for Applied Medical Research, Sydney, Australia

Daisuke Hasegawa
Division of Liver Diseases, Icahn School of Medicine at Mount Sinai, New York, NY, USA

Michael C. Wallace
Division of Liver Diseases, Icahn School of Medicine at Mount Sinai, New York, NY, USA; School of Medicine and Pharmacology, University of Western Australia, Perth, Australia

Scott L. Friedman
Division of Liver Diseases, Icahn School of Medicine at Mount Sinai, New York, NY, USA

David A. Brenner
Department of Medicine, University of California, San Diego, La Jolla, CA, USA

Alessandra Caligiuri
Dipartimento di Medicina Sperimentale e Clinica, University of Florence, Italy

Lara Campana
MRC/UoE Centre for Inflammation Research Queen's Medical Research Institute, Edinburgh, UK

Dianne H. Dapito
Institute of Human Nutrition, Columbia University, New York, NY, USA

Chandrashekhar R. Gandhi
Department of Pediatrics, Cincinnati Children's Hospital Medical Center, Department of Surgery, University of Cincinnati and Cincinnati VA Medical Center, Cincinnati, OH, USA

John Iredale
MRC/UoE Centre for Inflammation Research Queen's Medical Research Institute, Edinburgh, UK

Norifumi Kawada
Department of Hepatology, Graduate School of Medicine, Osaka City University, Osaka, Japan

Tatiana Kisseleva
Department of Surgery, University of California, San Diego, La Jolla, CA, USA

Yukinori Koyama
Department of Medicine University of California, San Diego, La Jolla, CA, USA; Department of Target Therapy Oncology, Graduate School of Medicine, Kyoto University, Kyoto, Japan

Sophie Lotersztajn
INSERM U1149, Center for Research on Inflammation, Paris, France; Université Paris Diderot, Paris, France; Laboratory of Excellence Inflammex, Paris, France

Ariane Mallat
AP-HP, Hôpital Henri Mondor-Albert Chenevier, Service d'Hépatologie, Créteil, France; INSERM U955, Créteil, France; Université Paris-Est, Créteil, France

Palash Mandal
Department of Biological Sciences, BITS PILANI Hyderabad Campus, Hyderabad, India

Derek A. Mann
Fibrosis Research Group, Institute of Cellular Medicine, Newcastle University, Newcastle upon Tyne, UK

Fabio Marra
Dipartimento di Medicina Sperimentale e Clinica, University of Florence, Italy

Laura E. Nagy
Center for Liver Disease Research, Department of Pathobiology, Cleveland Clinic, Cleveland, OH, USA, Departments of Molecular Medicine and Nutrition, Case Western Reserve University, Cleveland, OH, USA

Fiona Oakley
Fibrosis Research Group, Institute of Cellular Medicine, Newcastle University, Newcastle upon Tyne, UK

Maurizio Parola
Department of Clinical and Biological Sciences, Unit of Experimental Medicine and Clinical Pathology, School of Medicine, University of Torino, Torino, Italy

Massimo Pinzani
UCL Institute for Liver and Digestive Health, Division of Medicine, University College London, Royal Free Hospital, London, UK

Romano C. Pirola
Pancreatic Research Group, South Western Sydney Clinical School, University of New South Wales, Ingham Institute for Applied Medical Research, Sydney, Australia

Don C. Rockey
Department of Internal Medicine, Medical University of South Carolina, Charleston, SC, USA

Krista Rombouts
UCL Institute for Liver and Digestive Health, Division of Medicine, University College London, Royal Free Hospital, London, UK

Robert F. Schwabe
Institute of Human Nutrition, Columbia University, New York, NY, USA; Department of Medicine, Columbia University, New York, NY, USA

Ping Wang
Department of Medicine, University of California, San Diego, La Jolla, CA, USA; Department of Surgery, University of California, San Diego, La Jolla, CA, USA; Beijing Friendship Hospital, Capital Medical University, Beijing, China

Jeremy S. Wilson
Pancreatic Research Group, South Western Sydney Clinical School, University of New South Wales, Ingham Institute for Applied Medical Research, Sydney, Australia

FOREWORD

D. Montgomery Bissell MD
Professor, Division of Gastroenterology, University of California, San Francisco, CA, USA

Stellate cells were described several decades ago under a number of names: Ito cells, fat-storing cells, lipocytes, and perisinusoidal cells. In diseased liver, they were found together with collagen and postulated to be a source of fibrotic scar [1,2]. In the early 1980s, methods for the isolation and culture of hepatic stellate cells were developed that allowed direct study of their role in fibrosis [3,4]. The early work showed that the capacity for fibrogenesis is latent in normal stellate cells and emerges rapidly in culture [5]. This change in phenotype was termed *activation*. It occurred spontaneously in culture, becoming evident within the first few hours *ex vivo*. Similar changes were then documented in vivo [6]. At that time, because specific treatment for the most prevalent types of chronic liver injury was quite limited, interest in stellate cells grew rapidly. A number of laboratories took up the challenge of further defining the role of these cells.

WHAT HAVE WE LEARNED?

Progress has been impressive, as demonstrated by the chapters in this volume, each by an expert in the subject. The main focus has been stellate cell regulation in its several facets: cytokine secretion, matrix production, migration, proliferation, contraction, and immune function. Although the search for a general anti-fibrotic therapy goes on, several broad concepts have emerged that have changed the management of liver injury.

The specific composition of fibrotic scar varies dynamically and has biological effects. Prior to the current era, it was held that injury scar was essentially collagen, a proline-rich, rather featureless, helical protein, which lent tensile strength to the repairing wound but had little or no direct biological activity. The pathophysiology of severe scarring was ascribed solely to disruption of tissue structure by thick collagenous bands, which caused portal hypertension and its complications and were permanent. Based on this, it appeared that treatment for fibrosis, if feasible, would do little to restore liver function.

An alternative view of extracellular matrix (ECM) complexes began to emerge in the 1960s. Studies revealed that the ECM was highly heterogeneous at the molecular level, depending on location within an epithelium and the physiological context (embryonic, growing, nongrowing, injured, and so on). Moreover, a number of model systems indicated biological effects of matrix components [7,8]. The perisinusoidal space of the normal liver was shown to contain basement-membrane constituents including type IV collagen, which is structurally very different from type I collagen. The latter was found to be present but sparse in normal liver, present largely at vascular branch points [9]. In injury, however, collagen I along with collagen III and novel glycoproteins such as the EDA isoform of fibronectin [10] appeared in the subendothelial space. Associated with the 'injury' ECM were morphological events including effacement of hepatocyte microvilli and loss of sinusoidal fenestrae [11]. Finally, culture studies with normal primary hepatocytes indicated that expression of hepatocyte-specific function is strikingly dependent on the type of ECM in contact with the cell [9]. The data overall suggested that hepatocellular function depends on both the localization of injury matrix and its composition.

Fibrosis is dynamic and reversible. Historically cirrhosis was viewed as permanent. This was despite evidence to the contrary from experimental injury models and some human reports [12]. A sizable body of clinical data now has confirmed that fibrosis – and, occasionally, cirrhosis – resolves with elimination of the injury agent. In chronic liver disease of many years' duration, the signs of cirrhosis may persist, although functional improvement with treatment of the cause is seen and even impressive. This suggests that inflammation and acute fibrosis are as important as collagenous bands in their negative impact on hepatocellular function.

With this new understanding of fibrosis lability, the outlook for many patients with active cirrhosis has changed. Rather than being regarded as beyond help ('end-stage'), they have become prime candidates for treatment when a specific remedy exists.

Pathological fibrosis represents counterproductive (overactive, excessive) wound repair. The new ECM in injury, which derives from a number of cell types, constitutes the platform on which stellate cells become activated and proceed with repair. When the injury is limited in time (acute hepatitis A, for example), inflammation subsides as the noxious agent is controlled. In the final stages of repair, the platform is dismantled with takedown of the injury ECM and 'deactivation' or apoptosis of stellate cells [13]. The result is complete, or nearly complete, restoration of normal tissue structure. On the other hand, if the injury agent lingers and inflammation persists, recurrent cycles of repair lead to excess fibrosis, ultimately with symptoms of hepatic insufficiency.

THE OUTLOOK FOR ANTI-FIBROTIC TREATMENT

Wound repair is essential to life. Therefore, anti-fibrotic interventions that cripple or ablate the process are unlikely to be adopted because of safety concerns. A safer approach will be manipulations that decelerate the *normal* production of 'injury' matrix. This is effectively what happens with neutralization of an injury agent – recent examples being chronic hepatitis B [14]. and hepatitis C [15]. For diseases in which the pathogenic factor is unknown, the focus of treatment should be upstream initiators of wound repair, where intervention would be most likely to replicate the effect of removing the injury stimulus. Identifying the key initiators will involve linking immune cell activity and inflammatory signals to stellate cell activation. The chapters in this book are pointing the way.

REFERENCES

[1] Kent G, Gay S, Inouye T, Bahu R, Minick OT, Popper H. Vitamin A-containing lipocytes and formation of type III collagen in liver injury. Proc Natl Acad Sci U S A 1976;73:3719–22.

[2] Mak KM, Leo MA, Lieber CS. Alcoholic liver injury in baboons: transformation of lipocytes to transitional cells. Gastroenterology 1984;87:188–200.

[3] Knook DL, Seffelaar AM, de Leeuw AM. Fat-storing cells of the rat liver. their isolation and purification. Exp Cell Res 1982;139:468–71.

[4] de Leeuw AM, McCarthy SP, Geerts A, Knook DL. Purified rat liver fat-storing cells in culture divide and contain collagen. Hepatology 1984;4:392–403.

[5] Friedman SL, Roll FJ, Boyles J, Bissell DM. Hepatic lipocytes: the principal collagen-producing cells of normal rat liver. Proc Natl Acad Sci U S A 1985;82:8681–5.

[6] Maher JJ, McGuire RF. Extracellular matrix gene expression increases preferentially in rat lipocytes and sinusoidal endothelial cells during hepatic fibrosis in vivo. J Clin Invest 1990;86:1641–8.

[7] Meier L, Hay ED. Stimulation of corneal differentiation by interaction between cell surface and extracellular matrix. I. Morphometric analysis of transfilter "induction". J Cell Biol 1975;66:275–91.

[8] Bissell MJ, Hall HG, Parry G. How does the extracellular matrix direct gene expression? J Theor Biol 1982;99:31–68.

[9] Bissell DM, Arenson DM, Maher JJ, Roll FJ. Support of cultured hepatocytes by a laminin-rich gel. Evidence for a functionally significant subendothelial matrix in normal rat liver. J Clin Invest 1987;79:801–12.

[10] Jarnagin WR, Rockey DC, Koteliansky VE, Wang SS, Bissell DM. Expression of variant fibronectins in wound healing: cellular source and biological activity of the EIIIA segment in rat hepatic fibrogenesis. J Cell Biol 1994;127:2037–48.

[11] McGuire RF, Bissell DM, Boyles J, Roll FJ. Role of extracellular matrix in regulating fenestrations of sinusoidal endothelial cells isolated from normal rat liver. Hepatology 1992;15:989–97.

[12] Perez-Tamayo R. Cirrhosis of the liver: a reversible disease? Pathol Annu 1979;14(Pt 2):183–213.

[13] Gaca MD, Zhou X, Issa R, Kiriella K, Iredale JP, Benyon RC. Basement membrane-like matrix inhibits proliferation and collagen synthesis by activated rat hepatic stellate cells: evidence for matrix-dependent deactivation of stellate cells. Matrix Biol 2003;22:229–39.

[14] Yao FY, Terrault NA, Freise C, Maslow L, Bass NM. Lamivudine treatment is beneficial in patients with severely decompensated cirrhosis and actively replicating hepatitis B infection awaiting liver transplantation: a comparative study using a matched, untreated cohort. Hepatology 2001;34:411–6.

[15] D'Ambrosio R, Aghemo A, Rumi MG, et al. A morphometric and immunohistochemical study to assess the benefit of a sustained virological response in hepatitis C virus patients with cirrhosis. Hepatology 2012;56:532–43.

PREFACE

One and a half centuries ago von Kupffer identified an interesting group of cells that he initially called "perivascular cells of the connective tissue." These were subsequently identified as fat-storing cells or stellate cells. The field of stellate cells has expanded exponentially in the last four decades, primarily due to increasing evidence that they are the major cell type to cause liver fibrosis. Thus, hepatic stellate cells have become synonymous with liver fibrosis and targeting them is a realistic approach to treating fibrosis and cirrhosis of the liver.

Overwhelming interest in stellate cells research is evident with nearly 10,000 research articles including nearly 1,000 reviews (and several book chapters) published to date. From about 100 research articles before the 1970s, the number of publications quadrupled in the next decade and has consistently doubled in each of the next three decades. Another 2,000 articles have been published since 2011. With initial interest as a major cell type of storage and metabolism of retinoids (including vitamin A), and subsequently for its function in hepatic fibrosis, stellate cells continued to guard its secrets in regard to its role in other facets of hepatic pathophysiology. Emerging evidence in the last 15 years has revealed its role in liver cell (hepatocyte) survival and regeneration, hepatic inflammation, and immunobiology including transplantation tolerance and hepatocellular carcinoma.

All 16 chapters of this book have been contributed by leading experts in their respective fields, and cover the entire spectrum of stellate cell biology known to date. Thus they provide an up-to-date comprehensive treatise and perspectives of the established and putative functions of stellate cells in various pathophysiological conditions that affect the structure and functions of the liver. An independent chapter has also been devoted to characteristics of stellate cells in pancreatic pathobiology. This book is expected to serve as a reference to researchers in the field (cell and molecular biologists, hepatologists, gastroenterologists, and liver surgeons) as well as those who have an interest in hepatic biology and pathobiology. It is also intended to be used for graduate courses and for training residents and fellows in Gastroenterology and Hepatology.

Chandrashekhar Gandhi
Massimo Pinzani

CHAPTER 1

History and Early Work

Massimo Pinzani[1] and Chandrashekhar R. Gandhi[2]
[1]UCL Institute for Liver and Digestive Health, Division of Medicine, University College London, Royal Free Hospital, London, UK
[2]Department of Pediatrics, Cincinnati Children's Hospital Medical Center, Department of Surgery, University of Cincinnati and Cincinnati VA Medical Center, Cincinnati, OH, USA

1.1 DISCOVERY OF HEPATIC STELLATE CELLS

The discovery of hepatic stellate cells (HSCs) is credited to a German anatomist Karl Wilhelm von Kupffer (Figure 1.1) who, while studying the liver's nervous system, observed star-shaped, gold, chloride-stained cells within the hepatic lobules (Figures 1.2 and 1.3). Because these cells were localized around blood capillaries within the liver (now known as sinusoids), von Kupffer described them as *perivasculäre bindgewebszellen* (perivascular cells of the connective tissue) in an 1876 letter to his colleague Waldeyer [3]. While continuing his investigation of these "*perivasculäre bindgewebszellen*," von Kupffer observed uptake of intravenously injected India ink or sheep red blood cells by the rabbit liver cells stained with gold chloride and he changed his earlier opinion, concluding that they were phagocytes and also called them "specialized endothelial cells of the sinusoids" [4]. In a paper published in 1912, Pearce and Austin from the University of Pennsylvania, USA [5] referred to the liver cells that phagocytosed red blood cells in splenectomized dogs as "stellate endothelial cells" or "Kupfer's" (instead of Kupffer's) cells. Such a multitude of nomenclature continued for several more decades. In 1952, Toshio Ito of Keiko University, Japan, found that the cells in the perisinusoidal space contained abundant fat droplets (which were originally observed by von Kupffer), and were distinct from the phagocytosing cells in the sinusoids (now known as Kupffer cells). He called them "fat-storing cells" [6]. Almost two decades passed before Kenjiro Wake at Osaka University, Japan, using gold chloride and silver impregnation technique and vitamin A autofluorescence, confirmed that von Kupffer and Ito had described the same cells and further observed that they contained well-developed rough endoplasmic reticulum [7]. This observation in 1971 suggested the strong ability of stellate cells to synthesize proteins. In a subsequent study 3 years later, Wake reported the presence of cytosolic lipid droplets in stellate cells that were differentiated into two types: Type I droplets were electron dense, variable in size (up to 2 μm in diameter), membrane-bound, and appeared to be derived from multivesicular bodies, whereas Type II droplets were larger and uniform in size [8]. Type I lipid

Stellate Cells in Health and Disease
DOI: http://dx.doi.org/10.1016/B978-0-12-800134-9.00001-4

Figure 1.1 Karl von Kupffer.

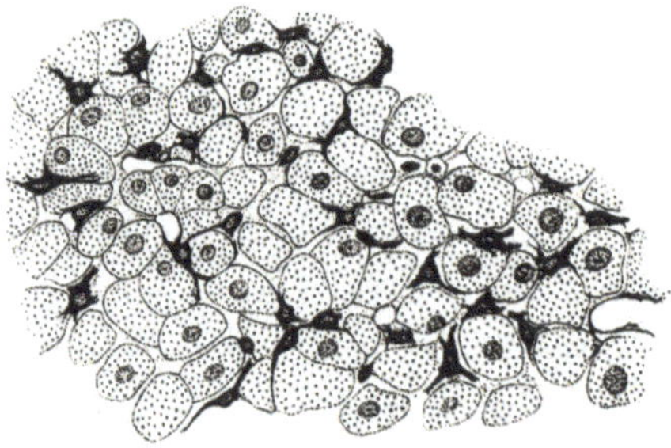

Figure 1.2 Reproduction of von Kupffer's drawing of the HSCs (Sterzellen) in dog liver. *From Ref. [1] (Courtesy of Prof. Kenjiro Wake.)*

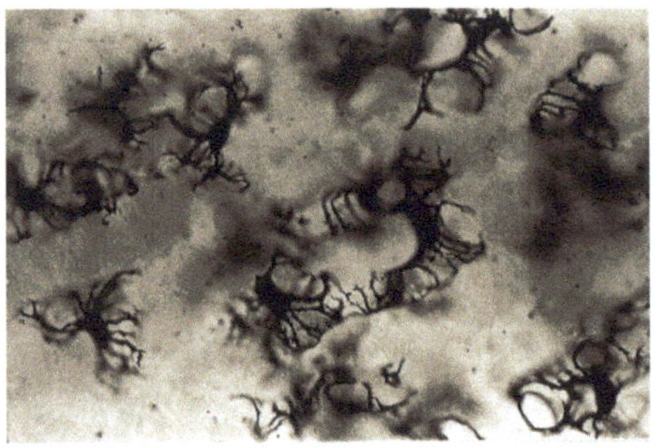

Figure 1.3 Stellate cells in the porcine liver. The long cytoplasmic processes encompass sinusoids at regular intervals. Golgi's silver method. × 840. *From Ref. [2] (Courtesy of Prof. Kenjiro Wake.)*

droplets were found to accumulate in the cells of animals with hypervitaminosis A in the intermediate and central zones of the liver lobule [1,8]. The number of stellate cells containing lipid droplets was estimated to be up to 75% both in rat and human [1,9]. The lipid droplets in isolated stellate cells contained high concentrations of both retinol and retinyl palmitate [10]. The observation of vitamin A stores (lipid droplets) in the 1980s led investigators to refer to stellate cells variously as fat-storing cells, lipocytes, perisinusoidal cells, and Ito cells (after Toshio Ito). Due to potential confusion arising from such different nomenclature, a consensus was reached in 1996 to call this cell type as HSC [11]. It is important to appreciate the efforts of von Kupffer in distinguishing these cells with the very basic techniques available at that time. Thus it was only apt that hepatic resident macrophages were named "Kupffer cells" in recognition of von Kupffer's original contribution to the field.

1.2 HSCs AND VITAMIN A HOMEOSTASIS

Wake's work in the 1970s established that HSCs were the major storage site of vitamin A (nearly 80% of the body's retinoids is stored in these cells). This was confirmed in 1982 by Knook and co-workers of the Institute of Gerontology, the Netherlands. They found that HSCs isolated from the livers of 12-month-old rats contained much higher levels of retinoids (32 μg retinol and 685 μg retinyl palmitate per mg cellular protein) as compared with the whole liver homogenate (1 μg retinol and 15 μg retinyl palmitate per mg protein) [12]. Further analysis revealed that HSCs contained 300-fold greater amount of retinyl esters than hepatocytes [13]. Approximately 42% retinyl ester, 28% triglycerides, 13% cholesterol, and 4% phospholipids were estimated to constitute the lipid droplets in HSCs [14]. In another study, lipid droplets isolated from HSCs were found to be much larger (up to 8 μm diameter) than those isolated from parenchymal cells (up to 2.5 μm diameter), but both contained similar concentration of neutral lipids [15]. Whereas retinyl- and cholesteryl esters comprised 65% and triglycerides 20% of the lipid droplet fractions of HSCs, lipid droplet fractions of the parenchymal cells comprised 62% triglycerides and up to 30% cholesteryl esters [15]. These findings in the 1980s opened up a major area of investigation to determine the role of HSCs in vitamin A homeostasis, which, as we know now, is critical in maintaining the liver's physiologic function [16]. The loss of retinoids by HSCs is paralleled by their transformation into a highly proliferative, contractile, and fibrogenic myofibroblastic phenotype that is responsible for liver fibrosis, cirrhosis, portal hypertension, and also implicated in playing a role in hepatocellular carcinoma (see Chapters 4, 6, 8, and 9).

A large amount of work on vitamin A homeostasis by HSCs has been performed by Blomhoff at the University of Oslo, Norway. In the intestine, retinol is esterified and retinyl esters are packaged in chylomicrons for delivery to the liver via the lymph. Hepatocytes are the primary cell type to endocytose retinyl esters, which are transferred to adjacent HSCs upon binding to specific retinol binding proteins [17]. However, HSCs were shown to possess an ability to directly take up retinol and its esters (e.g., retinyl acetate), as demonstrated by *in vitro* experiments [18,19]. Intracellular retinol-binding protein (CRBP), whose concentration in HSCs is almost

20 times greater than in hepatocytes, regulates the uptake, storage as well as the mobilization of retinoids to meet the peripheral demand [20,21]. In addition to CRBP, HSCs also contain other proteins related to retinoid metabolism such as cellular retinoic acid-binding protein, retinol palmitate hydrolase, bile salt-dependent and -independent retinol ester hydrolase, and acyl coenzyme A:retinol acyltransferase [22,23]. Furthermore, HSCs have been shown to express nuclear retinoid receptors including retinoic acid receptor (RAR) α, β, and γ [23,24], and retinoid X receptor (RXR) α and β [25]. Thus HSC is not just a cell type that simply stores and releases retinoids, but it also regulates their physiological and pathophysiological metabolism.

1.3 MORPHOLOGICAL CHARACTERISTICS OF HSCs

An elaborate electron microscopic examination of the bat liver performed by Tanuma and Ito in 1978 described the special arrangement of the three main nonparenchymal cell types, namely sinusoidal endothelial cells, Kupffer cells, and HSCs [26]. Accordingly, the cytoplasmic extensions of sinusoidal endothelial cells consist of continuous thicker parts (cytoplasmic processes) and discontinuous thinner parts (sieve plates) containing fenestrations that allow a free exchange of molecules between the blood flowing through the sinusoids and hepatocytes. Figure 1.4 (transmission electron micrograph of rat liver section) shows some of the characteristics of the interactions of HSC with other hepatic cells. The endothelium is single layered and devoid of basal lamina, with junctions between closely apposed margins of the cytosolic processes. Kupffer cells adhering to the sinusoidal walls protrude into the sinusoidal lumen and contain many microvillous pseudopods. HSCs localized in the space of Disse contain fat droplets, large Golgi complex and well developed rough endoplasmic reticulum filled with electron dense material. In

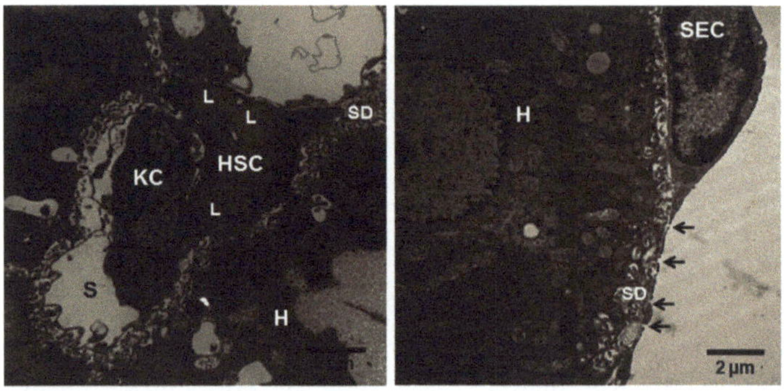

Figure 1.4 Transmission electron micrographs of rat liver. (Left) An HSC containing lipid droplets (L) is seen in close association with a Kupffer cell (KC), and making physical contact with the microvilli of a hepatocyte (H). S, sinusoid; SD, space of Disse. (Right) A sinusoidal cell (SEC) lining the sinusoid. Endothelial fenestrations (arrows) are clearly visible. The images were kindly provided by Dr. Donna Stolz, Center for Biologic Imaging, University of Pittsburgh.

normal human liver also HSCs located in the space of Disse were seen making direct physical contacts with hepatocytes and endothelial cells, and their nucleus-to-nucleus distance was estimated to be about 40 μm [27] (Figure 1.5).

Zonal distribution studies of human liver performed in 1981 by Giampieri et al. showed that the number of HSCs was higher in the centrilobular than in the periportal region [28]. HSCs in the periportal zone 1 were found to be relatively small with short cytosolic processes and contained smaller lipid droplets, those in zone 2 displayed longer branching processes and abundant vitamin A stores in the lipid droplets, and those in centrilobular zone 3 demonstrated dendritic appearance and progressively lower vitamin A content [27,29]. However, administration of excess vitamin A to rats caused an increase in the number and size of vitamin A-lipid droplets in HSCs located in the central zone, indicating that the intralobular heterogeneity may reflect differences in the metabolic handling of vitamin A. Such heterogeneity was also apparent with diffuse immunostaining for desmin (a marker for HSCs) in zone 1, intense staining in zone 2, and progressively diminishing staining moving toward the centrilobular vein [27,29]. HSCs were found to extend cytoplasmic processes ramifying beneath the endothelial lining, occasionally surrounding the sinusoids almost completely. These processes of HSCs seemed to reinforce the endothelial lining, potentially providing a regulation of the sinusoidal tone. In this regard, the subendothelial processes of HSCs were found to be equipped with massive actin-like filaments [1]. In fact, subsequent work demonstrated contractile properties of HSCs [30,31] and their role in contractile component of portal hypertension, which is almost always present in advanced cirrhosis (see Chapter 8). With their number (up to 10% of the total number of liver cells) [12,28] and ability to form physical association with all of the hepatic cell types, it became apparent that HSCs form a unique cell population organized in a three-dimensional network with the potential to closely regulate several important functions of the liver.

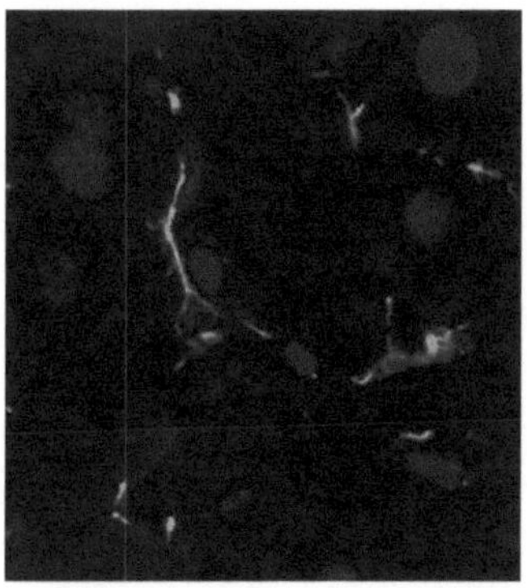

Figure 1.5 A mouse liver section stained for GFAP, a marker of HSC. With long cytosolic processes, one HSC can physically contact 1–2 hepatocytes as well as sinusoidal endothelial and Kupffer cells (40× magnification).

1.4 HSCs AND LIVER FIBROSIS

Although the focus of research in the 1970s was to understand the mechanisms of uptake, storage, composition, and release of retinoids by HSCs, evidence started accumulating that these cells could be a major source of extracellular matrix deposited in the liver during fibrogenesis [32–35]. McGee and Patrick (Royal Infirmary, Glasgow) in 1972 [32,33] observed massive hepatocellular necrosis that preceded fibrogenesis in rats injected with carbon tetrachloride, and suggested that HSCs were responsible for the latter event. Subsequently in 1977, Kent (University of Chicago, Illinois, USA), Popper (Mount Sinai School of Medicine, New York, USA) and co-workers demonstrated accumulation of transitional cells with morphological features of HSCs (lipocytes) and fibroblasts along the fibrous septa during carbon tetrachloride-induced liver injury in rats that also received subcutaneous injections of vitamin A [34]. Because collagen type III deposits were associated with these transitional cells, the authors postulated that they originated from HSCs. The relevance of these findings to humans became apparent later in a 1984 study with nonhuman primates. The hypothesis that HSCs were the precursors of the "transitional cells" that caused hepatic fibrosis was drawn based on the cellular depletion of lipid droplets and hypertrophy of endoplasmic reticulum observed in experiments performed in baboons maintained on a diet containing alcohol [35]. Investigation of HSCs as the primary fibrogenic cells during wound healing, hepatic fibrosis, and cirrhosis became, and continues to be, a major area of research described in Chapter 4. We now know that inflammatory mediators and reactive oxygen species produced by infiltrating blood cells (e.g., neutrophils and monocytes) as well as Kupffer cells and apoptotic bodies derived from dying hepatocytes are major stimuli for the transformation of HSCs into fibrogenic phenotype. It is interesting to note that McGee and Patrick [32,33] postulated a link between hepatocellular necrosis and fibrosis as early as 1972, although this reality remained obscure until the late 1980s.

1.5 ISOLATION AND CULTURE OF HSCs

Until the early 1980s, knowledge of the functions of HSCs was limited. Based on the ultrastructural characteristics, they resembled fibroblasts and there were strong indications for an active role of these cells in hepatic fibrosis [1,32–35]. Thus, a major breakthrough to determine precisely the functions of HSCs required isolation and culture of highly purified cells. In retrospect, this was critical because contaminating Kupffer cells, endothelial cells, or hepatocytes can influence the properties of HSCs. In the 1980s, attempts were made to isolate HSCs based on their relatively lower density (high lipid content) than other hepatic cells. In 1980, Otto and Veech at the National Institute on Alcohol Abuse and Alcoholism [10] reported isolation of HSCs from the livers of rats treated with vitamin A to increase the number and size of lipid droplets. The livers were digested with pronase, and total nonparenchymal cell fraction was separated by differential centrifugation; this was followed by purification of HSCs by centrifugation in a linear metrizamide gradient [10]. Based on the quickly fading green autofluorescence at 330 nm due to vitamin A, the procedure yielded a fourfold enrichment of HSCs in the cell fraction with gradient density of 1.043 g/mL. Shortly afterwards, Knook et al. [12] employed similar metrizamide density gradient centrifugation technique to isolate HSCs from rats without prior vitamin A treatment, and purified them from other nonparenchymal cells by centrifugal elutriation. The yield of HSCs from the livers of old animals (12 months of age) was much superior as compared with younger animals (3 months of age). Most investigators now use old rats and mice (retired breeders) for isolation of HSCs. However, even with the centrifugal elutriation procedure [12], the purity of HSCs was estimated to be only about 70–75%, the major contaminants being lymphocytes and endothelial cells. In 1985, Friedman et al. (Liver Center, University of California, San Francisco) achieved > 95% pure HSC preparation from rat liver [36]. The investigators digested the livers of rats pre-treated with retinyl acetate with pronase and collagenase; following separation of nonparenchymal cells by low-speed centrifugation, they used stractan (arabinogalactan) gradient (6%, 8%, 12%, and 20%) and obtained highly purified HSC fraction between the medium and 6% stractan [36]. These procedures or modifications thereof have become established techniques and are now routinely used by researchers to isolate HSCs from various species including human. Details of various procedures of HSC isolation and purification are addressed in Chapter 2.

1.6 ACTIVATION AND TRANSDIFFERENTIATION OF HSCs

In their follow-up investigation, Knook's group demonstrated successful culture of HSCs that divided, and could be established as cell lines for at least two passages [37]. The cells contained vimentin (indicating fibroblastic or myogenic origin), actin filaments, collagen type I and IV, and laminin in first passage and only collagen type I in the fourth passage. Because prolonged culture of HSCs was associated with gradual disappearance of vitamin A and phenotypic transition to more fibroblastic phenotype, the authors concluded that they might be the cell type involved in liver fibrosis, a view proposed much earlier (1972) by McGee and Patrick [32,33]. This was later confirmed by Friedman et al. using highly purified HSCs; they further demonstrated that the fibrogenic potential of HSCs, as assessed by collagen synthesis, was far superior to that of hepatocytes and endothelial cells [36]. It was further demonstrated in 1987 that highly purified desmin-positive rat HSCs lost retinoids progressively upon repeated sub-culture as they transdifferentiated into the myofibroblast-like phenotype [38]. *In vivo* experiments with carbon tetrachloride-treated rats also showed significant reduction in the hepatic vitamin A content in association with increased fibrosis, providing additional evidence that the loss of vitamin A might be a prerequisite for the transition of HSCs into the fibrogenic phenotype [39]. In fact, administration of vitamin A to rats was found to suppress carbon tetrachloride-induced fibrosis [39]. This issue (loss of retinoids and change of HSC phenotype) is addressed in detail in Chapter 4. Even in lower vertebrates, such as lamprey, evidence for the role of HSCs in hepatic fibrosis during biliary atresia has been presented [40].

In situ, HSCs have a small cell body from which cytosolic processes radiate and make physical contacts with surrounding cells (Figure 1.5). Upon isolation and culture, HSCs appear spherical and stain strongly for lipid droplets with oil red O (Figure 1.6A), but start flattening and generating cytosolic processes shortly in culture in association with the loss of lipid droplets (Figure 1.6B). Prolonged culture or passaging induces increase in size and myofibroblastic appearance with very low or absence of oil red O staining (Figure 1.6C and D). The process of the loss of retinoids and transformation of HSCs into myofibroblasts is known as "activation." Such spontaneous activation in culture indicated that HSCs produced factors that initiated and perpetuated their activation and proliferation, and provided an excellent model to discover extracellular factors that induce or accelerate the process of activation/transdifferentiation and the underlying intracellular signaling mechanisms. For culture-induced activation, it is essential to incubate HSCs in a medium supplemented by 5–10% serum. The cells at different stages of activation can be used to characterize their functions and relevance to the stages of liver injury, bearing in mind that HSCs are a highly heterogeneous population. It should also be considered that the findings derived from

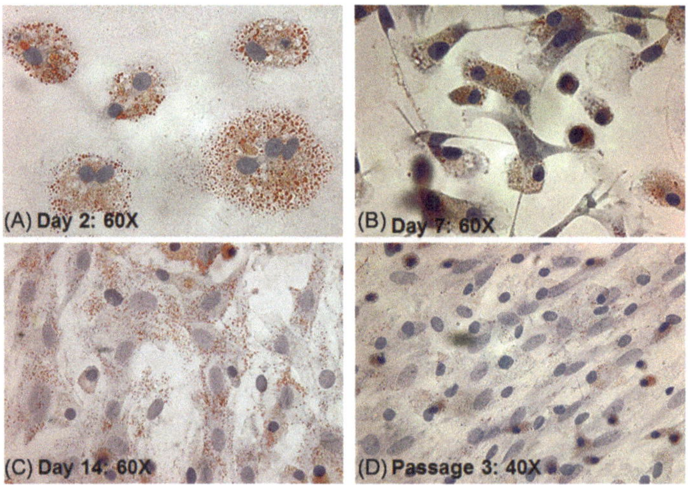

Figure 1.6 Rat stellate cells at indicated times in culture after isolation. The cells were stained with Oil Red O to show lipid stores. Note that with time in culture, HSCs progressively lose their lipid content and become flattened, gaining a myofibroblast-like phenotype.

experiments with HSCs from laboratory animals that are maintained in controlled pathogen-free environments may not necessarily be similar to the cells obtained from humans because humans are subjected to variable environmental conditions/stresses, food-derived antigens, drugs, xenobiotics, and microbial products that impact liver cell characteristics significantly.

Because HSCs isolated from normal liver rapidly begin to change their phenotype, it was necessary to develop a method to maintain them in the quiescent state to investigate their physiological characteristics and functions. Friedman et al. [41] found that HSCs retained their quiescent physiologic phenotype when cultured on basement membrane-like matrix (Matrigel™) derived from Engelbreth-Holm-Swarm murine tumor. These cells produced 70% less collagen (consisting only of type III collagen) than those cultured on plastic, which differentiated into myofibroblasts and produced large amounts of type I collagen. Although these findings provided key information on the relationship between HSCs and the surrounding three-dimensional extracellular matrix microenvironment in the process of HSC activation, this method has not been extensively employed in routine laboratory practice because of the inherent non-standardized content of growth factors and cytokines of Matrigel™ and its barrier effect toward exogenously added experimental agonists.

The search for exogenous factors that induce activation of HSCs directed attention to Kupffer cells, which were already known to produce reactive oxygen species and a variety of cytokines and growth mediators [42]. Indeed, medium conditioned by Kupffer cells induced proliferation of HSCs and increased their collagen production [43]. The mediator in Kupffer cell medium that caused proliferation was identified to be platelet-derived growth factor (PDGF) [43,44], which is perhaps the most potent mitogen for HSCs. Whereas PDGF induced proliferation of HSCs, TGFβ1, which is also produced by Kupffer cells, was identified to be a very potent cytokine to induce collagen production by HSCs [44].

1.7 MARKERS FOR HSCs

Until the mid-1980s, identification and quantification of HSCs largely depended on microscopic examination and ultrastructural analysis in combination with vitamin A autofluorescence [12,45]. Because a significant number of HSCs might not contain vitamin A [27,29], it was essential to identify specific markers. Knook et al. demonstrated that HSCs expressed vimentin [37], a type III intermediate filament protein that is expressed in mesenchymal cells. However, Kupffer cells as well as endothelial cells also express vimentin [38]. Although Yokoi et al. found desmin, a smooth muscle protein consisting of intermediate filaments, to be a reliable marker for rat HSCs [40], it was later determined that there was significant variability in desmin expression by HSCs within the liver lobule [27,29]. Rat HSCs were also reported to express glial fibrillary acidic protein (GFAP), the major component of intermediate filament in astrocytes [46]. This indicated an interesting possibility that HSCs may have neural origin during embryonic development. The accumulation of HSCs around necrotic area and their potential role in fibrosis was postulated in the 1970s [32–34], and was confirmed by Burt et al. in 1986 using anti-desmin antibody to identify HSCs in areas of hepatocellular necrosis caused by a single injection of carbon tetrachloride [45]. It is important to mention that human quiescent HSCs may not express desmin or GFAP [47]. Thus lipid deposits, vitamin A autofluorescence, and electron microscopy remain the primary mode of their identification. However, in the normal human liver some HSCs may express α-smooth muscle actin (α-sma), and regardless of the species all HSCs express α-sma upon activation.

As mentioned earlier, HSCs differentiate into proliferative and fibrogenic phenotype during liver injury and in culture. In addition to the loss of retinoids, the activation process is associated with the expression of α-sma, which is a highly reliable marker of HSC activation *in vivo* and *in vitro* (Figure 1.7). It has become a standard practice to assess the purity of freshly isolated HSCs by vitamin A autofluorescence and immunostaining for GFAP and desmin (HSC) in conjunction with markers for Kupffer cells, endothelial cells, and epithelial cells (see Chapter 2).

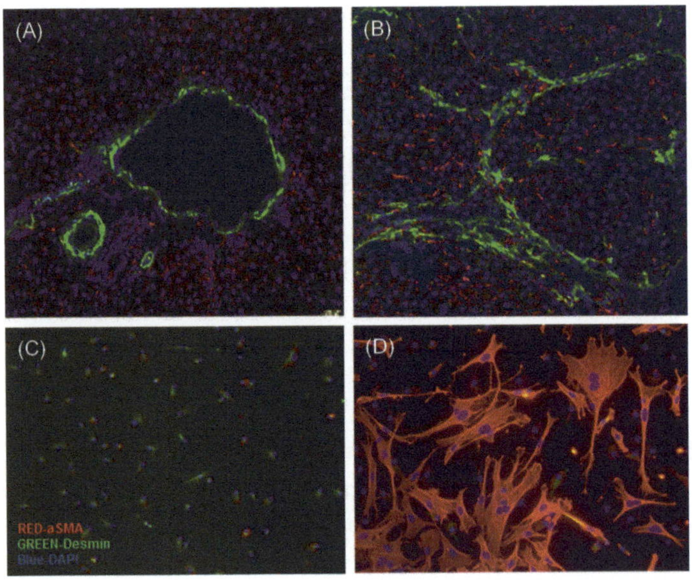

Figure 1.7 Stellate cell activation *in vivo* and in culture. Sections of control (A) and carbon tetrachloride-induced cirrhotic (B) rat livers stained for α-sma. The α-sma–positive cells can be seen only in the blood vessels of the control liver; in cirrhotic liver, in addition to blood vessels, α-sma–positive cells localize also to the fibrotic areas (activated stellate cells). Magnification 10 ×. Stellate cells on day 3 (C) and day 7 (D) of culture were stained for desmin (green) and α-sma (red). Note very weak staining for α-sma on day 3, and almost all cells acquired myofibroblast-like phenotype and demonstrated intense α-sma staining on day 7. Magnification 10×.

1.8 PERSPECTIVE

Since their discovery almost 140 years ago and somewhat slow-paced research for nearly a century, the stellate cell field has exploded since the 1980s with the demonstration that these cells are responsible for liver fibrosis. Apart from storing and releasing retinoids and changing its phenotype to repair liver injury, accumulating evidence indicates that stellate cell can be a major player in hepatic inflammation and immune regulation. Similar functions can now be attributed to stellate cells in other organs. Stellate cells can induce chemotaxis of inflammatory and immune cells, and interact, physically and via soluble mediators, with parenchymal cells (hepatocytes) and other nonparenchymal cells (Kupffer cells, sinusoidal endothelial cells, and biliary epithelial cells). These interactions are bidirectional and essential for hepatic homeostasis. The intent of this chapter was to provide a historical perspective recognizing the early research on stellate cells and put forth fundamental information that formed the foundation for several ensuing important discoveries. The subsequent chapters will describe in depth the current understanding of the various properties of stellate cell in the liver and other organs.

REFERENCES

[1] Wake K. Perisinusoidal stellate cells (fat-storing cells, interstitial cells, lipocytes), their related structure in and around the liver sinusoids, and vitamin A-storing cells in extrahepatic organs. Int Rev Cytol 1980;66:303–53.
[2] Wake K, Sato T. Cell Tissue Res 1993;273:227–37.
[3] von Kupffer K. Uber Sternzellen der Leber. Briefliche Mitteilung an Professor Waldeyer. Arch Mikr Anat 1876;12:353–8.
[4] von Kupffer C. Ueber die sogennanten Sternzellen der S–ugethierleber. Arch Mikr Anat 1899;54:254–88.
[5] Pearce RM, Austin JH. The relation of the spleen to blood destruction and regeneration and tp hemolytic jaundice:V. Changes in the endothelial cells of the lymph nodes and liver in splenectomized animals receiving hemolytic serum. J Exp Med 1912;16:780–8.
[6] Ito T, Nemoto M. Uber die Kupfferschen Sternzellen und die "Fettspeicherungszellen" (fat storing cells) in der Blutkapillarenwand der menschlichen Leber. Okajima Folia Anat Jpn 1952;24:243–58.
[7] Wake K. "Sternzellen" in the liver: perisinusoidal cells with special reference to storage of vitamin A. Am J Anat 1971;132:429–62.
[8] Wake K. Development of vitamin A-rich lipid droplets in multivesicular bodies of rat liver stellate cells. J Cell Biol 1974;63(2 Pt 1):683–91.
[9] Sztark F, Dubroca J, Latry P, Quinton A, Balabaud C, Bioulac-Sage P. Perisinusoidal cells in patients with normal liver histology. A morphometric study. J Hepatol 1986;2:358–69.
[10] Otto DA, Veech RL. Isolation of a lipocyte-rich fraction from rat liver nonparenchymal cells. Adv Exp Med Biol 1980;132:509–17.
[11] Hepatic stellate cell nomenclature. Hepatology 1996;23:193.
[12] Knook DL, Seffelaar AM, de Leeuw AM. Fat-storing cells of the rat liver. Their isolation and purification. Exp Cell Res 1982;139:468–71.
[13] Hendriks HF, Verhoofstad WA, Brouwer A, de Leeuw AM, Knook DL. Perisinusoidal fat-storing cells are the main vitamin A storage sites in rat liver. Exp Cell Res 1985;160:138–49.
[14] Yamada M, Blaner WS, Soprano DR, Dixon JL, Kjeldbye HM, Goodman DS. Biochemical characteristics of isolated rat liver stellate cells. Hepatology 1987;7:1224–9.

[15] Hendriks HF, Brekelmans PJ, Buytenhek R, Brouwer A, de Leeuw AM, Knook DL. Liver parenchymal cells differ from the fat-storing cells in their lipid composition. Lipids 1987;22:266–73.
[16] Senoo H, Kojima N, Sato M. Vitamin A-storing cells (stellate cells). Vitam Horm 2007;75:131–59.
[17] Blomhoff R, Berg T, Norum KR. Transfer of retinol from parenchymal to stellate cells in liver is mediated by retinol-binding protein. Proc Natl Acad Sci USA 1988;85:3455–8.
[18] Matsuura T, Nagamori S, Fujise K, Hasumura S, Homma S, Sujino H, et al. Retinol transport in cultured fat-storing cells of rat liver. Quantitative analysis by anchored cell analysis and sorting system. Lab Invest 1989;61:107–15.
[19] Pinzani M, Gentilini P, Abboud HE. Phenotypical modulation of liver fat-storing cells by retinoids. Influence on unstimulated and growth factor-induced cell proliferation. J Hepatol 1992;14:211–20.
[20] Blomhoff R, Rasmussen M, Nilsson A, Norum KR, Berg T, Blaner WS, et al. Hepatic retinol metabolism. Distribution of retinoids, enzymes, and binding proteins in isolated rat liver cells. J Biol Chem 1985;260:13560–5.
[21] Uchio K, Tuchweber B, Manabe N, Gabbiani G, Rosenbaum J, Desmouliere A. Cellular retinol-binding protein-1 expression and modulation during *in vivo* and *in vitro* myofibroblastic differentiation of rat hepatic stellate cells and portal fibroblasts. Lab Invest 2002;82:619–28.
[22] Blaner WS, Hendriks HF, Brouwer A, de Leeuw AM, Knook DL, Goodman DS. Retinoids, retinoid binding proteins and retinyl palmitate hydrolase distributions in different types of rat liver cells. J Lipid Res 1985;26:1241–51.
[23] Friedman SL, Wei S, Blaner WS. Retinol release by activated rat hepatic lipocytes: regulation by Kupffer cell-conditioned medium and PDGF. Am J Physiol 1993;264:G947–52.
[24] Weiner FR, Blaner WS, Czaja MJ, Shah A, Geerts A. Ito cell expression of a nuclear retinoic acid receptor. Hepatology 1992;15:336–42.
[25] Ulven SM, Natarajan V, Holven KB, Lovdal T, Berg T, Blomhoff R. Expression of retinoic acid receptor and retinoid X receptor subtypes in rat liver cells: implications for retinoid signalling in parenchymal, endothelial, Kupffer and stellate cells. Eur J Cell Biol 1998;77:111–6.
[26] Tanuma Y, Ito T. Electron microscope study on the hepatic sinusoidal wall and fat-storing cells in the bat. Arch Histol Jpn 1978;41:1–39.
[27] Wake K. Liver perivascular cells revealed by gold and silver impregnation methods and electron microscopy Motta P, editor. Biopathology of the liver, an ultrastructural approach. Dordrecht: Kluwer; 1988. p. 23–6.
[28] Giampieri MP, Jezequel AM, Orlandi F. The lipocytes in normal human liver. A quantitative study. Digestion 1981;22:165–9.
[29] Wake K. Hepatic stellate cells Tanikawa K, Ueno T, editors. Liver diseases and hepatic sinusoidal cells. Tokyo: Springer-Verlag; 1999. p. 56–65.
[30] Pinzani M, Failli P, Ruocco C, Casini A, Milani S, Baldi E, et al. Fat-storing cells as liver-specific pericytes. Spatial dynamics of agonist-stimulated intracellular calcium transients. J Clin Invest 1992;90:642–6.
[31] Rockey DC, Boyles JK, Gabbiani G, Friedman SL. Rat hepatic lipocytes express smooth muscle actin upon activation *in vivo* and in culture. J Submicrosc Cytol Pathol 1992;24:193–203.
[32] McGee JO, Patrick RS. The role of perisinusoidal cells in experimental hepatic fibrogenesis. J Pathol 1972;106(1): vi.
[33] McGee JO, Patrick RS. The role of perisinusoidal cells in hepatic fibrogenesis. An electron microscopic study of acute carbon tetrachloride liver injury. Lab Invest 1972;26:429–40.
[34] Kent G, Inouye T, Minick OT, Bahu RM, Minick OT, Popper H. Fat-storing cells (lipocytes) in the liver: their role in vitamin A storage and fibrogenesis. Med Chir Dig 1977;6:425–8.
[35] Mak KM, Leo MA, Lieber CS. Alcoholic liver injury in baboons: transformation of lipocytes to transitional cells. Gastroenterology 1984;87:188–200.
[36] Friedman SL, Roll FJ, Boyles J, Bissell DM. Hepatic lipocytes: the principal collagen-producing cells of normal rat liver. Proc Natl Acad Sci USA 1985;82:8681–5.
[37] de Leeuw AM, McCarthy SP, Geerts A, Knook DL. Purified rat liver fat-storing cells in culture divide and contain collagen. Hepatology 1984;4:392–403.

[38] Tsutsumi M, Takada A, Takase S. Characterization of desmin-positive rat liver sinusoidal cells. Hepatology 1987;7:277–84.
[39] Senoo H, Wake K. Suppression of experimental hepatic fibrosis by administration of vitamin A. Lab Invest 1985;52:182–94.
[40] Yokoi Y, Namihisa T, Kuroda H, Komatsu I, Miyazaki A, Watanabe S, et al. Immunocytochemical detection of desmin in fat-storing cells (Ito cells). Hepatology 1984;4:709–14.
[41] Friedman SL, Roll FJ, Boyles J, Arenson DM, Bissell DM. Maintenance of differentiated phenotype of cultured rat hepatic lipocytes by basement membrane matrix. J Biol Chem 1989;264:10756–62.
[42] Decker K. Biologically active products of stimulated liver macrophages (Kupffer cells). Eur J Biochem 1990;192:245–61.
[43] Friedman SL, Arthur MJ. Activation of cultured rat hepatic lipocytes by Kupffer cell conditioned medium. Direct enhancement of matrix synthesis and stimulation of cell proliferation via induction of platelet-derived growth factor receptors. J Clin Invest 1989;84:1780–5.
[44] Pinzani M, Gesualdo L, Sabbah GM, Abboud HE. Effects of platelet-derived growth factor and other polypeptide mitogens on DNA synthesis and growth of cultured rat liver fat-storing cells. J Clin Invest 1989;84:1786–93.
[45] Burt AD, Robertson JL, Heir J, MacSween RN. Desmin-containing stellate cells in rat liver; distribution in normal animals and response to experimental acute liver injury. J Pathol 1986;150:29–35.
[46] Gard AL, White FP, Dutton GR. Extra-neural glial fibrillary acidic protein (GFAP) immunoreactivity in perisinusoidal stellate cells of rat liver. J Neuroimmunol 1985;8:359–75.
[47] Geerts A. History, heterogeneity, developmental biology, and functions of quiescent hepatic stellate cells. Semin Liver Dis 2001;21:311–35.
[48] Matsuoka M, Tsukamoto H. Stimulation of hepatic lipocyte collagen production by Kupffer cell-derived transforming growth factor beta: implication for a pathogenetic role in alcoholic liver fibrogenesis. Hepatology 1990;11:599–605.

CHAPTER 2

Hepatic Stellate Cell Culture Models

Krista Rombouts
UCL Institute for Liver and Digestive Health, Division of Medicine, University College London, Royal Free Hospital, London, UK

2.1 ISOLATION OF HEPATIC STELLATE CELLS

In the past four decades several approaches such as density gradient separation [1–3], fluorescent cell sorting [4] and explant culture [5] have been employed to isolate hepatic stellate cells (HSCs). Separating HSCs from the total hepatic cell suspension allows culture of purified cell preparation to study, in a much defined way, the primary molecular signaling pathways associated with fibrogenesis. Explant culture and fluorescence-activated cell sorting (FACS) analysis are two cell isolation procedures that avoid the usage of density gradients. Nevertheless, density gradient centrifugation remains the method of preference for many investigators to isolate human and rodent HSCs. This method involves *ex vivo* digestion of the liver tissue by enzymes such as collagenase, pronase, and deoxyribonuclease (DNase) to dissociate the hepatic cells from the surrounding ECM. This is then followed by sequential centrifugation steps at different *g* force to separate HSCs from the other hepatic cell types. This is followed by a refined multi-step or single-step gradient centrifugation of the cell suspension through a density gradient of Nycodenz [6,7], Larex/Stractan/arabinogalactan [1,8], Optiprep [9], or Percoll [10,11] (Figure 2.1). Several alternatives have been optimized to isolate HSCs in combination with hepatic cells such as hepatocytes [12], liver sinusoidal endothelial cells (LSECs) [11], and Kupffer cells [13,14].

HSCs obtained thus can be further enriched by sorting based on high side scatter of incident light [15]. Moreover, immediately after the density gradient centrifugation, purity of HSCs can be analyzed by taking advantage of the quickly fading greenish autofluorescence containing vitamin A-enriched lipid droplets under 328-nm illumination. Over the years, several markers for HSCs have been identified to ascertain purity of the cell preparation as well as *in situ* identification including vimentin, desmin, and glial fibrillary acidic protein (GFAP), nestin, synaptophysin, nerve growth factor (NGF) receptor p75, and alpha-smooth muscle actin (α-sma) [16]. However, desmin and GFAP are the gold standard immediately after isolation or early culture, and upon activation almost all HSCs express α-sma and nestin, which makes it easier to evaluate purity of the cell preparation. In addition, staining for macrophages (ED1

Stellate Cells in Health and Disease
DOI: http://dx.doi.org/10.1016/B978-0-12-800134-9.00002-6

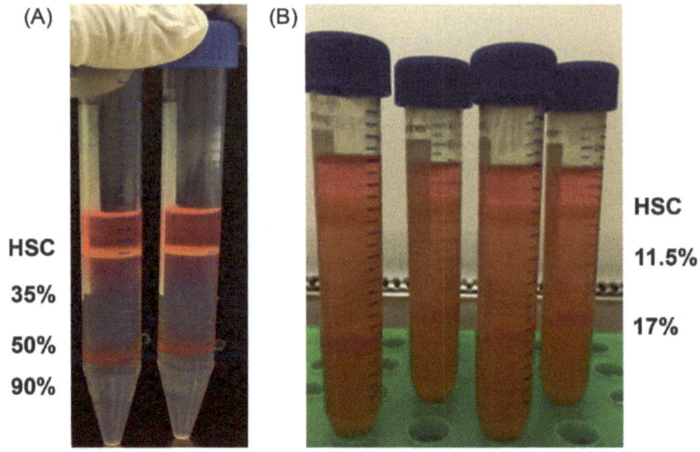

Figure 2.1 HSCs were isolated by employing an *ex vivo* enzymatic digestion of the liver tissue by enzymes such as collagenase, pronase, and DNase, which allows the dissociation of the hepatic cells from the surrounding ECM. This is then followed by sequential centrifugation steps at different *g* force to separate the various cell types from the HSC population. Next, a refined multistep or single-step gradient centrifugation allows to select and to purify HSCs. Different gradients and density solutions were used such as Percoll (35–50–90%) (A) and Optiprep (11.5–17%) (B).

and ED2 in rat and F4/80 in mouse), epithelial cells (cytokeratin 19), and endothelial cells (SE1) is performed to further assure the proportion of contaminating cells.

The principle of using the density gradient centrifugation to isolate HSCs from other hepatic cell types is based on the presence of intracellular vitamin A-containing lipid droplets in HSCs. Indeed, HSCs are a major storage site of retinoids including vitamin A and play a cardinal role in their storage and controlled release. These lipid droplets differ in number and diameter and vary between species and under different physiological conditions [16,17]. The presence or absence of the lipid droplets is of major importance as the "activation" of resting vitamin A-rich HSCs into myofibroblast-like phenotype observed in chronic liver diseases is associated with the loss of retinoids and an increase in ECM synthesis [18–20]. This should be taken into account when isolating HSCs from the diseased fibrotic liver, induced, for example, by bile duct ligation (BDL) or carbon tetrachloride (CCl_4) injections, and requires an increased collagenase and pronase concentration to digest the ECM and when using the density gradient centrifugation based upon the presence of retinoids lipid droplets [7,21].

The freshly isolated HSCs show prominent dendritic cytoplasmic processes, and the presence of lipid droplets. During culture over the following days, the morphology of HSCs gradually changes and displays a slightly more myofibroblast-like phenotype with heterogeneous retinoid droplet size (Figure 2.2).

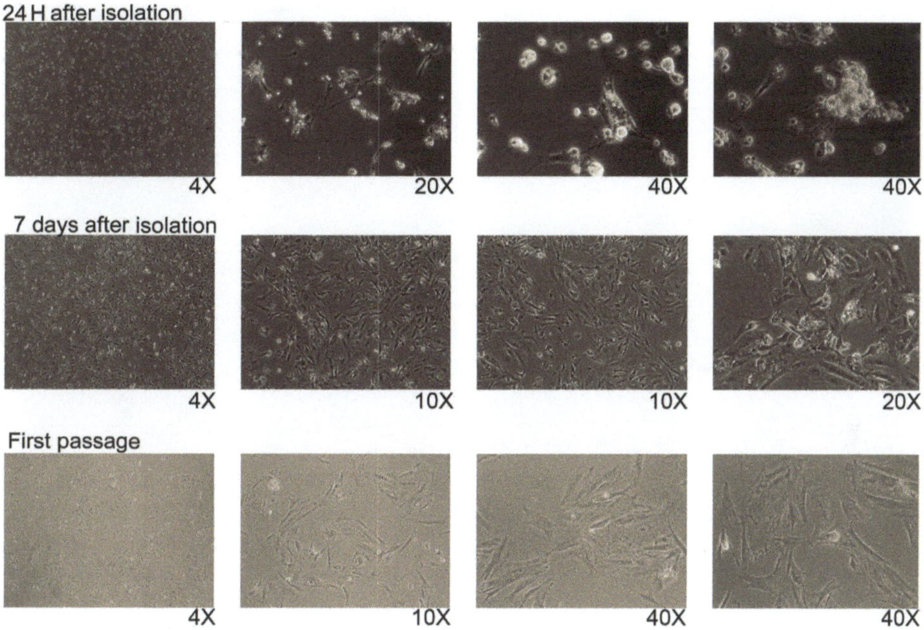

Figure 2.2 Phase contrast microscopic images of HSCs in culture. Twenty-four hours after isolation the freshly isolated HSCs show prominent dendritic cytoplasmic processes and the presence of lipid droplets. The HSC morphology gradually displays a slightly more myofibroblast-like phenotype during the subsequent days in culture. Images were taken after HSC isolation at 24 h, 7 days in culture, and after the first passage. Magnification 4 ×, 10 ×, 20 ×, and 40 ×.

2.2 SINGLE CELL CULTURE

Many cell culture models exist with different complexity. The most described and used model is the single monolayer culture of HSCs. The reason to use this *in vitro* model is that the primary quiescent HSCs found in the normal healthy liver spontaneously "transdifferentiate" into activated myofibroblast-like cells when cultured on non-coated plastic culture dishes in the presence of fetal bovine serum. Hence HSC "activation" refers to the transformation of the resting vitamin A-rich cell into a proliferating, fibrogenic, and highly contractile phenotype [22,23]. This *in vitro* model is well established and represents the process similar to that observed in chronic liver diseases. Thus, the single monolayer culture of quiescent HSCs has been used extensively to determine the role of specific proteins and genes during HSC activation *in vitro* (Figure 2.3). Moreover, this *in vitro* model enables researchers to investigate and compare the possible anti-fibrogenic effects of compounds between quiescent HSCs, transdifferentiated HSCs, and fully activated HSCs [6,24]. This information can then be extended/applied to the *in vivo* models [25]. Therefore, quiescent HSCs isolated from the normal, healthy liver and grown in a

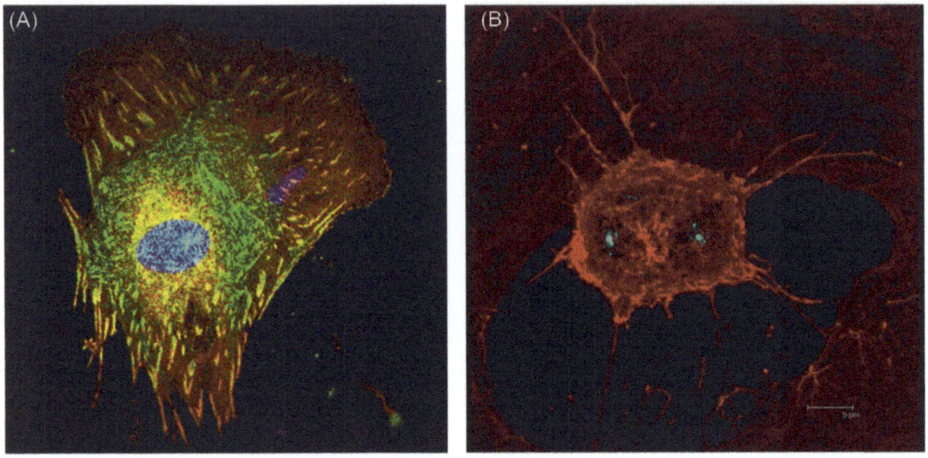

Figure 2.3 Confocal images of rat HSCs (A) and human HSCs (B) at different time points in culture. Cells were cultured on glass coverslip and stained with different antibodies. (A) Activated rat HSC demonstrates a strong presence of focal adhesion kinase (FAK, green color) and partial co-localization with vinculin (red color). (B) A mitotic human HSC is shown with B-actin localized in the cell body and small cellular processes, whereas phosphorylated myristoylated alanine-rich C-kinase substrate (MARCKS) is present as a centrosome protein. Both proteins are key players during mitosis in human HSC biology.

monolayer culture is an optimal tool to characterize the molecular mechanisms of the progression of HSC "transdifferentiation." As a consequence, this model has been used to determine the specific therapeutic target protein and signaling pathway(s) in HSCs and to determine which HSC phenotype (i.e., a quiescent vs. activated HSC) is targeted or is sensitive to the putative drug [6,26–28].

2.3 *IN VITRO-* VERSUS *IN VIVO*-ACTIVATED HSCs

Almost three decades ago, attempts were made to isolate *in vivo*-activated HSCs from the rodents in which fibrosis was induced by administration of CCl_4 injection or by BDL. Differences were found in the number of lipid droplets, the rate of collagen synthesis, and proliferation between HSCs isolated from the healthy versus the diseased liver [29,30]. These studies provided the initial evidence that HSCs may play a key role in liver fibrosis. Recently, with advanced technology such as gene expression microarray analysis and proteomics, clear differences have been demonstrated between *in vitro-* and *in vivo*-activated HSCs. Thus these findings favor the use of *in vivo*-activated HSCs isolated from a fibrotic liver over *in vitro*-activated HSCs [7,31,32]. The *in vivo*-activated HSCs reflect more the influence of the specific microenvironment on their behavior in the fibrotic liver. This model has now been adapted by several investigators to explore in greater detail the interplay between HSCs and the changing microenvironment in the diseased liver. Indeed, this concept has gained more attention and has been used to compare the characteristics of *in vitro*-activated HSCs with *in vivo*-activated HSCs in different animal models/chronic diseases [32–34]. Furthermore, fate tracking in combination with new markers in *in vivo* models has confirmed that activated HSCs are the major source of myofibroblasts and liver fibrosis [35,36].

2.4 SINGLE CELL CULTURE AND 2D: IMPORTANCE OF ADHESION, Arg–Gly–Asp, AND MATRIX COMPONENTS

HSCs, located in the perisinusoidal space of Disse, adhere to the endothelial cells on one hand and make contact with the parenchymal cells on the other hand. The normal perisinusoidal space contains type I, III, IV, V, and VI collagens, fibronectin, laminin, and proteoglycans. Upon liver injury, and especially during chronic liver damage, HSCs "transdifferentiate" and become activated by paracrine stimulation, which is then followed by autocrine amplification of the fibrogenic signals. As a result, HSCs produce and secrete large amounts of ECM components that interrupt the normal fine collagen type IV- and laminin-rich, low-density basement, membrane-like structure of the space of Disse, resulting in remodeled ECM with collagen type I- and III-rich fibrillar matrix [30,37–41]. These changes in collagen production can become so dramatic that the total content of collagens and the non-collagenous components increases three- to fivefold, indicating that the three-dimensional (3D) ECM in the space of Disse is a very dynamic component [23]. Not only are the different ECM components localized in the liver, but correlations have been made between their liver and serum concentrations [42,43]. Several investigations have demonstrated *in vitro* the remarkable plasticity of HSCs, which not only reflects various phenotypes but also mirrors different functions. In the 1990s, evidence was provided for the involvement of important signaling pathways of cellular adhesion molecules in HSC activation [43,44], which were found to differ depending upon whether HSCs are grown on plastic, collagen, or cultured on ECM [44–47]. The advanced understanding to control ECM composition and presentation has facilitated studies that explain the spatial distribution of integrin-mediated adhesion, cellular function, cell polarity, and proliferation. The most basic biological functionality in using two-dimensional (2D) cultures in combination with ECM is achieved by addition of the commonly used Arg–Gly–Asp (RGD) peptide as cell-adhesive ligands. Until today, the RGD peptide is a known adhesive domain of several ECM components, such as that involved in the binding of fibronectin [48], an important assembler of the collagen matrix, to the $\alpha 5\beta 1$ integrin receptor [49] with the presence of the intra-molecular RGD motifs in collagen [50].

HSCs are anchorage-dependent cells that sense the mechanics of their surroundings by pushing and pulling on the ECM, and change intracellular signals and cellular behavior [51]. Hence, investigating the mechanotransduction phenomenon and its effect on the HSC phenotype has led to the observation that depending on the hardness of the substrates used (i.e., soft vs. stiff substrates), the quiescent phenotype will be maintained or HSCs transform into activated myofibroblast-like cells. These studies are important to understand the complexity of interactions between HSCs and their ECM microenvironment in a healthy liver versus a diseased liver [52]. Notwithstanding all observations in 2D single culture on plastic culture dishes or ECM matrix components, the current *in vitro* models are still limited. This is due to the fact that the mechanical heterogeneity typical of fibrosis is absent. For instance, even culturing HSCs on collagen does not recapitulate the cellular and genetic behavior of the cells present in fibrous septa or liver nodule during the process of progressive liver fibrosis. Therefore, reductionist 2D monolayer cell culture systems with collagen or ECM substrates do not address some of the many complex multicellular processes that give shape to the normal healthy hepatic tissue or fibrotic liver tissue.

2.5 HSC CO-CULTURES WITH KUPFFER CELLS, HEPATOCYTES, LSEC, HCC, AND CC CELLS

One way to investigate the intercellular communication between HSCs and other hepatic cells is co-culturing HSCs with the cell type of interest such as Kupffer cells [53], hepatocytes, LSECs [54], CC [60], and hepatocellular carcinoma cell (HCC) lines [55]. Several experimental approaches can be used depending on what is being investigated. For example, paracrine effects can be determined when the two cell types are in direct contact, but this does not discriminate between the contact-dependent effects and those due to soluble mediators produced by one cell type on the other. On the other hand, one can investigate the paracrine interactions and measure the stimulation/reaction of a cell type when it is brought in contact with the secreted cytokines or growth factors produced by the other cell type (i.e., non-contact co-cultures). From a technical point of view, with different outcomes, one can use the cell culture inserts to have both cell types in co-culture but still physically separated. The trans-well insert contains pores of specific sizes, depending on the cell type used, that permits access to only growth factors/cytokines and the like to the lower compartment where HSCs are cultured. The effects of the soluble mediators on gene/protein expression of HSCs can then be investigated.

Another way to investigate the effect of a certain hepatic cell type on HSCs is by placing them in the conditioned medium produced by the other cell type (i.e., paracrine stimulation). This conditioned medium contains all cytokines, growth factors/soluble factors, exosomes, microparticles, and lysosomes produced and secreted by the hepatic cell [29,56–58]. Cells are typically cultured for up to 48 h in serum-free medium, which is collected and centrifuged at a predetermined *g*-force depending on which fraction contains the protein/growth factor/vesicle of interest. After culturing

HSCs in a particular conditioned medium, the morphological, functional, and genetic modifications in HSCs can be ascertained and quantified. On the other hand, several studies have investigated the effect of HSCs on signaling pathways and functions of cancer cells by exposing them to the conditioned medium collected from activated HSCs [59–62]. Investigation of the content of the conditioned medium will identify possible key players/mechanisms and their function in the interaction between HSCs and the other hepatic cells [36,57]. This approach can be applied to fully characterize the conditioned medium of HSCs isolated from patients with liver disease of different etiologies. This would identify secreted mediators, which are etiology-specific and important in the development of liver fibrosis.

2.6 *IN VITRO* 3D CULTURE SYSTEMS

It is becoming clear that the recreation of the liver microenvironment with cell–matrix interactions, cell–cell adhesion, and cellular signaling is essential in liver studies. Traditional 2D cell culture systems are proven to be valid in investigating possible mechanisms of cell behavior and screening for drugs to some extent. However, many previous investigations revealed a lack in translation towards animal models and, more importantly, into clinical studies. Therefore, the development of a well-defined 3D *in vitro* model, which mimics ECM structures as found *in vivo*, has gained strong interest. Indeed, matrix mechanics is a key parameter in regulating a range of cell behaviors such as cell proliferation, cell migration, and ECM production. In fact, it has been known for almost two decades that the behavior of HSCs is regulated by 3D ECM [63–67]. Initially, freshly isolated HSCs were cultured on thick Matrigel or in 3D type I collagen gel [67–69]. More recently, Lee et al. have investigated the paracrine effects of rat HSCs on primary hepatocytes in spheroid-based 3D culture by creating a chip with a cascade design in which the culture medium flows from the HSC culture towards the 3D-spheroids containing hepatocytes. Afterwards HSC- and hepatocyte-specific functions were measured for up to 9 days [70]. In contrast, Kostadinova et al. have inoculated primary hepatocytes into a pre-established non-parenchymal cell culture grown on 3D nylon scaffolds and demonstrated that this allows a culture time of up to 3 months while maintaining important hepatic functions [71]. Of course, both systems have several limitations such as standardized matrix stiffness and the presence of a uniform topography [72,73], which are likely two major variables that greatly contribute to an altered HSC phenotype in different liver diseases, thereby changing HSC behavior. Moreover, it has been shown to be very critical for promoting liver cell-specific functions *in vitro* to co-culture hepatocytes with non-parenchymal cells either in direct contact or via paracrine stimulation in these systems [70,71].

Given the structural diversity of the extracellular environment *in vivo* (i.e., healthy liver versus diseased liver) it is not surprising that research towards a more *in vivo*-like system is required. Thus, major efforts have been made to mimic the *in vivo* HSC–liver cell and HSC–ECM interactions that may more closely mimic the physiologic state. For example, precision-cut liver slices (PCLS) is an example of how a 3D *in vitro* model can be used to study the HSCs in a system that very closely reflects the *in vivo* situation with maintaining the intact hepatic architecture and cellular heterogeneity [72]. This 3D system can be obtained from rat [74], mouse [75], and human [76,77]. Liver slices or tissue cores are normally 8 mm in diameter, 250 μm thick, and contain 70–100 lobules. The PCLS is a dynamic *in vitro* 3D model as each PCLS is maintained in a vial and set on a roller platform and gently agitated at 37°C, with 5% CO_2 and 40% O_2 in a humidified incubator. In this way experiments can be performed over a time period of 24 h–1 week [78–80]. By using this 3D *in vitro* model, behavior of HSCs, endothelial cells, Kupffer cells, biliary epithelial cells, and portal fibroblasts has been investigated [78–82]. By using PCLS, the behavior of cells and the interactions of cells within their original ECM can be determined, and also can be compared between different pathologies [80]. It may be argued that the short incubation time is a major drawback of this system [75] and proteomics profiling may show different outcomes depending on which species has been investigated [76].

Despite major improvements in refining the isolation procedures and culture conditions in 2D and 3D, no efficient anti-fibrogenic therapy has been developed yet [26,83,84]. Many of the promising compounds have failed when translated from 2D *in vitro* culture into *in vivo* models owing to the lack of specific liver microenvironment. The absence of excellent and rapid 3D *in vitro* screening assays has slowed or arrested further investigation of possible lead compounds before they can be tested in animal models and phase I clinical trials. The Scar-in-a-Jar model may provide a new pathophysiological relevant *in vitro* screening assay to test novel anti-fibrotic compounds, which specifically target the collagen biosynthesis. This system is based on culturing fibroblasts in 24 wells with the addition of neutral or charged macromolecules (i.e., creating a 3D system). This enhances the collagen synthesis and induces a higher secretion rate compared to the classical 2D plastic culture dish [85]. Whether the profibrogenic behavior of HSCs and more specifically the screening of anti-fibrogenic compounds on HSCs will render this system favorable has yet to be proven.

During the last decade, much effort has been made to develop ECM scaffolds by using artificial scaffolds or rodent tissues such as whole rat kidney [86] and gastrointestinal tract [87] as well as rat liver tissue (as extensively reviewed elsewhere) [88–91]. Basically, by applying different perfusion or mechanical assays a decellularization process removes all cellular material and this generates a native ECM scaffold, which remains highly preserved (extensively described elsewhere [92]). In this context, the repopulation of the scaffold with one or more cell types represents a highly dynamic *in vitro* culture system, which reflects more realistically the hepatic 3D microenvironment. Indeed a recent study had demonstrated the use of human liver tissue as ECM 3D bioscaffold [93].

2.7 CONCLUSIONS

Optimization of the *in vitro* models will lead to new objectives and to a possible new era in the search for anti-fibrogenic compounds. It will be critical to improve the *in vitro* culture models that reproduce liver microenvironment. Therefore, efforts continue to validate/optimize and standardize 3D biosystems composed of single or co-culture platforms. In this regard, HSCs have been shown to be very important as secondary supportive or stromal cells to generate heterotypic interactions with the parenchymal cells. Indeed, non-parenchymal cells such as HSCs increase the viability and functionality of hepatocytes in the 3D biosystems composed of synthetic and biological-derived materials [94–96]. Finally, it will be critical to put more efforts on investigating the 3D *in vitro* models, making the 3D cell culture a routine method for researchers in academia and the pharmaceutical industries.

REFERENCES

[1] Pinzani M, Failli P, Ruocco C, et al. Fat-storing cells as liver-specific pericytes. Spatial dynamics of agonist-stimulated intracellular calcium transients. J Clin Invest 1992;90(2):642–6.
[2] Bataller R, Gasull X, Gines P, et al. *In vitro* and *in vivo* activation of rat hepatic stellate cells results in *de novo* expression of L-type voltage-operated calcium channels. Hepatology 2001;33(4):956–62.
[3] Rombouts K, Niki T, Wielant A, et al. Effect of aldosterone on collagen steady state levels in primary and subcultured rat hepatic stellate cells. J Hepatol 2001;34(2):230–8.
[4] Troeger JS, Mederacke I, Gwak GY, et al. Deactivation of hepatic stellate cells during liver fibrosis resolution in mice. Gastroenterology 2012;143(4):1073–83.
[5] Mallat A, Preaux AM, Serradeil-Le GC, et al. Growth inhibitory properties of endothelin-1 in activated human hepatic stellate cells: a cyclic adenosine monophosphate-mediated pathway. Inhibition of both extracellular signal-regulated kinase and c-Jun kinase and upregulation of endothelin B receptors. J Clin Invest 1996;98(12):2771–8.
[6] Niki T, Rombouts K, De BP, et al. A histone deacetylase inhibitor, trichostatin A, suppresses myofibroblastic differentiation of rat hepatic stellate cells in primary culture. Hepatology 1999;29(3):858–67.
[7] De MS, Seki E, Uchinami H, et al. Gene expression profiles during hepatic stellate cell activation in culture and *in vivo*. Gastroenterology 2007;132(5):1937–46.
[8] Friedman SL, Rockey DC, McGuire RF, et al. Isolated hepatic lipocytes and Kupffer cells from normal human liver: morphological and functional characteristics in primary culture. Hepatology 1992;15(2):234–43.
[9] Ichikawa S, Mucida D, Tyznik AJ, et al. Hepatic stellate cells function as regulatory bystanders. J Immunol 2011;186(10):5549–55.
[10] Yata Y, Takahara T, Furui K, et al. Expression of matrix metalloproteinase-13 and tissue inhibitor of metalloproteinase-1 in acute liver injury. J Hepatol 1999;30(3):419–24.
[11] Vrochides D, Papanikolaou V, Pertoft H, et al. Biosynthesis and degradation of hyaluronan by non-parenchymal liver cells during liver regeneration. Hepatology 1996;23(6):1650–5.
[12] Riccalton-Banks L, Bhandari R, Fry J, et al. A simple method for the simultaneous isolation of stellate cells and hepatocytes from rat liver tissue. Mol Cell Biochem 2003;248(1–2):97–102.
[13] Friedman SL, Roll FJ. Isolation and culture of hepatic lipocytes, Kupffer cells, and sinusoidal endothelial cells by density gradient centrifugation with Stractan. Anal Biochem 1987;161(1):207–18.
[14] Friedman SL. "Cuts both ways": collagenases, lipocyte activation and polyunsaturated lecithin. Hepatology 1992;15(3):549–51.
[15] Geerts A, Niki T, Hellemans K, et al. Purification of rat hepatic stellate cells by side scatter-activated cell sorting. Hepatology 1998;27(2):590–8.

[16] Geerts A. History, heterogeneity, developmental biology, and functions of quiescent hepatic stellate cells. Semin Liver Dis 2001;21(3):311–35.

[17] Geerts A. On the origin of stellate cells: mesodermal, endodermal or neuro-ectodermal? J Hepatol 2004;40(2):331–4.

[18] Beljaars L, Olinga P, Molema G, et al. Characteristics of the hepatic stellate cell-selective carrier mannose 6-phosphate modified albumin (M6P(28)-HSA). Liver 2001;21(5):320–8.

[19] Blaner WS, O'Byrne SM, Wongsiriroj N, et al. Hepatic stellate cell lipid droplets: a specialized lipid droplet for retinoid storage. Biochim Biophys Acta 2009;1791(6):467–73.

[20] Wongsiriroj N, Jiang H, Piantedosi R, et al. Genetic dissection of retinoid esterification and accumulation in the liver and adipose tissue. J Lipid Res 2014;55(1):104–14.

[21] Mannaerts I, Nuytten NR, Rogiers V, Vanderkerken K, van Grunsven LA, Geerts A. Chronic administration of valproic acid inhibits activation of mouse hepatic stellate cells *in vitro* and *in vivo*. Hepatology 2010;51(2):603–14.

[22] Friedman SL. Hepatic stellate cells: protean, multifunctional, and enigmatic cells of the liver. Physiol Rev 2008;88(1):125–72.

[23] Friedman SL. Molecular regulation of hepatic fibrosis, an integrated cellular response to tissue injury. J Biol Chem 2000;275(4):2247–50.

[24] Pinzani M, Rombouts K. Liver fibrosis: from the bench to clinical targets. Dig Liver Dis 2004;36(4):231–42.

[25] Liedtke C, Luedde T, Sauerbruch T, et al. Experimental liver fibrosis research: update on animal models, legal issues and translational aspects. Fibrogenesis Tissue Repair 2013;6:19.

[26] Lim MP, Devi LA, Rozenfeld R. Cannabidiol causes activated hepatic stellate cell death through a mechanism of endoplasmic reticulum stress-induced apoptosis. Cell Death Dis 2011;2:e170.

[27] Maubach G, Lim MC, Chen J, et al. miRNA studies in *in vitro* and *in vivo* activated hepatic stellate cells. World J Gastroenterol 2011;17(22):2748–73.

[28] Iizuka M, Murata T, Hori M, et al. Increased contractility of hepatic stellate cells in cirrhosis is mediated by enhanced Ca^{2+}-dependent and Ca^{2+}-sensitization pathways. Am J Physiol Gastrointest Liver Physiol 2011;300(6):G1010–21.

[29] Shiratori Y, Geerts A, Ichida T, et al. Kupffer cells from CCl_4-induced fibrotic livers stimulate proliferation of fat-storing cells. J Hepatol 1986;3(3):294–303.

[30] Shiratori Y, Ichida T, Geerts A, et al. Modulation of collagen synthesis by fat-storing cells, isolated from CCl_4- or vitamin A-treated rats. Dig Dis Sci 1987;32(11):1281–9.

[31] Kristensen DB, Kawada N, Imamura K, et al. Proteome analysis of rat hepatic stellate cells. Hepatology 2000;32(2):268–77.

[32] Gandhi CR, Kuddus RH, Uemura T, Rao AS. Endothelin stimulates transforming growth factor-β1 and collagen synthesis in stellate cells from control but not cirrhotic rat liver. Eur J Pharmacol 2000;406:311–8.

[33] Nevzorova YA, Hu W, Cubero FJ, et al. Overexpression of c-myc in hepatocytes promotes activation of hepatic stellate cells and facilitates the onset of liver fibrosis. Biochim Biophys Acta 2013;1832(10):1765–75.

[34] Scarpa M, Grillo AR, Brun P, et al. Snail1 transcription factor is a critical mediator of hepatic stellate cell activation following hepatic injury. Am J Physiol Gastrointest Liver Physiol 2011;300(2):G316–26.

[35] Mederacke I, Hsu CC, Troeger JS, et al. Fate tracing reveals hepatic stellate cells as dominant contributors to liver fibrosis independent of its aetiology. Nat Commun 2013;4:2823.

[36] Iwaisako K, Jiang C, Zhang M, et al. Origin of myofibroblasts in the fibrotic liver in mice. Proc Natl Acad Sci USA 2014;111(32):E3297–305.

[37] Verbruggen LA, Geerts A, Wisse E, et al. Serum and synovial fluid antibodies to collagen in rheumatic diseases: a review. Clin Rheumatol 1986;5(4):440–4.

[38] Geerts A, Vrijsen R, Rauterberg J, et al. *In vitro* differentiation of fat-storing cells parallels marked increase of collagen synthesis and secretion. J Hepatol 1989;9(1):59–68.

[39] van Oortmarssen GJ, Habbema JD, van der Maas PJ, et al. A model for breast cancer screening. Cancer 1990;66(7):1601–12.

[40] Hayasaka A, Koch J, Schuppan D, et al. The serum concentrations of the aminoterminal propeptide of procollagen type III and the hepatic content of mRNA for the alpha 1 chain of procollagen type III in carbon tetrachloride-induced rat liver fibrogenesis. J Hepatol 1991;13(3):328–38.

[41] Geerts A, Geuze HJ, Slot JW, et al. Immunogold localization of procollagen III, fibronectin and heparan sulfate proteoglycan on ultrathin frozen sections of the normal rat liver. Histochemistry 1986;84(4–6):355–62.
[42] Schuppan D, Dumont JM, Kim KY, et al. Serum concentration of the aminoterminal procollagen type III peptide in the rat reflects early formation of connective tissue in experimental liver cirrhosis. J Hepatol 1986;3(1):27–37.
[43] Carloni V, Defranco RM, Caligiuri A, et al. Cell adhesion regulates platelet-derived growth factor-induced MAP kinase and PI-3 kinase activation in stellate cells. Hepatology 2002;36(3):582–91.
[44] Olsen AL, Sackey BK, Marcinkiewicz C, et al. Fibronectin extra domain-A promotes hepatic stellate cell motility but not differentiation into myofibroblasts. Gastroenterology 2012;142(4):928–37.
[45] Gaca MD, Zhou X, Benyon RC. Regulation of hepatic stellate cell proliferation and collagen synthesis by proteinase-activated receptors. J Hepatol 2002;36(3):362–9.
[46] Mazzocca A, Carloni V, Sciammetta S, et al. Expression of transmembrane 4 superfamily (TM4SF) proteins and their role in hepatic stellate cell motility and wound healing migration. J Hepatol 2002;37(3):322–30.
[47] Guvendiren M, Perepelyuk M, Wells RG, et al. Hydrogels with differential and patterned mechanics to study stiffness-mediated myofibroblastic differentiation of hepatic stellate cells. J Mech Behav Biomed Mater 2014;38:198–208.
[48] Xu G, Niki T, Virtanen I, et al. Gene expression and synthesis of fibronectin isoforms in rat hepatic stellate cells. Comparison with liver parenchymal cells and skin fibroblasts. J Pathol 1997;183(1):90–8.
[49] Kawelke N, Vasel M, Sens C, et al. Fibronectin protects from excessive liver fibrosis by modulating the availability of and responsiveness of stellate cells to active TGF-beta. PLoS One 2011;6(11):e28181.
[50] Birukawa NK, Murase K, Sato Y, et al. Activated hepatic stellate cells are dependent on self-collagen, cleaved by membrane type 1 matrix metalloproteinase for their growth. J Biol Chem 2014;289(29):20209–21.
[51] Hinz B, Phan SH, Thannickal VJ, et al. The myofibroblast: one function, multiple origins. Am J Pathol 2007;170(6):1807–16.
[52] Rombouts K, Carloni V. The fibrotic microenvironment as a heterogeneity facet of hepatocellular carcinoma. Fibrogenesis Tissue Repair 2013;6(1):17.
[53] Milosevic N, Schawalder H, Maier P. Kupffer cell-mediated differential down-regulation of cytochrome P450 metabolism in rat hepatocytes. Eur J Pharmacol 1999;368(1):75–87.
[54] Scholzel K, Schildberg FA, Welz M, et al. Transfer of MHC-class-I molecules among liver sinusoidal cells facilitates hepatic immune surveillance. J Hepatol 2014.
[55] Coulouarn C, Corlu A, Glaise D, et al. Hepatocyte-stellate cell cross-talk in the liver engenders a permissive inflammatory microenvironment that drives progression in hepatocellular carcinoma. Cancer Res 2012;72(10):2533–42.
[56] Friedman SL, Arthur MJ. Activation of cultured rat hepatic lipocytes by Kupffer cell conditioned medium. Direct enhancement of matrix synthesis and stimulation of cell proliferation via induction of platelet-derived growth factor receptors. J Clin Invest 1989;84(6):1780–5.
[57] Chen L, Charrier A, Zhou Y, et al. Epigenetic regulation of connective tissue growth factor by MicroRNA-214 delivery in exosomes from mouse or human hepatic stellate cells. Hepatology 2014;59(3):1118–29.
[58] Povero D, Eguchi A, Niesman IR, et al. Lipid-induced toxicity stimulates hepatocytes to release angiogenic microparticles that require Vanin-1 for uptake by endothelial cells. Sci Signal 2013;6(296):ra88.
[59] Rombouts K, Carloni V, Mello T, et al. Myristoylated Alanine-Rich protein Kinase C Substrate (MARCKS) expression modulates the metastatic phenotype in human and murine colon carcinoma *in vitro* and *in vivo*. Cancer Lett 2013.
[60] Gentilini A, Rombouts K, Galastri S, et al. Role of the stromal-derived factor-1 (SDF-1)-CXCR4 axis in the interaction between hepatic stellate cells and cholangiocarcinoma. J Hepatol 2012;57(4):813–20.
[61] Amann T, Bataille F, Spruss T, et al. Activated hepatic stellate cells promote tumorigenicity of hepatocellular carcinoma. Cancer Sci 2009;100(4):646–53.
[62] Badiola I, Olaso E, Crende O, et al. Discoidin domain receptor 2 deficiency predisposes hepatic tissue to colon carcinoma metastasis. Gut 2012;61(10):1465–72.

[63] Senoo H, Hata R. Extracellular matrix regulates cell morphology, proliferation, and tissue formation. Kaibogaku Zasshi 1994;69(6):719–33.
[64] Senoo H, Imai K, Matano Y, et al. Molecular mechanisms in the reversible regulation of morphology, proliferation and collagen metabolism in hepatic stellate cells by the three-dimensional structure of the extracellular matrix. J Gastroenterol Hepatol 1998;13(Suppl.):S19–32.
[65] Miura M, Sato M, Toyoshima I, et al. Extension of long cellular processes of hepatic stellate cells cultured on extracellular type I collagen gel by microtubule assembly: observation utilizing time-lapse video-microscopy. Cell Struct Funct 1997;22(4):487–92.
[66] Awata R, Sawai H, Imai K, et al. Morphological comparison and functional reconstitution of rat hepatic parenchymal cells on various matrices. J Gastroenterol Hepatol 1998;13(Suppl.):S55–61.
[67] Imai K, Sato T, Senoo H. Adhesion between cells and extracellular matrix with special reference to hepatic stellate cell adhesion to three-dimensional collagen fibers. Cell Struct Funct 2000;25(6):329–36.
[68] Takahara T, Zhang LP, Yata Y, et al. Modulation of matrix metalloproteinase-9 in hepatic stellate cells by three-dimensional type I collagen: its activation and signaling pathway. Hepatol Res 2003;26(4):318–26.
[69] Takahra T, Smart DE, Oakley F, et al. Induction of myofibroblast MMP-9 transcription in three-dimensional collagen I gel cultures: regulation by NF-kappaB, AP-1 and Sp1. Int J Biochem Cell Biol 2004;36(2):353–63.
[70] Lee SA, No da Y, Kang E, et al. Spheroid-based three-dimensional liver-on-a-chip to investigate hepatocyte–hepatic stellate cell interactions and flow effects. Lab Chip 2013;13(18):3529–37.
[71] Kostadinova R, Boess F, Applegate D, et al. A long-term three dimensional liver co-culture system for improved prediction of clinically relevant drug-induced hepatotoxicity. Toxicol Appl Pharmacol 2013;268(1):1–16.
[72] Hakkinen KM, Harunaga JS, Doyle AD, et al. Direct comparisons of the morphology, migration, cell adhesions, and actin cytoskeleton of fibroblasts in four different three-dimensional extracellular matrices. Tissue Eng Part A 2011;17(5–6):713–24.
[73] Harunaga JS, Yamada KM. Cell-matrix adhesions in 3D. Matrix Biol 2011;30(7–8):363–8.
[74] Westra IM, Oosterhuis D, Groothuis GM, et al. Precision-cut liver slices as a model for the early onset of liver fibrosis to test antifibrotic drugs. Toxicol Appl Pharmacol 2014;274(2):328–38.
[75] Szalowska E, Stoopen G, Groot MJ, et al. Treatment of mouse liver slices with cholestatic hepatotoxicants results in down-regulation of Fxr and its target genes. BMC Med Genomics 2013;6:39.
[76] van Swelm RP, Hadi M, Laarakkers CM, et al. Proteomic profiling in incubation medium of mouse, rat and human precision-cut liver slices for biomarker detection regarding acute drug-induced liver injury. J Appl Toxicol 2014;34(9):993–1001.
[77] Olinga P, Meijer DK, Slooff MJ, et al. Liver slices in *in vitro* pharmacotoxicology with special reference to the use of human liver tissue. Toxicol In Vitro 1997;12(1):77–100.
[78] van de Bovenkamp M, Groothuis GM, Draaisma AL, et al. Precision-cut liver slices as a new model to study toxicity-induced hepatic stellate cell activation in a physiologic milieu. Toxicol Sci 2005;85(1):632–8.
[79] Clouzeau-Girard H, Guyot C, Combe C, et al. Effects of bile acids on biliary epithelial cell proliferation and portal fibroblast activation using rat liver slices. Lab Invest 2006;86(3):275–85.
[80] Guyot C, Lepreux S, Combe C, et al. Fibrogenic cell phenotype modifications during remodelling of normal and pathological human liver in cultured slices. Liver Int 2010;30(10):1529–40.
[81] Vickers AE, Saulnier M, Cruz E, et al. Organ slice viability extended for pathway characterization: an *in vitro* model to investigate fibrosis. Toxicol Sci 2004;82(2):534–44.
[82] Olinga P, Groen K, Hof IH, et al. Comparison of five incubation systems for rat liver slices using functional and viability parameters. J Pharmacol Toxicol Methods 1997;38(2):59–69.
[83] Friedman SL. Preface. Hepatic fibrosis: pathogenesis, diagnosis, and emerging therapies. Clin Liver Dis 2008;12(4):13–24.
[84] Schuppan D, Pinzani M. Anti-fibrotic therapy: lost in translation? J Hepatol 2012;56(Suppl. 1):S66–74.
[85] Chen CZ, Peng YX, Wang ZB, et al. The Scar-in-a-Jar: studying potential antifibrotic compounds from the epigenetic to extracellular level in a single well. Br J Pharmacol 2009;158(5):1196–209.

[86] Bonandrini B, Figliuzzi M, Papadimou E, et al. Recellularization of well-preserved acellular kidney scaffold using embryonic stem cells. Tissue Eng Part A 2014;20(9–10):1486–98.
[87] Orlando G. Regenerative medicine technology applied to gastroenterology: current status and future perspectives. World J Gastroenterol 2012;18(47):6874–5.
[88] Faulk DM, Johnson SA, Zhang L, et al. Role of the extracellular matrix in whole organ engineering. J Cell Physiol 2014;229(8):984–9.
[89] Godoy P, Hewitt NJ, Albrecht U, et al. Recent advances in 2D and 3D *in vitro* systems using primary hepatocytes, alternative hepatocyte sources and non-parenchymal liver cells and their use in investigating mechanisms of hepatotoxicity, cell signaling and ADME. Arch Toxicol 2013;87(8):1315–530.
[90] Mao TK, Kimura Y, Kenny TP, et al. Elevated expression of tyrosine kinase DDR2 in primary biliary cirrhosis. Autoimmunity 2002;35(8):521–9.
[91] Hammond JS, Gilbert TW, Howard D, et al. Scaffolds containing growth factors and extracellular matrix induce hepatocyte proliferation and cell migration in normal and regenerating rat liver. J Hepatol 2011;54(2):279–87.
[92] Khan AA, Vishwakarma SK, Bardia A, et al. Repopulation of decellularized whole organ scaffold using stem cells: an emerging technology for the development of neo-organ. J Artif Organs 2014;17(4):291–300.
[93] Mazza G, Rombouts K, Malago M, Dhar D, Hall A, Dhillon P, et al. 3D biological scaffolds obtained from discarded human livers as a platform for tissue engineering and regenerative medicine. J Hepatol 2014;60(1):S69.
[94] Bhatia SN, Underhill GH, Zaret KS, et al. Cell and tissue engineering for liver disease. Sci Transl Med 2014;6(245):245sr2.
[95] Kasuya J, Sudo R, Mitaka T, et al. Hepatic stellate cell-mediated three-dimensional hepatocyte and endothelial cell triculture model. Tissue Eng Part A 2011;17(3–4):361–70.
[96] Kasuya J, Sudo R, Mitaka T, et al. Spatio-temporal control of hepatic stellate cell-endothelial cell interactions for reconstruction of liver sinusoids *in vitro*. Tissue Eng Part A 2012;18(9–10):1045–56.

CHAPTER 3

Hepatic Fibrosis: A Global Clinical Problem

Massimo Pinzani
UCL Institute for Liver and Digestive Health, Division of Medicine, University College London, Royal Free Hospital, London, UK

3.1 INTRODUCTION

Progressive accumulation of fibrillar extracellular matrix (ECM) in the liver is the consequence of reiterated liver tissue damage due to infectious (mostly hepatitis B virus [HBV] and hepatitis C virus [HCV]), toxin/drug-induced, metabolic, and autoimmune causes and the relative chronic activation of the wound-healing reaction. The process may result in clinically evident liver cirrhosis and hepatic failure. Cirrhosis is defined as an advanced stage of fibrosis, characterized by the formation of regenerative nodules of liver parenchyma that are separated by and encapsulated in fibrotic septa and associated with major angio-architectural changes [1].

Independent of the etiology, liver cirrhosis is globally the most common non-neoplastic cause of death among hepatobiliary and digestive diseases. In addition, this condition is largely associated with hepatocellular carcinoma (HCC), with a further increase in the relative mortality rate [2]. HCV infection is associated with the highest HCC incidence in persons with cirrhosis, occurring twice as commonly in Japan as in the West (5-year cumulative incidence, 30% and 17%, respectively), followed by hereditary hemochromatosis (5-year cumulative incidence, 21%). In HBV-related cirrhosis, the 5-year cumulative HCC risk is 15% in high endemic areas and 10% in the West. In the absence of HCV and HBV infection, the HCC incidence is lower in alcoholic cirrhotics (5-year cumulative risk, 8%) and subjects with advanced primary biliary cirrhosis (PBC) (5-year cumulative risk, 4%) [2]. Increasing evidence suggests a high but not yet clearly quantified incidence in nonalcoholic fatty liver disease (NAFLD) also in patients without cirrhosis [3].

Regardless of the etiology, the following clinical features have been shown to be predictors of the development of significant fibrosis, or at least, of an accelerated progression to cirrhosis: (i) male gender (for groups of age <50 years); (ii) age at infection (HBV and HCV); (iii) obesity, diabetes mellitus and, more in general, the clinical features of the so-called "metabolic syndrome"; (iv) excessive daily alcohol intake; and (v) elevated hepatic iron content. In addition, individual (genetic and epigenetic) factors

Stellate Cells in Health and Disease
DOI: http://dx.doi.org/10.1016/B978-0-12-800134-9.00003-8

are likely to affect several aspects of the fibrogenic process (i.e., differences in handling a metabolic/toxic load, in the immune system reactions toward infectious agents and autoantigens, and in the evolution of the chronic wound-healing reaction). These individual differences are responsible for markedly different rates of fibrosis progression (i.e., slow, fast, and intermediate "fibrosers") for apparently similar clinical features [4,5].

The incidence and the impact of hepatic fibrosis on public health are closely associated with the prevalence of the various causes of chronic liver damage. Classically, excessive alcohol consumption, HBV, and HCV have been the main causes of cirrhosis. In recent years, prevalence of obesity and metabolic syndrome has resulted in increasing incidence of cirrhosis secondary to NAFLD, especially in developed countries [6]. Regardless, it is increasingly evident that NAFLD is also an emerging phenomenon in developing countries exposed to the industrial Western diet rich in fat and fructose [7].

3.2 CHRONIC VIRAL HEPATITIS

Worldwide, an estimated 130–170 million people have HCV infection. HCV prevalence is highest in Egypt at >10% of the general population and China has the most people with HCV (29.8 million) [8,9]. Differences in past HCV incidence and current HCV prevalence, together with the generally protracted nature of HCV disease progression, has led to considerable diversity in the burden of advanced liver disease in different countries around the globe. Countries with a high incidence of HCV or peak incidence in the recent past will have further escalations in HCV-related cirrhosis and HCC over the next two decades [10]. Indeed, morbidity and mortality rates from chronic HCV infection will increase because the infection incidence peaked in the mid-1980s and because the liver disease progresses slowly and is clinically silent to cirrhosis and end-stage liver disease over a 15- to 20-year time period for 15–20% of chronically infected individuals. Acute HCV infection is difficult to detect because of the generally asymptomatic nature of the disease and the marginalization of at-risk populations especially in large industrialized cities. Around 25% of the patients with acute HCV infection undergo spontaneous clearance, with increased rates among those with favorable *IL28B* genotypes, acute symptoms, and in women. The remaining 75% of patients progress to chronic HCV infection and are subsequently at risk of progression to hepatic fibrosis, cirrhosis, and HCC [3]. Chronic hepatitis C generally progresses slowly in the initial two decades, but can be accelerated during this time as a result of advancing age and co-factors such as heavy alcohol intake, HIV co-infection, and obesity/diabetes.

HBV is the most common among those hepatitis viruses that cause chronic infections of the liver in humans, and it represents a global public health problem. Chronic hepatitis caused by HBV is the leading cause of cirrhosis worldwide and approximately one-quarter of patients with cirrhosis develop decompensated liver disease within

5 years [11,12]. In addition, chronic HBV infection is the major cause of HCC worldwide [13]. Approximately 15 million of the estimated 350 million individuals with chronic HBV infection have evidence of exposure to hepatitis D (delta) virus (HDV), which requires hepatitis B surface antigen for transmission and packaging [14,15]. HBV/HDV co-infection is associated with more severe acute hepatitis and higher mortality than acute HBV mono-infection [16]. Chronic co-infection is clearly associated with a higher risk of cirrhosis and decompensated liver disease. Hepatitis B and all of the complications resulting from it, as well HDV and its complications, are globally preventable by hepatitis B vaccination, and therefore elimination of HBV transmission and of new acute and chronic infections appear to be a feasible goal.

In intravenous drug abusers, the co-infection of HIV with HBV and/or HCV is common because of shared modes of transmission [16]. It is known that HIV accelerates the progression of liver fibrosis and results in increased morbidity and mortality associated with viral hepatitis, but it is less clear if viral hepatitis has a direct effect on HIV. Currently, the association of HIV and HCV has the most frequent occurrence and an estimated 5–10 million individuals in the Western world alone are infected with both viruses [17,18]. Treatment of viral hepatitis improves the outcome and should be considered in all HIV-infected patients. Treatment of HBV without concurrent treatment of HIV is risky because resistance can occur in both viruses if regimens are not carefully chosen.

The incidence of liver fibrosis consequent to chronic HBV and HCV infection is supposed to slowly decline with the introduction of effective antiviral agents. Indeed, the recent development and introduction of antiviral regimens with an extremely high eradication efficacy for HCV infection has signed a turning point in the fight against liver fibrosis. However, the very high cost for an individual treatment will limit the use of these drugs to high-income industrialized countries for at least the next decade [19,20].

3.2.1 Alcohol-induced liver fibrosis

The incidence of alcohol-induced (acute and chronic) liver disease (ALD) is still very high in both industrialized and developing countries. The association between alcohol intake and alcoholic liver disease has been well documented, although cirrhosis of the liver develops in only a relatively small proportion of heavy drinkers [21]. The risk of cirrhosis increases proportionally with consumption of more than 30 g of alcohol per day, and the highest risk is associated with consumption of more than 120 g per day [21]. The prevalence of cirrhosis is 1% in persons drinking 30–60 g of alcohol per day and up to 5.7% in those consuming 120 g daily. It is presumed that other factors, such as gender, genetic and epigenetic characteristics, and presence of co-morbidities (including metabolic syndrome and chronic viral infection), play an important role in the genesis of ALD [22–25].

Chronic alcohol use may cause several types of liver injury. Regular alcohol abuse, even for just a few days, can result in steatosis, a disorder in which hepatocytes accumulate macrovesicular droplets containing triglycerides. Although alcoholic fatty liver resolves with abstinence, steatosis predisposes people who continue to consume alcohol to develop chronic alcoholic steatohepatitis (ASH), hepatic fibrosis, and cirrhosis [25]. A typical feature of advanced alcoholic liver disease is the manifestation of episodes of acute hepatitis over a chronic background of fibrosis or cirrhosis (acute on chronic).

An alarming aspect of alcohol abuse, particularly among adolescent and young adults, is the so-called "binge drinking" (i.e., excessive alcohol intake over a short period of time, in general during the weekend) with intervals of abstinence. Recent studies from experimental animals and from humans indicate that binge drinking has profound effects on immunological, signaling, and epigenetic parameters of the liver. This is in addition to the known metabolic effects of acute alcohol intake. Indeed, binge drinking alters the levels of several cellular components and dramatically amplifies liver injury in chronically ethanol exposed liver [26].

3.3 NAFLD AND NASH

NAFLD has evolved as a serious public health problem in both industrialized and developing countries. At present, 14–27% of the general population in the industrialized world has NAFLD. Moreover, nonalcoholic steatohepatitis (NASH)—the most serious form of NAFLD—is predicted to become the leading cause of liver transplantation in the United States and likely in Europe by the year 2020 [7].

NAFLD covers cases of a wide spectrum of severity, ranging from bland fatty liver without any inflammation and with little or no tendency to progress all the way to NASH with inflammatory reactions and hepatocyte damage, with or without fibrosis. Five to twenty percent of patients with NAFLD develop NASH, which undergoes a further transition to higher-grade fibrosis in 10–20% of cases [7]. In fewer than 5% of cases, fibrosis progresses to cirrhosis. These approximate figures lead to an estimate of 0.05–0.3% for the prevalence of cirrhosis in the general population. Moreover, NAFLD can progress to HCC also in the absence of apparent cirrhosis [3,27]. The diagnosis of fatty liver disease can be suspected initially on the basis of abnormally high aspartate aminotransferase (AST) and/or alanine aminotransferase (ALT) levels and ultrasonographic findings of fatty liver [6].

The increasing burden of NAFLD and the current lack of effective treatments to halt or reverse disease progression in patients with NASH have led to intense investigation into the basic mechanisms involved in the development and progression of the disease. The most accepted concept outlining the pathogenesis of NAFLD implies the occurrence of parallel events (hits) that involve a complex interaction and crosstalk among environmental factors, host genetics, and gut microflora [28]. This interaction might promote isolated steatosis, cell death or progressive liver damage, innate immune activation, inflammation, and fibrosis.

3.4 AUTOIMMUNE HEPATITIS, PRIMARY BILIARY CIRRHOSIS, AND PRIMARY SCLEROSING CHOLANGITIS

Autoimmune hepatitis (AIH) is a chronic inflammatory disease of the liver that occurs worldwide with a low but probably underestimated prevalence. Although AIH typically affects young and middle-aged women, it can occur in both sexes and across all age groups [29]. Prognosis of severe AIH is poor if untreated. The pathogenesis is complex, combining environmental factors (external chemical or infectious triggers) and host genetic susceptibility.

PBC is a chronic, progressive, cholestatic, organ-specific autoimmune disease of unknown etiology. It predominantly affects middle-aged women and is characterized by autoimmune-mediated destruction of small- and medium-size intrahepatic bile ducts, portal inflammation, and progressive scarring, which without proper treatment can ultimately lead to biliary fibrosis and hepatic failure. In recent years, advanced biochemical assays, further delineation of specific liver histological findings, more effective serum auto-antibody detection methods, and improved diagnostic abilities have led to higher prevalence estimates worldwide [30,31]. Currently, it is believed that PBC is likely to be triggered by a combination of environmental factors including infection in a genetically susceptible individual. This hypothesis is supported by the high concordance rate of PBC among first-degree relatives and in homozygous twins (~60%) [32,33].

Primary sclerosing cholangitis (PSC) is a chronic cholestatic disease of the liver that is characterized by progressive inflammation, fibrosis, and stricturing of the intrahepatic and extrahepatic bile ducts [34,35]. It is progressive in most patients and leads to cirrhosis. It is a rare disease, mostly affecting people of northern European descent and males more than females. Inflammatory bowel disease (IBD) is present in ~75% of the patients with PSC, mostly ulcerative colitis (~85% of the cases). In addition to biliary fibrosis and cirrhosis, complications of PSC include dominant strictures of the bile ducts, cholangitis, and cholangiocarcinoma. PSC likely occurs in genetically susceptible individuals, perhaps after exposure to environmental triggers. These could initiate a series of events that involve complex interactions between the innate and adaptive immune systems, ultimately leading to lymphocyte migration, cholangiocyte damage, and progressive fibrosis. Several important observations, coupled with the strong association between certain human leukocyte antigen (HLA) haplotypes and frequency of concurrent extrahepatic autoimmune disorders, support the concept that PSC is an immune-mediated phenomenon.

3.5 ETIOLOGY-DRIVEN LIVER FIBROSIS

Although cirrhosis is the common result of progressive fibrogenesis, there are distinct patterns of fibrotic development related to the underlying disorders causing the fibrosis and the prevalent type of tissue insult and relative damage [1,36].

The pattern of fibrosis development typical of chronic viral hepatitis or of AIH is considered the result of portal–central (vein) bridging necrosis, thus originating in portal–central septa. In addition, this form of fibrogenic evolution is characterized by the presence of "interface" hepatitis and development of portal to portal septa and septa ending blind in the parenchyma, and by rapid derangement of the vascular connections with the portal system (early portal hypertension) [37]. Chronic viral hepatitis and AIH are characterized primarily by chronic hepatocellular necrosis and apoptosis. Consequently, the main driver of fibrogenesis is a typical chronic wound-healing reaction. This process, which is highly efficient in the presence of single acute tissue insult, leads to progressive scarring when tissue damage is chronic. In other words, deposition of fibrillar matrix rather than organized tissue regeneration becomes the best option in order to maintain tissue continuity. In addition to chronic hepatocellular necrosis and apoptosis, hepatic fibrogenesis due to the chronic activation of the wound-healing reaction is characterized by the following key features: (i) a complex inflammatory infiltrate including mononuclear cells and immunocompetent cells; (ii) the activation of different types of ECM-producing cells (HSC and portal myofibroblasts) with marked proliferative, synthetic, and contractile features; and (iii) marked changes in the quality and quantity of the hepatic ECM associated with very limited or absent possibilities of remodeling and regeneration [36,38].

Several lines of evidence suggest that viral proteins may directly exert a pro-fibrogenic effect. Incubation of activated but not quiescent human HSC with recombinant HCV core and nonstructural protein 3 (NS3) protein increased intracellular calcium concentration and reactive oxygen species (ROS) production [39]. Expression of the core HCV protein in HSC by employing adenoviral gene transfer induced increased cell proliferation in a Ras/ERK- and PI3K/AKT-dependent manner, while the expression of the NS3–NS5 protein expression preferentially induced proinflammatory actions, such as increased chemokine secretion and expression of intercellular cell adhesion molecule type 1 (ICAM-1) through the NF-κB and c-Jun N-terminal kinase pathways. In addition, the interaction between recombinant HCV envelope E2 glycoprotein and CD81, a tetraspanin highly expressed on the activated HSC membrane, has been shown to induce a time-dependent increase in the synthesis and activity of matrix metallopeptidase-2 (MMP-2; gelatinase A), a major enzyme involved in the degradation of normal hepatic ECM [40]. Activation of MMP-2 leads to the degradation of the normal hepatic ECM in areas where HCV is concentrated, favoring the penetration of the inflammatory infiltrate in the space of Disse along chemotactic gradients, leading to further parenchymal damage. During HIV infection, target cells are exposed to viral gene products expressed at the surface of infected cells, or released in the microenvironment. In particular, the human immunodeficiency virus 1 (HIV-1) envelope protein gp120 is present in tissues and circulates in the blood on the surface of virions or as a free protein. Exposure of human HSC to recombinant gp120 effectively induced a significant increase in cell migration and augmented the secretion of monocyte chemotactic protein 1 (MCP-1), a proinflammatory chemokine, and interleukin 6 (IL-6). Overall these data suggest that gp120 exposed by infected mononuclear cells leads to further MCP-1 production by HSC thus facilitating additional HSC and inflammatory cell recruitment in the areas of active tissue remodeling [41].

The general pathological features of liver fibrosis developing in ASH and NASH are very similar, if not indistinguishable (Figure 3.1). Typically, fibrosis is initially limited to lobular zone 3 and is concentrated around the sinusoids (capillarization) and around groups of hepatocytes (chicken-wire pattern) on a general background of hepatocellular steatosis/ballooning with different degrees of lobular inflammation. This pericellular fibrosis eventually forms septa isolating regenerating nodules leading to a picture typical of cirrhosis. Although the mechanisms underlying fibrosis in ALD and NAFLD are in part common to those observed in other CLDs ("core pathways" of hepatic fibrogenesis), there are specific "regulatory pathways" for ASH and NASH. In both

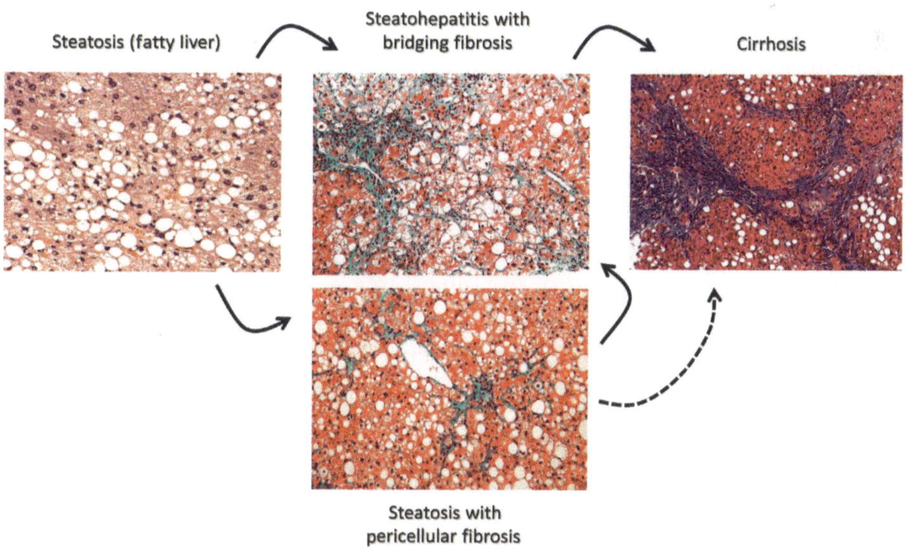

Figure 3.1 Fibrogenic evolution of alcoholic and nonalcoholic fatty liver diseases. Simple fatty liver can originate a progressive hepatic fibrogenic disorders with the appearance of inflammation and cell necrosis (steatohepatitis) leading to cirrhosis. An alternative to this "direct" pathway to cirrhosis is the slow deposition of fibrotic tissue within sinusoidal structure with eventual evolution to cirrhosis.

conditions, oxidative stress-induced inflammation and fibrosis play a central role in the progression of the disease [42]. In alcoholic liver injury, for example, acetaldehyde, the main metabolite of ethanol, is able to increase gene transcription and synthesis of different ECM components in activated hepatic stellate cells (HSCs) [43]. In addition to acetaldehyde, products of lipid peroxidation generated by exposure to ethanol or the production of iron overload may also perpetuate HSC activation. Along these lines, stimulation of lipid peroxidation or exposure to 4-hydroxynonenal (4-HNE), a highly reactive aldehydic end-product of lipid peroxidation, increases procollagen I gene expression in activated human HSC [44,45].

In NASH, the altered pattern of circulating adipokines, oxidative stress generation, and the hormonal profile associated with the metabolic syndrome clearly have a specific role for the induction of fibrogenesis in this condition. Expansion of adipose tissue, and in particular of visceral fat has been associated with a decreased release of insulin-sensitizing and antiinflammatory cytokines, and increased expression of proinflammatory and profibrogenic molecules, particularly TNF-α, leptin, and resistin. In addition, the expansion of visceral fat leads to an elevated free fatty acid (FFA) release. FFA-related lipotoxicity has been shown to play a key role in the genesis of liver fibrosis typical of NASH. Other key profibrogenic mechanisms typical of both ASH and NASH include methionine metabolism abnormalities, autophagy, hepatocyte apoptosis, and increased intestinal permeability with endotoxin (lipopolysaccharide)-induced activation of innate immunity mechanisms including the activation of toll-like receptors and of the so-called inflammasome [7].

Biliary fibrosis, typical of PBC and PSC, is characterized by the co-proliferation of reactive bile ductules and periductular (myo)fibroblast-like cells at the portal–parenchymal interface and develops following a portal to portal direction. This leads to the formation of portal–portal septa surrounding liver nodules, where the central vein and its connections with the portal tract are preserved until late stages (Figure 3.2). The histologic features of PSC are often nonspecific and prone to sampling variations due to the heterogeneous involvement of the biliary tree. Unfortunately, the classic description of concentric ductal fibrosis (onion skinning) involving bile ducts within portal tract areas is rarely encountered in clinical practice [34,35].

Finally, the central to central (vein) form of fibrogenic evolution is in general secondary to venous outflow problems (e.g., Budd-Chiari syndrome, chronic heart failure) and is characterized by the development of central to central septa and "reversed lobulation" [5].

The difference in the patterns of evolution underlines the prevalence of different fibrogenic mechanisms and multiple cellular effectors of fibrogenesis. Accordingly, the natural history of disease associated with a cirrhotic liver caused by chronic alcohol abuse could be markedly different when compared to a cirrhotic liver caused by chronic viral hepatitis. In addition, the potential for disease progression and regression

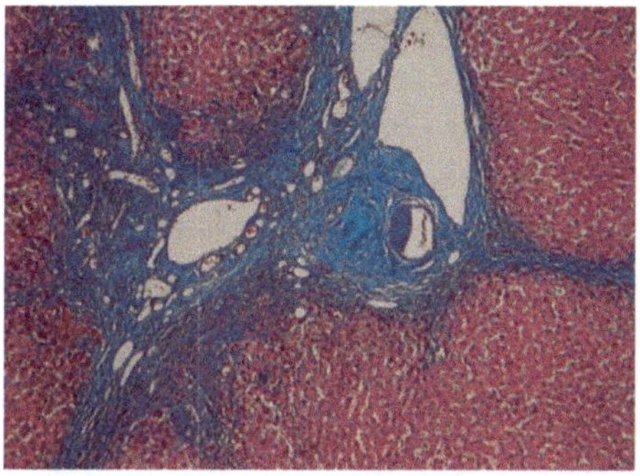

Figure 3.2 Example of biliary fibrosis typical of PBC and PSC. The co-proliferation of reactive bile ductules and periductular (myo) fibroblast-like cells at the portal–parenchymal interface leads to a fibrotic pattern developing along a portal to portal direction. This leads to the formation of portal–portal septa surrounding liver nodules, where the central vein and its connections with the portal tract are preserved until late stages.

following abstinence and treatment will also be different. Along these lines, the results of a study [46] analyzing fibrosis distribution in explanted cirrhotic liver indicate that a case of alcoholic cirrhosis requiring liver transplantation is characterized by a collagen content, assessed with recently introduced collagen proportionate area (CPA) method [47], which is on average double of that of a case of HCV cirrhosis. These observations affirm the idea that "cirrhosis," or more precisely advanced-stage CLD, should be primarily classified according to etiology: it is indeed rare in modern medicine to find a term equivalent to cirrhosis that serves as a blanket term to denote "end-stage fibrous disease" of other organs.

3.5.1 Perspectives

Increasing knowledge of the pathogenesis of hepatic fibrosis has led to important changes in the clinical interpretation of this phenomenon. The awareness that liver fibrosis has different patterns of evolution depending on the etiology of CLD and several prevailing mechanisms, will lead to the development of distinct diagnostic and therapeutic modalities. This will ultimately improve the clinical management of chronic fibrogenic liver diseases. In addition, the identification of genetic and epigenetic factors involved in the progression of liver fibrosis is expected to lead to the establishment of more accurate prognostic markers and to a better stratification of patients for targeted therapeutic treatments.

It is predictable that the incidence of advanced CLD due to chronic HBV and HCV infection will slowly decrease due to increased coverage of HBV vaccination and increasingly effective antiviral treatments. However, due to inherent high social costs, it is likely that these improvements will be limited to wealthy industrialized countries and will cause a further global disparity. The incidence of alcohol-induced liver damage and fibrosis, while slowly decreasing due to increased public awareness, is facing new patterns of abuse (i.e., increased incidence of binge drinking in the younger generations). Overall, it is likely that liver fibrosis developing as consequence of NASH and the metabolic syndrome will represent the most important clinical burden in the following decades.

REFERENCES

[1] Friedman SL. Evolving challenges in hepatic fibrosis. Nat Rev Gastroenterol Hepatol 2010;7:425–36.
[2] El-Serag HB. Epidemiology of viral hepatitis and hepatocellular carcinoma. Gastroenterology 2012;142:1264–73.
[3] Baffy G, Brunt EM, Caldwell SH. Hepatocellular carcinoma in non-alcoholic fatty liver disease: an emerging menace. J Hepatol 2013;56:1384–91.
[4] Poynard T, Bedossa P, Opolon P. Natural history of liver fibrosis progression in patients with chronic hepatitis C. The OBSVIRC, METAVIR, CLINIVIR, and DOSVIRC groups. Lancet 1997;349:825–32.
[5] Pinzani M, Rombouts K, Colagrande S. Fibrosis in chronic liver diseases: diagnosis and management. J Hepatol 2005;42(Suppl. 1):S22–36.
[6] Ong JP, Younossi ZM. Epidemiology and natural history of NAFLD and NASH. Clin Liver Dis 2007;11(1):1–16.
[7] Wree A, Broderick L, Canbay A, Hoffman HM, Feldstein AE. From NAFLD to NASH to cirrhosis—new insights into disease mechanisms. Nat Rev Gastroenterol Hepatol 2013;10:627–36.
[8] Shepard CW, Finelli L, Alter MJ. Global epidemiology of hepatitis C virus infection. Lancet Infect Dis 2005;5:558–67.
[9] Gower E, Estes C, Blach S, Razavi-Shearer K, Razavi H. Global epidemiology and genotype distribution of the hepatitis C virus infection. J Hepatol 2014 pii:S0168-8278(14)00526-1. http://dx.doi.org/10.1016/j.jhep.2014.07.027 [Epub ahead of print].
[10] Aghemo A, Colombo M. Hepatocellular carcinoma in chronic hepatitis C: from bench to bedside. Semin Immunopathol 2013;35:111–20.
[11] McMahon BJ. Chronic hepatitis B virus infection. Med Clin North Am 2014;98:39–54. http://dx.doi.org/10.1016/j.mcna.2013.08.004 [Epub ahead of print].
[12] Iavarone M, Colombo M. HBV infection and hepatocellular carcinoma. Clin Liver Dis 2013;17:375–97.
[13] Rizzetto M, Alavian SM. Hepatitis delta: the rediscovery. Clin Liver Dis 2013;17:475–87.
[14] Rizzetto M, Ciancio A. Epidemiology of hepatitis D. Semin Liver Dis 2012;32:211–9.
[15] Farci P, Niro GA. Clinical features of hepatitis D. Semin Liver Dis 2012;32:228–36.
[16] Petty LA, Steinbeck JL, Pursell K, Jensen DM. Human immunodeficiency virus and coinfection with hepatitis B and C. Infect Dis Clin North Am 2014;28:477–99.
[17] Clausen LN, Lundbo LF, Benfield T. Hepatitis C virus infection in the human immunodeficiency virus infected patient. World J Gastroenterol 2014;20:12132–43.
[18] Nelson PK, Mathers BM, Cowie B, Hagan H, Des Jarlais D, Horyniak D, et al. Global epidemiology of hepatitis B and hepatitis C in people who inject drugs: results of systematic reviews. Lancet 2011;378:571–83.
[19] Toy M. Cost-effectiveness of viral hepatitis B and C treatment. Best Pract Res Clin Gastroenterol 2013;27:973–85.
[20] Zeuzem S. Decade in review-HCV: hepatitis C therapy-a fast and competitive race. Nat Rev Gastroenterol Hepatol 2014. http://dx.doi.org/10.1038/nrgastro.2014.164 [Epub ahead of print].
[21] Bellentani S, Saccoccio G, Costa G, Tiribelli C, Manenti F, Sodde M, et al. Drinking habits as cofactors of risk for alcohol induced liver damage. Gut 1997;41:845–50.
[22] Becker U, Deis A, Sorensen TI, Grønbaek M, Borch-Johnsen K, Müller CF, et al. Prediction of risk of liver disease by alcohol intake, sex, and age: a prospective population study. Hepatology 1996;23:1025–9.
[23] de Alwis W, Day CP. Genetics of alcoholic liver disease and nonalcoholic fatty liver disease. Semin Liver Dis 2007;27:44–54.

[24] Nishiguchi S, Kuroki T, Yabusako T, Seki S, Kobayashi K, Monna T, et al. Detection of hepatitis C virus antibodies and hepatitis C virus RNA in patients with alcoholic liver disease. Hepatology 1991;14:985–9.
[25] Teli MR, Day CP, Burt AD, Bennett MK, James OF. Determinants of progression to cirrhosis or fibrosis in pure alcoholic fatty liver. Lancet 1995;346:987–90.
[26] Shukla SD, Pruett SB, Szabo G, Arteel GE. Binge ethanol and liver: new molecular developments. Alcohol Clin Exp Res 2013;37:550–7.
[27] Ertle J, Dechêne A, Sowa JP, Penndorf V, Herzer K, Kaiser G, et al. Non-alcoholic fatty liver disease progresses to hepatocellular carcinoma in the absence of apparent cirrhosis. Int J Cancer 2011;128:2436–43.
[28] Tilg H, Moschen AR. Evolution of inflammation in nonalcoholic fatty liver disease: the multiple parallel hits hypothesis. Hepatology 2010;52:1836–46.
[29] Heneghan MA, Yeoman AD, Verma S, Smith AD, Longhi MS. Autoimmune hepatitis. Lancet 2013;382:1433–44.
[30] Selmi C, Invernizzi P, Zuin M, Podda M, Gershwin ME. Genetics and geoepidemiology of primary biliary cirrhosis: following the footprints to disease etiology. Semin Liver Dis 2005;25:265–80.
[31] Prince MI, James OF. The epidemiology of primary biliary cirrhosis. Clin Liver Dis 2003;7:795–819.
[32] Selmi C, Mayo MJ, Bach N, Ishibashi H, Invernizzi P, Gish RG, et al. Primary biliary cirrhosis in monozygotic and dizygotic twins: genetics, epigenetics, and environment. Gastroenterology 2004;127:485–92.
[33] Lazaridis KN, Juran BD, Boe GM, Slusser JP, de Andrade M, Homburger HA, et al. Increased prevalence of antimitochondrial antibodies in first-degree relatives of patients with primary biliary cirrhosis. Hepatology 2007;46:785–92.
[34] Karlsen TH, Boberg KM. Update on primary sclerosing cholangitis. J Hepatol 2013;59:571–82.
[35] Yimam KK, Bowlus CL. Diagnosis and classification of primary sclerosing cholangitis. Autoimmun Rev 2014;13:445–50.
[36] Pinzani M, Macias-Barragan J. Update on the pathophysiology of liver fibrosis. Expert Rev Gastroenterol Hepatol 2010;4:459–72.
[37] Bosch J, Garcia-Pagan GC, Berzigotti A, Abraldres G. Measurement of portal pressure and its role in the management of chronic liver disease. Semin Liver Dis 2006;26:348–62.
[38] Mehal WZ, Iredale J, Friedman SL. Scraping fibrosis: expressway to the core of fibrosis. Nat Med 2011;17:552–3.
[39] Bataller R, Paik YH, Lindquist JN, Lemasters JJ, Brenner DA. Hepatitis C virus core and nonstructural proteins induce fibrogenic effects in hepatic stellate cells. Gastroenterology 2004;126:529–40.
[40] Mazzocca A, Sciammetta SC, Carloni V, et al. Binding of hepatitis C virus envelope protein E2 to CD81 up-regulates matrix metalloproteinase-2 in human hepatic stellate cells. J Biol Chem 2005;280:11329–39.
[41] Bruno R, Galastri S, Sacchi P, et al. The HIV envelope protein GP120 modulates the biology of human hepatic stellate cells: a link between HIV infection and liver fibrogenesis. Gut 2010;59:513–20.
[42] Novo E, Parola M. Redox mechanisms in hepatic chronic wound healing and fibrogenesis. Fibrogenesis Tissue Repair 2008;1(1):5.
[43] Casini A, Cunningham M, Rojkind M, Lieber CS. Acetaldehyde increases procollagen type I and fibronectin gene transcription in cultured rat fat-storing cells through a protein synthesis-dependent mechanism. Hepatology 1991;13:758–65.
[44] Casini A, Ceni E, Salzano R, Biondi P, Parola M, Galli A, et al. Neutrophil-derived superoxide anion induces lipid peroxidation and stimulate collagen synthesis in human hepatic stellate cells. Role of nitric oxide. Hepatology 1997;25:361–7.
[45] Parola M, Robino G, Marra F, Pinzani M, Bellomo G, Leonarduzzi G, et al. HNE interacts directly with JNK isoforms in human hepatic stellate cells. J Clin Invest 1998;102:1942–50.
[46] Hall A, Germani G, Isgrò G, Burroughs AK, Dhillon AP. Fibrosis distribution in explanted cirrhotic livers. Histopathology 2012;60:270–7.
[47] Calvaruso V, Burroughs AK, Standish K, Manousou P, Grillo F, Leandro G, et al. Computer-assisted image analysis of liver collagen: relationship to Ishak scoring and hepatic venous pressure gradient. Hepatology 2009;49:1236–44.

CHAPTER 4

Stellate Cells and Hepatic Fibrosis

Daisuke Hasegawa*[,1], Michael C. Wallace*[,1,2] and Scott L. Friedman[1]

[1]Division of Liver Diseases, Icahn School of Medicine at Mount Sinai, New York, NY, USA
[2]School of Medicine and Pharmacology, University of Western Australia, Perth, Australia

4.1 INTRODUCTION

In the United States in 2010, chronic liver disease and cirrhosis together were the 12th leading cause of mortality, accounting for 33,642 or 1.3% of all deaths [1]. Hepatic fibrosis is a reversible, protective response to virtually all forms of liver injury (inherited, metabolic, inflammatory, infectious, vascular, toxins, and drugs) [2]. While initially advantageous, when the injury becomes chronic, the ongoing cellular response becomes dysregulated and the fibrosis may eventually progress to cirrhosis and its associated complications [2]. Although the progression rate is variable and can be disease-specific, hepatic fibrosis may advance to cirrhosis in <10 years in some patients; once cirrhosis is established, the median 1-year survival in decompensated disease is <50% [3].

Fibrosis and cirrhosis are strictly defined pathological entities whose histological features have been extensively characterized, first in 1978 by the World Health Organization [4]. Fibrosis is defined as the presence of excess collagen within the extracellular matrix (ECM), produced by myofibroblasts that have been activated in response to injury [2,4,5]. Importantly, mild hepatic fibrosis typically leads to negligible clinical symptoms or disturbance of liver cell function until cirrhosis has developed. In the developed world, up to 45% of deaths are associated with fibrotic disease [6]. Increasing efforts to identify the common pathogenic underpinnings of these fibrotic processes, not only in liver but also across most other organs including lung, kidney, heart, pancreas, skin, and bone marrow, will yield great public health benefit with the anticipated arrival of anti-fibrotic therapies.

Recent advances in hepatic fibrosis have built upon the fundamental tenet that hepatic stellate cells (HSCs) orchestrate a complex and tightly regulated network of cellular and molecular responses, culminating in ECM accumulation [7]. In addition, genetic lineage tracing experiments in mouse models have firmly established HSCs as

*Both authors contributed equally to this manuscript

Stellate Cells in Health and Disease
DOI: http://dx.doi.org/10.1016/B978-0-12-800134-9.00004-X

the dominant source of myofibroblasts in experimental injury [8–10]. The cell's roles have expanded beyond fibrosis development to critical cellular responses during fibrosis resolution, as well as contributions to hepatic stem cell amplification, liver regeneration, immunoregulation, and hepatocellular carcinoma (HCC) development (explored in other chapters of this book) [5,7]. Finally, recent studies uncover a deepening complexity in HSC activation pathways. In this chapter, we provide a concise review of vital contributions of HSCs to hepatic fibrosis, and we highlight advances that may eventually lead to novel anti-fibrotic therapies.

4.2 PATHOGENESIS OF HEPATIC FIBROSIS

4.2.1 Pathological and clinical associations of hepatic fibrosis

HSCs reside in the perisinusoidal space of Disse. This space separates hepatocytes, which are lined with a low-density basal membrane, from endothelial cells, which form the walls of the hepatic sinusoids. Loss of the characteristic pores of endothelial cells, or fenestrations, is an early event in hepatic injury and is accompanied by secretion of ECM into the space of Disse by HSCs [11,12]. This loss of fenestrations or "capillarization" of the sinusoids interferes with the critical metabolic, bidirectional homeostatic flow between mixed blood (from the hepatic artery and portal vein) and the functional cell of the liver, the hepatocyte [2], and underlies the link between fibrosis and organ dysfunction. As more ECM is deposited, its components progressively infiltrate into hepatic parenchyma, first as an expansion of portal tracts with fibrotic tissue, followed by vascularized bridging septae between portal tracts, and eventually into frank cirrhosis with extensive replacement of parenchyma with bands of fibrotic tissue interspersed with characteristic regenerative nodules [13]. This gross distortion of the hepatic parenchyma and vasculature undermines the synthetic, detoxification, and filtration capacity of the liver and elevates portal pressure, often leading to liver failure and death. Additionally, the resultant fibrotic microenvironment is fertile soil for malignant transformation of hepatocytes [14]. Indeed, up to 90% of all HCCs arise within a cirrhotic liver [15,16]. In addition to the more than 30,000 deaths in the United States due to cirrhosis and chronic liver disease in 2010, 21,608 people died from liver cancers, the vast majority of which were HCC that arose in the setting of chronic liver disease [1]. Figure 4.1 demonstrates cellular and ECM alterations between normal and fibrotic liver.

4.2.2 ECM composition and matrix homeostasis

More than an inert support structure, the ECM is a highly dynamic and well organized functional system that both responds to and influences the local microenvironment [17]. ECM homeostasis is maintained by a constant, balanced turnover of ECM components, mediated by matrix metalloproteinases (MMPs; which degrade ECM

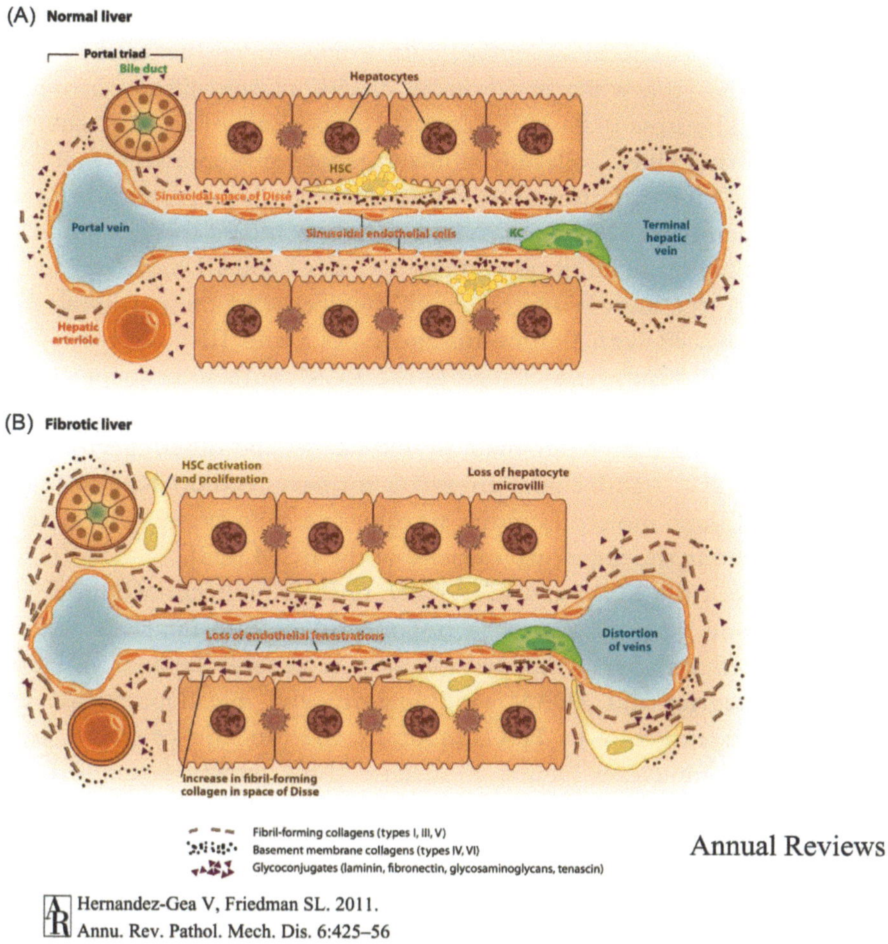

Figure 4.1 ***Cellular and matrix alterations between normal and fibrotic liver.*** The liver is a complex three-dimensional structure consisting of parenchymal (hepatocytes) and non-parenchymal cells including HSCs, which reside in the perisinusoidal space of Disse, Kupffer cells (KCs), which reside within sinusoids, and endothelial cells, which form the sinusoidal lining.

A. *Normal liver:* Hepatic sinusoids receive blood from both the hepatic arteriole and the portal vein tributary. Bidirectional flow across fenestrated endothelial cells and the low-density basement membrane-like matrix in the subendothelial space of Disse promotes efficient metabolite exchange between the sinusoidal blood and hepatocytes.
B. *Fibrotic liver:* Early events in liver fibrosis include deposition of fibrillar matrix within the space of Disse, loss of endothelial cell fenestrations, leading to impaired metabolic exchange between hepatocytes and sinusoidal blood. HSCs activate, proliferate, and may migrate from the space of Disse into hepatic parenchyma where they further deposit matrix.

Reproduced with permission from Ref. [2].

components) and tissue inhibitors of metalloproteinases (TIMPs; which inhibit MMPs) [18,19]. In healthy liver, ECM comprises just a fraction of the total liver weight (approximately 0.5%) [20], while in advanced disease, fibrotic tissue can replace the bulk of the organ. The structural and functional components of ECM in liver include several families of molecules: collagens, noncollagenous glycoproteins, proteoglycans, glycosaminoglycans, growth factors, integrins, and matricellular proteins [5,21]. In the normal liver, collagen types I, III, V, and XI are largely confined to the capsule and the portal triad, with only scattered fibrils containing types I and III in the subendothelial space [22]. In injury, the low-density basement membrane of the perisinusoidal space is disrupted and replaced with fibronectin and types I and III collagen, resulting in both a qualitative and quantitative change in the ECM, with subsequent interference of the bidirectional flow between hepatocytes and the sinusoids. Components of the ECM, such as decorin and biglycan, bind growth factors and other effector molecules in their latent form (e.g., transforming growth factor beta [TGFβ]). Release and subsequent activation of these molecules results in paracrine stimulation of local cells, such as HSCs, demonstrating the functional capacity of the ECM [2,5].

Signaling between the ECM and nearby cell types is bidirectional; the ECM may both respond to and generate signals in response to cellular injury. Many of these signals may be mediated by membrane adhesion receptors, the most important of which include integrins and ADAM (a disintegrin and metalloproteinase domain) molecules. Integrins expressed by HSCs modulate the function of several key profibrotic pathways and molecules such as connective tissue growth factor (CTGF) [23], Hedgehog [24], TGFβ [2], and are central to the procarcinogenic milieu of fibrotic liver [14]. ADAM molecules are expressed by HSCs and endothelial cells, and mediate profibrotic signals such as those from epidermal growth factor receptor (EGFR), among others [25–27]. In disease, the ECM may generate feed-forward loops; changes in stiffness generate more activation signals for HSCs, which in turn increases fibrosis development [28–30].

4.2.3 Cellular sources of ECM

Myofibroblasts are the mesenchymal cell type responsible for wound healing and tissue repair across all organs and various physiological states, including cancer [31]. In the liver, the dominant source of myofibroblasts are HSCs, which, in response to liver injury, transdifferentiate from quiescent, lipid-laden cells into fibrogenic and contractile myofibroblasts. While some myofibroblasts may also be derived from portal myofibroblasts and epithelial-to-mesenchymal transition (EMT), fate tracing studies suggest that myofibroblasts are derived primarily from HSCs across a range of disease states, with the possible exception of cholestatic disease [8–10]. This concept is further detailed in Chapter 6.

4.3 HSCs AND HEPATIC FIBROSIS

HSC activation is a dynamic, programmed event in which resting, lipid-laden HSCs transdifferentiate into fibrogenic, contractile, and proliferative myofibroblasts under the direction of multiple signaling pathways, both paracrine and autocrine. As the HSCs are activated, they not only respond to signals but also generate them; resulting in an intricate, bidirectional signaling network of cells and mediators, which culminates in an accumulation of ECM and hepatic fibrosis. Conceptually, this event may be divided into two major phases: *initiation* and *perpetuation*. Depending upon the nature and course of the injury, a third phase, *regression*, may follow. *Initiation* is a range of genetic and phenotypic changes that render the HSCs more responsive to a range of cellular stimuli. It can thus be considered a "priming" step. *Perpetuation* is a continuous process, which encompasses those initial cellular events, plus autocrine and paracrine loops, and responses to alterations in the local ECM that amplify and expand the activated phenotype. *Regression* will only occur with the cessation of the injurious stimuli and involves activated HSCs undergoing apoptosis, senescence, or converting to a separate "inactivated" phenotype. This section is devoted to well established or "classic" pathways of HSC activation.

4.3.1 Initiation

When the liver is injured, HSCs receive paracrine signals from damaged neighboring cells such as hepatocytes, KCs, leukocytes, platelets, and sinusoidal endothelial cells. Damaged hepatocytes release reactive oxygen species (ROS), lipid peroxides, and generate apoptotic bodies in response to liver injury, all of which are potent HSC activation signals. For example, ethanol metabolized via cytochrome P450 2E1 (CYP2E1) [32,33] and lipid peroxides in nonalcoholic fatty liver disease (NAFLD) may both stimulate HSCs [34]. ROS generated by hepatocytes may also increase collagen synthesis in HSCs, which can be attenuated by antioxidants, CYP2E1 inhibitors, and hydrogen peroxidase [32]. Fas and tumor necrosis factor-related apoptosis inducing ligand (TRAIL)-mediated hepatocyte apoptosis and the generation of apoptotic bodies produce profibrogenic stimuli [35,36]. Furthermore, ROS can induce endoplasmic reticulum (ER) stress in HSCs, which in turn induces autophagy and contributes to HSC activation by providing a fuel source in the form of free fatty acids (FFAs) (see later for further details) [37–39].

Sinusoidal endothelial cells, by virtue of their close proximity to both HSCs and the blood supply of the liver, are important in the early stages of HSC initiation. Fenestrated endothelial cells ("differentiated" endothelial cells) are critical for maintaining a quiescent HSC phenotype as evidenced by the co-culture of HSCs with differentiated (but not capillarized) endothelial cells, which prevents spontaneous

HSC activation [40]. Indeed, the capillarization of endothelial cells precedes fibrosis development in several liver diseases including toxic, alcoholic, and nonalcoholic injuries [12]. This effect is likely mediated through vascular endothelial growth factor (VEGF)-mediated nitric oxide (NO) production. In genetic models of VEGF deletion, endothelial cells lose their fenestrations [41,42]. In animal models, pharmacological modulation that restores the differentiated endothelial cell phenotype promotes reversion of activated HSCs to quiescence [41]. Endothelial cells are a source of cellular (alternative spliced domain A [EDA]) fibronectin, which may promote early migration of HSCs to the site of injury, and convert the latent form of TGFβ, both of which promote cellular activation [5].

Platelets are a major source of the potent HSC mitogen platelet-derived growth factor B (PDGFβ), as well as secreting TGFβ and epidermal growth factor (EGF) [5]; despite this, their exact role in hepatic fibrogenesis and HSC activation remains to be clarified. Evidence documenting their role in fibrogenic pathways includes the following: (i) PDGF-B protein levels in fibrotic mice can be reduced to normal with an anti-platelet antibody, suggesting they are a dominant source of this mitogen; (ii) platelets tend to localize in areas of fibrotic tissue in close proximity to activated HSCs; and (iii) platelet depletion leads to a reduction in alpha smooth muscle actin (α-sma) and fibrogenic gene expression [43]. On the other hand, there is evidence to suggest platelets may inhibit HSC activation, including: (i) HSC activation in culture is blocked by platelet-derived adenosine triphosphate (ATP) [44]; (ii) hepatocyte growth factor (HGF) released by activated platelets *in vitro* reduces the production of collagen by HSCs [45]; and (iii) transgenic mice with thrombocytopenia develop advanced fibrosis in response to cholestasis [45]. These divergent studies addressing the role of platelets in HSC activation and fibrosis have not yet been reconciled.

4.3.2 Perpetuation

If the initial insult is sustained, ongoing paracrine signaling from neighboring cells and the surrounding ECM, plus autocrine signals generated from within HSCs, perpetuate and amplify the activated phenotype. Continued ECM deposition leads to fibrosis development. Perpetuation of HSC activation is an organized process including a number of functional outcomes that can conceptually be divided into retinoid loss, fibrogenesis, matrix turnover, chemotaxis, proliferation, contractility, and immunoregulation.

4.3.2.1 Retinoid loss

Between 50% and 80% of retinoids (vitamin A compounds) are stored in the characteristic cytoplasmic droplets of HSCs as retinyl esters, primarily retinyl palmitate; these droplets also contain FFAs, cholesterol, and triglycerides [46,47]. Lecithin retinyl acyltransferase (LRAT) esterifies retinoids and is the sole retinyl acyltransferase in the liver

[47,48]. As HSCs activate, they lose their characteristic droplets, an event traditionally considered a hallmark of HSC activation [7]; whether this droplet loss is *critical* for activation remains controversial [49–53]. More recent evidence indicates that retinoid loss is a consequence, not a cause of HSC activation and fibrogenesis [54]; when LRAT is genetically deleted, HSCs lack retinyl ester-containing droplets, yet develop the same amount of fibrosis as wild type mice following experimental injury due to bile duct occlusion or toxic injury. Interestingly, these knockout mice are protected from hepatocarcinogenesis [54]. There is renewed interest and additional contradictory findings in retinoid metabolism and hepatic fibrogenesis, which is discussed later in Section 4.4.

4.3.2.2 Fibrogenesis

Production and secretion of ECM components, particularly collagen type I, is a defining characteristic of both HSC activation and development of hepatic fibrosis. TGFβ, the most potent fibrogenic cytokine, is secreted in its latent form by several cell types but requires activation to be fibrogenic [2,5]. Once activated, TGFβ signals via intracellular Smad proteins that subsequently induces transcription of collagens type I and III [55,56]. TGFβ activity is inhibited by peroxisome proliferator-activated receptor gamma (PPARγ) by interacting with JunD and inhibiting Smad3 [57,58]. The complexity of TGFβ signaling is increasingly evident. For example, TGFβ activity may be modulated by osteopontin [59] by the expression of its pseudoreceptor, bone morphogenetic protein, and activin membrane-bound inhibitor (BAMBI) via nuclear factor kappa-light-chain-enhancer of activated B cells (NF-κB) signaling [60]; by crosstalk between immune cells and pro-inflammatory cytokines such as interleukin 17A (IL-17A) [61]; both transcriptionally and post-transcriptionally via PDGFβ receptor (βPDGFR) signaling [62]; and through crosstalk with vitamin D [63]. Other important profibrogenic molecules include CTGF [64,65], osteopontin [59,66,67], and several other cytokines and chemokines [68]. The role of cytokines in HSCs will be covered in detail in Chapter 5.

4.3.2.3 Matrix turnover

Rather than an uncontrolled, continuous process of ECM secretion and deposition, hepatic fibrosis is characterized by a shift in the *balance* between deposition and degradation. The major pathological outcome is a replacement of the normal basement membrane, which consists primarily of collagen type IV with collagen type I [5]. This shift is achieved by alterations in the quantities and activity of the MMPs and their inhibitors, TIMPs, which bind and inactivate MMPs (see Section 4.2.2) [5,7]. There is increasing interest in the intracellular regulation of collagen and other matrix components via autophagy and the unfolded protein response (UPR) in tissues including the kidney [69,70]. This has not as yet been demonstrated in the liver but merits further investigation. The functions of MMPs and their inhibitors are detailed in Chapter 7.

4.3.2.4 Chemotaxis

Chemotaxis of activated HSCs leads to accumulation at sites of injury, often beyond the space of Disse within developing vascularized fibrous septae. PDGFβ is the prototypical HSC chemoattractant. Elongated HSC protrusions, which contain cytoskeletal components, respond to PDGFβ through a series of morphological events including cell spreading at the tip closest to the PDGFβ, movement of the cell body toward the PDGFβ, and retraction of the trailing protrusion [71]. HSCs migrate toward several classes of molecules and local environmental conditions including chemoattractants, ROS, hypoxia, ECM components, and the hyaluronic acid receptor [5]. Other than PDGFβ, major HSC chemoattractants include VEGF [72], EGF [73], and monocyte chemoattractant protein-1 (MCP-1) [74], all of which engage the Ras/ERK or PI3K/Akt pathways in promoting migration [75]. ROS produced by damaged hepatocytes, inflammatory cells, or generated by nicotinamide adenine dinucleotide phosphate (NADPH)-oxidase within HSCs can lead to JNK1/2 mediated chemotaxis, even in the absence of chemoattractants [75,76]. Local tissue hypoxia may also generate ROS, which leads to a biphasic HSC migration pattern. The early phase is a mitochondrial driven, ROS-dependent pathway that activates the ERK1/2 and JNK1/2 cascades, while the delayed phase is a hypoxia-inducible factor 1-alpha (HIF-1α) and VEGF-upregulation pathway [77]. The response of HSCs to changes in local ECM is one among many feed-forward loops in hepatic fibrosis pathogenesis. HSCs may migrate toward collagen type I and MMP-2, further amplifying the fibrotic response [73].

4.3.2.5 Proliferation

A hallmark of hepatic fibrosis is both qualitative and quantitative changes in HSCs. HSCs proliferate in response to a host of molecules including PDGFβ, which is considered the most potent of the HSC mitogens, tumor necrosis factor alpha (TNFα), thrombin, EGF, and VEGF, among others [5,7,78]. At the onset of hepatic injury there is both a rapid increase in βPDGF receptor (βPDGFR) and PDGFβ expression, the result of which is signaling through the Ras-MAPK and phosphoinositide 3-kinase (PI3K) pathways that ultimately lead to a proliferative, profibrotic transcriptional program [79–81]. The presence and crosstalk with TNFα appears to be critical for PDGFβ-mediated HSC proliferation [82]. Angiogenesis allows for the proliferation of HSCs by providing important nutrients and vascularizing the fibrous septae in which the HSCs will reside. In fact, HSCs may contribute directly to angiogenesis through the stabilization of new vessels [83]. VEGF is a central mediator of both angiogenesis and HSC proliferation [84–87]. Fascinating functions of vitamin D in hepatic biology and fibrosis are emerging. Patients with alcoholic liver disease and low 25(OH)D levels have more severe hepatic damage and worse outcomes [88], and patients with the highest baseline 25(OH)D levels have a 50% reduction in HCC risk irrespective of underlying liver disease [89]. It is feasible that some of this effect may be due to the anti-proliferative effect of the activated form of vitamin D, 1,25$(OH)_2D_3$, which when added to primary rat HSCs suppresses HSC proliferation, increases MMP-9, and decreases TIMP-1 expression [90]. Vitamin D receptor ligands have been recently proposed as an endocrine checkpoint in hepatic fibrosis [91], and treatment with 1,25$(OH)_2D_3$ attenuates fibrosis in a chronic thioacetamide (TAA) rodent model [90].

4.3.2.6 Contractility

Disruption of the normal architecture of the liver, primarily through remodeling of the hepatic sinusoids and development of fibrous septae replete with contractile HSCs, is the major determinant of portal hypertension, which in turn leads to a cascade of further clinical complications. The mechanisms by which HSCs become contractile are primarily via calcium-dependent pathways [92], although calcium-independent pathways have been described [93,94]. The force generated by the contraction of activated HSCs is sufficient to contract sinusoids [95]. The major inducers of HSC contraction and relaxation are endothelin-1 (ET-1) and NO, respectively, although several other mediators have been implicated, including eicosanoids, carbon monoxide, adenosine, atrial natriuretic peptide, and angiotensin II [5,7]. ET-1, the receptors of which are found primarily on smooth muscle cells, HSCs and endothelial cells, acts in a paracrine and autocrine fashion within the liver. It is released in part from HSCs in response to liver injury and leads to a constriction of the sinusoids and a reduction in flow [96,97]. NO induces HSC relaxation; in hepatic injury, sinusoidal endothelial cells produce less NO, which contributes to a contractile phenotype [98]. Modulation of HSC contractility is an evolving treatment concept for portal hypertension [99]. The contribution of HSCs to portal hypertension will be discussed further in Chapter 8.

4.3.2.7 Immunoregulation

In the setting of sustained liver injury, HSCs attract immune cells to the site of injury, encourage their maturation, and subsequently respond to signals generated by the infiltrating cells. This feed-forward loop and bidirectional signaling are two important tenets of HSC biology. The interactions between HSCs and the immune system are detailed in Chapters 12 and 14.

4.3.3 Regression

The reversibility of advanced hepatic fibrosis, perhaps even early cirrhosis, has been conclusively demonstrated across a range of diseases [100–102]. This realization, which is in contrast to earlier notions of hepatic fibrosis as a progressive, irreversible condition [103,104], has provided the foundation for attempts at anti-fibrotic drug development by exploiting these regression pathways, a concept that will be explored in Chapter 11. Chemical cross-linking may prevent the resolution of more advanced, macronodular cirrhosis, while micronodular cirrhosis may spontaneously resolve should the injury be removed, at least in animal model [105], and anecdotally in humans [102]. The mechanisms by which fibrosis may reverse remain to be completely characterized, but it is clear that "regression" of activated HSCs and myofibroblasts is a critical step. The current understanding of fate decisions for activated HSCs describes three mechanisms by which they may regress. These are apoptosis, senescence, and reversion to a distinct, "inactivated" phenotype. Teasing out how and why activated HSCs move along one of these pathways is likely to be a critical next phase in understanding HSC biology.

4.3.3.1 Apoptosis

The contribution of myofibroblast apoptosis to resolution of fibrosis has been established in animal models, although evidence in humans remains elusive. In both models of cholestasis (BDL followed by bilio-jejunal anastomosis), and toxic liver injury (e.g., chronic carbon tetrachloride [CCl_4]), in which the insult is removed after the development of fibrosis, there is resolution of injury and fibrosis with essentially normal liver histology within 6 weeks, and marked apoptosis of myofibroblasts [106,107]. It is important to note that human HSCs are relatively resistant to apoptosis compared to rodents due to increased Bcl-2 expression and thus caution must be taken in extrapolating these rodent findings to humans [108]. The mechanisms by which apoptosis is regulated are not entirely clear but several mediators and cell populations have been implicated: (i) NF-κB, which protects HSCs from apoptosis [109], and inhibition of which accelerates fibrosis resolution in CCl_4 treated rodents [110]; (ii) nerve growth factor (NGF), which is secreted by hepatocytes and may promote apoptosis via inhibition of NF-κB [111]; (iii) natural killer (NK) cells, which induce the apoptosis of activated HSCs when they express NKG2D/RAE1 [112]; (iv) members of the Hsp70 heat-shock family of proteins, which protect against stress-induced apoptosis [113] and whose genetic deletion results in rapid recovery from fibrosis and disappearance of activated HSCs in animal models [114]; and (v) ER stress, which has been demonstrated in several recent studies, but whose mechanisms remain unclear [115–118]. A recent fate mapping study in rodents identified just 2.6% of activated HSCs undergoing apoptosis 7 days after CCl_4 cessation, which was the peak apoptotic activity, and thus the overall contribution of apoptosis to reduced numbers of activated HSCs during fibrosis regression is uncertain [114].

4.3.3.2 Senescence

Cellular senescence is characterized by stable cell-cycle arrest; in addition to lack of proliferation, senescent HSCs limit production of profibrogenic proteins, degrade those that are present and up regulate immune surveillance and clearance of activated HSCs. In this way, senescent HSCs act as a brake on fibrosis development and are an important homeostatic regulator of the hepatic wound-healing response [119]. Genetic deletion of the p53 senescence program increases hepatic fibrosis development and malignant transformation of epithelial cells [119,120]. The signals that drive activated HSCs to senesce remain to be clarified. IL-22 promotes HSC senescence, protects against hepatic fibrosis, and may shift the fate of HSCs away from apoptosis [121]. A recent finding may suggest that there are yet to be established context- and disease-specific implications of HSC cellular senescence. The acquisition of an obesity and gut microbiome-associated HSC senescence secretory program may promote the development of HCC, as suggested by a rodent model [122]. However, in this model there was no fibrosis development and thus the study's implications for human disease are not certain.

4.3.3.3 Reversion to an inactivated phenotype

Two related *in vivo* genetic fate mapping studies have demonstrated that activated HSCs may revert to an "inactivated" phenotype following cessation of experimental liver injury, one that is a distinct cellular type from a quiescent HSC [114,123]. Critically, these HSCs are able to escape apoptosis, down regulate their fibrotic transcriptional program and acquire a phenotype that is similar to quiescent HSCs, with the important difference that they are "primed" to more readily reactivate when exposed to repeat injury [114,123]. One month after cessation of liver injury, approximately 40–50% of the liver's HSCs may have this activated phenotype, with the remainder consisting of residual quiescent HSCs or newly developed cells from a precursor population [114,123]. As with senescence, the mediators that allow this escape from apoptosis and reversion to an inactivated state are unknown. One study demonstrated up-regulation of heat shock proteins early in the resolution phase [114], while another did not [123]. In a study of cultured HSCs, fructose-1,6-bisphosphate mediated PPARγ activation in cultured HSCs to reduce collagen production and increase lipid content; it is unclear whether this response represents the inactivated phenotype seen in animal models [124].

4.4 NEW AND EMERGING PATHWAYS OF HSC ACTIVATION

In the last 5 years, our appreciation of the complexity of the pathways leading to activation of HSCs has expanded. The "classic pathway" of HSC activation detailed in the previous section remains as a useful model; however, the discovery of several important molecules and effectors that were previously thought unrelated to HSC biology may challenge this dogma and eventually lead to a deeper understanding of the processes that lead to fibrosis development. This section outlines some of these emerging pathways that may shape the future of research into HSCs and hepatic fibrosis.

The identification of a central regulatory hub in HSCs that controls cellular activation would represent an attractive anti-fibrotic drug target. Recently, a Gα-Interacting Vesicle (GIV; also known as Girdin) has been proposed as such a regulatory hub

[125]. GIV is positioned downstream of several important pro-activation pathways (TGFβ, βPDGFR, chemokines, and Toll-like receptors [TLRs]). In response to liver injury, GIV is up regulated and shifts the balance of several downstream pathways within HSCs to a profibrotic phenotype, with increased PI3K and Smad signaling and decreased cyclic adenosine monophosphate (cAMP). In this way, HSCs are programmed to proliferate and increase collagen production, are protected from apoptosis, and migrate to sites of injury. Furthermore, GIV activity molds the innate immune response in liver injury, primarily through TLR responses and KC activation. The presence of GIV mRNA but not protein in quiescent HSCs suggests that its activity is regulated at several steps, possibly including via signal transducer and activator of transcription 3 (STAT3). The authors propose interference at the GIV:Gαi interface as a novel and attractive anti-fibrotic therapy. Finally, high GIV expression in formalin-fixed paraffin embedded liver tissue in those with chronic hepatitis C but with low degrees of fibrosis may predict subsequent rapid progression to fibrosis and cirrhosis [125]. While this finding was reported in relatively small number of individuals (21 males) and will require further validation in other diseases, the implications may be significant in prioritizing development of GIV antagonists as potential anti-fibrotics, provided that the effects are restricted to HSCs.

The distinctions between transcriptional and epigenetic control of important pro-activation pathways, exemplified by GIV and STAT3, is an emerging field that has been reviewed recently [126–128]. Recent advances include the identification of several transcriptional control programs mediated by absent, small, or homeotic disc 1 (ASH1) [129] and myocardin-related transcription factor A (MRTF-A) [130]. ASH1 is a histone-modifying enzyme that in response to liver injury binds directly to the promoter regions of the profibrotic and HSC activation genes TIMP-1, ASMA, collagen type I and TGFβ, increases levels of histone H3 lysine 4 trimethylation and ultimately results in increased gene transcription [129]. Upstream of ASH1 is methyl-CpG binding protein 2 (MeCP2), which is another transcriptional regulator and part of a so-called "epigenetic relay pathway" [131]. MeCP2 promotes myofibroblast transdifferentiation in other organs including the heart and lungs, and may be a core fibrogenic regulator across multiple disease states and tissue types [132,133]. MRTF-A can modulate the chromatin structure of several genes crucial to HSC activation in response to liver injury [130]. As with MeCP2, MRTF-A is conserved across tissue types and is central to the fibrotic response in heart and lung [134,135], and is implicated in HCC development [136]. Multiple micro RNAs (miRNAs) have been identified as critical mediators in liver fibrosis at all levels, including HSC activation and fibrogenesis and there is likely to be an increasingly sophisticated appreciation of their role in the coming years [131].

A central component of HSC activation is proliferation (see Section 4.3.2.5) with engagement of cell cycle machinery. Rather than these processes simply responding to pro-survival signals in a non-autonomous fashion, they may operate as regulatory molecules in their own right. For example, c-myc is a central driver of cellular proliferation [137] and its amplification is commonly found in HCC [138]. It has recently been shown to be overexpressed in hepatocytes of human liver samples with advanced fibrosis and promotes a "profibrotic tissue environment" via pre-activation of HSCs, which primes them for a second profibrotic insult [139]. This effect is mediated in part by an inhibition of PPARγ and reduction of lipogenesis, plus increased PDGF-B expression in hepatocytes and subsequent paracrine crosstalk with HSCs. Interestingly, the increase in c-myc may result either from a direct inflammatory response to liver injury or from injury-induced gene amplification [139]. C-myc may be a target for HCC chemoprevention and fibrosis modulation. In a rodent model of liver fibrosis and HCC, short-acting RNA therapy was used to increase the hepatocyte expression of albumin, which resulted in a significant reduction in tumor burden compared to the WT mice (80%) [140]. Using the same technique in HepG2 cells resulted in significant reductions in c-myc expression; unfortunately, it is not clear if c-myc levels were assessed in the *in vivo* models. However, in view of how HSCs and a fibrotic microenvironment promote HCC [14], some of the observed results in the *in vivo* model may have been due to reduced c-myc expression and attenuated HSC activation, but again, these parameters were not directly measured. Cyclin E1 is known to be under the control of c-myc and is critical for HSC activation [141]. As with c-myc, cyclin E1 is highly expressed in cirrhotic human liver tissue and not only promotes the transdifferentiation of quiescent HSCs but drives the proliferation of activated myofibroblasts. Interestingly, in this model, c-myc was not required for cyclin E1 mediated HSC activation. Several other levels of control may exist for cyclin E1, as it appears to be negatively regulated by cyclin E2 [141] and miR-195 [142]. In a human immortalized HSC line (LX-2) in culture, spontaneous activation is associated with a reduction in miR-195, an increase in cyclin E1, and entry into the cell cycle with subsequent proliferation [142]. In hepatocytes, cyclin E1 and cyclin dependent kinase 2 are indispensable for liver regeneration. Conditional knock out of these molecules in the rodent model of partial hepatectomy completely blocks the cell cycle and impaired parenchymal regeneration [143].

The metabolic demands in fibrosis are significant, and HSCs rely upon autophagy to liberate FFAs from cytoplasmic droplets to meet these needs. When crucial components of the autophagy machinery are genetically deleted within HSCs in rodent models of liver injury, the WT mice develop more fibrosis, presumably due to the inability to meet the necessary energy requirements [38,144,145]. These important observations have paved the way for in-depth analyses of the association between autophagy and HSC activation. The UPR is frequently seen in activated HSCs, and is generated in response to paracrine signals from neighboring cells exposed to ROS and ER stress [5,146]. Recent work has highlighted inositol requiring enzyme 1 alpha (IRE1α) and X-box transcription factor 1 (XBP1s) expression as a link among UPR,

ER stress, autophagy, and HSC activation [39]. Modulation of autophagy for the purposes of attenuating fibrosis will require specific cellular targeting given its crucial roles in normal homeostasis [144].

Several other metabolic pathways besides autophagy have been linked to HSC activation. Broadly, several molecules that are responsible for metabolic homeostasis in a quiescent HSC may transition to pro-activation functions during liver injury. This metabolic reprogramming is not unique to HSCs and has been proposed as a new hallmark of cancer [147]. Rev-erbα, a protein previously considered to function mainly as a transcriptional repressor, serves a unique bifunctional role in HSC biology [148]. In normal liver, Rev-erbα helps maintain the adipogenic, quiescent HSC phenotype, but when exposed to fibrogenic stimuli it acquires new functions and promotes transdifferentiation and increased contractility of HSCs. Other examples of metabolic reprogramming in HSCs include: (i) PPARγ, a master inhibitor of HSC activation that affects lipid metabolism and maintains an adipogenic phenotype [57]; (ii) Hedgehog ligands, which induce aerobic glycolysis, suppression of normal glucose homeostatic functions, and HSC activation [149]; (iii) deletion of the HSC nuclear receptor, liver X receptor (LXR), which results in deranged lipid metabolism, HSC activation, and fibrosis [150]; and (iv) the nuclear receptor, farnesoid X receptor (FXR), which is central to bile acid, lipid, and carbohydrate metabolism in the liver as well as a negative regulator of HSC activation [151]. Overall, tight control of metabolic homeostasis promotes a quiescent phenotype; any signals that interrupt this balance may favor transdifferentiation of HSCs into an activated myofibroblast. This concept may be exploited by anti-fibrotic therapy. For example, obeticholic acid, an FXR agonist, improves portal hypertension and protects against fibrosis in rodents [152,153], in addition to promising results in humans with nonalcoholic steatohepatitis (NASH).

As mentioned in Section 4.2, the significance of retinoid metabolism in HSC activation and hepatic fibrogenesis remains unclear, with recent evidence suggesting it is more consequence than causative [54]. The discovery that alcohol dehydrogenase III (ADH3) has diverse functions in retinoid metabolism in HSCs and NK cell crosstalk has re-opened the debate. The ADH enzymes within HSCs do not contribute meaningfully to ethanol metabolism within the liver [7]. ADH3, which is the dominant subtype within HSCs, is the major enzyme responsible for the oxidation of retinols released from cytoplasmic droplets, is a positive regulator of HSC activation, and is a negative regulator of NK cells in the setting of liver fibrosis [154]. It is uncertain whether these functions are co-regulated, as ADH3 KO mice develop similar degrees of fibrosis as WT mice [154]. Genome-wide association studies and population-based studies have identified an association between a common polymorphism in PNPLA3 (I148M) and the severity of liver disease including fibrosis and steatosis, initially in NAFLD/NASH but now across a range of liver diseases. There is no clear mechanistic link between this polymorphism and the biology of NASH or fibrosis as yet [155]. Recent work has suggested patatin-like phosphatase domain-containing 3 (PNPLA3) may be a "retinol-gatekeeper"; the wild type but not the polymorphic form has potent retinyl palmitate lipase activity, and hence those with the polymorphism may have retinol retention, in addition to more severe liver disease. The link between PNPLA3 and HSC activation is uncertain but merits further investigation [156].

NAFLD is the hepatic manifestation of the metabolic syndrome. Increasing attention has been focused upon the inherent metabolic disturbances in NAFLD and their effects upon HSCs. Obese, diabetic mice accumulate free cholesterol (FC) in the liver [157]. Rodent models have demonstrated the pro-activation effects of FC accumulation within HSCs, through a number of interrelated steps [158–160]. FC interferes with the endo-somal-lysosomal degradation pathway of TLR4; subsequently, levels of the membrane bound TGFβ pseuodreceptor, BAMBI, are increased, which sensitizes HSCs to profibro-genic stimuli. Furthermore, the disruption of cholesterol homeostasis suppresses PPARγ activity [158–160]. The addition of commonly used lipid-lowering drugs may reverse this phenotype [161].

Evidence has accumulated that the Hedgehog pathway contributes to HSC responses and fibrogenesis. The Hedgehog pathway is a series of ligands and receptors that have been implicated in a host of homeostatic and pathogenic liver processes including regeneration, fibrogenesis, inflammation, vascular remodeling, and cancer [162]. More recently, Hedgehog has been proposed as an important intermediate regulatory node, which facilitates paracrine crosstalk between damaged hepatocytes and hepatic progenitor cells that expand and contribute to fibrogenesis and myofibroblast activation [163]. The PIVENS trial, which randomized NASH patients to placebo, pioglitazone, or vitamin E, suggested histological response was strongly correlated with loss of the important Hedgehog ligand, Sonic hedgehog [164].

4.5 CONCLUSIONS

This last 5 years have seen substantial increases in our understanding of the role of HSCs in hepatic fibrosis. However, the "classic" pathway of HSC activation remains as an important framework for understanding how quiescent HSCs may transdifferentiate to an activated myofibroblast. Despite knowledge of other sources of myofibroblasts in hepatic fibrosis (see Chapter 6), recent fate tracing studies have placed HSCs back at the very forefront and thus they represent the most logical target for anti-fibrotic therapy (see Chapter 11) [8–10]. The search for a regulatory hub in HSC activation and hepatic fibrosis [125] is the first step toward a deeper appreciation of the myriad pathways that contribute to this potentially fatal condition; the ultimate result of which will hopefully be a safe and effective way of slowing or reversing this progressive disease.

REFERENCES

[1] Murphy SL, Xu J, Kochanek KD. Deaths: final data for 2010. Natl Vital Stat Rep 2013;61(4):1–117.
[2] Hernandez-Gea V, Friedman SL. Pathogenesis of liver fibrosis. Annu Rev Pathol 2011;6:425–56.
[3] Asrani SK, Kamath PS. Natural history of cirrhosis. Curr Gastroenterol Rep 2013;15(2):308.
[4] Anthony PP, Ishak KG, Nayak NC, et al. The morphology of cirrhosis. Recommendations on definition, nomenclature, and classification by a working group sponsored by the World Health Organization. J Clin Pathol 1978;31(5):395–414.
[5] Puche JE, Saiman Y, Friedman SL. Hepatic stellate cells and liver fibrosis. Compr Physiol 2013;3(4):1473–92.
[6] Wynn TA. Cellular and molecular mechanisms of fibrosis. J Pathol 2008;214(2):199–210.
[7] Friedman SL. Hepatic stellate cells: protean, multifunctional, and enigmatic cells of the liver. Physiol Rev 2008;88(1):125–72.
[8] Iwaisako K, Jiang C, Zhang M, et al. Origin of myofibroblasts in the fibrotic liver in mice. Proc Natl Acad Sci USA 2014;111(32):E3297–305.
[9] Mederacke I, Hsu CC, Troeger JS, et al. Fate tracing reveals hepatic stellate cells as dominant contributors to liver fibrosis independent of its aetiology. Nat Commun 2013;4:2823.
[10] Michelotti GA, Xie G, Swiderska M, et al. Smoothened is a master regulator of adult liver repair. J Clin Invest 2013;123(6):2380–94.
[11] Schaffner F, Poper H. Capillarization of hepatic sinusoids in man. Gastroenterology 1963;44:239–42.
[12] DeLeve LD. Liver sinusoidal endothelial cells in hepatic fibrosis. Hepatology 2014. http://dx.doi.org/10.1002/hep.27376.
[13] Ferrell L. Liver pathology: cirrhosis, hepatitis, and primary liver tumors. Update and diagnostic problems. Mod Pathol 2000;13(6):679–704.
[14] Zhang DY, Friedman SL. Fibrosis-dependent mechanisms of hepatocarcinogenesis. Hepatology 2012;56(2):769–75.
[15] Chiesa R, Donato F, Tagger A, et al. Etiology of hepatocellular carcinoma in Italian patients with and without cirrhosis. Cancer Epidemiol Biomarkers Prev 2000;9(2):213–6.
[16] Simonetti RG, Camma C, Fiorello F, et al. Hepatocellular carcinoma. A worldwide problem and the major risk factors. Dig Dis Sci 1991;36(7):962–72.
[17] Lu P, Takai K, Weaver VM, Werb Z. Extracellular matrix degradation and remodeling in development and disease. Cold Spring Harb Perspect Biol 2011;3:12.
[18] Knittel T, Mehde M, Kobold D, et al. Expression patterns of matrix metalloproteinases and their inhibitors in parenchymal and non-parenchymal cells of rat liver: regulation by TNF-alpha and TGF-beta1. J Hepatol 1999;30(1):48–60.
[19] Arthur MJ, Friedman SL, Roll FJ, Bissell DM. Lipocytes from normal rat liver release a neutral metalloproteinase that degrades basement membrane (type IV) collagen. J Clin Invest 1989;84(4):1076–85.
[20] Rojkind M, Giambrone MA, Biempica L. Collagen types in normal and cirrhotic liver. Gastroenterology 1979;76(4):710–9.
[21] Schuppan D, Ruehl M, Somasundaram R, Hahn EG. Matrix as a modulator of hepatic fibrogenesis. Semin Liver Dis 2001;21(3):351–72.
[22] Friedman SL, Roll FJ, Boyles J, Bissell DM. Hepatic lipocytes: the principal collagen-producing cells of normal rat liver. Proc Natl Acad Sci USA 1985;82(24):8681–5.
[23] Pi L, Robinson PM, Jorgensen M, et al. Connective tissue growth factor and integrin alphavbeta6: a new pair of regulators critical for ductular reaction and biliary fibrosis. Hepatology 2014. http://dx.doi.org/10.1002/hep.27425.
[24] Omenetti A, Porrello A, Jung Y, et al. Hedgehog signaling regulates epithelial–mesenchymal transition during biliary fibrosis in rodents and humans. J Clin Invest 2008;118(10):3331–42.
[25] Kesteloot F, Desmouliere A, Leclercq I, et al. ADAM metallopeptidase with thrombospondin type 1 motif 2 inactivation reduces the extent and stability of carbon tetrachloride-induced hepatic fibrosis in mice. Hepatology 2007;46(5):1620–31.
[26] Niiya M, Uemura M, Zheng XW, et al. Increased ADAMTS-13 proteolytic activity in rat hepatic stellate cells upon activation *in vitro* and *in vivo*. J Thromb Haemost 2006;4(5):1063–70.

[27] Oikawa H, Maesawa C, Tatemichi Y, et al. A disintegrin and metalloproteinase 17 (ADAM17) mediates epidermal growth factor receptor transactivation by angiotensin II on hepatic stellate cells. Life Sci 2014;97(2):137–44.

[28] Wells RG. The role of matrix stiffness in hepatic stellate cell activation and liver fibrosis. J Clin Gastroenterol 2005;39(4 Suppl. 2):S158–61.

[29] Georges PC, Hui JJ, Gombos Z, et al. Increased stiffness of the rat liver precedes matrix deposition: implications for fibrosis. Am J Physiol Gastrointest Liver Physiol 2007;293(6):G1147–54.

[30] Iredale JP. Models of liver fibrosis: exploring the dynamic nature of inflammation and repair in a solid organ. J Clin Invest 2007;117(3):539–48.

[31] Kalluri R, Zeisberg M. Fibroblasts in cancer. Nat Rev Cancer 2006;6(5):392–401.

[32] Nieto N, Friedman SL, Cederbaum AI. Stimulation and proliferation of primary rat hepatic stellate cells by cytochrome P450 2E1-derived reactive oxygen species. Hepatology 2002;35(1):62–73.

[33] Nieto N, Friedman SL, Cederbaum AI. Cytochrome P450 2E1-derived reactive oxygen species mediate paracrine stimulation of collagen I protein synthesis by hepatic stellate cells. J Biol Chem 2002;277(12):9853–64.

[34] Leclercq IA, Farrell GC, Field J, et al. CYP2E1 and CYP4A as microsomal catalysts of lipid peroxides in murine nonalcoholic steatohepatitis. J Clin Invest 2000;105(8):1067–75.

[35] Canbay A, Higuchi H, Bronk SF, et al. Fas enhances fibrogenesis in the bile duct ligated mouse: a link between apoptosis and fibrosis. Gastroenterology 2002;123(4):1323–30.

[36] Canbay A, Taimr P, Torok N, et al. Apoptotic body engulfment by a human stellate cell line is profibrogenic. Lab Invest 2003;83(5):655–63.

[37] Hernandez-Gea V, Friedman SL. Autophagy fuels tissue fibrogenesis. Autophagy 2012;8(5):849–50.

[38] Hernandez-Gea V, Ghiassi-Nejad Z, Rozenfeld R, et al. Autophagy releases lipid that promotes fibrogenesis by activated hepatic stellate cells in mice and in human tissues. Gastroenterology 2012;142(4):938–46.

[39] Hernandez-Gea V, Hilscher M, Rozenfeld R, et al. Endoplasmic reticulum stress induces fibrogenic activity in hepatic stellate cells through autophagy. J Hepatol 2013;59(1):98–104.

[40] Deleve LD, Wang X, Guo Y. Sinusoidal endothelial cells prevent rat stellate cell activation and promote reversion to quiescence. Hepatology 2008;48(3):920–30.

[41] Xie G, Wang X, Wang L, et al. Role of differentiation of liver sinusoidal endothelial cells in progression and regression of hepatic fibrosis in rats. Gastroenterology 2012;142(4):918–27. e916.

[42] May D, Djonov V, Zamir G, et al. A transgenic model for conditional induction and rescue of portal hypertension reveals a role of VEGF-mediated regulation of sinusoidal fenestrations. PLoS One 2011;6(7):e21478.

[43] Yoshida S, Ikenaga N, Liu SB, et al. Extra-hepatic PDGFB, delivered by platelets, promotes activation of hepatic stellate cells and biliary fibrosis in mice. Gastroenterology 2014. http://dx.doi.org/10.1053/j.gastro.2014.08.038.

[44] Ikeda N, Murata S, Maruyama T, et al. Platelet-derived adenosine 5′-triphosphate suppresses activation of human hepatic stellate cell: *in vitro* study. Hepatol Res 2012;42(1):91–102.

[45] Kodama M, Kitadai Y, Sumida T, et al. Expression of platelet-derived growth factor (PDGF)-B and PDGF-receptor beta is associated with lymphatic metastasis in human gastric carcinoma. Cancer Sci 2010;101(9):1984–9.

[46] Blomhoff R, Green MH, Berg T, Norum KR. Transport and storage of vitamin A. Science 1990;250(4979):399–404.

[47] O'Byrne SM, Blaner WS. Retinol and retinyl esters: biochemistry and physiology. J Lipid Res 2013;54(7):1731–43.

[48] O'Byrne SM, Wongsiriroj N, Libien J, et al. Retinoid absorption and storage is impaired in mice lacking lecithin:retinol acyltransferase (LRAT). J Biol Chem 2005;280(42):35647–57.

[49] Davis BH, Kramer RT, Davidson NO. Retinoic acid modulates rat Ito cell proliferation, collagen, and transforming growth factor beta production. J Clin Invest 1990;86(6):2062–70.

[50] Geubel AP, De Galocsy C, Alves N, Rahier J, Dive C. Liver damage caused by therapeutic vitamin A administration: estimate of dose-related toxicity in 41 cases. Gastroenterology 1991;100(6):1701–9.

[51] Hellemans K, Verbuyst P, Quartier E, et al. Differential modulation of rat hepatic stellate phenotype by natural and synthetic retinoids. Hepatology 2004;39(1):97–108.

[52] Okuno M, Moriwaki H, Imai S, et al. Retinoids exacerbate rat liver fibrosis by inducing the activation of latent TGF-beta in liver stellate cells. Hepatology 1997;26(4):913–21.
[53] Senoo H, Wake K. Suppression of experimental hepatic fibrosis by administration of vitamin A. Lab Invest 1985;52(2):182–94.
[54] Kluwe J, Wongsiriroj N, Troeger JS, et al. Absence of hepatic stellate cell retinoid lipid droplets does not enhance hepatic fibrosis but decreases hepatic carcinogenesis. Gut 2011;60(9):1260–8.
[55] Inagaki Y, Okazaki I. Emerging insights into transforming growth factor beta Smad signal in hepatic fibrogenesis. Gut 2007;56(2):284–92.
[56] Breitkopf K, Godoy P, Ciuclan L, Singer MV, Dooley S. TGF-beta/Smad signaling in the injured liver. Z Gastroenterol 2006;44(1):57–66.
[57] Hazra S, Xiong S, Wang J, et al. Peroxisome proliferator-activated receptor gamma induces a phenotypic switch from activated to quiescent hepatic stellate cells. J Biol Chem 2004;279(12):11392–401.
[58] Lin J, Chen A. Activation of peroxisome proliferator-activated receptor-gamma by curcumin blocks the signaling pathways for PDGF and EGF in hepatic stellate cells. Lab Invest 2008;88(5):529–40.
[59] Wang X, Lopategi A, Ge X, et al. Osteopontin induces ductular reaction contributing to liver fibrosis. Gut 2014. http://dx.doi.org/10.1136/gutjnl-2013-306373.
[60] Liu C, Chen X, Yang L, et al. Transcriptional repression of the transforming growth factor beta (TGF-beta) Pseudoreceptor BMP and activin membrane-bound inhibitor (BAMBI) by Nuclear Factor kappaB (NF-kappaB) p50 enhances TGF-beta signaling in hepatic stellate cells. J Biol Chem 2014;289(10):7082–91.
[61] Fabre T, Kared H, Friedman SL, Shoukry NH. IL-17A enhances the expression of profibrotic genes through upregulation of the TGF-beta receptor on hepatic stellate cells in a JNK-dependent manner. J Immunol 2014. http://dx.doi.org/10.4049/jimmunol.1400861.
[62] Liu C, Li J, Xiang X, et al. PDGF receptor alpha promotes TGF-beta signaling in hepatic stellate cells via transcriptional and post transcriptional regulation of TGF-beta receptors. Am J Physiol Gastrointest Liver Physiol 2014. http://dx.doi.org/10.1152/ajpgi.00138.2014.
[63] Beilfuss A, Sowa JP, Sydor S, et al. Vitamin D counteracts fibrogenic TGF-beta signalling in human hepatic stellate cells both receptor-dependently and independently. Gut 2014. http://dx.doi.org/10.1136/gutjnl-2014-307024.
[64] Huang G, Brigstock DR. Regulation of hepatic stellate cells by connective tissue growth factor. Front Biosci (Landmark Ed) 2012;17:2495–507.
[65] Gao R, Brigstock DR. Connective tissue growth factor (CCN2) induces adhesion of rat activated hepatic stellate cells by binding of its C-terminal domain to integrin alpha(v)beta(3) and heparan sulfate proteoglycan. J Biol Chem 2004;279(10):8848–55.
[66] Urtasun R, Lopategi A, George J, et al. Osteopontin, an oxidant stress sensitive cytokine, up-regulates collagen-I via integrin alpha(V)beta(3) engagement and PI3K/pAkt/NFkappaB signaling. Hepatology 2012;55(2):594–608.
[67] Leung TM, Wang X, Kitamura N, Fiel MI, Nieto N. Osteopontin delays resolution of liver fibrosis. Lab Invest 2013;93(10):1082–9.
[68] Sahin H, Trautwein C, Wasmuth HE. Functional role of chemokines in liver disease models. Nat Rev Gastroenterol Hepatol 2010;7(12):682–90.
[69] Kim SI, Na HJ, Ding Y, et al. Autophagy promotes intracellular degradation of type I collagen induced by transforming growth factor (TGF)-beta1. J Biol Chem 2012;287(15):11677–88.
[70] Ishida Y, Yamamoto A, Kitamura A, et al. Autophagic elimination of misfolded procollagen aggregates in the endoplasmic reticulum as a means of cell protection. Mol Biol Cell 2009;20(11):2744–54.
[71] Melton AC, Yee HF. Hepatic stellate cell protrusions couple platelet-derived growth factor-BB to chemotaxis. Hepatology 2007;45(6):1446–53.
[72] Novo E, Cannito S, Zamara E, et al. Proangiogenic cytokines as hypoxia-dependent factors stimulating migration of human hepatic stellate cells. Am J Pathol 2007;170(6):1942–53.
[73] Yang C, Zeisberg M, Mosterman B, et al. Liver fibrosis: insights into migration of hepatic stellate cells in response to extracellular matrix and growth factors. Gastroenterology 2003;124(1):147–59.
[74] Marra F, Romanelli RG, Giannini C, et al. Monocyte chemotactic protein-1 as a chemoattractant for human hepatic stellate cells. Hepatology 1999;29(1):140–8.

[75] Novo E, Busletta C, Bonzo LV, et al. Intracellular reactive oxygen species are required for directional migration of resident and bone marrow-derived hepatic pro-fibrogenic cells. J Hepatol 2011;54(5):964–74.
[76] Paik YH, Iwaisako K, Seki E, et al. The nicotinamide adenine dinucleotide phosphate oxidase (NOX) homologues NOX1 and NOX2/gp91(phox) mediate hepatic fibrosis in mice. Hepatology 2011;53(5):1730–41.
[77] Novo E, Povero D, Busletta C, et al. The biphasic nature of hypoxia-induced directional migration of activated human hepatic stellate cells. J Pathol 2012;226(4):588–97.
[78] Lee UE, Friedman SL. Mechanisms of hepatic fibrogenesis. Best Pract Res Clin Gastroenterol 2011;25(2):195–206.
[79] Wong L, Yamasaki G, Johnson RJ, Friedman SL. Induction of beta-platelet-derived growth factor receptor in rat hepatic lipocytes during cellular activation *in vivo* and in culture. J Clin Invest 1994;94(4):1563–9.
[80] Kelly JD, Haldeman BA, Grant FJ, et al. Platelet-derived growth factor (PDGF) stimulates PDGF receptor subunit dimerization and intersubunit trans-phosphorylation. J Biol Chem 1991;266(14):8987–92.
[81] Failli P, Ruocco C, De Franco R, et al. The mitogenic effect of platelet-derived growth factor in human hepatic stellate cells requires calcium influx. Am J Physiol 1995;269(5 Pt 1):C1133–9.
[82] Tarrats N, Moles A, Morales A, et al. Critical role of tumor necrosis factor receptor 1, but not 2, in hepatic stellate cell proliferation, extracellular matrix remodeling, and liver fibrogenesis. Hepatology 2011;54(1):319–27.
[83] Lee JS, Semela D, Iredale J, Shah VH. Sinusoidal remodeling and angiogenesis: a new function for the liver-specific pericyte? Hepatology 2007;45(3):817–25.
[84] Kaur S, Tripathi D, Dongre K, et al. Increased number and function of endothelial progenitor cells stimulate angiogenesis by resident liver sinusoidal endothelial cells (SECs) in cirrhosis through paracrine factors. J Hepatol 2012;57(6):1193–8.
[85] Ishikawa K, Mochida S, Mashiba S, et al. Expressions of vascular endothelial growth factor in non-parenchymal as well as parenchymal cells in rat liver after necrosis. Biochem Biophys Res Commun 1999;254(3):587–93.
[86] Zhao Y, Wang Y, Wang Q, et al. Hepatic stellate cells produce vascular endothelial growth factor via phospho-p44/42 mitogen-activated protein kinase/cyclooxygenase-2 pathway. Mol Cell Biochem 2012;359(1–2):217–23.
[87] Yoshiji H, Kuriyama S, Yoshii J, et al. Vascular endothelial growth factor and receptor interaction is a prerequisite for murine hepatic fibrogenesis. Gut 2003;52(9):1347–54.
[88] Trepo E, Ouziel R, Pradat P, et al. Marked 25-hydroxyvitamin D deficiency is associated with poor prognosis in patients with alcoholic liver disease. J Hepatol 2013;59(2):344–50.
[89] Fedirko V, Duarte-Salles T, Bamia C, et al. Prediagnostic circulating vitamin D levels and risk of hepatocellular carcinoma in European populations: a nested case-control study. Hepatology 2014. http://dx.doi.org/10.1002/hep.27079.
[90] Abramovitch S, Dahan-Bachar L, Sharvit E, et al. Vitamin D inhibits proliferation and profibrotic marker expression in hepatic stellate cells and decreases thioacetamide-induced liver fibrosis in rats. Gut 2011;60(12):1728–37.
[91] Ding N, Yu RT, Subramaniam N, et al. A vitamin D receptor/SMAD genomic circuit gates hepatic fibrotic response. Cell 2013;153(3):601–13.
[92] Iizuka M, Murata T, Hori M, Ozaki H. Increased contractility of hepatic stellate cells in cirrhosis is mediated by enhanced Ca^{2+}-dependent and Ca^{2+}-sensitization pathways. Am J Physiol Gastrointest Liver Physiol 2011;300(6):G1010–21.
[93] Saiman Y, Agarwal R, Hickman DA, et al. CXCL12 induces hepatic stellate cell contraction through a calcium-independent pathway. Am J Physiol Gastrointest Liver Physiol 2013;305(5):G375–82.
[94] Melton AC, Datta A, Yee Jr. HF. $[Ca^{2+}]_i$-independent contractile force generation by rat hepatic stellate cells in response to endothelin-1. Am J Physiol Gastrointest Liver Physiol 2006;290(1):G7–G13.
[95] Thimgan MS, Yee Jr. HF. Quantitation of rat hepatic stellate cell contraction: stellate cells' contribution to sinusoidal resistance. Am J Physiol 1999;277(1 Pt 1):G137–43.

[96] Shao R, Yan W, Rockey DC. Regulation of endothelin-1 synthesis by endothelin-converting enzyme-1 during wound healing. J Biol Chem 1999;274(5):3228–34.
[97] Housset C, Rockey DC, Bissell DM. Endothelin receptors in rat liver: lipocytes as a contractile target for endothelin 1. Proc Natl Acad Sci USA 1993;90(20):9266–70.
[98] Loureiro-Silva MR, Cadelina GW, Groszmann RJ. Deficit in nitric oxide production in cirrhotic rat livers is located in the sinusoidal and postsinusoidal areas. Am J Physiol Gastrointest Liver Physiol 2003;284(4):G567–74.
[99] Fallowfield JA, Hayden AL, Snowdon VK, et al. Relaxin modulates human and rat hepatic myofibroblast function and ameliorates portal hypertension *in vivo*. Hepatology 2014;59(4):1492–504.
[100] Brown A, Goodman Z. Hepatitis B-associated fibrosis and fibrosis/cirrhosis regression with nucleoside and nucleotide analogs. Expert Rev Gastroenterol Hepatol 2012;6(2):187–98.
[101] Ramachandran P, Iredale JP. Liver fibrosis: a bidirectional model of fibrogenesis and resolution. QJM 2012;105(9):813–7.
[102] Ellis EL, Mann DA. Clinical evidence for the regression of liver fibrosis. J Hepatol 2012;56(5):1171–80.
[103] Popper H, Schaffner F. Hepatic cirrhosis. A problem in communication. Isr J Med Sci 1968;4(1):1–7.
[104] Popper H, Uenfriend S. Hepatic fibrosis. Correlation of biochemical and morphologic investigations. Am J Med 1970;49:707–21.
[105] Issa R, Zhou X, Constandinou CM, et al. Spontaneous recovery from micronodular cirrhosis: evidence for incomplete resolution associated with matrix cross-linking. Gastroenterology 2004;126(7):1795–808.
[106] Issa R, Williams E, Trim N, et al. Apoptosis of hepatic stellate cells: involvement in resolution of biliary fibrosis and regulation by soluble growth factors. Gut 2001;48(4):548–57.
[107] Iredale JP, Benyon RC, Pickering J, et al. Mechanisms of spontaneous resolution of rat liver fibrosis. Hepatic stellate cell apoptosis and reduced hepatic expression of metalloproteinase inhibitors. J Clin Invest 1998;102(3):538–49.
[108] Novo E, Marra F, Zamara E, et al. Overexpression of Bcl-2 by activated human hepatic stellate cells: resistance to apoptosis as a mechanism of progressive hepatic fibrogenesis in humans. Gut 2006;55(8):1174–82.
[109] Watson MR, Wallace K, Gieling RG, et al. NF-kappaB is a critical regulator of the survival of rodent and human hepatic myofibroblasts. J Hepatol 2008;48(4):589–97.
[110] Oakley F, Meso M, Iredale JP, et al. Inhibition of inhibitor of kappaB kinases stimulates hepatic stellate cell apoptosis and accelerated recovery from rat liver fibrosis. Gastroenterology 2005;128(1):108–20.
[111] Oakley F, Trim N, Constandinou CM, et al. Hepatocytes express nerve growth factor during liver injury: evidence for paracrine regulation of hepatic stellate cell apoptosis. Am J Pathol 2003;163(5):1849–58.
[112] Radaeva S, Sun R, Jaruga B, et al. Natural killer cells ameliorate liver fibrosis by killing activated stellate cells in NKG2D-dependent and tumor necrosis factor-related apoptosis-inducing ligand-dependent manners. Gastroenterology 2006;130(2):435–52.
[113] Gabai VL, Yaglom JA, Volloch V, et al. Hsp72-mediated suppression of c-Jun N-terminal kinase is implicated in development of tolerance to caspase-independent cell death. Mol Cell Biol 2000;20(18):6826–36.
[114] Kisseleva T, Cong M, Paik Y, et al. Myofibroblasts revert to an inactive phenotype during regression of liver fibrosis. Proc Natl Acad Sci USA 2012;109(24):9448–53.
[115] Huang Y, Li X, Wang Y, et al. Endoplasmic reticulum stress-induced hepatic stellate cell apoptosis through calcium-mediated JNK/P38 MAPK and Calpain/Caspase-12 pathways. Mol Cell Biochem 2014;394(1–2):1–12.
[116] De Minicis S, Candelaresi C, Agostinelli L, et al. Endoplasmic reticulum stress induces hepatic stellate cell apoptosis and contributes to fibrosis resolution. Liver Int 2012;32(10):1574–84.
[117] Lim MP, Devi LA, Rozenfeld R. Cannabidiol causes activated hepatic stellate cell death through a mechanism of endoplasmic reticulum stress-induced apoptosis. Cell Death Dis 2011;2:e170.
[118] Zhu Y, Men R, Wen M, et al. Blockage of TRPM7 channel induces hepatic stellate cell death through endoplasmic reticulum stress-mediated apoptosis. Life Sci 2014;94(1):37–44.
[119] Krizhanovsky V, Yon M, Dickins RA, et al. Senescence of activated stellate cells limits liver fibrosis. Cell 2008;134(4):657–67.

[120] Lujambio A, Akkari L, Simon J, et al. Non-cell-autonomous tumor suppression by p53. Cell 2013;153(2):449–60.

[121] Kong X, Feng D, Wang H, et al. Interleukin-22 induces hepatic stellate cell senescence and restricts liver fibrosis in mice. Hepatology 2012;56(3):1150–9.

[122] Yoshimoto S, Loo TM, Atarashi K, et al. Obesity-induced gut microbial metabolite promotes liver cancer through senescence secretome. Nature 2013;499(7456):97–101.

[123] Troeger JS, Mederacke I, Gwak GY, et al. Deactivation of hepatic stellate cells during liver fibrosis resolution in mice. Gastroenterology 2012;143(4):1073–83. e1022.

[124] de Mesquita FC, Bitencourt S, Caberlon E, et al. Fructose-1,6-bisphosphate induces phenotypic reversion of activated hepatic stellate cell. Eur J Pharmacol 2013;720(1–3):320–5.

[125] Lopez-Sanchez I, Dunkel Y, Roh YS, et al. GIV/Girdin is a central hub for profibrogenic signalling networks during liver fibrosis. Nat Commun 2014;5:4451.

[126] Mann J, Mann DA. Transcriptional regulation of hepatic stellate cells. Adv Drug Deliv Rev 2009;61(7–8):497–512.

[127] Zeybel M, Mann DA, Mann J. Epigenetic modifications as new targets for liver disease therapies. J Hepatol 2013;59(6):1349–53.

[128] Mann DA. Epigenetics in liver disease. Hepatology 2014;60(4):1418–25.

[129] Perugorria MJ, Wilson CL, Zeybel M, et al. Histone methyltransferase ASH1 orchestrates fibrogenic gene transcription during myofibroblast transdifferentiation. Hepatology 2012;56(3):1129–39.

[130] Tian W, Hao C, Fan Z, et al. Myocardin related transcription factor A programs epigenetic activation of hepatic stellate cells. J Hepatol 2014. http://dx.doi.org/10.1016/j.jhep.2014.07.029.

[131] Mann J, Chu DC, Maxwell A, et al. MeCP2 controls an epigenetic pathway that promotes myofibroblast transdifferentiation and fibrosis. Gastroenterology 2010;138(2):705–14. e701–704.

[132] Hu B, Gharaee-Kermani M, Wu Z, Phan SH. Essential role of MeCP2 in the regulation of myofibroblast differentiation during pulmonary fibrosis. Am J Pathol 2011;178(4):1500–8.

[133] Katare R, Riu F, Mitchell K, et al. Transplantation of human pericyte progenitor cells improves the repair of infarcted heart through activation of an angiogenic program involving micro-RNA-132. Circ Res 2011;109(8):894–906.

[134] Luchsinger LL, Patenaude CA, Smith BD, Layne MD. Myocardin-related transcription factor-A complexes activate type I collagen expression in lung fibroblasts. J Biol Chem 2011;286(51):44116–25.

[135] Small EM, Thatcher JE, Sutherland LB, et al. Myocardin-related transcription factor-a controls myofibroblast activation and fibrosis in response to myocardial infarction. Circ Res 2010;107(2):294–304.

[136] Hampl V, Martin C, Aigner A, et al. Depletion of the transcriptional coactivators megakaryoblastic leukaemia 1 and 2 abolishes hepatocellular carcinoma xenograft growth by inducing oncogene-induced senescence. EMBO Mol Med 2013;5(9):1367–82.

[137] Meyer N, Penn LZ. Reflecting on 25 years with MYC. Nat Rev Cancer 2008;8(12):976–90.

[138] Takahashi Y, Kawate S, Watanabe M, et al. Amplification of c-myc and cyclin D1 genes in primary and metastatic carcinomas of the liver. Pathol Int 2007;57(7):437–42.

[139] Nevzorova YA, Hu W, Cubero FJ, et al. Overexpression of c-myc in hepatocytes promotes activation of hepatic stellate cells and facilitates the onset of liver fibrosis. Biochim Biophys Acta 2013;1832(10):1765–75.

[140] Reebye V, Saetrom P, Mintz PJ, et al. Novel RNA oligonucleotide improves liver function and inhibits liver carcinogenesis *in vivo*. Hepatology 2014;59(1):216–27.

[141] Nevzorova YA, Bangen JM, Hu W, et al. Cyclin E1 controls proliferation of hepatic stellate cells and is essential for liver fibrogenesis in mice. Hepatology 2012;56(3):1140–9.

[142] Sekiya Y, Ogawa T, Iizuka M, et al. Down-regulation of cyclin E1 expression by microRNA-195 accounts for interferon-beta-induced inhibition of hepatic stellate cell proliferation. J Cell Physiol 2011;226(10):2535–42.

[143] Hu W, Nevzorova YA, Haas U, et al. Concurrent deletion of cyclin E1 and cyclin-dependent kinase 2 in hepatocytes inhibits DNA replication and liver regeneration in mice. Hepatology 2014;59(2):651–60.

[144] Czaja MJ, Ding WX, Donohue Jr. TM, et al. Functions of autophagy in normal and diseased liver. Autophagy 2013;9(8):1131–58.

[145] Thoen LF, Guimaraes EL, Dolle L, et al. A role for autophagy during hepatic stellate cell activation. J Hepatol 2011;55(6):1353–60.

[146] Tanjore H, Lawson WE, Blackwell TS. Endoplasmic reticulum stress as a pro-fibrotic stimulus. Biochim Biophys Acta 2013;1832(7):940–7.
[147] Ward PS, Thompson CB. Metabolic reprogramming: a cancer hallmark even warburg did not anticipate. Cancer Cell 2012;21(3):297–308.
[148] Li T, Eheim AL, Klein S, et al. Novel role of nuclear receptor Rev-erbalpha in hepatic stellate cell activation: potential therapeutic target for liver injury. Hepatology 2014;59(6):2383–96.
[149] Chen Y, Choi SS, Michelotti GA, et al. Hedgehog controls hepatic stellate cell fate by regulating metabolism. Gastroenterology 2012;143(5):1319–29. e1311.
[150] Beaven SW, Wroblewski K, Wang J, et al. Liver X receptor signaling is a determinant of stellate cell activation and susceptibility to fibrotic liver disease. Gastroenterology 2011;140(3):1052–62.
[151] Calkin AC, Tontonoz P. Transcriptional integration of metabolism by the nuclear sterol-activated receptors LXR and FXR. Nat Rev Mol Cell Biol 2012;13(4):213–24.
[152] Fiorucci S, Antonelli E, Rizzo G, et al. The nuclear receptor SHP mediates inhibition of hepatic stellate cells by FXR and protects against liver fibrosis. Gastroenterology 2004;127(5):1497–512.
[153] Verbeke L, Farre R, Trebicka J, et al. Obeticholic acid, a farnesoid X receptor agonist, improves portal hypertension by two distinct pathways in cirrhotic rats. Hepatology 2014;59(6):2286–98.
[154] Yi HS, Lee YS, Byun JS, et al. Alcohol dehydrogenase III exacerbates liver fibrosis by enhancing stellate cell activation and suppressing natural killer cells in mice. Hepatology 2014. http://dx.doi.org/10.1002/hep.27137.
[155] Boursier J, Diehl AM. PNPLA3 and liver disease opportunities to unravel mechanisms underlying statistical associations. Hepatology 2014. http://dx.doi.org/10.1002/hep.27445.
[156] Pirazzi C, Valenti L, Motta BM, et al. PNPLA3 has retinyl-palmitate lipase activity in human hepatic stellate cells. Hum Mol Genet 2014;23(15):4077–85.
[157] Van Rooyen DM, Larter CZ, Haigh WG, et al. Hepatic free cholesterol accumulates in obese, diabetic mice and causes nonalcoholic steatohepatitis. Gastroenterology 2011;141(4):1393–403. e1391–1395.
[158] Teratani T, Tomita K, Suzuki T, et al. A high-cholesterol diet exacerbates liver fibrosis in mice via accumulation of free cholesterol in hepatic stellate cells. Gastroenterology 2012;142(1):152–64. e110.
[159] Tomita K, Teratani T, Suzuki T, et al. Free cholesterol accumulation in hepatic stellate cells: mechanism of liver fibrosis aggravation in nonalcoholic steatohepatitis in mice. Hepatology 2014;59(1):154–69.
[160] Tomita K, Teratani T, Suzuki T, et al. Acyl-CoA:cholesterol acyltransferase 1 mediates liver fibrosis by regulating free cholesterol accumulation in hepatic stellate cells. J Hepatol 2014;61(1):98–106.
[161] Van Rooyen DM, Gan LT, Yeh MM, et al. Pharmacological cholesterol lowering reverses fibrotic NASH in obese, diabetic mice with metabolic syndrome. J Hepatol 2013;59(1):144–52.
[162] Omenetti A, Choi S, Michelotti G, Diehl AM. Hedgehog signaling in the liver. J Hepatol 2011;54(2):366–73.
[163] Grzelak CA, Martelotto LG, Sigglekow ND, et al. The intrahepatic signalling niche of hedgehog is defined by primary cilia positive cells during chronic liver injury. J Hepatol 2014;60(1):143–51.
[164] Guy CD, Suzuki A, Abdelmalek MF, Burchette JL, Diehl AM. Treatment response in the PIVENS trial is associated with decreased hedgehog pathway activity. Hepatology 2014. http://dx.doi.org/10.1002/hep.27235.

CHAPTER 5

Cytokine Production and Signaling in Stellate Cells

Fabio Marra and Alessandra Caligiuri
Dipartimento di Medicina Sperimentale e Clinica, University of Florence, Italy

5.1 INTRODUCTION

Hepatic stellate cells (HSCs) were initially identified and characterized for their ability to secrete extracellular matrix components, participating in the pathogenesis of fibrosis. It soon became apparent that the role of HSCs in the hepatic wound-healing response is far more complex than initially thought, and that these cells are able to regulate a number of different biological functions ranging from immunity to angiogenesis and liver regeneration. Most of the activities to which HSCs contribute are regulated through the secretion of soluble mediators. In this group, cytokines represent a significantly relevant factor capable of affecting numerous functions in which HSCs are implicated. On the other hand, HSCs express a number of different cytokine receptors, and their activation induces various signaling pathways that ultimately results in changes in their gene expression and biological functions.

A characteristic of the cytokine network is its ability to regulate the various functions of target cells. For this reason, it is difficult to group cytokines and cytokine receptors based on their predominant functions. Thus, this chapter is organized around the different groups of cytokines and their receptors rather than according to the ultimate effect(s) exerted on HSCs.

5.2 PLATELET-DERIVED GROWTH FACTOR

Platelet-derived growth factor (PDGF) has been characterized as one of the most potent mitogens for HSCs isolated from rat, mouse, or human liver [1]. Among the different isoforms of PDGF, those containing the A- or B-chains have been extensively characterized and better defined. PDGF isoforms are secreted by various resident and infiltrating cells within the liver, including macrophages and endothelial cells. However, HSCs are also a source of PDGF A- and B-chains, and although their contribution is quantitatively minor, it may generate autocrine and paracrine loops [2]. Inducers of PDGF B-chain include PDGF itself and other factors such as HBV-derived proteins [2,3].

Stellate Cells in Health and Disease
DOI: http://dx.doi.org/10.1016/B978-0-12-800134-9.00005-1

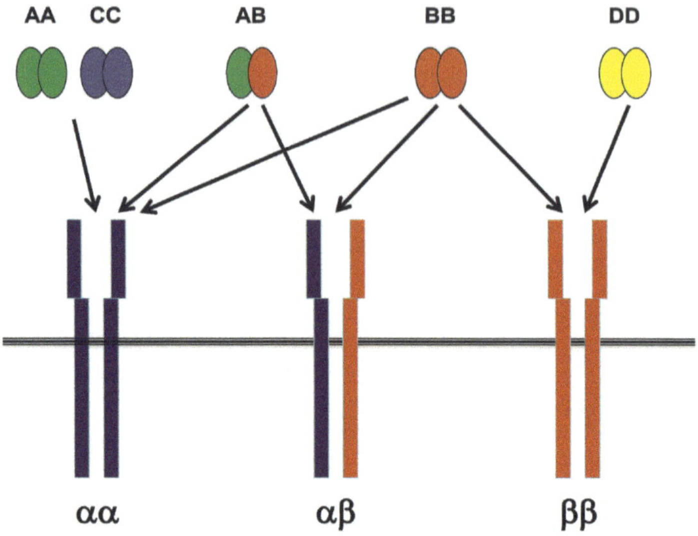

Figure 5.1 ***Interaction of the different isoforms of PDGF with PDGF receptors.*** The different PDGF chains forming homo- or heterodimers are indicated in different colors. They interact with PDGF-R homo- or heterodimers of α- and/or β-chain, as indicated. Low affinity interactions are not depicted.

The effects of secreted PDGF-B may also have an impact on cancer biology, such as in the case of cholangiocarcinoma, where it promotes cell survival via activation of the Hedgehog pathway [4].

PDGF-BB is a potent agonist for proliferation and migration of HSCs and interacts with all PDGF receptors containing alpha or beta subunits (Figure 5.1). However, the receptor predominantly expressed in these cells is the beta isoform, which is markedly upregulated upon HSC activation [5,6]. While PDGF chains co-distribute with cells expressing PDGF receptor *in vivo* during fibrogenesis [7], PDGF stimulation of HSCs is a relatively minor inducer of the expression of genes related to the extracellular matrix such as procollagens. However, the role of the beta subunit of the PDGF receptor is indicated by recent studies where an adenovirally expressed soluble PDGF-beta receptor subunit was found to ameliorate fibrosis in a rat model [8]. Furthermore, PDGF causes upregulation of various factors that contribute to angiogenesis, a well-established component of the hepatic wound-healing response.

The pivotal role played by PDGF in the modulation of relevant biologic activities of HSCs, such as proliferation and chemotaxis, has spurred a number of studies directed to the understanding of the intracellular signaling pathways activated by this growth factor in HSCs and their role in different biologic activities. Recently, two additional PDGF polypeptide chains were identified, namely PDGF-C and PDGF-D, suggesting that PDGF-mediated signaling is more complex than previously anticipated.

The biologic relevance of the newly discovered PDGF isoforms is evident from the development of fibrosis and hepatocellular carcinoma in a transgenic mouse model expressing PDGF-C [9]. Interactions of PDGF isoforms with different receptors is illustrated in Figure 5.1.

In regard to the relevance of this polypeptide growth factor in stellate cell growth and chemotaxis, considerable efforts have been made in understanding the intracellular signaling events elicited by the interaction of PDGF with its receptor. As for many other receptors, interaction with a cognate ligand leads to generation of both positive and negative signaling pathways, allowing a fine modulation of the effects of receptor activation to a finite duration.

The PDGF receptors belong to the family of tyrosine kinases receptors, and activation is provided by cross-phosphorylation consequent to receptor dimerization [10]. Once activated, the PDGF receptor associates with multiple intracellular proteins that contribute to the downstream transmission of the signal. Extracellular signal-regulated kinase (ERK) belongs to the mitogen-activated protein kinase family and its activation is pivotal in generating proliferative responses in HSC exposed to PDGF [11]. As a result, inhibitors of ERK 1/2 efficiently inhibit cell growth and limit the further activation of molecules dependent on ERK activation, such as c-fos and AP-1 [11]. Other pathways regulated by ERK 1/2 in response to PDGF include STAT-1 and phosphatidic acid, a lipid second messenger that provides additional proliferative signals [12,13]. More recently, human antigen R, an RNA-binding protein, has been identified as a target of PDGF receptor activation and ERK signaling. Human antigen R (HuR) contributes to fibrogenesis, as its silencing attenuates fibrosis development *in vivo* and reduces liver damage and oxidative stress [14].

Activation of ERK in HSCs has been demonstrated in a number of conditions associated with liver damage and/or fibrosis such as the acute intoxication with CCl_4 [11]. Interestingly, ERK activation was found to precede the proliferation of HSCs, and was associated with a time point when PDGF is particularly abundant, suggesting that this growth factor may play a major role in inducing ERK activation *in vivo* [11]. In other studies, modulation of the ERK pathway has been achieved by interfering with upstream signaling events. For example, accumulation of intracellular cyclic adenosine monophosphate (cAMP) upon treatment with phosphodiesterase inhibitors results in a marked inhibition of PDGF-induced ERK activation. Pentoxifylline is the first compound of this class to be described for its ability to modulate PDGF-induced ERK phosphorylation [15]. More recently, rats treated with the PDE4-specific inhibitor rolipram showed reduced expression of inflammatory and profibrotic cytokines and fibrosis, indicating that this class of compounds may deserve further consideration in the treatment of fibrosis [16]. Prostaglandin (PG) E_2 also causes an increase in intracellular cAMP, thus leading to inhibition of the ERK pathway [17]. Interestingly, PDGF is one of the stimuli that increases PGE_2 synthesis, illustrating the existence of an autocrine pathway of self-limiting the downstream effects of PDGF [17]. Another group has reported that the COX-2 inhibitor celecoxib inhibits PDGF-induced HSC proliferation, and that PGE_2 and PGI_2 induce ERK activation in HSCs causing their proliferation [18]. Although these observations together suggest that PGE_2 contributes to chemokine secretion [19], the significance of the prostaglandin system in HSC biology remains controversial.

Phosphatidylinositol 3 kinase (PI3K) is another important molecule recruited upon activation of PDGF receptor, and its downstream effectors include Akt, also originally known as protein kinase B. This pathway contributes to the generation of mitogenic and migratory signals in HSCs exposed to PDGF [20]. The *in vivo* relevance of this finding is indicated by the observation that the p85 subunit of PI3K is bound by the PDGF-receptor and activation of PI3K occurs following acute CCl_4-induced liver damage in rat [20]. Accordingly, inhibitors of PI3K activation block proliferation and chemotaxis of HSCs in response to PDGF. In addition, a role of interaction between the PDGF receptor and the axonal guidance protein, neuropilin-1, has been reported [21]. Knockdown of NRP-1 attenuates PDGF-induced chemotaxis, while its overexpression increases cell motility and collagen production.

The PI3K/Akt pathway is also implicated in the control of HSC survival. Remarkably, differential effects in response to cytokines that activate PI3K may be observed by comparing proliferation and survival. Specifically, PDGF is a much stronger mitogen than insulin-like growth factor-1 (IGF-1), a factor induced by stimulation with PDGF [22]. IGF-1, however, acts as a survival factor rather than a mitogenic growth factor for HSCs [23]. Of note, the action of IGF-1 is also regulated by the different binding proteins that associate with this cytokine. The repertoire of binding proteins depends not only on the types secreted by HSCs, but also on those produced by the neighboring cells [24,25]. The regulation of this system in HSCs has been recently shown to be dependent on selected miRNAs, such as miR-29b, which inhibits the expression of IGF-1 and PDGF-C [26].

PDGF-signaling in HSCs is also dependent on changes in intracellular calcium $[Ca^{2+}]_i$ [27], consisting of a peak increase followed by a plateau phase due to the influx from the external medium. Induction of replicative competence by PDGF is dependent on the maintenance of sustained increase in $[Ca^{2+}]_i$ due to calcium entry rather than its release from the intracellular stores. In addition, PDGF induces the activity of the Na^+/H^+ exchanger, and changes in intracellular pH are essential for the full biologic activity of PDGF. Inhibition of the Na^+/H^+ exchanger reduces PDGF-induced HSC proliferation *in vitro* [28], and a Na^+/H^+ exchanger inhibitor amiloride blocks HSC proliferation and collagen deposition *in vivo* [29]. Another protein recently found to be affected by PDGF is the myristoylated alanine rich protein kinase C substrate (MARCKS), a protein kinase C substrate regulated by calcium. MARCKS regulates cell motility and upon phosphorylation induced by PDGF, it detaches from the plasma membrane favoring cell migration [30]. ERK5 is a less studied member of MAPK family. Recently, ERK5 was found to be a target of the PDGF receptor in HSCs. Its role in PDGF-stimulated HSC biology is demonstrated by inhibition of cell proliferation, and expression and activation of c-Jun by ERK5 knockdown [31].

Several observations indicate that PDGF signaling interacts with products of oxidative stress in HSCs. Reactive aldehydes such as 4-hydroxy-2,3-nonenal (HNE), exert direct profibrogenic effects via nuclear translocation of c-Jun NH_2-terminal kinases (JNKs), upregulation of c-Jun and increased AP-1 binding [32]. However, HNE abolishes PDGF-BB mitogenic signaling in human HSCs through specific inhibition of the intrinsic tyrosine kinase activity [33].

5.3 ANGIOGENIC CYTOKINES

Formation of new blood vessels is an integral part of the wound-healing response and has been documented in chronic liver diseases, where it is associated with fibrosis. An initial connection was found between cancer and the angiogenic potential of HSCs, as melanoma supernatants stimulated vascular endothelial growth factor (VEGF) expression in these cells. Moreover, these effects were potentiated by hypoxia, indicating that HSCs are a part of the network of cytokines that modulate angiogenesis during reduced oxygen availability [34]. Of note, VEGF is not the only factor produced by HSCs, as angiopoietins and apelin that have the ability to modulate the angiogenic process were also found to be secreted by HSCs [35–37]. Factors that modulate expression of this group of cytokines include hypoxia and several soluble factors, including PDGF and leptin [36,38]. Importantly, several stimuli potentially relevant during chronic liver injury, including hypoxia, nitric oxide, and oxidative stress-related conditions, can upregulate VEGFR1 [39].

Most of the factors involved in angiogenesis regulate the biology of HSCs. For example, fibroblast growth factor-1 exerts potent mitogenic action on HSCs, and a profibrogenic role of this factor has been documented in animal models [1]. Activation of HSCs is associated with an increased expression of VEGF receptors, and VEGF induces cell growth [39]. Interestingly, at later stages of HSC activation, expression of VEGFR1 increases progressively whereas VEGFR2 expression decreases, and VEGF reduces the contractile properties of HSCs and their α-sma expression [39]. VEGF, together with angiopoietin 1 (Ang-1), also induces HSC migration via activation of the Ras/ERK pathway. These effects are mediated by Ang-1 receptor Tie-2, VEGFR-1, and VEGFR-2 [40]. The role of Ang-1 has been more extensively evaluated under *in vivo* conditions, where adenoviral vectors expressing the extracellular domain of Tie-2 were used to block angiopoietin signaling [37], resulting in inhibition of angiogenesis and liver fibrosis.

Apelin is another factor related to angiogenesis, that is principally induced by angiotensin II and endothelin-1 in HSCs. Exposure to recombinant apelin increased synthesis of collagen-I, and this effect was reduced upon blocking apelin receptor [35]. Another factor linking angiogenesis and HSC biology is placental growth factor (PIGF), which is upregulated in a rodent model of fibrosis [41]. Mice deficient in PIGF have less inflammation and fibrosis. Moreover, PIGF induced proliferation and migration of cultured HSCs via activation of ERK1/2. Another study indicating the role of angiogenesis in the regulation of the fibrogenic process investigated the effects of everolimus, which inhibits collagen production by activated HSCs, as well as their contraction and transition toward an activated status [42]. The mechanism of angiogenesis is also tightly regulated through angiogenesis inhibitors. A recent study investigating the role of vasohibin-1 in samples from cirrhotic patients and in rodent models of liver fibrosis demonstrated that vasohibin-1 regulates fibrogenesis- and portal hypertension-associated pathological angiogenesis [43].

5.4 TGF-β SUPERFAMILY

TGF-β, a potent cytokine that regulates the stellate cell phenotype [44], is produced by circulating cells as well as several cell types in the liver including HSCs [45]. TGF-β is secreted in a latent form that requires proteolysis, and is associated with a latent TGF-β binding protein. While integrins, fibrinogen, and urokinase-type plasminogen activator are factors involved in TGF-β activation [46–48], TGF-β may be inactivated by the proteoglycan decorin [49]. In HSCs, ADAMTS1 induces TGF-β activation through an interaction of the latency-associated peptide with TGF-β, and its downregulation decreases the release of TGF-β [50].

Expression of TGF-β is increased in models of fibrosis and in patients with chronic liver disease [51,52]. A major effect of TGF-β is the induction of the deposition of extracellular matrix, and several strategies that reduce the biologic action of this cytokine in the target tissue result in an amelioration of liver fibrosis [51]. TGF-β is the most potent stimulus for production extracellular matrix by HSCs, and it induces both qualitative and quantitative changes via differential stimulation of each component [44]. A parallel effect is the inhibition of matrix degradation, via actions on matrix metalloproteinases and induction of tissue inhibitors of metalloproteinases (TIMPs) and plasminogen activator inhibitor (PAI).

TGF-β signals via a complex receptor system. Quiescent HSCs are poorly responsive to TGF-β [53], while TGF-β receptors are present after activation of HSCs that respond to TGF-β. Interestingly, in non-activated cells, mRNA transcripts for TGF-β receptor II (TβRII) and TβRIII (betaglycan) were actually greater than after activation on uncoated plastic [53], and it has been suggested that the ratio of type II to type I receptors may be critical to mediate the biologic effects of TGF-β. The leading role played by TGF-β in hepatic fibrosis has generated interest in identifying strategies to block its actions as an anti-fibrotic agent. Along these lines, soluble TGF-β receptors

[51] increasing the levels of small mother against decapentaplegic (Smad)7 [54], or a variety of other approaches [55] have been proposed. Recently, it was also shown that the vasodilator-stimulated phosphoprotein (VASP) contributes to the activation process of HSCs in response to TGF-β, forming a complex with TβRII and Rab11, a Ras-like small GTPase [56]. Semaphorin 7A is another factor that is increased in response to TGF-β in activated HSCs, leading to enhanced expression of fibrogenic and inflammatory factors [57]. In contrast, other proteins counteract the fibrogenic actions of TGF-β in HSCs. The IQ motif containing GTPase activating protein 1 (IQGAP1) suppresses TβRII-mediated signaling in HSCs, whereas IQGAP1 knockdown stabilizes the effects of TGF-β1 on HSC transdifferentiation into myofibroblasts [58]. Similarly, Bone Morphogenetic Protein and Activin Membrane-Bound Inhibitor (Bambi) is a TGF-β coreceptor that downregulates downstream signaling of TGF-β, and may be modulated by profibrogenic factors such as TLR-4 [59]. Other factors that modulate TGF-β signaling include some miRNAs such as miR-133a, which is specifically downregulated in HSCs during fibrogenesis. Overexpression of this miRNA in HSCs leads to decreased expression of signaling through blunting of TGF-β activity.

Similar to other TGF-β-responsive cells, exposure of HSCs to TGF-β results in phosphorylation of Smad2 and Smad3 [60]. Nonetheless, fully activated myofibroblasts show reduced Smad activation, a finding associated with a reduced ability of this cytokine to induce its biologic actions [60]. Interaction between activated Smads and other factors contribute to the cellular specificity of TGF-β actions, and Sp1 binding to the specific promoter mediates the increased expression of the a2(I) collagen gene in HSC [61], whereas in hepatocytes Sp3 binds the same element, but with little transactivating activity. Moreover activated Smad3 interacts with Sp1 in HSCs, but not in hepatocytes, thus providing a link between TGF-β signaling and matrix upregulation [62]. To confirm the role of this pathway in the regulation of extracellular matrix, constitutive phosphorylation and nuclear localization of Smad3 has been described in a clone of HSC exhibiting high levels of collagen and PAI-1 [63]. Another aspect where TGF-β and Smads are involved is in the interaction with extracellular matrix [64]. Antagonistic Smads include Smad7, which interferes with many aspects of HSC activation [54]. TGF-β signaling may also be counteracted by different mechanisms including exposure of HSCs to vitamin D, which limits TGF-β's action via vitamin D receptor (VDR)-dependent and independent mechanisms [65]. The action of VDR appears to be particularly intriguing, and in a recent elegant study, it has been shown that TGF-β1 signaling causes a redistribution of genome-wide VDR-binding sites in HSCs. This facilitates VDR binding at Smad3 target genes, reducing Smad3 occupancy at these sites, thus inhibiting fibrosis [66]. Although Smads represent a critical pathway in TGF-β signaling, this cytokine has also been shown to activate the ERK1/2 pathway, with potential involvement in extracellular matrix synthesis [67].

Exposure of the cells to reactive oxygen intermediates also contributes to the induction of procollagen α1(I) in response to TGF-β, an effect mediated, at least in part, by activation of C/EBPβ [68]. It has been proposed that exposure to TGF-β is associated with high intracellular levels of H_2O_2, that could be efficiently removed by catalase, yielding high levels of glutathione and allowing autocrine secretion of TGF-β [69].

HSCs also produce activin and respond to recombinant activin. Activin A increased mRNA levels for type I collagen synergistically with TGF-β, thus establishing functionality of the activin receptors [70]. Activin A also stimulates CCL2 expression in human HSCs [71]. A natural inhibitor of activin is follistatin, whose expression decreases during HSC activation, leading to more unopposed activin activity [72]. However, the dynamics of activin receptor expression during HSC activation and the specific role of activin receptor-dependent signaling pathways have not yet been evaluated.

Connective tissue growth factor (CTGF) (also known as CCN2) is a growth factor modulator protein that promotes fibrogenesis in skin, lung, and kidney [73,74]. CTGF is strongly expressed during hepatic fibrosis, and HSCs appear to be a source of this cytokine in the liver [75]. A recent report indicates that its expression in HSCs is TGF-β and Smad 2/3 independent, in contrast to hepatocytes where CTGF regulation is TGF-β dependent. These findings are particularly interesting as they emphasize that regulation of the same cytokine can be completely divergent between two resident liver cell populations, and point to the possibility of cell- and pathway-specific inhibition of a factor that is widely expressed [76].

5.5 CHEMOKINES

The chemokine system is a major modulator of many critical functions both in physiologic and pathologic conditions, including inflammation, development, leukocyte trafficking, angiogenesis, and cancer [77]. The observation that HSCs are able to secrete different chemokines highlighted the concept that they may influence recruitment of cells of the immune system, regulating inflammation and creating a crosstalk between the inflammatory and reparative phase of the wound-healing response [77].

CCL2 is one of the best characterized chemokines during hepatic fibrogenesis, and HSCs are major contributors of its expression in the chronically inflamed liver [78]. CCL2 promotes the recruitment of monocytes expressing the cognate receptor CCR2, and several models of chronic liver injury have shown the relevance of the CCL2–CCR2 pathway in different models of fibrosis, employing CCL2- or CCR2-deficient mice, or pharmacologic interference with this system [77]. Interestingly, HSCs do not express the receptor CCR2, yet incubation of HSCs with CCL2 induces their chemotaxis [79], which is associated with activation of intracellular signaling. Thus, HSC are not only producers of CCL2 and other chemokines, but also a target of their action. Table 5.1 illustrates the different chemokines and chemokine receptors expressed by HSCs.

Table 5.1 Major Chemokines Expressed by HSC and Their Predominant Function

Chemokine	Receptor	Major function
CCL2	CCR2	Profibrotic
CCL5	CCR1, CCR5	Profibrotic
CCL20	CCR6	Profibrotic
CXCL8	CXCR1, CXCR2	Profibrotic
CXCL9	CXCR3	Anti-fibrotic
CXCL10	CXCR3	Profibrotic
CX3CL1	CX3CR1	Anti-fibrotic

HSCs have also been shown to express and secrete members of the chemokine system, which results in chemoattraction of neutrophils, such as CXCL1 or CXCL8 [80,81]. Several factors stimulate the secretion of this group of cytokines, including molecules derived from Gram-negative or positive bacteria [82], adipokines such as resistin [83], or the core protein of HCV [84]. More recently, CCL20, the ligand of CCR6, has been identified as a chemokine secreted by HSCs [85]. CCL20 expression was increased in animal models of liver injury, and macrophages and HSCs were identified as the main cell types expressing CCL20. Moreover, CCL20 induced proinflammatory and profibrogenic effects in cultured primary HSCs. On the other hand, in murine experimental fibrosis, CCL20 favors the recruitment of gamma-delta T cells, limiting inflammation and fibrosis [86].

Chemokine receptor CXCR3, which binds the ligands CXCL10, CXCL9, and CXCL11, is one of the first chemokine receptors described in HSCs [87]. Its activation is associated with increased cell migration and leads to activation of the Ras/ERK cascade, and to activation of PI3K. The three CXCR3 ligands are upregulated in patients with chronic liver diseases, and their serum levels correlate with the severity of fibrosis [77]. Recent findings report that among these, CXCL9 exerts anti-fibrogenic and antiangiogenic functions whereas CXCL10 appears to be profibrogenic, directly affecting HSCs [77,88–90].

Expression of other functional chemokine receptors have been recently demonstrated in cultured HSCs. CCR7 mediates cell migration and secretion of other chemokines [91]. Similarly, CCR5 has been found on the surface of activated HSCs, which are also an important source of its ligand CCL5 indicating the existence of an autocrine loop of action [92]. The profibrogenic role of the CCR5/CCL5 pathway is indicated by several lines of evidence, including that from mice with genetic deletion of CCR5 or CCL5 [93,94]. It is important to highlight that in experiments carried out in chimeric mice, the major profibrogenic role was attributed to the expression of CCR5 in HSCs, rather than in hematopoietic cells [93]. Of note, CCR5 also mediates the direct profibrogenic action of the gp120 HIV envelope protein on HSCs [95].

Genetic deletion of CCL3 also abrogates experimental hepatic fibrosis in mice through a direct inhibitory action on HSCs [77].

HSCs also express the chemokine receptor CXCR4 that is activated by stimulation of its cognate ligand CXCL12 [96]. CXCR4 has also been shown to interfere with HIV envelope protein gp120 [97]. In addition, CXCL12 (SDF-1) was recently recognized as a crucial factor balancing regenerative and fibrogenic response by sinusoidal endothelial cells, and activation of CXCR4 in sinusoidal endothelial cells was found to induce profibrogenic response in HSCs [98]. Additional chemokine receptors that modulate the profibrogenic features of HSC include CCR9, the activation of which, in response to CCL25, functions as direct stimulus for HSCs activation [99].

5.6 TUMOR NECROSIS FACTOR SUPERFAMILY

The receptors for tumor necrosis factor (TNF) include molecules binding to known cytokines and orphan receptors [100]. Based on the presence or absence of a "death domain," molecules of this superfamily may be divided into two groups, including TNFR1, Fas, and the p75 receptor for nerve growth factor with "death domain" on the one hand, and TNFR2, and CD40 that lack the "death domain" on the other [101]. TNF activates a complex series of pathways that regulate inflammation, and others leading to cell death. The complex network of signals downstream of TNFR-1 has been reviewed by Walczak [102]. The inflammatory pathway is mediated by downstream activation of molecules such as NF-κB and JNK, while cell death is mediated by interaction of TRADD with FADD that eventually leads to activation of the caspase cascade.

TNF has been shown to exert relevant actions in HSCs and to modulate several aspects of the fibrogenic response. Early data indicate that TNF participates in the activation process [103], while it inhibits expression and synthesis of fibrillar collagen [104] and proliferation of HSC [105]. On the other hand, TNF critically regulates the "proinflammatory" phenotype of HSCs, upregulating the expression and secretion of different cytokines and chemokines (reviewed in Ref. [106]). While TNFR1 is expressed at the mRNA level in quiescent HSC, its signaling capabilities appear to be somehow limited because it does not result in NF-κB activation [107]. TNF upregulates NF-κB target molecules such as ICAM-1 only in activated cells. Activation of NF-κB is critical to induce proinflammatory actions on HSCs in response to TNF and other cytokines. This is clearly indicated by the effects of inhibition of NF-κB activation, which results in a marked reduction of the expression and secretion of a number of factors involved in inflammation and secreted by HSCs, including interleukin-6, CCL2, CXCL1, CXCL2, and ICAM-1 [19,108]. TNF also causes upregulation of Rel-like proteins, which ultimately participate in assembly of the NF-κB DNA binding complex in activated HSCs [109], and contribute to the induction of cyclo-oxygenase

2 (COX-2) [19,105]. Upon inhibition of NF-κB, exposure of HSCs to TNF results in apoptosis [110], while in the presence of an intact NF-κB pathway, TNF actually favors survival of these cells, highlighting the complex role of this factor [111].

Stress-activated protein kinases, such as JNK or p38, are additional targets of the action of NF-κB, and have been involved in the modulation of the activated phenotype of HSCs [112]. JNK is required for the induction of prepro-endothelin-1 (precursor of endothelin-1, a potent vasoconstrictor and multifunctional peptide) expression in activated HSCs [113]. Moreover, activation of JNK or of p38MAPK was necessary for TNF-mediated induction of CCL2 [71]. Remarkably, activation of p38 was also necessary for the release of several cytokines including TNF-α by activated HSCs, identifying a mechanism for an autocrine loop involving TNF in these cells [114].

The inhibition of type I collagen expression by TNF is mediated by a complex mechanism. Preincubation of HSCs with pertussis toxin abolishes this effect, while ceramide mimics the action of TNF [115]. Moreover, increased binding of C/EBPβ and C/EBPδ, and reduced binding of Sp1 have been shown to be involved in the inhibitory effects of TNF [116–118]. An interesting crosstalk involving TNF and the peroxisome proliferator-activated receptor-γ (PPAR-γ) has been described in HSCs, where TNF reduced the transcriptional activity of PPAR-γ via activation of ERK and JNK [119]. As PPAR-γ plays a pivotal role in the maintenance of the quiescent phenotype of HSC [120], these data indicate a mechanism for the induction of HSC activation in response to TNF. Recent investigation has focused on the contribution of the different TNF receptors in HSC biology. TNF was found to regulate HSC proliferation and extracellular matrix remodeling via TNFR1 [121], and accordingly, liver damage and fibrogenesis caused by bile-duct ligation were reduced in TNFR1 knockout mice.

HSCs express Fas (CD95), which increases during the activation process in parallel with HSC apoptosis [122]. Moreover, HSCs express Fas-ligand [111], which is down-regulated by TNF-α. The expression of receptors for TRAIL, another receptor belonging to the TNF superfamily, has been shown in LX-2 cells, where it mediated apoptosis following exposure to TRAIL [123] or co-incubation with Kupffer cells [123,124]. Recently, TRAIL was found to inhibit Hsp47 expression in HSCs, and to lead to a decrease in collagen expression through this mechanism [125].

The p75 nerve growth factor receptor also possesses a death domain and binds NGF with low affinity as compared to TrkA [126]. In activated HSCs, p75 was demonstrated *in vivo* and in culture, and exposure to NGF caused apoptosis [127]. Interestingly, while mature NGF is proapoptotic for HSCs, proNGF protects hepatic myofibroblasts from apoptosis, and during the resolution of fibrosis a decrease in proNGF parallels loss of hepatic myofibroblasts by apoptosis [128]. Other studies indicated that deletion of p75NTR in HSCs blocks differentiation and does not support hepatocyte proliferation during liver regeneration [129].

5.7 INTERLEUKINS AND INTERFERONS

Interleukin-1 has a potent proinflammatory action in HSCs and activates NF-κB in activated but not in quiescent cells [107,108]. IL-1 also inhibits the expression of procollagen α1(I) at a post-transcriptional level [104] and stimulates HSC proliferation [130]. However, in a more recent study IL-1 was found to induce collagen synthesis via the JNK and p38/AP-1 pathways [131]. Altogether, IL-1 exerts profibrogenic actions, as indicated by the anti-fibrogenic effects of IL-1 receptor antagonist [132]. This is supported by the fact that IL-1 also promotes survival of activated HSCs *in vitro* and *in vivo* and induces fibrosis, as indicated by the experiments involving genetic ablation of IL-1 receptors [133]. HSCs are also the target of IL-1β produced during activation of the inflammasome system [134].

IL-17A is critical in inflammation and stimulates fibrogenesis. IL-17 directly induces production of collagen in HSCs and activates Stat3 [135]. Moreover, reduced inflammation and fibrosis were observed in IL-17RA–deficient mice [136]. Recent evidence for a modulation of HSC functions by IL-27 is provided by the observation that HSCs express receptors for this cytokine, with actions similar to the ones induced by interferon-γ [137]. HSCs also respond to IL-10, IL-6, and oncostatin M [138–140]. IL-6 is also produced by HSCs, and has been implicated in the regenerative response and, on the other hand, in the profibrogenic actions of glycotoxins [141,142]. IL-10 inhibits procollagen α(1) expression at the transcriptional level, and may have important anti-fibrogenic properties [139,143], as also indicated by the anti-fibrogenic actions of transgenic IL-10 expression in hepatocytes [144]. Bone marrow-derived cells may be an additional source of IL-10, resulting in a limitation of fibrosis [145]. HSCs are an additional source of IL-10, and release of this cytokine ameliorates experimental alcoholic liver disease following activation of toll-like receptor-3 [141].

Receptors for interferons are ubiquitous, and IFN-γ downregulates activation, matrix synthesis, and proliferation of HSCs both *in vitro* and *in vivo* [146–148]. These actions may be exploited by targeting IFN-γ toward HSCs to limit experimental fibrosis [149]. IFN-γ upregulates the expression of proinflammatory factors such as CCL2 in HSCs [78], and this effect is differentially regulated as compared to other cytokines [19]. Surprisingly, when human HSCs are incubated with IFN-γ for prolonged periods and exposed to mitogens such as serum or PDGF, an increase in cell proliferation is observed [150]. An interesting novel role of HSCs is their immunomodulatory activity, as evidenced in islet cell transplantation. HSC cotransplantation protects allografts from rejection, an effect that requires an intact IFN-γ signaling pathway in HSCs [145].

5.8 ADIPOKINES AND OTHER CYTOKINES RELATED TO METABOLISM

Adipokines belong to a relatively novel class of cytokines whose major site of production is the adipose tissue [151]. Accumulating evidence over the past 10 years has indicated that the adipose tissue is not merely a storage site for excess energy in the

Table 5.2 Metabolic Actions of Leptin and Adiponectin

Leptin
• Inhibition of food intake • Stimulation of energy expenditure • Suppression of hepatic glucose production • Suppression of fatty acid biosynthesis (inhibition of stearoyl coenzyme A desaturase-1 (SCD-1)) • Stimulation of fatty acid oxidation in the liver and skeletal muscle • Stimulation of glucose uptake in skeletal muscle • Stimulation of insulin secretion • Stimulation of adiponectin expression • Stimulation of proinflammatory cytokines • Suppression of resistin expression • Suppression of RBP4 expression
Adiponectin
• Suppression of hepatic glucose production • Suppression of hepatic lipogenesis • Stimulation of glucose uptake by skeletal muscle • Stimulation of fatty acid oxidation in the liver and skeletal muscle • Stimulation of insulin secretion • Modulation of food intake and energy expenditure • Inhibition of proinflammatory cytokines • Induced by PPAR-agonists (thiazolidinediones)

form of triglycerides, but rather a dynamic tissue playing a critical role in the regulation of metabolism (Table 5.2). Leptin is the product of the *ob* gene, and its expression is acquired by HSCs along the activation process [152]. Leptin release is increased in obese patients, and its circulating levels correlate with the fat mass. Thus, leptin reaching the liver through the systemic and portal circulation, together with that released by HSC and other liver cells, may activate target receptors on different cell types.

Work conducted by different groups has indicated that leptin stimulates the fibrogenic process with different mechanisms operating on multiple cellular targets, including HSCs. The expression of leptin receptors at the gene and protein levels has been shown in primary rodent or human HSCs, and in cell lines such as LX-2 cells [151]. HSCs express the functional long form of the receptor (ObRb), which is responsible for activation of protein tyrosine kinases of the Jak family and activation of Stat3 and the MAPK cascade [153]. In terms of biologic effects (Figure 5.2), leptin induces the expression of type I procollagen in HSCs, and upregulates expression of the TIMP-1 [151]. Moreover, leptin induces proinflammatory and proangiogenic actions via upregulated secretion of CCL2, VEGF, and Ang-1. Interestingly, leptin increases the abundance of hypoxia-inducible factor 1α (HIF-1α), and of the mammalian target of rapamycin (TOR) pathway, resulting in expression of angiogenic cytokines [38]. The downstream

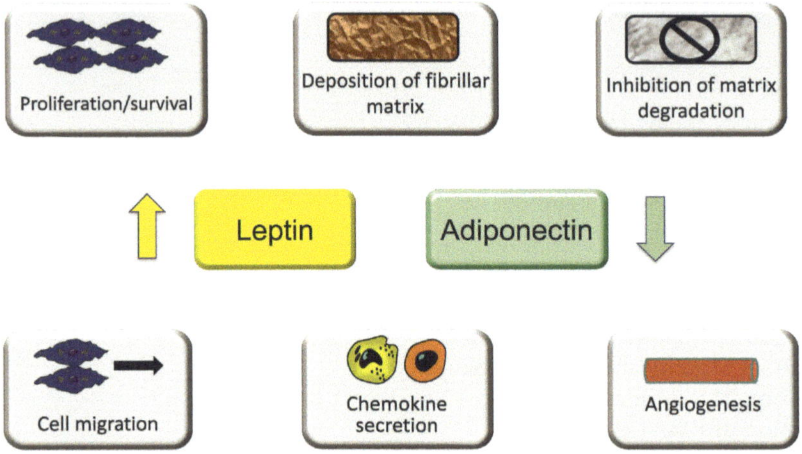

Figure 5.2 ***Opposing actions of leptin and adiponectin on HSC biology.*** **Different biologic actions of myofibroblastic HSC, relevant to the fibrogenic process are shown. Leptin and adiponectin exert opposing effects on these actions, as indicated by the arrows.**

effectors of leptin signaling also include activation of the Hedgehog pathway [154], which contributes to confer to these cells a myofibroblastic phenotype, and activation of the Sp1 and Sp3 transcription factors, modulating collagen expression [155]. More recently, the β-catenin pathway has been indicated as an additional target of leptin contributing to liver fibrogenesis [106].

Adiponectin (Figure 5.2) may be considered a counter-regulator of leptin's actions in HSCs. Circulating levels of this cytokine are inversely related to adiposity, and decrease in patients with the metabolic syndrome. Adiponectin is also expressed by HSCs and limits the development of liver fibrosis in different experimental models [151]. Both receptors for adiponectin, Adipo R1 and Adipo R2, have been identified in HSCs, and after transition to myofibroblasts, expression of Adipo R1 predominates [156]. Exposure of HSCs to adiponectin activates adenosine monophosphate-activated protein kinase (AMPK), a negative modulator of different characteristics of myofibroblastic HSCs [156,157]. Additionally, adiponectin induces the expression of suppressor of cytokine signaling-3 and protein tyrosine phosphatase 1B, negative regulators of JAK/STAT, a pathway necessary for leptin's fibrogenic signaling [158]. Other mechanisms of the anti-fibrogenic actions of adiponectin include the disassembly of focal adhesion complexes, which may be relevant to resolution of fibrosis [159]. Interestingly, adiponectin has been proposed as a mediator of the anti-fibrogenic effects of thiazolidinediones on HSCs [159].

Resistin has been indicated as a possible mediator of the detrimental effects of obesity on metabolism, although its biology is not totally comparable in rodents and humans [160,161]. Resistin is upregulated in humans in conditions of severe hepatic damage, and a proinflammatory action of this adipokine has been shown in cultured human HSCs [83]. Moreover, resistin activates the p38MAPK pathway in HSCs, and induces proliferation and migration of these cells [162]. Apelin, which interacts with the G protein-coupled receptor APJ and is highly expressed by HSCs, exerts pro-angiogenic actions, and contributes to hepatic fibrosis [163]. Exposure to recombinant apelin results in increased synthesis of type I collagen and PDGFβ receptor in HSCs [35]. Expression of apelin is upregulated by angiotensin II and endothelin-1, and the apelin receptor APJ is modulated by LPS and hypoxia [35,164].

Ghrelin is an orexigenic hormone with pleiotropic properties, mainly produced by the gastric mucosa, but also by other organs including the liver. Ghrelin was found to be expressed in normal and diseased liver and to exert anti-fibrotic actions, preventing hepatocyte death, reducing the accumulation of activated HSCs, and decreasing collagen deposition [165].

5.9 OSTEOPONTIN

Osteopontin (OPN) is a phosphorylated acidic glycoprotein that functions as an immune modulator, and affects wound healing [166]. Using a cDNA microarray, OPN was identified as one of the genes upregulated in activated HSCs, where incubation with recombinant OPN induced proliferative and migratory effects [167]. Activation of the OPN system was also found to depend on the Hedgehog pathway, as in cultured HSCs, Hedgehog agonists upregulated OPN expression, and OPN promoted fibrogenic responses [168]. More recently, activation of SOX9 was indicated as a downstream mediator of Hedgehog responsible for induction of OPN expression [169]. Other studies indicate that OPN exerts a direct control over collagen production. Overexpression of OPN in HSCs resulted in increased collagen-I production through engagement of integrin αvβ3 and activation of the PI3k/Akt pathway [170]. OPN is also a critical molecule in the crosstalk between biliary epithelial cells, progenitor cells, and HSCs. Biliary cells induced upregulation of OPN in HSCs, resulting in expression of TGF-β1 and enhanced collagen-I synthesis by HSCs [171]. This pathway could be implicated in the profibrogenic effect of ductular reaction.

5.10 PERSPECTIVE

The past 20 years have witnessed an explosion of information on the role of cytokines in the regulation of HSC biology. Data on the different types of molecules have been accompanied by elucidation of the different signaling pathways regulating the biology of these cells. It emerges as a complex network that, in association with other soluble mediators, determines a fine-tuning of the hepatic wound-healing response, and of the fibrogenic and fibrolytic processes. More important, elucidation of the pattern of regulation of the cytokine network involving HSCs, and of the different signaling pathways, has allowed identification of multiple candidates to be tested in the clinical setting as therapeutic targets and/or biomarkers.

REFERENCES

[1] Pinzani M, Gesualdo L, Sabbah GM, Abboud HE. Effects of platelet-derived growth factor and other polypeptide mitogens on DNA synthesis and growth of cultured rat liver fat-storing cells. J Clin Invest 1989;84:1786–93.

[2] Marra F, Choudhury GG, Pinzani M, Abboud HE. Regulation of platelet-derived growth factor secretion and gene expression in human liver fat-storing cells. Gastroenterology 1994;107:1110–7.

[3] Bai Q, An J, Wu X, You H, Ma H, Liu T, et al. HBV promotes the proliferation of hepatic stellate cells via the PDGF-B/PDGFR-beta signaling pathway *in vitro*. Int J Mol Med 2012;30:1443–50.

[4] Fingas CD, Bronk SF, Werneburg NW, Mott JL, Guicciardi ME, Cazanave SC, et al. Myofibroblast-derived PDGF-BB promotes Hedgehog survival signaling in cholangiocarcinoma cells. Hepatology 2011;54:2076–88.

[5] Friedman SL, Arthur MJ. Activation of cultured rat hepatic lipocytes by Kupffer cell conditioned medium. Direct enhancement of matrix synthesis and stimulation of cell proliferation via induction of platelet-derived growth factor receptors. J Clin Invest 1989;84:1780–5.

[6] Pinzani M, Gentilini A, Caligiuri A, De Franco R, Pellegrini G, Milani S, et al. Transforming growth factor-beta 1 regulates platelet-derived growth factor receptor beta subunit in human liver fat-storing cells. Hepatology 1995;21:232–9.

[7] Pinzani M, Milani S, Herbst H, DeFranco R, Grappone C, Gentilini A, et al. Expression of platelet-derived growth factor and its receptors in normal human liver and during active hepatic fibrogenesis. Am J Pathol 1996;148:785–800.

[8] Reichenbach V, Fernandez-Varo G, Casals G, Oro D, Ros J, Melgar-Lesmes P, et al. Adenoviral dominant-negative soluble PDGFRbeta improves hepatic collagen, systemic hemodynamics, and portal pressure in fibrotic rats. J Hepatol 2012;57:967–73.

[9] Czochra P, Klopcic B, Meyer E, Herkel J, Garcia-Lazaro JF, Thieringer F, et al. Liver fibrosis induced by hepatic overexpression of PDGF-B in transgenic mice. J Hepatol 2006;45:419–28.

[10] Claesson-Welsh L. Platelet-derived growth factor receptor signals. J Biol Chem 1994;269:32023–6.

[11] Marra F, Arrighi MC, Fazi M, Caligiuri A, Pinzani M, Romanelli RG, et al. Extracellular signal-regulated kinase activation differentially regulates platelet-derived growth factor's actions in hepatic stellate cells, and is induced by *in vivo* liver injury in the rat. Hepatology 1999;30:951–8.

[12] Benitez-Rajal J, Lorite MJ, Burt AD, Day CP, Thompson MG. Phospholipase D and extracellular signal-regulated kinase in hepatic stellate cells: effects of platelet-derived growth factor and extracellular nucleotides. Am J Physiol Gastrointest Liver Physiol 2006;291:G977–86.

[13] Reeves HL, Thompson MG, Dack CL, Burt AD, Day CP. The role of phosphatidic acid in platelet-derived growth factor-induced proliferation of rat hepatic stellate cells. Hepatology 2000;31:95–100.

[14] Woodhoo A, Iruarrizaga-Lejarreta M, Beraza N, Garcia-Rodriguez JL, Embade N, Fernandez-Ramos D, et al. Human antigen R contributes to hepatic stellate cell activation and liver fibrosis. Hepatology 2012;56:1870–82.

[15] Mallat A, Gallois C, Tao J, Habib A, Maclouf J, Mavier P, et al. Platelet-derived growth factor-BB and thrombin generate positive and negative signals for human hepatic stellate cell proliferation. Role of a prostaglandin/cyclic AMP pathway and cross-talk with endothelin receptors. J Biol Chem 1998;273:27300–5.

[16] Gobejishvili L, Barve S, Breitkopf-Heinlein K, Li Y, Zhang J, Avila DV, et al. Rolipram attenuates bile duct ligation-induced liver injury in rats: a potential pathogenic role of PDE4. J Pharmacol Exp Ther 2013;347:80–90.

[17] Mallat A, Fouassier L, Preaux AM, Gal CS, Raufaste D, Rosenbaum J, et al. Growth inhibitory properties of endothelin-1 in human hepatic myofibroblastic Ito cells. An endothelin B receptor-mediated pathway. J Clin Invest 1995;96:42–9.

[18] Paik YH, Kim JK, Lee JI, Kang SH, Kim DY, An SH, et al. Celecoxib induces hepatic stellate cell apoptosis through inhibition of Akt activation and suppresses hepatic fibrosis in rats. Gut 2009;58:1517–27.

[19] Efsen E, Bonacchi A, Pastacaldi S, Valente AJ, Wenzel UO, Tosti-Guerra C, et al. Agonist-specific regulation of monocyte chemoattractant protein-1 expression by cyclooxygenase metabolites in hepatic stellate cells. Hepatology 2001;33:713–21.

[20] Marra F, Gentilini A, Pinzani M, Choudhury GG, Parola M, Herbst H, et al. Phosphatidylinositol 3-kinase is required for platelet-derived growth factor's actions on hepatic stellate cells. Gastroenterology 1997;112:1297–306.
[21] Cao S, Yaqoob U, Das A, Shergill U, Jagavelu K, Huebert RC, et al. Neuropilin-1 promotes cirrhosis of the rodent and human liver by enhancing PDGF/TGF-beta signaling in hepatic stellate cells. J Clin Invest 2010;120:2379–94.
[22] Pinzani M, Abboud HE, Aron DC. Secretion of insulin-like growth factor-I and binding proteins by rat liver fat-storing cells: regulatory role of platelet-derived growth factor. Endocrinology 1990;127:2343–9.
[23] Gentilini A, Marra F, Gentilini P, Pinzani M. Phosphatidylinositol-3 kinase and extracellular signal-regulated kinase mediate the chemotactic and mitogenic effects of insulin-lice growth factor-I in human hepatic stellate cells. J Hepatol 2000;32:227–34.
[24] Gentilini A, Feliers D, Pinzani M, Woodruff K, Abboud S. Characterization and regulation of insulin-like growth factor binding proteins in human hepatic stellate cells. J Cell Physiol 1998;174:240–50.
[25] Scharf JG, Knittel T, Dombrowski F, Muller L, Saile B, Braulke T, et al. Characterization of the IGF axis components in isolated rat hepatic stellate cells. Hepatology 1998;27:1275–84.
[26] Kwiecinski M, Elfimova N, Noetel A, Tox U, Steffen HM, Hacker U, et al. Expression of platelet-derived growth factor-C and insulin-like growth factor I in hepatic stellate cells is inhibited by miR-29. Lab Invest 2012;92:978–87.
[27] Failli P, De FR, Caligiuri A, Gentilini A, Romanelli RG, Marra F, et al. Nitrovasodilators inhibit platelet-derived growth factor-induced proliferation and migration of activated human hepatic stellate cells. Gastroenterology 2000;119:479–92.
[28] Di Sario A, Bendia E, Svegliati Baroni G, Ridolfi F, Bolognini L, Feliciangeli G, et al. Intracellular pathways mediating Na^+/H^+ exchange activation by platelet-derived growth factor in rat hepatic stellate cells. Gastroenterology 1999;116:1155–66.
[29] Benedetti A, Di Sario A, Casini A, Ridolfi F, Bendia E, Pigini P, et al. Inhibition of the NA(+)/H(+) exchanger reduces rat hepatic stellate cell activity and liver fibrosis: an *in vitro* and *in vivo* study. Gastroenterology 2001;120:545–56.
[30] Rombouts K, Lottini B, Caligiuri A, Liotta F, Mello T, Carloni V, et al. MARCKS is a downstream effector in platelet-derived growth factor-induced cell motility in activated human hepatic stellate cells. Exp Cell Res 2008;314:1444–54.
[31] Rovida E, Navari N, Caligiuri A, Dello Sbarba P, Marra F. ERK5 differentially regulates PDGF-induced proliferation and migration of hepatic stellate cells. J Hepatol 2008;48:107–15.
[32] Parola M, Robino G, Marra F, Pinzani M, Bellomo G, Leonarduzzi G, et al. HNE interacts directly with JNK isoforms in human hepatic stellate cells. J Clin Invest 1998;102:1942–50.
[33] Robino G, Parola M, Marra F, Caligiuri A, De Franco RM, Zamara E, et al. Interaction between 4-hydroxy-2,3-alkenals and the platelet-derived growth factor-beta receptor. Reduced tyrosine phosphorylation and downstream signaling in hepatic stellate cells. J Biol Chem 2000;275:40561–7.
[34] Olaso E, Salado C, Egilegor E, Gutierrez V, Santisteban A, Sancho-Bru P, et al. Proangiogenic role of tumor-activated hepatic stellate cells in experimental melanoma metastasis. Hepatology 2003;37:674–85.
[35] Melgar-Lesmes P, Casals G, Pauta M, Ros J, Reichenbach V, Bataller R, et al. Apelin mediates the induction of profibrogenic genes in human hepatic stellate cells. Endocrinology 2010;151:5306–14.
[36] Aleffi S, Petrai I, Bertolani C, Parola M, Colombatto S, Novo E, et al. Upregulation of proinflammatory and proangiogenic cytokines by leptin in human hepatic stellate cells. Hepatology 2005;42:1339–48.
[37] Taura K, De Minicis S, Seki E, Hatano E, Iwaisako K, Osterreicher CH, et al. Hepatic stellate cells secrete angiopoietin 1 that induces angiogenesis in liver fibrosis. Gastroenterology 2008;135:1729–38.
[38] Aleffi S, Navari N, Delogu W, Galastri S, Novo E, Rombouts K, et al. Mammalian target of rapamycin mediates the angiogenic effects of leptin in human hepatic stellate cells. Am J Physiol Gastrointest Liver Physiol 2011;301:G210–9.
[39] Ankoma-Sey V, Matli M, Chang KB, Lalazar A, Donner DB, Wong L, et al. Coordinated induction of VEGF receptors in mesenchymal cell types during rat hepatic wound healing. Oncogene 1998;17:115–21.

[40] Novo E, Cannito S, Zamara E, Valfre di Bonzo L, Caligiuri A, Cravanzola C, et al. Proangiogenic cytokines as hypoxia-dependent factors stimulating migration of human hepatic stellate cells. Am J Pathol 2007;170:1942–53.
[41] Van Steenkiste C, Ribera J, Geerts A, Pauta M, Tugues S, Casteleyn C, et al. Inhibition of placental growth factor activity reduces the severity of fibrosis, inflammation, and portal hypertension in cirrhotic mice. Hepatology 2011;53:1629–40.
[42] Piguet AC, Majumder S, Maheshwari U, Manjunathan R, Saran U, Chatterjee S, et al. Everolimus is a potent inhibitor of activated hepatic stellate cell functions *in vitro* and *in vivo*, while demonstrating anti-angiogenic activities. Clin Sci (Lond) 2014;126:775–84.
[43] Coch L, Mejias M, Berzigotti A, Garcia-Pras E, Gallego J, Bosch J, et al. Disruption of negative feedback loop between vasohibin-1 and vascular endothelial growth factor decreases portal pressure, angiogenesis, and fibrosis in cirrhotic rats. Hepatology 2014;60:633–47.
[44] Puche JE, Saiman Y, Friedman SL. Hepatic stellate cells and liver fibrosis. Compr Physiol 2013;3:1473–92.
[45] Marra F, Bonewald LF, Park-Snyder S, Park IS, Woodruff KA, Abboud HE. Characterization and regulation of the latent transforming growth factor-beta complex secreted by vascular pericytes. J Cell Physiol 1996;166:537–46.
[46] Botella LM, Sanchez-Elsner T, Sanz-Rodriguez F, Kojima S, Shimada J, Guerrero-Esteo M, et al. Transcriptional activation of endoglin and transforming growth factor-beta signaling components by cooperative interaction between Sp1 and KLF6: their potential role in the response to vascular injury. Blood 2002;100:4001–10.
[47] Racine-Samson L, Rockey DC, Bissell DM. The role of alpha1beta1 integrin in wound contraction. A quantitative analysis of liver myofibroblasts *in vivo* and in primary culture. J Biol Chem 1997;272:30911–7.
[48] Smedsrod B, Le Couteur D, Ikejima K, Jaeschke H, Kawada N, Naito M, et al. Hepatic sinusoidal cells in health and disease: update from the 14th International symposium. Liver Int 2009;29:490–501.
[49] Baghy K, Iozzo RV, Kovalszky I. Decorin-TGFbeta axis in hepatic fibrosis and cirrhosis. J Histochem Cytochem 2012;60:262–8.
[50] Bourd-Boittin K, Bonnier D, Leyme A, Mari B, Tuffery P, Samson M, et al. Protease profiling of liver fibrosis reveals the ADAM metallopeptidase with thrombospondin type 1 motif, 1 as a central activator of transforming growth factor beta. Hepatology 2011;54:2173–84.
[51] George J, Roulot D, Koteliansky VE, Bissell DM. *In vivo* inhibition of rat stellate cell activation by soluble transforming growth factor beta type II receptor: a potential new therapy for hepatic fibrosis. PNAS 1999;96:12719–24.
[52] Nakatsukasa H, Evarts RP, Hsia CC, Thorgeirsson SS. Transforming growth factor-beta 1 and type I procollagen transcripts during regeneration and early fibrosis of rat liver. Lab Invest 1990;63:171–80.
[53] Friedman SL, Yamasaki G, Wong L. Modulation of transforming growth factor beta receptors of rat lipocytes during the hepatic wound healing response. Enhanced binding and reduced gene expression accompany cellular activation in culture and *in vivo*. J Biol Chem 1994;269:10551–8.
[54] Dooley S, Hamzavi J, Breitkopf K, Wiercinska E, Said HM, Lorenzen J, et al. Smad7 prevents activation of hepatic stellate cells and liver fibrosis in rats. Gastroenterology 2003;125:178–91.
[55] Liu X, Hu H, Yin JQ. Therapeutic strategies against TGF-beta signaling pathway in hepatic fibrosis. Liver Int 2006;26:8–22.
[56] Tu K, Li J, Verma VK, Liu C, Billadeau DD, Lamprecht G, et al. Vasodilator-stimulated phosphoprotein promotes activation of hepatic stellate cells by regulating Rab11-dependent plasma membrane targeting of transforming growth factor beta receptors. Hepatology 2015;61:361–74.
[57] De Minicis S, Rychlicki C, Agostinelli L, Saccomanno S, Trozzi L, Candelaresi C, et al. Semaphorin 7A contributes to TGF-beta-mediated liver fibrogenesis. Am J Pathol 2013;183:820–30.
[58] Liu C, Billadeau DD, Abdelhakim H, Leof E, Kaibuchi K, Bernabeu C, et al. IQGAP1 suppresses TbetaRII-mediated myofibroblastic activation and metastatic growth in liver. J Clin Invest 2013;123:1138–56.
[59] Seki E, De Minicis S, Osterreicher CH, Kluwe J, Osawa Y, Brenner DA, et al. TLR4 enhances TGF-beta signaling and hepatic fibrosis. Nat Med 2007;13:1324–32.

[60] Dooley S, Delvoux B, Lahme B, Mangasser-Stephan K, Gressner AM. Modulation of transforming growth factor beta response and signaling during transdifferentiation of rat hepatic stellate cells to myofibroblasts. Hepatology 2000;31:1094–106.

[61] Inagaki Y, Truter S, Ramirez F. Transforming growth factor-beta stimulates alpha 2(I) collagen gene expression through a cis-acting element that contains an Sp1-binding site. J Biol Chem 1994;269:14828–34.

[62] Inagaki Y, Nemoto T, Nakao A, Dijke P, Kobayashi K, Takehara K, et al. Interaction between GC box binding factors and Smad proteins modulates cell lineage-specific alpha 2(I) collagen gene transcription. J Biol Chem 2001;276:16573–9.

[63] Inagaki Y, Mamura M, Kanamaru Y, Greenwel P, Nemoto T, Takehara K, et al. Constitutive phosphorylation and nuclear localization of Smad3 are correlated with increased collagen gene transcription in activated hepatic stellate cells. J Cell Physiol 2001;187:117–23.

[64] Wells RG, Discher DE. Matrix elasticity, cytoskeletal tension, and TGF-beta: the insoluble and soluble meet. Sci Signal 2008;1 pe13.

[65] Beilfuss A, Sowa JP, Sydor S, Beste M, Bechmann LP, Schlattjan M, et al. Vitamin D counteracts fibrogenic TGF-beta signalling in human hepatic stellate cells both receptor-dependently and independently. Gut 2014.

[66] Ding N, Yu RT, Subramaniam N, Sherman MH, Wilson C, Rao R, et al. A vitamin D receptor/SMAD genomic circuit gates hepatic fibrotic response. Cell 2013;153:601–13.

[67] Reimann T, Hempel U, Krautwald S, Axmann A, Scheibe R, Seidel D, et al. Transforming growth factor-beta1 induces activation of Ras, Raf-1, MEK and MAPK in rat hepatic stellate cells. FEBS Lett 1997;403:57–60.

[68] Garcia-Trevijano ER, Iraburu MJ, Fontana L, Dominguez-Rosales JA, Auster A, Covarrubias-Pinedo A, et al. Transforming growth factor beta1 induces the expression of alpha1(I) procollagen mRNA by a hydrogen peroxide-C/EBPbeta-dependent mechanism in rat hepatic stellate cells. Hepatology 1999;29:960–70.

[69] De Bleser PJ, Xu G, Rombouts K, Rogiers V, Geerts A. Glutathione levels discriminate between oxidative stress and transforming growth factor-beta signaling in activated rat hepatic stellate cells. J Biol Chem 1999;274:33881–7.

[70] Sugiyama M, Ichida T, Sato T, Ishikawa T, Matsuda Y, Asakura H. Expression of activin A is increased in cirrhotic and fibrotic rat livers. Gastroenterology 1998;114:550–8.

[71] Marra F, Delogu W, Petrai I, Pastacaldi S, Bonacchi A, Efsen E, et al. Differential requirement of members of the MAPK family for CCL2 expression by hepatic stellate cells. Am J Physiol Gastrointest Liver Physiol 2004;287:G18–26.

[72] Patella S, Phillips DJ, Tchongue J, de Kretser DM, Sievert W. Follistatin attenuates early liver fibrosis: effects on hepatic stellate cell activation and hepatocyte apoptosis. Am J Physiol Gastrointest Liver Physiol 2006;290:G137–44.

[73] Gao R, Brigstock DR. Connective tissue growth factor (CCN2) induces adhesion of rat activated hepatic stellate cells by binding of its C-terminal domain to integrin alpha(v)beta(3) and heparan sulfate proteoglycan. J Biol Chem 2004;279:8848–55.

[74] Brigstock DR. The CCN family: a new stimulus package. J Endocrinol 2003;178:169–75.

[75] Paradis V, Perlemuter G, Bonvoust F, Dargere D, Parfait B, Vidaud M, et al. High glucose and hyperinsulinemia stimulate connective tissue growth factor expression: a potential mechanism involved in progression to fibrosis in nonalcoholic steatohepatitis. Hepatology 2001;34:738–44.

[76] Gressner OA, Lahme B, Demirci I, Gressner AM, Weiskirchen R. Differential effects of TGF-beta on connective tissue growth factor (CTGF/CCN2) expression in hepatic stellate cells and hepatocytes. J Hepatol 2007;47:699–710.

[77] Marra F, Tacke F. Roles for chemokines in liver disease. Gastroenterology 2014;147:577–94. e1.

[78] Marra F, Valente AJ, Pinzani M, Abboud HE. Cultured human liver fat-storing cells produce monocyte chemotactic protein-1. Regulation by proinflammatory cytokines. J Clin Invest 1993;92:1674–80.

[79] Marra F, Romanelli RG, Giannini C, Failli P, Pastacaldi S, Arrighi MC, et al. Monocyte chemotactic protein-1 as a chemoattractant for human hepatic stellate cells. Hepatology 1999;29:140–8.

[80] Maher JJ, Lozier JS, Scott MK. Rat hepatic stellate cells produce cytokine-induced neutrophil chemoattractant in culture and *in vivo*. Am J Physiol 1998;275:G847–53.

[81] Paik YH, Schwabe RF, Bataller R, Russo MP, Jobin C, Brenner DA. Toll-like receptor 4 mediates inflammatory signaling by bacterial lipopolysaccharide in human hepatic stellate cells. Hepatology 2003;37:1043–55.
[82] Paik YH, Lee KS, Lee HJ, Yang KM, Lee SJ, Lee DK, et al. Hepatic stellate cells primed with cytokines upregulate inflammation in response to peptidoglycan or lipoteichoic acid. Lab Invest 2006;86:676–86.
[83] Bertolani C, Sancho-Bru P, Failli P, Bataller R, Aleffi S, DeFranco R, et al. Resistin as an intrahepatic cytokine: overexpression during chronic injury and induction of proinflammatory actions in hepatic stellate cells. Am J Pathol 2006;169:2042–53.
[84] Clement S, Pascarella S, Conzelmann S, Gonelle-Gispert C, Guilloux K, Negro F. The hepatitis C virus core protein indirectly induces alpha-smooth muscle actin expression in hepatic stellate cells via interleukin-8. J Hepatol 2010;52:635–43.
[85] Affo S, Morales-Ibanez O, Rodrigo-Torres D, Altamirano J, Blaya D, Dapito DH, et al. CCL20 mediates lipopolysaccharide induced liver injury and is a potential driver of inflammation and fibrosis in alcoholic hepatitis. Gut 2014;63:1782–92.
[86] Hammerich L, Bangen JM, Govaere O, Zimmermann HW, Gassler N, Huss S, et al. Chemokine receptor CCR6-dependent accumulation of gammadelta T cells in injured liver restricts hepatic inflammation and fibrosis. Hepatology 2014;59:630–42.
[87] Bonacchi A, Romagnani P, Romanelli RG, Efsen E, Annunziato F, Lasagni L, et al. Signal transduction by the chemokine receptor CXCR3: activation of Ras/ERK, Src, and phosphatidylinositol 3-kinase/Akt controls cell migration and proliferation in human vascular pericytes. J Biol Chem 2001;276:9945–54.
[88] Sahin H, Borkham-Kamphorst E, Kuppe C, Zaldivar MM, Grouls C, Al-samman M, et al. Chemokine Cxcl9 attenuates liver fibrosis-associated angiogenesis in mice. Hepatology 2012;55:1610–9.
[89] Berres ML, Trautwein C, Schmeding M, Eurich D, Tacke F, Bahra M, et al. Serum chemokine CXC ligand 10 (CXCL10) predicts fibrosis progression after liver transplantation for hepatitis C infection. Hepatology 2011;53:596–603.
[90] Wasmuth HE, Lammert F, Zaldivar MM, Weiskirchen R, Hellerbrand C, Scholten D, et al. Antifibrotic effects of CXCL9 and its receptor CXCR3 in livers of mice and humans. Gastroenterology 2009;137:309–19. 19 e1–3.
[91] Bonacchi A, Petrai I, Defranco RM, Lazzeri E, Annunziato F, Efsen E, et al. The chemokine CCL21 modulates lymphocyte recruitment and fibrosis in chronic hepatitis C. Gastroenterology 2003;125:1060–76.
[92] Schwabe RF, Bataller R, Brenner DA. Human hepatic stellate cells express CCR5 and RANTES to induce proliferation and migration. Am J Physiol Gastrointest Liver Physiol 2003;285:G949–58.
[93] Seki E, De Minicis S, Gwak GY, Kluwe J, Inokuchi S, Bursill CA, et al. CCR1 and CCR5 promote hepatic fibrosis in mice. J Clin Invest 2009;119:1858–70.
[94] Berres ML, Koenen RR, Rueland A, Zaldivar MM, Heinrichs D, Sahin H, et al. Antagonism of the chemokine Ccl5 ameliorates experimental liver fibrosis in mice. J Clin Invest 2010;120:4129–40.
[95] Bruno R, Galastri S, Sacchi P, Cima S, Caligiuri A, DeFranco R, et al. gp120 modulates the biology of human hepatic stellate cells: a link between HIV infection and liver fibrogenesis. Gut 2010;59:513–20.
[96] Hong F, Tuyama A, Lee TF, Loke J, Agarwal R, Cheng X, et al. Hepatic stellate cells express functional CXCR4: role in stromal cell-derived factor-1alpha-mediated stellate cell activation. Hepatology 2009;49:2055–67.
[97] Tuyama AC, Hong F, Saiman Y, Wang C, Ozkok D, Mosoian A, et al. Human immunodeficiency virus (HIV)-1 infects human hepatic stellate cells and promotes collagen I and monocyte chemoattractant protein-1 expression: implications for the pathogenesis of HIV/hepatitis C virus-induced liver fibrosis. Hepatology 2010;52:612–22.
[98] Ding BS, Cao Z, Lis R, Nolan DJ, Guo P, Simons M, et al. Divergent angiocrine signals from vascular niche balance liver regeneration and fibrosis. Nature 2014;505:97–102.
[99] Chu PS, Nakamoto N, Ebinuma H, Usui S, Saeki K, Matsumoto A, et al. C-C motif chemokine receptor 9 positive macrophages activate hepatic stellate cells and promote liver fibrosis in mice. Hepatology 2013;58:337–50.

[100] Ashkenazi A, Dixit VM. Death receptors: signaling and modulation. Science 1998;281:1305–8.
[101] Ledgerwood EC, Pober JS, Bradley JR. Recent advances in the molecular basis of TNF signal transduction. Lab Invest 1999;79:1041–50.
[102] Walczak H. TNF and ubiquitin at the crossroads of gene activation, cell death, inflammation, and cancer. Immunol Rev 2011;244:9–28.
[103] Knittel T, Muller L, Saile B, Ramadori G. Effect of tumour necrosis factor-alpha on proliferation, activation and protein synthesis of rat hepatic stellate cells. J Hepatol 1997;27:1067–80.
[104] Armendariz-Borunda J, Katayama K, Seyer JM. Transcriptional mechanisms of type I collagen gene expression are differentially regulated by interleukin-1 beta, tumor necrosis factor alpha, and transforming growth factor beta in Ito cells. J Biol Chem 1992;267:14316–21.
[105] Gallois C, Habib A, Tao J, Moulin S, Maclouf J, Mallat A, et al. Role of NF-kappaB in the antiproliferative effect of endothelin-1 and tumor necrosis factor-alpha in human hepatic stellate cells. Involvement of cyclooxygenase-2. J Biol Chem 1998;273:23183–90.
[106] Zhai X, Yan K, Fan J, Niu M, Zhou Q, Zhou Y, et al. The beta-catenin pathway contributes to the effects of leptin on SREBP-1c expression in rat hepatic stellate cells and liver fibrosis. Br J Pharmacol 2013;169:197–212.
[107] Hellerbrand C, Jobin C, Licato LL, Sartor RB, Brenner DA. Cytokines induce NF-kappaB in activated but not in quiescent rat hepatic stellate cells. Am J Physiol 1998;275:G269–78.
[108] Marra F. Hepatic stellate cells and the regulation of liver inflammation. J Hepatol 1999;31:1120–30.
[109] Elsharkawy AM, Wright MC, Hay RT, Arthur MJ, Hughes T, Bahr MJ, et al. Persistent activation of nuclear factor-kappaB in cultured rat hepatic stellate cells involves the induction of potentially novel Rel-like factors and prolonged changes in the expression of IkappaB family proteins. Hepatology 1999;30:761–9.
[110] Lang A, Schoonhoven R, Tuvia S, Brenner DA, Rippe RA. Nuclear factor kappaB in proliferation, activation, and apoptosis in rat hepatic stellate cells. J Hepatol 2000;33:49–58.
[111] Saile B, Matthes N, Knittel T, Ramadori G. Transforming growth factor beta and tumor necrosis factor alpha inhibit both apoptosis and proliferation of activated rat hepatic stellate cells. Hepatology 1999;30:196–202.
[112] Reeves HL, Dack CL, Peak M, Burt AD, Day CP. Stress-activated protein kinases in the activation of rat hepatic stellate cells in culture. J Hepatol 2000;32:465–72.
[113] Zhan S, Rockey DC. Tumor necrosis factor alpha stimulates endothelin-1 synthesis in rat hepatic stellate cells in hepatic wound healing through a novel IKK/JNK pathway. Exp Cell Res 2011;317:1040–8.
[114] Thirunavukkarasu C, Watkins SC, Gandhi CR. Mechanisms of endotoxin-induced NO, IL-6, and TNF-alpha production in activated rat hepatic stellate cells: role of p38 MAPK. Hepatology 2006;44:389–98.
[115] Hernandez-Munoz I, de la Torre P, Sanchez-Alcazar JA, Garcia I, Santiago E, Munoz-Yague MT, et al. Tumor necrosis factor alpha inhibits collagen alpha 1(I) gene expression in rat hepatic stellate cells through a G protein. Gastroenterology 1997;113:625–40.
[116] Hernandez I, de la Torre P, Rey-Campos J, Garcia I, Sanchez JA, Munoz R, et al. Collagen alpha1(I) gene contains an element responsive to tumor necrosis factor-alpha located in the 5' untranslated region of its first exon. DNA Cell Biol 2000;19:341–52.
[117] Houglum K, Buck M, Kim DJ, Chojkier M. TNF-alpha inhibits liver collagen-alpha 1(I) gene expression through a tissue-specific regulatory region. Am J Physiol 1998;274:G840–7.
[118] Iraburu MJ, Dominguez-Rosales JA, Fontana L, Auster A, Garcia-Trevijano ER, Covarrubias-Pinedo A, et al. Tumor necrosis factor alpha down-regulates expression of the alpha1(I) collagen gene in rat hepatic stellate cells through a p20C/EBPbeta- and C/EBPdelta-dependent mechanism. Hepatology 2000;31:1086–93.
[119] Sung CK, She H, Xiong S, Tsukamoto H. Tumor necrosis factor-alpha inhibits peroxisome proliferator-activated receptor gamma activity at a posttranslational level in hepatic stellate cells. Am J Physiol Gastrointest Liver Physiol 2004;286:G722–9.
[120] Marra F, Efsen E, Romanelli RG, Caligiuri A, Pastacaldi S, Batignani G, et al. Ligands of peroxisome proliferator-activated receptor gamma modulate profibrogenic and proinflammatory actions in hepatic stellate cells. Gastroenterology 2000;119:466–78.

[121] Tarrats N, Moles A, Morales A, Garcia-Ruiz C, Fernandez-Checa JC, Mari M. Critical role of tumor necrosis factor receptor 1, but not 2, in hepatic stellate cell proliferation, extracellular matrix remodeling, and liver fibrogenesis. Hepatology 2011;54:319–27.

[122] Saile B, Knittel T, Matthes N, Schott P, Ramadori G. CD95/CD95L-mediated apoptosis of the hepatic stellate cell. A mechanism terminating uncontrolled hepatic stellate cell proliferation during hepatic tissue repair. Am J Pathol 1997;151:1265–72.

[123] Taimr P, Higuchi H, Kocova E, Rippe RA, Friedman S, Gores GJ. Activated stellate cells express the TRAIL receptor-2/death receptor-5 and undergo TRAIL-mediated apoptosis. Hepatology 2003;37:87–95.

[124] Fischer R, Cariers A, Reinehr R, Haussinger D. Caspase 9-dependent killing of hepatic stellate cells by activated Kupffer cells. Gastroenterology 2002;123:845–61.

[125] Park SJ, Sohn HY, Park SI. TRAIL regulates collagen production through HSF1-dependent Hsp47 expression in activated hepatic stellate cells. Cell Signal 2013;25:1635–43.

[126] Bonacchi A, Taddei ML, Petrai I, Efsen E, Defranco R, Nosi D, et al. Nuclear localization of TRK-A in liver cells. Histol Histopathol 2008;23:327–40.

[127] Trim N, Morgan S, Evans M, Issa R, Fine D, Afford S, et al. Hepatic stellate cells express the low affinity nerve growth factor receptor p75 and undergo apoptosis in response to nerve growth factor stimulation. Am J Pathol 2000;156:1235–43.

[128] Kendall TJ, Hennedige S, Aucott RL, Hartland SN, Vernon MA, Benyon RC, et al. p75 Neurotrophin receptor signaling regulates hepatic myofibroblast proliferation and apoptosis in recovery from rodent liver fibrosis. Hepatology 2009;49:901–10.

[129] Passino MA, Adams RA, Sikorski SL, Akassoglou K. Regulation of hepatic stellate cell differentiation by the neurotrophin receptor p75NTR. Science 2007;315:1853–6.

[130] Matsuoka M, Pham NT, Tsukamoto H. Differential-effects of interleukin-1-alpha, tumor necrosis factor-alpha, and transforming growth factor-beta-1 on cell-proliferation and collagen formation by cultured fat-storing cells. Liver 1989;9:71–8.

[131] Zhang Y, Yao X. Role of c-Jun N-terminal kinase and p38/activation protein-1 in interleukin-1beta-mediated type I collagen synthesis in rat hepatic stellate cells. APMIS 2012;120:101–7.

[132] Mancini R, Benedetti A, Jezequel AM. An interleukin-1 receptor antagonist decreases fibrosis induced by dimethylnitrosamine in rat liver. Virchows Arch 1994;424:25–31.

[133] Pradere JP, Kluwe J, De Minicis S, Jiao JJ, Gwak GY, Dapito DH, et al. Hepatic macrophages but not dendritic cells contribute to liver fibrosis by promoting the survival of activated hepatic stellate cells in mice. Hepatology 2013;58:1461–73.

[134] Wree A, Eguchi A, McGeough MD, Pena CA, Johnson CD, Canbay A, et al. NLRP3 inflammasome activation results in hepatocyte pyroptosis, liver inflammation, and fibrosis in mice. Hepatology 2014;59:898–910.

[135] Meng F, Wang K, Aoyama T, Grivennikov SI, Paik Y, Scholten D, et al. Interleukin-17 signaling in inflammatory, Kupffer cells, and hepatic stellate cells exacerbates liver fibrosis in mice. Gastroenterology 2012;143:765–76. e1–3.

[136] Tan Z, Qian X, Jiang R, Liu Q, Wang Y, Chen C, et al. IL-17A plays a critical role in the pathogenesis of liver fibrosis through hepatic stellate cell activation. J Immunol 2013;191:1835–44.

[137] Schoenherr C, Weiskirchen R, Haan S. Interleukin-27 acts on hepatic stellate cells and induces signal transducer and activator of transcription 1-dependent responses. Cell Commun Signal 2010;8:19.

[138] Greenwel P, Iraburu MJ, Reyes-Romero M, Meraz-Cruz N, Casado E, Solis-Herruzo JA, et al. Induction of an acute phase response in rats stimulates the expression of alpha 1(I) procollagen messenger ribonucleic acid in their livers. Possible role of interleukin-6. Lab Invest 1995;72:83–91.

[139] Wang SC, Ohata M, Schrum L, Rippe RA, Tsukamoto H. Expression of interleukin-10 by *in vitro* and *in vivo* activated hepatic stellate cells. J Biol Chem 1998;273:302–8.

[140] Levy MT, Trojanowska M, Reuben A. Oncostatin M: a cytokine upregulated in human cirrhosis, increases collagen production by human hepatic stellate cells. J Hepatol 2000;32:218–26.

[141] Byun JS, Suh YG, Yi HS, Lee YS, Jeong WI. Activation of toll-like receptor 3 attenuates alcoholic liver injury by stimulating Kupffer cells and stellate cells to produce interleukin-10 in mice. J Hepatol 2013;58:342–9.

[142] Leung C, Herath CB, Jia Z, Goodwin M, Mak KY, Watt MJ, et al. Dietary glycotoxins exacerbate progression of experimental fatty liver disease. J Hepatol 2014;60:832–8.
[143] Mathurin P, Xiong SG, Kharbanda KK, Veal N, Miyahara T, Motomura K, et al. IL-10 receptor and coreceptor expression in quiescent and activated hepatic stellate cells. Am J Physiol Gastrointest Liver Physiol 2002;282:G981–90.
[144] Safadi R, Ohta M, Alvarez CE, Fiel MI, Bansal M, Mehal WZ, et al. Immune stimulation of hepatic fibrogenesis by CD8 cells and attenuation by transgenic interleukin-10 from hepatocytes. Gastroenterology 2004;127:870–82.
[145] Suh YG, Kim JK, Byun JS, Yi HS, Lee YS, Eun HS, et al. CD11b(+) Gr1(+) bone marrow cells ameliorate liver fibrosis by producing interleukin-10 in mice. Hepatology 2012;56:1902–12.
[146] Rockey DC, Chung JJ. Interferon gamma inhibits lipocyte activation and extracellular matrix mRNA expression during experimental liver injury: implications for treatment of hepatic fibrosis. J Investig Med 1994;42:660–70.
[147] Rockey DC, Chung JJ. Inducible nitric oxide synthase in rat hepatic lipocytes and the effect of nitric oxide on lipocyte contractility. J Clin Invest 1995;95:1199–206.
[148] Mallat A, Preaux AM, Blazejewski S, Rosenbaum J, Dhumeaux D, Mavier P. Interferon alfa and gamma inhibit proliferation and collagen synthesis of human Ito cells in culture. Hepatology 1995;21:1003–10.
[149] Bansal R, Prakash J, Post E, Beljaars L, Schuppan D, Poelstra K. Novel engineered targeted interferon-gamma blocks hepatic fibrogenesis in mice. Hepatology 2011;54:586–96.
[150] Marra F, Choudhury GG, Abboud HE. Interferon-gamma-mediated activation of STAT1alpha regulates growth factor-induced mitogenesis. J Clin Invest 1996;98:1218–30.
[151] Marra F, Bertolani C. Adipokines in liver diseases. Hepatology 2009;50:957–69.
[152] Potter JJ, Womack L, Mezey E, Anania FA. Transdifferentiation of rat hepatic stellate cells results in leptin expression. Biochem Biophys Res Commun 1998;244:178–82.
[153] Cao Q, Mak KM, Ren C, Lieber CS. Leptin stimulates tissue inhibitor of metalloproteinase-1 in human hepatic stellate cells: respective roles of the JAK/STAT and JAK-mediated H_2O_2-dependant MAPK pathways. J Biol Chem 2004;279:4292–304.
[154] Choi SS, Syn WK, Karaca GF, Omenetti A, Moylan CA, Witek RP, et al. Leptin promotes the myofibroblastic phenotype in hepatic stellate cells by activating the hedgehog pathway. J Biol Chem 2010;285:36551–60.
[155] Garcia-Ruiz I, Gomez-Izquierdo E, Diaz-Sanjuan T, Grau M, Solis-Munoz P, Munoz-Yague T, et al. Sp1 and Sp3 transcription factors mediate leptin-induced collagen alpha1(I) gene expression in primary culture of male rat hepatic stellate cells. Endocrinology 2012;153:5845–56.
[156] Caligiuri A, Bertolani C, Guerra CT, Aleffi S, Galastri S, Trappoliere M, et al. Adenosine monophosphate-activated protein kinase modulates the activated phenotype of hepatic stellate cells. Hepatology 2008;47:668–76.
[157] Adachi M, Brenner DA. High molecular weight adiponectin inhibits proliferation of hepatic stellate cells via activation of adenosine monophosphate-activated protein kinase. Hepatology 2008;47:677–85.
[158] Handy JA, Fu PP, Kumar P, Mells JE, Sharma S, Saxena NK, et al. Adiponectin inhibits leptin signalling via multiple mechanisms to exert protective effects against hepatic fibrosis. Biochem J 2011;440:385–95.
[159] Kumar P, Smith T, Rahman K, Mells JE, Thorn NE, Saxena NK, et al. Adiponectin modulates focal adhesion disassembly in activated hepatic stellate cells: implication for reversing hepatic fibrosis. FASEB J 2014.
[160] Savage DB, Sewter CP, Klenk ES, Segal DG, Vidal-Puig A, Considine RV, et al. Resistin/Fizz3 expression in relation to obesity and peroxisome proliferator-activated receptor-gamma action in humans. Diabetes 2001;50:2199–202.
[161] Patel L, Buckels AC, Kinghorn IJ, Murdock PR, Holbrook JD, Plumpton C, et al. Resistin is expressed in human macrophages and directly regulated by PPAR gamma activators. Biochem Biophys Res Commun 2003;300:472–6.
[162] Dong ZX, Su L, Brymora J, Bird C, Xie Q, George J, et al. Resistin mediates the hepatic stellate cell phenotype. World J Gastroenterol 2013;19:4475–85.

[163] Principe A, Melgar-Lesmes P, Fernandez-Varo G, del Arbol LR, Ros J, Morales-Ruiz M, et al. The hepatic apelin system: a new therapeutic target for liver disease. Hepatology 2008;48:1193–201.
[164] Melgar-Lesmes P, Pauta M, Reichenbach V, Casals G, Ros J, Bataller R, et al. Hypoxia and proinflammatory factors upregulate apelin receptor expression in human stellate cells and hepatocytes. Gut 2011;60:1404–11.
[165] Moreno M, Chaves JF, Sancho-Bru P, Ramalho F, Ramalho LN, Mansego ML, et al. Ghrelin attenuates hepatocellular injury and liver fibrogenesis in rodents and influences fibrosis progression in humans. Hepatology 2010;51:974–85.
[166] Konno S, Kurokawa M, Uede T, Nishimura M, Huang SK. Role of osteopontin, a multifunctional protein, in allergy and asthma. Clin Exp Allergy 2011;41:1360–6.
[167] Lee SH, Seo GS, Park YN, Yoo TM, Sohn DH. Effects and regulation of osteopontin in rat hepatic stellate cells. Biochem Pharmacol 2004;68:2367–78.
[168] Syn WK, Choi SS, Liaskou E, Karaca GF, Agboola KM, Oo YH, et al. Osteopontin is induced by hedgehog pathway activation and promotes fibrosis progression in nonalcoholic steatohepatitis. Hepatology 2011;53:106–15.
[169] Pritchett J, Harvey E, Athwal V, Berry A, Rowe C, Oakley F, et al. Osteopontin is a novel downstream target of SOX9 with diagnostic implications for progression of liver fibrosis in humans. Hepatology 2012;56:1108–16.
[170] Urtasun R, Lopategi A, George J, Leung TM, Lu Y, Wang X, et al. Osteopontin, an oxidant stress sensitive cytokine, up-regulates collagen-I via integrin alpha(V)beta(3) engagement and PI3K/pAkt/NFkappaB signaling. Hepatology 2012;55:594–608.
[171] Wang X, Lopategi A, Ge X, Lu Y, Kitamura N, Urtasun R, et al. Osteopontin induces ductular reaction contributing to liver fibrosis. Gut 2014;63:1805–18.

CHAPTER 6

Stellate Cells, Portal Myofibroblasts, and Epithelial-to-Mesenchymal Transition

Yukinori Koyama[1,3], Ping Wang[1,2,4], David A. Brenner[1] and Tatiana Kisseleva[2]
[1]Department of Medicine, University of California, San Diego, La Jolla, CA, USA
[2]Department of Surgery, University of California, San Diego, La Jolla, CA, USA
[3]Department of Target Therapy Oncology, Graduate School of Medicine, Kyoto University, Kyoto, Japan
[4]Beijing Friendship Hospital, Capital Medical University, Beijing, China

6.1 INTRODUCTION

Hepatic fibrosis is an outcome of many chronic liver diseases, including hepatitis B virus (HBV), hepatitis C virus (HCV), alcoholic liver disease, and nonalcoholic steatohepatitis (NASH) [1]. Chronic liver injury often leads to hepatic fibrosis, a condition associated with increased levels of circulating transforming growth factor-β1 (TGF-β1) and lipopolysaccharide (LPS), activation of myofibroblasts, and extensive deposition of extracellular matrix (ECM), mostly collagen Type I, leading to scar formation. Several early events play an important role in the pathogenesis of liver fibrosis, including: (i) damage to hepatic epithelial and endothelial cells; (ii) release of TGF-β1, the major fibrogenic cytokine; (iii) increase of intestinal permeability producing increased bacterial products in the blood; (iv) recruitment of inflammatory cells; (v) induction of reactive oxygen species (ROS); and (vi) generation of ECM-producing myofibroblasts [1,2]. Immunophenotypically, myofibroblasts are characterized by a spindle or stellate shape and expression of abundant intracellular matrix proteins (vimentin, α-smooth muscle actin [α-SMA], non-muscle myosin, fibronectin) [3]. Ultrastructurally, myofibroblasts are identified by a rough endoplasmic reticulum (rER), a Golgi apparatus producing collagen, peripheral myofilaments, fibronexus (no lamina), and gap junctions [3]. The origin of myofibroblasts in the damaged liver has been extensively studied, and several sources of myofibroblasts have been identified [4–7]. In the fibrotic liver, hepatic stellate cells (HSCs) have been reported to contribute to the majority of the collagen Type I-producing myofibroblasts [1]. Hepatic myofibroblasts may also originate from portal fibroblasts (PFs), bone marrow (BM)-derived mesenchymal cells, and fibrocytes [8] (Figure 6.1). Two additional mechanisms implicated in fibrogenesis are the epithelial-to-mesenchymal transition (EMT), when epithelial cells acquire features of mesenchymal cells and may give rise to fully differentiated myofibroblasts [9,10],

Stellate Cells in Health and Disease
DOI: http://dx.doi.org/10.1016/B978-0-12-800134-9.00006-3

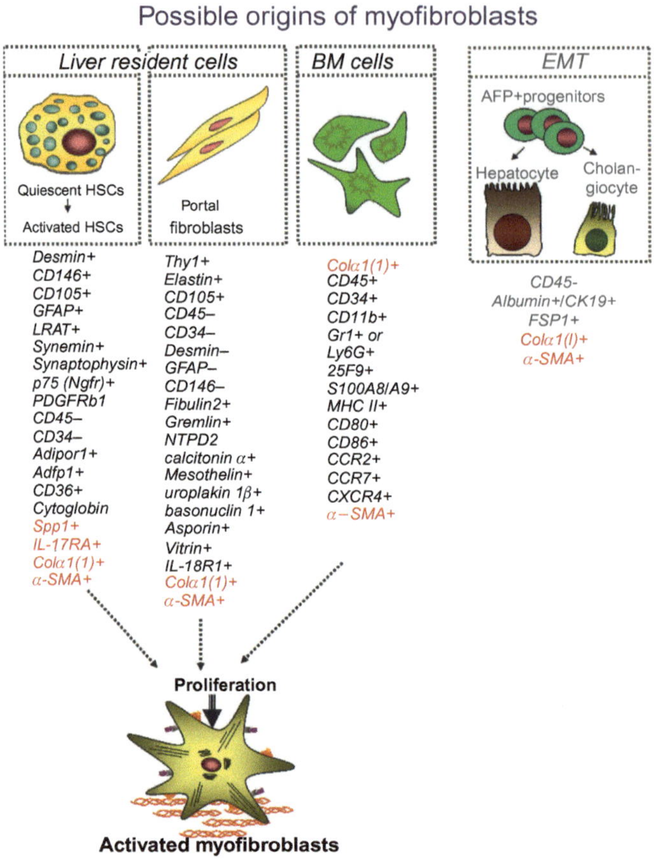

Figure 6.1 Three sources of myofibroblasts have been proposed in fibrotic liver: resident cells (HSCs and PFs); BM-derived cells (fibrocytes), and cells originated by EMT or EndMT. Cell specific markers are listed for each cell type. *Modified from Ref. [2].*

and endothelial-to-mesenchymal transition (EndMT), when endothelial cells undergo a similar phenotypic change [11,12].

The goal of this chapter is to summarize the cellular origin and the fate of fibrogenic myofibroblasts contributing to the pathogenesis of liver fibrosis. Here we discuss the contribution of (i) HSCs, (ii) PFs, and (iii) EMT to liver fibrosis of different etiologies, such as (a) postnecrotic (carbon tetrachloride, CCl_4), (b) cholestatic (surgical ligation of the common bile duct, BDL), (c) metabolic (choline deficient amino acid supplemented diet, CDAA diet), and (d) alcoholic (Tsukamoto-French model of intragastric gavage).

All human and experimental models of hepatic fibrosis are characterized by the excessive production of ECM proteins forming a fibrous scar by myofibroblasts. In fact, immunohistochemical staining with α-SMA is now used clinically to identify myofibroblasts in liver biopsies (reviewed in Ref. [1]). However, the origin of these fibrogenic myofibroblasts is not always known, and may vary between different diseases and different experimental models. Understanding the mechanism(s) of myofibroblast differentiation is critical for providing new targets for anti-fibrotic therapy.

6.2 HEPATIC STELLATE CELLS

6.2.1 Contribution of HSCs to liver fibrosis

6.2.1.1 HSCs/Myofibroblasts are the primary target of anti-fibrotic therapy

Under physiological conditions, HSCs exhibit a quiescent phenotype (qHSCs). qHSCs reside in the Disse space, express neural markers (glial fibrillar acidic protein [GFAP], synemin, synaptophysin [1], NGF receptor p75 [13,14]), and desmin, and serve as a major storage of vitamin A [15–17] (Figure 6.1). In response to injury, qHSCs downregulate levels of vitamin A, and activate into collagen Type I-producing myofibroblasts [4,7]. Using collagen-α1(I)-GFP, reporter mice activated HSCs (aHSCs) can be identified by GFP (green fluorescent protein) and vitamin A expression. In response to injury, qHSCs decrease vitamin A storage and peroxisome proliferator-activated receptor gamma (PPARγ) expression and activate into collagen Type I and α-SMA expressing myofibroblasts [1].

In response to fibrogenic injury, HSCs undergo initiation and perpetuation of activation [18]. The earliest changes in HSC phenotype and gene expression occur initially in response to paracrine signaling, cytokines, and other stimuli. These include altered composition of the ECM and exposure to lipid peroxides and products of damaged hepatocytes [18]. In response to injury, damaged hepatocytes produce fibrogenic ROS, which in turn facilitate HSC activation and migration. Furthermore, dying hepatocytes form apoptotic bodies and release TGF-β1, which causes upregulation of pro-collagen Type l, α-SMA, tissue inhibitor of matrix metalloproteinase-1 (TIMP-1), and TGF-βRI expression in neighboring HSCs, and trigger activation and flux of inflammatory cells into the injured liver [1]. Activation of Kupffer cells plays a critical role in HSC activation. Kupffer cells are the major source of TGF-β1, the major profibrogenic cytokine [19]. In addition, they produce other proinflammatory cytokines interleukin-6 (IL-6), IL-1β, and TNF-β that are strongly upregulated during development of liver fibrosis. Other factors such as platelet-derived growth factor (PDGF) and epidermal growth factor (EGF) also contribute to HSCs activation [4,20]. Recent studies suggest that IL-17A plays a significant role in pathogenesis of liver fibrosis by regulating TGF-β production by Kupffer cells, and directly inducing collagen Type I expression in HSCs [21]. The outcome of these changes is the replacement of "normal" ECM, with "scar" matrix [18].

Liver resident HSCs and PFs give rise to fibrogenic myofibroblasts in response to fibrogenic injury. However, PFs play a lesser role in the pathogenesis of liver fibrosis than HSCs because their activation occurs mostly in response to cholestatic liver injury [22]. Therefore, we may conclude that HSCs are the major source of hepatic myofibroblasts because HSCs activate and contribute to deposition of ECM during development of liver fibrosis of different etiologies [23]. These findings are confirmed by Ghiassi-Nejad et al. [24], who developed a new model for depleting mouse HSCs, which is based on overexpression of herpes simplex virus thymidine kinase gene (HSV-Tk) under control of the mouse GFAP promoter. Depletion of HSCs in the liver can be achieved in these mice upon the administration of ganciclovir (GCV). This experiment clarified their contributions to hepatic injury and fibrosis in response to damage caused by combination of CCl_4 (centrilobular injury) with allyl alcohol (AA) (periportal injury), as well as in a BDL model. Depletion of HSCs by 50–65% resulted in significant attenuation of liver fibrosis [24]. This powerful technique can be applied to other organs and systems, although extrahepatic $GFAP^+$ cells could be also affected by this model.

6.2.1.2 Reversibility of liver fibrosis

It was previously believed that hepatic fibrosis was irreversible [25]. However, sequential liver biopsies have documented that removing the underlying etiological agent may result in regression of hepatic fibrosis in patients with secondary biliary fibrosis [26], hepatitis C [27], hepatitis B [28], NASH [29], and autoimmune hepatitis [30]. Furthermore, in experimental models of CCl_4 [31,32] and BDL [33] induced liver fibrosis, removal of the etiological agent results in reversal of fibrosis [1]. Regression is characterized by a decrease of inflammatory cytokines and TGF-β1 [34], death of HSCs [31], and upregulation of matrix-metalloproteinases [7,35] relative to their inhibitors TIMPs [36]. Liver associated natural killer (NK) cells and NKT cells may play a critical role in HSC apoptosis [37]. Drugs that induce apoptosis in activated HSCs (gliotoxin, sulfasalazine, the inhibitors of IKK (the IκB kinase that is a part of upstream NF-κB signal transduction cascade), and anti-TIMP antibodies) cause liver fibrosis to regress [1]. Although HSC death by apoptosis [31] and senescence [38] during the regression of liver fibrosis is well documented, its quantitative contribution was unknown [39]. Theoretically, activated HSCs/myofibroblasts were suggested to transdifferentiate into another phenotype or revert to quiescent HSCs.

Although the cellular population of quiescent HSCs is restored in mice recovering from fibrosis, the etiology of these quiescent-like HSCs remained unknown until recently. The development of the Cre-lox-system [40] provided a unique tool to monitor specific cellular populations and their progeny in mice [41]. This system is based on the ability of the small 38-kDa bacteriophage protein Cre-recombinase (Cre) to recognize and excise inverted 13 base pairs *loxP* sequences and the deoxyribonucleic acid (DNA) it flanks [40,42]. Cre-*loxP* systems are also widely used for cell-specific gene deletion. Some of the data obtained in HSCs are discussed in Section 6.2.1.3. In addition, Cre-*LoxP* systems are used for cell fate mapping. Genetic labeling of aHSCs can be achieved by crossing mice expressing Cre under control of a myofibroblast-specific promoters (Collagen-α1(I)-Cre mice, Collagen-α2(I)-Cre mice, Collagen-α2(I)-Er-Cre mice, vimentin-Er-Cre mice) or HSC-specific promoters (Lrat-Cre mice, GFAP-Cre mice, PDGFRb-Cre mice) with reporter $Rosa26^{f/f\text{-}YFP}$ mice (or other reporter mice), ubiquitously expressing the *yfp* gene, which transcription is blocked by a *floxed* Stop cassette [42]. Upon genetic recombination, *floxed*-Stop-*floxed* cassette is removed from genomic DNA, causing irreversible labeling of HSCs and their progenies by expression of reporter YFP protein [43] (reviewed in Ref. [44]). Phenotypical changes of aHSCs and the mechanism of their inactivation have been studied during regression of liver fibrosis using models of hepatotoxin-induced liver fibrosis (such as CCl_4 or intragastric alcohol feeding).

6.2.1.3 Inactivation of HSCs (iHSCs)

Regression of liver fibrosis is associated with the disappearance of hepatic myofibroblasts. The previous concept was that on the basis of the documented senescence and apoptosis some aHSCs/myofibroblasts disappear during reversal of fibrosis (see Section 6.2.2.1). However, until recently it was not clear if all aHSCs apoptose during regression of liver fibrosis because this phenomenon is short-lived and difficult to quantify. We [45], and subsequently others [46], have used genetic labeling to demonstrate an alternative pathway in which myofibroblasts revert to a quiescent-like phenotype in CCl_4-induced liver injury and experimental alcoholic liver disease. The fate of these genetically marked myofibroblasts is quantitated in experimental models of fibrosis and its reversal. Following induction of liver injury in these mice, aHSCs and their progeny are permanently labeled by YFP expression even if the cells undergo a phenotypic change [45]. We demonstrated that half of the myofibroblasts escape apoptosis during regression of liver fibrosis, downregulate fibrogenic genes (Col1a1, Col1a2, α-SMA, TIMP1, TGFβ-RI) and acquire a phenotype similar to, but distinct from, qHSCs [45].

Inactivated HSCs acquire a novel phenotype that has not been previously described and is now designated iHSCs [47]. In particular, iHSCs more rapidly reactivate into myofibroblasts in response to fibrogenic stimuli and more effectively contribute to liver fibrosis. Inactivated HSCs downregulate α-SMA, Col1a1, Col1a2, TIMP1, TGFβ-RI, and obtain features of quiescent-like HSCs due to upregulation of PPAR-γ, CREBP, and BAMBI [45]. Our findings in mice *in vivo* support previous *in vitro* studies demonstrating the importance of PPAR-γ for maintaining and reestablishing the quiescent phenotype (qHSCs) [48,49]. Meanwhile, other quiescent-associated genes such as GFAP, Adipor1, Adpf, and Dbp are not re-expressed in iHSCs, suggesting that despite the similarities between qHSCs and iHSCs and their lack of fibrogenic gene expression, iHSCs possess properties distinct from qHSCs. Based on comparison of the global gene expression in qHSCs, aHSCs, and iHSCs, several genes that are differentially expressed in HSCs depending on their phenotype, have been identified and can be used to distinguish iHSCs from qHSCs and aHSCs [45]. This is of particular importance for identifying inactivation of human HSCs. Comparative analysis of phenotype-specific HSC signature genes may provide further insight in inactivation of HSCs *in vivo*. Specifically, we identified 7 qHSC-specific genes (that are expressed only in qHSCs: Nr1d2, Adipor1, Adpf, GFAP, Sbp, Prei4, Foxj1), 12 aHSC-specific genes (expressed only in aHSCs: α-SMA, Col1α1, TIMP-1, Crlf1, ILr1, Itga5, Spp1, Lox, LoxL2, IL-17ra, Fosl1, Folr1), 12 iHSC-specific genes (upregulated in iHSCs: Rnd1, Csf1R, IL-7R, Cntfr, Csf2rb, Cx3cr1, Ly86, Cxcl2, Fos, Egr1, Spic, Atf3), and 12 genes that are upregulated in both qHSCs and iHSCs (and therefore are associated with nonfibrogenic properties of HSCs: Mrc1, Agtr1a, Calcr1, Grap, Ngfr, Insig1, PPARγ, CD36, IL-10RA, Foxo1, Tal1, BAMBI). Furthermore, the genes Nr1d2, Adipor1, IL17ra, Itga5, Egfr, Crlf1, IL1r1, Rnd1, Csf1R, IL7R, Cntfr, Csf2rb, Cx3cr1, IL10RA, Ly86, CD36, Mrc1, Agtr1a, Calcr1, Grap, and Ngfr are cell surface receptors, so that their expression could be used to detect different HSC phenotypes using flow cytometry.

6.2.2 Other functions of HSCs

6.2.2.1 Distinct functions of senescent HSCs

Senescence is a form of proliferative arrest that is associated with the stable repression of E2F target genes [50]. Senescent cells also downregulate the production of ECM proteins, upregulate ECM-degrading enzymes, and secrete a variety of biologically active factors and cytokines [51]. The role of the senescence phenotype in regulation of HSC function will be discussed later in this section. Krizhanovsky et al. [38] observed that during resolution of liver fibrosis activated HSCs senesce, and a p53-dependent senescence is a mechanism by which HSCs regulate tissue homeostasis and fibrosis and fibrosis resolution. Consistently, mice in which p53 was specifically deleted in HSCs displayed exaggerated fibrosis in response to chronic injury. Other factors were also implicated in regulation of senescent phenotype in HSCs. HSCs were shown to express high levels of IL-22R1 and IL-10R2 and respond to IL-22 signaling [52]. Overexpression of IL-22 *in vivo* or treatment with IL-22 *in vitro* promotes HSC senescence and apoptosis. The anti-fibrotic effects of IL-22 are mediated via the activation of STAT3 and activation of suppressor of cytokine signaling 3 (SOCS3), which induces HSC senescence [52].

CCN1 is matricellular protein of the CCN (CYR61/CTGF/NOV) family [53]. CCN1 expression is upregulated in hepatocytes in response to fibrogenic liver injury (CCl_4 and BDL). CCN1 acts on aHSCs by engaging integrin α6β1 to induce ROS accumulation through the Rac1-NADPH oxidase 1 enzyme complex causing HSC senescence and regression of liver fibrosis. Mice with hepatocyte-specific CCN1 deletion lack senescence and develop exacerbated fibrosis, whereas overexpression of hepatic CCN1 reduces liver fibrosis with enhanced senescence [54]. A2A receptor regulates HSC fate by inducing HSC proliferation, protecting from apoptosis and senescence by decreasing p53 and Rb through cAMP-PKA/Rac1/p38 MAPK pathway, and promote liver fibrosis [55].

Cellular senescence has been also implicated in protection against tumorigenesis [56] caused by telomerase shortening, DNA damage, and oncogene activation via induction of senescence-induced cell-cycle arrest and apoptosis. HSCs do not undergo malignant transformation, but senescence is critical for apoptosis of activated HSCs during regression of liver fibrosis. Recently, a novel mechanism has been described that links senescent HSCs with development of hepatocellular carcinoma (HCC). Briefly, obesity was shown to be associated with alterations of gut microbiota, and subsequent increased production of DNA-damaging gut bacterial metabolite deoxycholic acid (DCA). The circulating DCA induces senescence-associated secretory (SASP) phenotype in HSCs (upregulation of 53BP1 foci, p21), and promotes production and secretion of various inflammatory and tumor-promoting factors (IL-6, Gro-α, CXCL9) in the liver [51] known to facilitate HCC development in mice after exposure to carcinogen. Blocking DCA production or reducing gut bacteria prevents HCC development in obese mice. Similar results were also observed in mice lacking an SASP inducer or depleted of senescent HSCs, implying that the DCA-SASP axis in HSCs may play a role in obesity-associated HCC development. Although this study provides new insights into the mechanism of HCC development, it appears to contrast previous reports that identified hepatic myeloid cells as the major source of proinflammatory cytokines in the course of HCC development [57–59]. Furthermore, Lujambio et al. [60] reported an opposing observation that senescent HSCs attenuate HCC development. In this study, HSC-specific p53-deficient mice, in which HSC do not undergo senescence, are more susceptible to liver fibrosis and HCC, indicating that HSC senescence suppresses liver fibrosis and HCC development [60].

6.2.2.2 HSC as regulatory bystanders

Despite continuous exposure to bacterial and dietary antigens through the portal vein blood flow [61], the liver is considered to be immunologically quiescent; it mediates immunosuppressive functions and is responsible for systemic tolerance [62,63]. On the other hand, perhaps this makes the liver more susceptible to chronic viral infections such as hepatitis virus B and C [64,65]. A sub-population of Treg cells, which is characterized by expression of Forkhead box P3 (Foxp3), was shown to suppress immune responses in the liver [66] and regulate liver-mediated tolerance [67]. In response to specific priming and cytokine stimulations, Foxp3 + Treg (iTreg) cells differentiate from activated CD4 T cells (which also give rise to other T helper (Th) subtypes, including Th1, Th2, and Th17 effector cells) and mediate antiinflammatory responses [68]. TGF-β1 triggers iTreg differentiation [69]. On the other hand, the combination of IL-6 and TGF-β1 drive differentiation of proinflammatory Th17 cells [70,71]. This raises the question of how TGF-β1 can generate such distinct outcomes. Recently, the vitamin A metabolite retinoic acid (RA) has been identified as a key regulator of TGF-β1-dependent immune responses, capable of inhibiting the IL-6-driven induction of proinflammatory T(H)17 cells and promoting antiinflammatory Treg cell differentiation [72,73]. Recent studies have suggested that HSCs play a critical role in regulation of iTreg differentiation, although the underlying mechanism is not well understood. HSCs are located in the space of Disse in close proximity to hepatocytes and liver endothelial cells (LCEC), and functionally act as liver pericytes. With these unique features, quiescent HSCs may directly contact immunocompetent cells and present antigens to naive T cells, or may mediate their activation indirectly by producing biologically active factors and promote iTreg differentiation. Several studies have proposed that HSCs have a capability to function as antigen presenting cells (APC) for MHC class II restricted T cells and affect their functions [74,75]. However, a recent study by Ichikawa et al. [76], which used cell sorting (for vitamin A + , desmin + , and GFAP+ cells) to isolate qHSCs, could not confirm these results. In contrast to previous reports, this group demonstrated that highly purified HSCs did not express co-stimulatory molecules and only upregulated MHC class II after *in vitro* culture in the presence of exogenous interferon-γ (IFN-γ). Consistent with an insufficient APC function, HSCs failed to stimulate naïve OT-II TCR transgenic (OT-II) $CD4^+$ T cells and only moderately stimulated α-GalCer primed invariant NKT (iNKT) cells, indicating that HSCs do not present antigen to naïve MHC class II restricted CD4 T cells. To investigate whether the RA-producing HSCs are part of the liver tolerance mechanism, Ichikawa et al. tested if HSCs can indirectly provide suppressive effects as RA-secreting regulatory bystanders. They showed that purified HSCs express high level of RDH1 and RALDH1 enzymes that are responsible for vitamin A metabolism and RA production, and release RA during priming of T cells by professional APC [76]. Overall, these finding describe a novel and rather unexpected role for HSCs as regulatory bystanders in RA-mediated and TGF-β1-dependent Treg induction. This result was supported by the notion that HSCs from vitamin A-deficient (VAD) mice failed to enhance Foxp3 induction on DC primed T cells, and their regulatory bystander function was completely abolished because of the absence of RA [76].

These findings are further confirmed in a more recent report, which investigated the molecular interactions among murine HSCs, dendritic cells, and naive $CD4^+$ T cells [77]. The authors demonstrated that HSCs alone do not present antigens to naive $CD4^+$ T cells, but in the presence of dendritic cells and TGF-β1, preferentially induce functional $Foxp3^+$ Tregs via production of all-trans retinoic acid, and therefore play a role in the tolerogenic nature of the liver [77].

6.2.2.3 HSCs as regulators of hepatocellular damage

Retinoids (vitamin A and its metabolites) are potent natural regulators of cellular activities, including cell proliferation, differentiation, and apoptosis [78,79]. Retinoic acid, the major transcriptionally active retinoid species, regulates more than 500 genes [80,81]. The all-*trans*- and 9-*cis*-isomers of retinoic acid regulate transcription upon binding to one of their six cognate nuclear receptors, the retinoic acid receptors (RARα, RARβ, and RARγ) and the retinoid X receptors (RXRα, RXRβ, and RXRγ) [82,83]. Most of this hepatic retinoid (80–90%) is stored as retinyl ester within the lipid droplets of HSCs [84,85]. The loss of lipid droplets and retinoids is a hallmark of HSC activation, but it remains unclear whether this loss promotes HSC activation, liver fibrogenesis, or carcinogenesis. Lecithin-retinol acyltransferase (LRAT) is the enzyme responsible for retinyl ester formation and storage of retinoids in HSCs [86]. LRAT-deficient mice ($LRAT^{-/-}$ mice) have been generated [86]. These mice are genetically unable to accumulate hepatic retinoids, and LRAT-deficient HSCs lack retinoid-containing lipids droplets. Surprisingly, when $LRAT^{-/-}$ mice were subjected to spontaneous and experimental fibrosis, they developed the same degree of liver fibrosis as the wild type mice [87]. Activation of LRAT-deficient HSCs in response to fibrogenic stimuli was comparable to that observed in wild type HSCs. These data demonstrated that profibrogenic functions of activated HSCs, such as ECM deposition, do not depend on lipid droplet content and retinoid metabolism. However, expression of LRAT in wild type HSCs is rapidly suppressed upon activation, suggesting that retinoid metabolism plays an important role in HSC biology. In an attempt to evaluate LRAT-mediated functions of HSCs, $LRAT^{-/-}$ mice were subjected to diethylnitrosamine-induced (DEN) tumorigenesis. Interestingly, $LRAT^{-/-}$ mice were protected from DEN-induced tumorigenesis [87] with reduced liver injury and tumor load [87,88]. This may result from high levels of free circulating retinoids, increased retinoid signaling, induction of the retinoic acid responsive genes Cyp26a1, RARb, and p21, and subsequent p21-mediated inhibition of hepatocyte proliferation [87]. This observation was unexpected because dietary retinoid intake in humans and retinoic acid administration to cultured cancer cells or to mice subjected to experimental cancer models have often been found to provide chemoprotection against cancer development [81,89]. To assess whether this hepatic response in the total absence of retinoid stores is specific to DEN or whether it can be more generalized to other hepatic toxins, Shmarakov et al. [81] undertook studies of thioacetamide (TAA)-induced hepatic toxicity in $LRAT^{-/-}$ mice. While wild type mice developed liver injury (evidenced by focal necrotic areas and increases in serum ALT activity and myeloperoxidase activity in hepatic parenchyma) within 48 h of TAA-treatment, remarkably none of these TAA-induced effects were observed in $LRAT^{-/-}$ mice [81]. Loss of hepatoprotection in $LRAT^{-/-}$ mice administered retinyl acetate over 48 h suggested that TAA-dependent hepatotoxicity depends on availability of retinoid metabolites. Overall, these studies established that endogenous hepatic retinoid stores can modulate the toxicity of TAA in mice by increasing oxidative stress and the activities of the cytochrome P450 (CYP) and flavin-containing monooxygenase (FMO) systems in the damaged liver [81].

6.2.2.4 The role of HSCs in liver regeneration

All-trans retinoic acid (RA) is a potent inducer of regeneration. Liu et al. [90] have shown that RA signals via retinoic acid receptor β (RARβ) and retinoid X receptor α (RXRα) to activate transcription of target cell cycle genes Cdk1, Cdk2, Cyclin B, Cyclin E, and Cdc25a. Consistently, wild type mice treated with a single dose of RA developed hepatomegaly due to increased number of Ki67-positive cells and induced expression of cell cycle genes in the liver. Experimental liver regeneration is induced in mice by partial hepatectomy (PH), which results in substantial depletion of hepatic retinoid stores (70%), rapid and synchronous induction of hepatocyte cell DNA replication and mitosis, and proliferation of normally quiescent HSCs [81]. Mice that received RA prior to PH demonstrated accelerated liver regeneration caused by early upregulation of RARβ, Aldh1a2, Crabp1, and Crbp1, and cell cycle genes. Postoperative administration of RA or synthetic retinoid NIK33 also enhanced hepatocyte proliferation in the regenerating rat liver [91]. Although the results on how RA affects hepatocyte proliferation at different time points after PH are rather conflicting, LRAT-deficient mice have reduced RA levels and impaired liver regeneration in response to PH, suggesting that hepatic retinoid storage is required for normal liver regeneration [81]. Taken together, HSCs regulate hepatocyte cell cycle progression and liver regeneration by producing and releasing biologically active retinoid metabolites.

However, the effect of HSCs on hepatocyte proliferation seems to differ in the injured liver. In response to chronic injury, failure of hepatocytes to regenerate is accompanied by deposition of ECM proteins that replace the damaged cells [92]. The mechanism that controls the balance between cell regeneration and fibrogenesis is not well understood. Ebrahimkhani et al. [93] demonstrated that activated HSCs in diseased livers inhibit hepatocyte proliferation by producing TGF-β1, a potent profibrogenic cytokine that also strongly affects hepatocyte proliferation. Selective depletion of activated HSCs increases hepatocyte proliferation in a biliary fibrosis model. Furthermore, the authors suggested that production of TGF-β1 is, at least in part, regulated in HSCs via serotonin-mediated activation of 5-hydroxytryptamine 2B receptor (5-HT2B), which is selectively expressed in activated HSCs and induces a signaling cascade that results in TGF-β1 expression. Serotonin (5-hydroxytryptamine, 5-HT), a well-known neurotransmitter, has been recently implicated in promoting tissue remodeling [94] and TGF-β1 production. Consistently, mice lacking 5-HT2B receptors and normal mice treated with a specific 5-HT2B receptor antagonist (SB-204741) demonstrated decreased TGF-β1 production and increased hepatocyte proliferation in response to PH. These findings indicated that regenerative response is one of many overlapping pathways involved in reconstituting normal and diseased livers [92,93].

6.3 PORTAL FIBROBLASTS

6.3.1 Characterization of PFs

PFs are spindle-shaped cells of mesenchymal origin that express vimentin, elastin, and Thy1 (a glycophosphatidylinositol-linked glycoprotein) [95,96] (Figure 6.1). Under physiological conditions, PFs comprise a small population of cells that surround the portal vein to maintain integrity of the portal tract [97]. PFs were implicated to pathogenesis of cholestatic liver injury [98]. In response to the BDL model of cholestatic injury in mice, PFs proliferate, differentiate into α-SMA-expressing myofibroblasts, and synthesize ECM [97–99]. PFs have been also implicated in pathogenesis of liver injury in Mdr2$^{-/-}$ mice, which is characterized by disruption of tight junctions and basement membranes of bile ducts, bile leakage to the portal tract, and formation of periportal biliary fibrosis [100,101]. Both aPFs and aHSCs were shown to contribute to collagen Type I deposition in Mdr2$^{-/-}$ mice [100,102]. However, the direct contribution of aPFs to ECM deposition and liver fibrosis (compared to aHSCs) in Mdr2$^{-/-}$ mice has not been assessed [101].

The contribution of PFs to liver fibrosis of different etiologies is not well understood, mainly because of difficulties with PF and myofibroblast isolation. The most widely used method of PF isolation from rats is based on liver perfusion with enzymatic digestion followed by size selection [103]. Cell outgrowth from dissected bile segments is still used to isolate murine PFs [104], and after 10–14 days in culture PFs undergo progressive myofibroblastic differentiation [97,105]. Another method of PF isolation and culturing is based on a precision-cut liver slice (PCLS), and is designed to maintain cell–cell and cell–matrix interactions and mimic natural microenvironment of PFs. These techniques were instrumental in identification of several markers of PFs that are now available to identify this specific population in the myofibroblast population: gremlin, Thy1 [106], fibulin 2 [106], IL-6, elastin [107], the ecto-AT-Pase nucleoside triphosphate diphosphohydrolase-2 (NTPD2) [108], and coffilin 1 [109]. In addition, the lack of desmin, cytoglobin, α2-macroglobulin [106], neural proteins (GFAP, NGFRp75, and synaptophysin) expression and absence of lipid droplets distinguishes PFs from HSCs [1,97,110]. Recently, we used a flow cytometry method to purify activated PFs. This method utilizes the reporter Collagen-α1(I)-GFP mice (Col-GFP mice, that expresses GFP upon activation of collagen-α1(I) promoter/enhancer) [99], which are widely used to study activation of myofibroblasts in fibrotic liver [111–113]. Thus, isolated aPFs can be identified as GFP^{+} Thy1^{+} Elastin^{+} cells activated in response to cholestatic liver injury. Isolated aPFs were subjected to gene expression profiling. In concordance with previous studies, aPFs lacked expression of cytoglobin [109,114], but expressed Thy1, elastin, Gremlin 1, Fibulin 2, and NTPD2 mRNAs [97,115]. In addition, several novel PF-specific markers were identified. We identified that aPFs uniquely expressed calcitonin α, mesothelin, uroplakin 1β, basonuclin 1, asporin, proteoglycan, glipican 3, and CD200. High expression of these genes specifically in aPFs (but not in quiescent or activated HSCs, endothelial cells, Kupffer cells, hepatocytes, or cholangiocytes) was confirmed. Some of these genes (including *basonuclin 1, glycoprotein m6a, uroplakin 3b and 1b, mesothelin, IL-18R, calcitonin-related peptides*, and *vitrin*) were reported as signature genes of murine hepatic mesothelial [116] and epicardial cells [117], supporting the theory that PFs originate from mesothelial cells [118,119]. However, the study did not discriminate between PFs at different stages of activation and recently identified mesothelial precursor cells located in the liver capsule [120].

The aPFs functionally differ from aHSCs. For example, aPFs respond rapidly to TGF-β1 [121] by upregulation of Col-α1(I), α-SMA, TIMP1, TGF-β2 [122], plasminogen activator inhibitor PAI-1, elastin [97], fibronectin [106,123], and fibulin-2 ECM components [106,123], and of CD73 ecto-enzyme [123,124], while specifically downregulating expression of CD39L1/NTPDase2 ecto-enzyme [125]. In addition, TGF-β1 signaling was shown to be associated with downregulation of PPAR-γ, induction of NFκB, and extracellular signal-regulated kinase (ERK) [126]. Bile acids were shown to induce aPF activation and proliferation [22,127]. They mediate their functions via activation of nuclear receptors, such as farnesoid X receptor (FXR), vitamin D receptor (VDR), pregnane X receptor (PXR), constitutive androstane receptor (CAR), and small heterodimer partner (SHP), and TGR5 (the membrane G protein-coupled receptor) [102,110]. However, it remains controversial, if aPFs express bile acid receptors, as it has been proposed in rat aPFs [110,128]. Recent studies reported that aPFs do not express FXR protein, but upregulate VDR [102,129]. Some bile acids can activate collagen Type I expression in aPFs, however, it remains unclear if this is a direct effect [22]. Interestingly, the same study had demonstrated that activation of PFs precedes activation of HSCs in response to cholestatic liver injury. At the onset of BDL, the number of aPFs exceeds the number of aHSCs (70% vs. 30%) [22]. We can speculate that there is a direct correlation between cholestatic injury, proliferation of cholangiocytes, and activation of PFs. Furthermore, responses of aPFs to fibrogenic stimuli differ from that observed in aHSCs. In contrast to aHSCs, aPFs do not respond to activation by PDGF or NGF. Another characteristic of aPFs is that they express IL-25R (IL-17ReceptorB). The aPFs respond to IL-25 stimulation by induction of IL-13, (similar to IL-13 induction by IL-25-treated macrophages [130] and fibroblasts [131]), which in turn stimulates activation of aHSCs *in vitro* [132,133] by inducing connective tissue growth factor (CTCF), Co1la1, α-SMA, and TIMP1 mRNAs, suggesting that aPFs may locally facilitate HSC activation. The aPFs also secrete TGF-β1 [122], suggesting that aPFs may facilitate HSC activation by secreting IL-13 and TGF-β1.

6.4 EPITHELIAL-TO-MESENCHYMAL TRANSITION

In theory, hepatic epithelial cells (cholangiocytes and hepatocytes) and hepatic sinusoidal endothelial cells may transition to myofibroblasts in response to sustained liver injury. Cell specific promoters (cytokeratin-19 [CK-19, K19], albumin, and Tie-1, respectively), driving expression of Cre for each of these cell types, are used to irreversibly mark these cell populations on the genetic level.

EMT [9,10] is a process where fully differentiated epithelial cells undergo phenotypic transition to fully differentiated mesenchymal cells (fibroblasts or myofibroblasts) [134]. During EMT, epithelial cells detach from the epithelial layer, lose their polarity, stop expressing epithelial markers (e.g., CK-19, CK-7, E-cadherin) and tight junction proteins (zonula occludens-1, ZO-1), increase their motility, and obtain a (myo) fibroblast-like phenotype [135]. Epithelial cells transitioning into (myo)fibroblasts are proposed to express fibroblast specific protein-1 (FSP-1), which is used as a universal marker of EMT in fibrogenesis and cancer [135,136]. Pathological EMT has been best characterized in fibrosing kidneys [137] in which more than 30% of renal interstitial fibroblasts are reported to be derived EMT [136]. EMT of hepatocytes and cholangiocytes has been reported in patients and in mice with liver fibrosis [138–141]. However, more recent studies using rigorous cell fate mapping have detected only minimal or no contribution of hepatocyte EMT to myofibroblasts was detected *in vivo* [11,142].

Studies have also claimed that endothelial cells may also transition to mesenchymal cells (EndMT), giving rise to (myo)fibroblasts in response to fibrogenic injury. EndMT has been reported to contribute to cardiac [11] and renal [12] fibrosis. EndMT is identified by expression of myofibroblast genes [135] in endothelial cells that are expressing, or have a history of expressing, PECAM-1/CD31, Tie1 [11], Tie-2, and CD34 [4,12]. Although endothelial cell injury [2] and neovascularization play a critical role in liver fibrosis, the role of EndMT in liver fibrosis is still not definitively resolved.

Genetic cell fate mapping *in vivo* provides powerful advantages over previous approaches (*in vitro* differentiation studies or dual immunostaining for original and new markers) to follow changes in cellular phenotype throughout the study period. Recent studies that investigated EMT in the liver were based on genetic fate mapping of specific cell populations and monitoring their transdifferentiation into myofibroblasts in response to sustained fibrogenic injury. These studies demonstrated that EMT in hepatocytes, cholangiocytes and their bi-potential precursors does not contribute to liver fibrosis. Thus, contribution of EMT of hepatocytes to collagen Type 1-expressing cells was studied in triple transgenic *Alb*^{LacZ}/*Col-GFP* mice, generated by crossing Col-GFP mice (expressing GFP in all collagen Type I^{+} cells) [99] with Alb^{LacZ} (Albumin-Cre mice × ROSA26$^{f/f\text{-}LacZ}$ reporter mice); Alb^{LacZ}/Col-GFP mice were treated with CCl_4 to induce fibrosis. Staining of liver with X-gal revealed that none of the genetically labeled hepatocytes co-expressed GFP *in vivo*, suggesting that EMT in hepatocytes does not contribute to myofibroblasts. This contrasts with studies of liver parenchymal cells in primary culture, in which TGF-β1 treatment induces phenotypic changes consistent with EMT [140,143]. To study EMT in cholangiocytes, tamoxifen-inducible CK-19^{YFP} mice were crossed with Rosa$26^{f/f\text{-}YFP}$ reporter mice, and subjected to liver injury. Again, genetically labeled cholangiocytes did not transdifferentiate into myofibroblasts. The final evidence was provided by a study published by Chu et al. [144]. To assess whether EMT is active during several experimental models of liver fibrosis, the authors used Alfp-Cre × Rosa26-YFP mice, in which the epithelial cells of the liver (hepatocytes, cholangiocytes, and their bi-potential progenitors) are genetically labeled with YFP. Primary cholangiocytes isolated from this reporter strain were able to undergo EMT *in vitro* when treated with TGF-β1, but failed to upregulate mesenchymal markers S100A4, vimentin, α-SMA, or pro-collagen 1α2 in response to BDL, CCl_4, and 3,5-diethoxycarbonyl-1,4-dihydrocollidine (DDC) models. Overall, these data strongly suggest that EMT does not contribute to hepatic myofibroblasts in fibrotic liver, supporting the concept that HSCs and aPFs are the primary source of collagen Type I-producing myofibroblasts.

6.5 PERSPECTIVE

Despite extensive research, the origin of myofibroblasts in different types of fibrotic liver diseases has been a subject of debate. Overwhelming evidence has established that activated HSCs are the major source of myofibroblasts in a variety of cholestatic and hepatotoxic murine models of liver injury, also supported strongly by human data. However, other sources of hepatic myofibroblasts such as activated PFs and BM-derived fibrocytes have also been identified. Recent studies suggest that the majority of hepatic myofibroblasts originate from liver resident mesenchymal cells, although composition of these myofibroblasts varies depending upon the etiology of liver fibrosis. Future investigations are expected to determine the dynamic interactions between myofibroblasts and microenvironment specific for a type of liver injury and provide a further insight into precise targets for anti-fibrotic therapy.

REFERENCES

[1] Bataller R, Brenner DA. Liver fibrosis. J Clin Invest 2005;115:209–18.
[2] Kisseleva T, Brenner DA. Mechanisms of fibrogenesis. Exp Biol Med (Maywood) 2008;233:109–22.
[3] Eyden B. The myofibroblast: phenotypic characterization as a prerequisite to understanding its functions in translational medicine. J Cell Mol Med 2008;12:22–37.
[4] Kisseleva T, Brenner DA. Hepatic stellate cells and the reversal of fibrosis. J Gastroenterol Hepatol 2006;21(Suppl. 3):S84–7.
[5] Kalluri R, Neilson EG. Epithelial–mesenchymal transition and its implications for fibrosis. J Clin Invest 2003;112:1776–84.

[6] Gomperts BN, Strieter RM. Fibrocytes in lung disease. J Leukoc Biol 2007;82:449–56.
[7] Fallowfield JA, Mizuno M, Kendall TJ, Constandinou CM, Benyon RC, Duffield JS, et al. Scar-associated macrophages are a major source of hepatic matrix metalloproteinase-13 and facilitate the resolution of murine hepatic fibrosis. J Immunol 2007;178:5288–95.
[8] Kisseleva T, Brenner DA. Fibrogenesis of parenchymal organs. Proc Am Thorac Soc 2008;5:338–42.
[9] Kalluri R. EMT: when epithelial cells decide to become mesenchymal-like cells. J Clin Invest 2009;119:1417–9.
[10] Choi SS, Diehl AM. Epithelial-to-mesenchymal transitions in the liver. Hepatology 2009;50(6):2007–13.
[11] Zeisberg EM, Tarnavski O, Zeisberg M, Dorfman AL, McMullen JR, Gustafsson E, et al. Endothelial-to-mesenchymal transition contributes to cardiac fibrosis. Nat Med 2007;13:952–61.
[12] Zeisberg EM, Potenta SE, Sugimoto H, Zeisberg M, Kalluri R. Fibroblasts in kidney fibrosis emerge via endothelial-to-mesenchymal transition. J Am Soc Nephrol 2008;19:2282–7.
[13] Sachs BD, Baillie GS, McCall JR, Passino MA, Schachtrup C, Wallace DA, et al. p75 neurotrophin receptor regulates tissue fibrosis through inhibition of plasminogen activation via a PDE4/cAMP/PKA pathway. J Cell Biol 2007;177:1119–32.
[14] Kendall TJ, Hennedige S, Aucott RL, Hartland SN, Vernon MA, Benyon RC, et al. p75 Neurotrophin receptor signaling regulates hepatic myofibroblast proliferation and apoptosis in recovery from rodent liver fibrosis. Hepatology 2009;49:901–10.
[15] Iredale JP. Models of liver fibrosis: exploring the dynamic nature of inflammation and repair in a solid organ. J Clin Invest 2007;117:539–48.
[16] Geerts A. History, heterogeneity, developmental biology, and functions of quiescent hepatic stellate cells. Semin Liver Dis 2001;21:311–35.
[17] Senoo H, Kojima N, Sato M. Vitamin A-storing cells (stellate cells). Vitam Horm 2007;75:131–59.
[18] Kocabayoglu P, Friedman SL. Cellular basis of hepatic fibrosis and its role in inflammation and cancer. Front Biosci (Schol Ed) 2013;5:217–30.
[19] Seki E, De Minicis S, Osterreicher CH, Kluwe J, Osawa Y, Brenner DA, et al. TLR4 enhances TGF-beta signaling and hepatic fibrosis. Nat Med 2007;13:1324–32.
[20] DeLeve LD, Jaeschke H, Kalra VK, Asahina K, Brenner DA, Tsukamoto H. 15th international symposium on cells of the hepatic sinusoid, 2010. Liver Int 2011;31:762–72.
[21] Meng F, Wang K, Aoyama T, Grivennikov SI, Paik Y, Scholten D, et al. Interleukin-17 signaling in inflammatory, Kupffer cells, and hepatic stellate cells exacerbates liver fibrosis in mice. Gastroenterology 2012;143:765–76. e761–3.
[22] Iwaisako K, Jiang C, Zhang M, Cong M, Moore-Morris TJ, Park TJ, et al. Origin of myofibroblasts in the fibrotic liver in mice. Proc Natl Acad Sci USA 2014;111:E3297–305.
[23] Mederacke I, Hsu CC, Troeger JS, Huebener P, Mu X, Dapito DH, et al. Fate tracing reveals hepatic stellate cells as dominant contributors to liver fibrosis independent of its aetiology. Nat Commun 2013;4:2823.
[24] Ghiassi-Nejad Z, Hernandez-Gea V, Woodrell C, Lang UE, Dumic K, Kwong A, et al. Reduced hepatic stellate cell expression of Kruppel-like factor 6 tumor suppressor isoforms amplifies fibrosis during acute and chronic rodent liver injury. Hepatology 2013;57:786–96.
[25] Ramachandran P, Iredale JP. Reversibility of liver fibrosis. Ann Hepatol 2009;8:283–91.
[26] Hammel P, Couvelard A, O'Toole D, Ratouis A, Sauvanet A, Flejou JF, et al. Regression of liver fibrosis after biliary drainage in patients with chronic pancreatitis and stenosis of the common bile duct. N Engl J Med 2001;344:418–23.
[27] Arthur MJ. Reversibility of liver fibrosis and cirrhosis following treatment for hepatitis C. Gastroenterology 2002;122:1525–8.
[28] Kweon YO, Goodman ZD, Dienstag JL, Schiff ER, Brown NA, Burchardt E, et al. Decreasing fibrogenesis: an immunohistochemical study of paired liver biopsies following lamivudine therapy for chronic hepatitis B. J Hepatol 2001;35:749–55.
[29] Dixon JB, Bhathal PS, Hughes NR, O'Brien PE. Nonalcoholic fatty liver disease: improvement in liver histological analysis with weight loss. Hepatology 2004;39:1647–54.
[30] Czaja AJ, Carpenter HA. Decreased fibrosis during corticosteroid therapy of autoimmune hepatitis. J Hepatol 2004;40:646–52.

[31] Iredale JP, Benyon RC, Pickering J, McCullen M, Northrop M, Pawley S, et al. Mechanisms of spontaneous resolution of rat liver fibrosis. Hepatic stellate cell apoptosis and reduced hepatic expression of metalloproteinase inhibitors. J Clin Invest 1998;102:538–49.

[32] Issa R, Zhou X, Constandinou CM, Fallowfield J, Millward-Sadler H, Gaca MD, et al. Spontaneous recovery from micronodular cirrhosis: evidence for incomplete resolution associated with matrix cross-linking. Gastroenterology 2004;126:1795–808.

[33] Issa R, Williams E, Trim N, Kendall T, Arthur MJ, Reichen J, et al. Apoptosis of hepatic stellate cells: involvement in resolution of biliary fibrosis and regulation by soluble growth factors. Gut 2001;48:548–57.

[34] Ueberham E, Low R, Ueberham U, Schonig K, Bujard H, Gebhardt R. Conditional tetracycline-regulated expression ofTGF-beta1 in liver of transgenic mice leads to reversible intermediary fibrosis. Hepatology 2003;37:1067–78.

[35] Uchinami H, Seki E, Brenner DA, D'Armiento J. Loss of MMP 13 attenuates murine hepatic injury and fibrosis during cholestasis. Hepatology 2006;44:420–9.

[36] Pinzani M, Vizzutti F. Fibrosis and cirrhosis reversibility: clinical features and implications. Clin Liver Dis 2008;12:901–13.

[37] Radaeva S, Sun R, Jaruga B, Nguyen VT, Tian Z, Gao B. Natural killer cells ameliorate liver fibrosis by killing activated stellate cells in NKG2D-dependent and tumor necrosis factor-related apoptosis-inducing ligand-dependent manners. Gastroenterology 2006;130:435–52.

[38] Krizhanovsky V, Yon M, Dickins RA, Hearn S, Simon J, Miething C, et al. Senescence of activated stellate cells limits liver fibrosis. Cell 2008;134:657–67.

[39] Varga J, Brenner D, Phan SE. Fibrosis research. Methods and protocols. Totowa, NJ: Humana Press; 2005. p. 293.

[40] Ramirez-Solis R, Liu P, Bradley A. Chromosome engineering in mice. Nature 1995;378:720–4.

[41] Perkins AS. Functional genomics in the mouse. Funct Integr Genomics 2002;2:81–91.

[42] Sternberg N, Austin S, Hamilton D, Yarmolinsky M. Analysis of bacteriophage P1 immunity by using lambda-P1 recombinants constructed *in vitro*. Proc Natl Acad Sci USA 1978;75:5594–8.

[43] Sauer B. Inducible gene targeting in mice using the Cre/lox system. Methods 1998;14:381–92.

[44] Xu J, Liu X, Koyama Y, Wang P, Lan T, Kim IG, et al. The types of hepatic myofibroblasts contributing to liver fibrosis of different etiologies. Front Pharmacol 2014;5:167.

[45] Kisseleva T, Cong M, Paik Y, Scholten D, Jiang C, Benner C, et al. Myofibroblasts revert to an inactive phenotype during regression of liver fibrosis. Proc Natl Acad Sci USA 2012;109:9448–53.

[46] Troeger JS, Mederacke I, Gwak GY, Dapito DH, Mu X, Hsu CC, et al. Deactivation of hepatic stellate cells during liver fibrosis resolution in mice. Gastroenterology 2012;143(4):73–83. e22.

[47] Cassiman D, Libbrecht L, Desmet V, Denef C, Roskams T. Hepatic stellate cell/myofibroblast subpopulations in fibrotic human and rat livers. J Hepatol 2002;36:200–9.

[48] She H, Xiong S, Hazra S, Tsukamoto H. Adipogenic transcriptional regulation of hepatic stellate cells. J Biol Chem 2005;280:4959–67.

[49] Tsukamoto H. Adipogenic phenotype of hepatic stellate cells. Alcohol Clin Exp Res 2005;29:132S–3S.

[50] Collado M, Serrano M. Senescence in tumours: evidence from mice and humans. Nat Rev Cancer 2010;10:51–7.

[51] Coppe JP, Patil CK, Rodier F, Sun Y, Munoz DP, Goldstein J, et al. Senescence-associated secretory phenotypes reveal cell-nonautonomous functions of oncogenic RAS and the p53 tumor suppressor. PLoS Biol 2008;6:2853–68.

[52] Kong X, Feng D, Mathews S, Gao B. Hepatoprotective and anti-fibrotic functions of interleukin-22: therapeutic potential for the treatment of alcoholic liver disease. J Gastroenterol Hepatol 2013;28(Suppl. 1):56–60.

[53] Jun JI, Lau LF. Taking aim at the extracellular matrix: CCN proteins as emerging therapeutic targets. Nat Rev Drug Discov 2011;10:945–63.

[54] Kim KH, Chen CC, Monzon RI, Lau LF. Matricellular protein CCN1 promotes regression of liver fibrosis through induction of cellular senescence in hepatic myofibroblasts. Mol Cell Biol 2013;33:2078–90.

[55] Ahsan MK, Mehal WZ. Activation of adenosine receptor A2A increases HSC proliferation and inhibits its death and senescence by down-regulation of p53 and Rb. Front Pharmacol 2014;5:69.

[56] Kuilman T, Peeper DS. Senescence-messaging secretome: SMS-ing cellular stress. Nat Rev Cancer 2009;9:81–94.
[57] Park EJ, Lee JH, Yu GY, He G, Ali SR, Holzer RG, et al. Dietary and genetic obesity promote liver inflammation and tumorigenesis by enhancing IL-6 and TNF expression. Cell 2010;140:197–208.
[58] Seki E. Microbiome-obesity-liver cancer interaction: senescence of hepatic stellate cells and bile acids play new roles. Gastroenterology 2014;146:860–1.
[59] Naugler WE, Sakurai T, Kim S, Maeda S, Kim K, Elsharkawy AM, et al. Gender disparity in liver cancer due to sex differences in MyD88-dependent IL-6 production. Science 2007;317:121–4.
[60] Lujambio A, Akkari L, Simon J, Grace D, Tschaharganeh DF, Bolden JE, et al. Non-cell-autonomous tumor suppression by p53. Cell 2013;153:449–60.
[61] Lumsden AB, Henderson JM, Kutner MH. Endotoxin levels measured by a chromogenic assay in portal, hepatic and peripheral venous blood in patients with cirrhosis. Hepatology 1988;8:232–6.
[62] Yu S, Nakafusa Y, Flye MW. Portal vein administration of donor cells promotes peripheral allospecific hyporesponsiveness and graft tolerance. Surgery 1994;116:229–34.
[63] Thomson AW, Knolle PA. Antigen-presenting cell function in the tolerogenic liver environment. Nat Rev Immunol 2010;10:753–66.
[64] Shoukry NH, Cawthon AG, Walker CM. Cell-mediated immunity and the outcome of hepatitis C virus infection. Annu Rev Microbiol 2004;58:391–424.
[65] Thimme R, Lohmann V, Weber F. A target on the move: innate and adaptive immune escape strategies of hepatitis C virus. Antiviral Res 2006;69:129–41.
[66] Sakaguchi S, Wing K, Miyara M. Regulatory T cells—a brief history and perspective. Eur J Immunol 2007;37(Suppl. 1):S116–23.
[67] He F, Chen Z, Xu S, Cai M, Wu M, Li H, et al. Increased $CD4^{+}CD25^{+}Foxp3^{+}$ regulatory T cells in tolerance induced by portal venous injection. Surgery 2009;145:663–74.
[68] Zhu J, Yamane H, Paul WE. Differentiation of effector CD4 T cell populations (*). Annu Rev Immunol 2010;28:445–89.
[69] Chen W, Jin W, Hardegen N, Lei KJ, Li L, Marinos N, et al. Conversion of peripheral $CD4^{+}CD25^{-}$ naive T cells to $CD4^{+}CD25^{+}$ regulatory T cells by TGF-beta induction of transcription factor Foxp3. J Exp Med 2003;198:1875–86.
[70] Mangan PR, Harrington LE, O'Quinn DB, Helms WS, Bullard DC, Elson CO, et al. Transforming growth factor-beta induces development of the T(H)17 lineage. Nature 2006;441:231–4.
[71] Bettelli E, Carrier Y, Gao W, Korn T, Strom TB, Oukka M, et al. Reciprocal developmental pathways for the generation of pathogenic effector TH17 and regulatory T cells. Nature 2006;441:235–8.
[72] Mucida D, Cheroutre H. TGFbeta and retinoic acid intersect in immune-regulation. Cell Adh Migr 2007;1:142–4.
[73] Mucida D, Park Y, Kim G, Turovskaya O, Scott I, Kronenberg M, et al. Reciprocal TH17 and regulatory T cell differentiation mediated by retinoic acid. Science 2007;317:256–60.
[74] Winau F, Hegasy G, Weiskirchen R, Weber S, Cassan C, Sieling PA, et al. Ito cells are liver-resident antigen-presenting cells for activating T cell responses. Immunity 2007;26:117–29.
[75] Eksteen B, Mora JR, Haughton EL, Henderson NC, Lee-Turner L, Villablanca EJ, et al. Gut homing receptors on CD8 T cells are retinoic acid dependent and not maintained by liver dendritic or stellate cells. Gastroenterology 2009;137:320–9.
[76] Ichikawa S, Mucida D, Tyznik AJ, Kronenberg M, Cheroutre H. Hepatic stellate cells function as regulatory bystanders. J Immunol 2011;186:5549–55.
[77] Dunham RM, Thapa M, Velazquez VM, Elrod EJ, Denning TL, Pulendran B, et al. Hepatic stellate cells preferentially induce $Foxp3^{+}$ regulatory T cells by production of retinoic acid. J Immunol 2013;190:2009–16.
[78] Gudas LJ. Emerging roles for retinoids in regeneration and differentiation in normal and disease states. Biochim Biophys Acta 2012;1821:213–21.
[79] Noy N. Between death and survival: retinoic acid in regulation of apoptosis. Annu Rev Nutr 2010;30:201–17.
[80] Balmer JE, Blomhoff R. Gene expression regulation by retinoic acid. J Lipid Res 2002;43:1773–808.
[81] Shmarakov IO, Jiang H, Yang KJ, Goldberg IJ, Blaner WS. Hepatic retinoid stores are required for normal liver regeneration. J Lipid Res 2013;54:893–908.

[82] Chawla A, Repa JJ, Evans RM, Mangelsdorf DJ. Nuclear receptors and lipid physiology: opening the X-files. Science 2001;294:1866–70.

[83] Duong V, Rochette-Egly C. The molecular physiology of nuclear retinoic acid receptors. From health to disease. Biochim Biophys Acta 2011;1812:1023–31.

[84] Blaner WS, O'Byrne SM, Wongsiriroj N, Kluwe J, D'Ambrosio DM, Jiang H, et al. Hepatic stellate cell lipid droplets: a specialized lipid droplet for retinoid storage. Biochim Biophys Acta 2009;1791:467–73.

[85] Goodman DW, Huang HS, Shiratori T. Tissue distribution and metabolism of newly absorbed vitamin A in the rat. J Lipid Res 1965;6:390–6.

[86] O'Byrne SM, Wongsiriroj N, Libien J, Vogel S, Goldberg IJ, Baehr W, et al. Retinoid absorption and storage is impaired in mice lacking lecithin: retinol acyltransferase (LRAT). J Biol Chem 2005;280:35647–57.

[87] Kluwe J, Wongsiriroj N, Troeger JS, Gwak GY, Dapito DH, Pradere JP, et al. Absence of hepatic stellate cell retinoid lipid droplets does not enhance hepatic fibrosis but decreases hepatic carcinogenesis. Gut 2011;60:1260–8.

[88] Shirakami Y, Gottesman ME, Blaner WS. Diethylnitrosamine-induced hepatocarcinogenesis is suppressed in lecithin: retinol acyltransferase-deficient mice primarily through retinoid actions immediately after carcinogen administration. Carcinogenesis 2012;33:268–74.

[89] Tang XH, Gudas LJ. Retinoids, retinoic acid receptors, and cancer. Annu Rev Pathol 2011;6:345–64.

[90] Liu HX, Ly I, Hu Y, Wan YJ. Retinoic acid regulates cell cycle genes and accelerates normal mouse liver regeneration. Biochem Pharmacol 2014;91:256–65.

[91] Kimura M, Watanabe M, Ishibashi N, Yanagida S, Ogihara M. Acyclic retinoid NIK-333 accelerates liver regeneration and lowers serum transaminase activities in 70% partially hepatectomized rats, *in vivo*. Eur J Pharmacol 2010;643:267–73.

[92] Chung C, Iwakiri Y. Activated hepatic stellate cells: negative regulators of hepatocyte proliferation in liver diseases. Hepatology 2012;56:389–91.

[93] Ebrahimkhani MR, Oakley F, Murphy LB, Mann J, Moles A, Perugorria MJ, et al. Stimulating healthy tissue regeneration by targeting the 5-HT(2)B receptor in chronic liver disease. Nat Med 2011;17:1668–73.

[94] Seuwen K, Pouyssegur J. Serotonin as a growth factor. Biochem Pharmacol 1990;39:985–90.

[95] Dudas J, Mansuroglu T, Batusic D, Saile B, Ramadori G. Thy-1 is an *in vivo* and *in vitro* marker of liver myofibroblasts. Cell Tissue Res 2007;329:503–14.

[96] Yovchev MI, Zhang J, Neufeld DS, Grozdanov PN, Dabeva MD. Thymus cell antigen-1-expressing cells in the oval cell compartment. Hepatology 2009;50:601–11.

[97] Dranoff JA, Wells RG. Portal fibroblasts: underappreciated mediators of biliary fibrosis. Hepatology 2010;51:1438–44.

[98] Desmouliere A, Darby I, Costa AM, Raccurt M, Tuchweber B, Sommer P, et al. Extracellular matrix deposition, lysyl oxidase expression, and myofibroblastic differentiation during the initial stages of cholestatic fibrosis in the rat. Lab Invest 1997;76:765–78.

[99] Yata Y, Scanga A, Gillan A, Yang L, Reif S, Breindl M, et al. DNase I-hypersensitive sites enhance alpha1(I) collagen gene expression in hepatic stellate cells. Hepatology 2003;37:267–76.

[100] Jinnin M, Ihn H, Asano Y, Yamane K, Trojanowska M, Tamaki K. Upregulation of tenascin-C expression by IL-13 in human dermal fibroblasts via the phosphoinositide 3-kinase/Akt and the protein kinase C signaling pathways. J Invest Dermatol 2006;126:551–60.

[101] McHedlidze T, Waldner M, Zopf S, Walker J, Rankin AL, Schuchmann M, et al. Interleukin-33-dependent innate lymphoid cells mediate hepatic fibrosis. Immunity 2013;39:357–71.

[102] Fickert P, Fuchsbichler A, Moustafa T, Wagner M, Zollner G, Halilbasic E, et al. Farnesoid X receptor critically determines the fibrotic response in mice but is expressed to a low extent in human hepatic stellate cells and periductal myofibroblasts. Am J Pathol 2009;175:2392–405.

[103] Wen JW, Olsen AL, Perepelyuk M, Wells RG. Isolation of rat portal fibroblasts by in situ liver perfusion. J Vis Exp 2012;64:3669.

[104] Uchio K, Tuchweber B, Manabe N, Gabbiani G, Rosenbaum J, Desmouliere A. Cellular retinol-binding protein-1 expression and modulation during *in vivo* and *in vitro* myofibroblastic differentiation of rat hepatic stellate cells and portal fibroblasts. Lab Invest 2002;82:619–28.

[105] Kruglov EA, Jain D, Dranoff JA. Isolation of primary rat liver fibroblasts. J Investig Med 2002;50:179–84.
[106] Knittel T, Kobold D, Saile B, Grundmann A, Neubauer K, Piscaglia F, et al. Rat liver myofibroblasts and hepatic stellate cells: different cell populations of the fibroblast lineage with fibrogenic potential. Gastroenterology 1999;117:1205–21.
[107] Goodpaster T, Legesse-Miller A, Hameed MR, Aisner SC, Randolph-Habecker J, Coller HA. An immunohistochemical method for identifying fibroblasts in formalin-fixed, paraffin-embedded tissue. J Histochem Cytochem 2008;56:347–58.
[108] Dranoff JA, Kruglov EA, Robson SC, Braun N, Zimmermann H, Sevigny J. The ecto-nucleoside triphosphate diphosphohydrolase NTPDase2/CD39L1 is expressed in a novel functional compartment within the liver. Hepatology 2002;36:1135–44.
[109] Bosselut N, Housset C, Marcelo P, Rey C, Burmester T, Vinh J, et al. Distinct proteomic features of two fibrogenic liver cell populations: hepatic stellate cells and portal myofibroblasts. Proteomics 2010;10:1017–28.
[110] Fausther M, Dranoff JA. New insights on the pathogenesis of biliary cirrhosis provided by studies in FXR knockout mice. J Hepatol 2011;55:939–40.
[111] Magness ST, Bataller R, Yang L, Brenner DA. A dual reporter gene transgenic mouse demonstrates heterogeneity in hepatic fibrogenic cell populations. Hepatology 2004;40:1151–9.
[112] Magness ST, Brenner DA. Targeted disruption of the mouse ferrochelatase gene producing an exon 10 deletion. Biochim Biophys Acta 1999;1453:161–74.
[113] Bataller R, Schwabe RF, Choi YH, Yang L, Paik YH, Lindquist J, et al. NADPH oxidase signal transduces angiotensin II in hepatic stellate cells and is critical in hepatic fibrosis. J Clin Invest 2003;112:1383–94.
[114] Kawada N, Kristensen DB, Asahina K, Nakatani K, Minamiyama Y, Seki S, et al. Characterization of a stellate cell activation-associated protein (STAP) with peroxidase activity found in rat hepatic stellate cells. J Biol Chem 2001;276:25318–23.
[115] Forbes SJ, Parola M. Liver fibrogenic cells. Best Pract Res Clin Gastroenterol 2011;25:207–17.
[116] Onitsuka I, Tanaka M, Miyajima A. Characterization and functional analyses of hepatic mesothelial cells in mouse liver development. Gastroenterology 2010;138:1525–35. 1535 e1521–6.
[117] Bochmann L, Sarathchandra P, Mori F, Lara-Pezzi E, Lazzaro D, Rosenthal N. Revealing new mouse epicardial cell markers through transcriptomics. PLoS One 2010;5:e11429.
[118] Asahina K. Hepatic stellate cell progenitor cells. J Gastroenterol Hepatol 2012;27(Suppl. 2):80–4.
[119] Asahina K, Tsai SY, Li P, Ishii M, Maxson Jr. RE, Sucov HM, et al. Mesenchymal origin of hepatic stellate cells, submesothelial cells, and perivascular mesenchymal cells during mouse liver development. Hepatology 2009;49:998–1011.
[120] Rinkevich Y, Mori T, Sahoo D, Xu PX, Bermingham Jr. JR, Weissman IL. Identification and prospective isolation of a mesothelial precursor lineage giving rise to smooth muscle cells and fibroblasts for mammalian internal organs, and their vasculature. Nat Cell Biol 2012;14:1251–60.
[121] Liu C, Gaca MD, Swenson ES, Vellucci VF, Reiss M, Wells RG. Smads 2 and 3 are differentially activated by transforming growth factor-beta (TGF-beta) in quiescent and activated hepatic stellate cells. Constitutive nuclear localization of Smads in activated cells is TGF-beta-independent. J Biol Chem 2003;278:11721–8.
[122] Wells RG, Kruglov E, Dranoff JA. Autocrine release of TGF-beta by portal fibroblasts regulates cell growth. FEBS Lett 2004;559:107–10.
[123] Ji J, Yu F, Ji Q, Li Z, Wang K, Zhang J, et al. Comparative proteomic analysis of rat hepatic stellate cell activation: a comprehensive view and suppressed immune response. Hepatology 2012;56:332–49.
[124] Fausther M, Sheung N, Saiman Y, Bansal MB, Dranoff JA. Activated hepatic stellate cells upregulate transcription of ecto-5′-nucleotidase/CD73 via specific SP1 and SMAD promoter elements. Am J Physiol Gastrointest Liver Physiol 2012;303:G904–14.
[125] Li Z, Dranoff JA, Chan EP, Uemura M, Sevigny J, Wells RG. Transforming growth factor-beta and substrate stiffness regulate portal fibroblast activation in culture. Hepatology 2007;46:1246–56.
[126] Meurer SK, Alsamman M, Sahin H, Wasmuth HE, Kisseleva T, Brenner DA, et al. Overexpression of endoglin modulates TGF-beta1-signalling pathways in a novel immortalized mouse hepatic stellate cell line. PLoS One 2013;8:e56116.

[127] Clouzeau-Girard H, Guyot C, Combe C, Moronvalle-Halley V, Housset C, Lamireau T, et al. Effects of bile acids on biliary epithelial cell proliferation and portal fibroblast activation using rat liver slices. Lab Invest 2006;86:275–85.

[128] Machado MV, Diehl AM. Liver renewal: detecting misrepair and optimizing regeneration. Mayo Clin Proc 2014;89:120–30.

[129] Kleaveland KR, Moore BB, Kim KK. Paracrine functions of fibrocytes to promote lung fibrosis. Expert Rev Respir Med 2014;8:163–72.

[130] Fort MM, Cheung J, Yen D, Li J, Zurawski SM, Lo S, et al. IL-25 induces IL-4, IL-5, and IL-13 and Th2-associated pathologies *in vivo*. Immunity 2001;15:985–95.

[131] Gregory LG, Jones CP, Walker SA, Sawant D, Gowers KH, Campbell GA, et al. IL-25 drives remodelling in allergic airways disease induced by house dust mite. Thorax 2012;68(1):82–90.

[132] Wynn TA. IL-13 effector functions. Annu Rev Immunol 2003;21:425–56.

[133] Liu Y, Meyer C, Muller A, Herweck F, Li Q, Mullenbach R, et al. IL-13 induces connective tissue growth factor in rat hepatic stellate cells via TGF-beta-independent Smad signaling. J Immunol 2011;187:2814–23.

[134] Zavadil J, Bitzer M, Liang D, Yang YC, Massimi A, Kneitz S, et al. Genetic programs of epithelial cell plasticity directed by transforming growth factor-beta. Proc Natl Acad Sci USA 2001;98:6686–91.

[135] Zeisberg M, Neilson EG. Biomarkers for epithelial-mesenchymal transitions. J Clin Invest 2009;119:1429–37.

[136] Iwano M, Plieth D, Danoff TM, Xue C, Okada H, Neilson EG. Evidence that fibroblasts derive from epithelium during tissue fibrosis. J Clin Invest 2002;110:341–50.

[137] Kalluri R, Weinberg RA. The basics of epithelial–mesenchymal transition. J Clin Invest 2009;119:1420–8.

[138] Nitta T, Kim JS, Mohuczy D, Behrns KE. Murine cirrhosis induces hepatocyte epithelial mesenchymal transition and alterations in survival signaling pathways. Hepatology 2008;48:909–19.

[139] Zeisberg M, Yang C, Martino M, Duncan MB, Rieder F, Tanjore H, et al. Fibroblasts derive from hepatocytes in liver fibrosis via epithelial to mesenchymal transition. J Biol Chem 2007;282:23337–47.

[140] Omenetti A, Porrello A, Jung Y, Yang L, Popov Y, Choi SS, et al. Hedgehog signaling regulates epithelial–mesenchymal transition during biliary fibrosis in rodents and humans. J Clin Invest 2008;118:3331–42.

[141] Syn WK, Jung Y, Omenetti A, Abdelmalek M, Guy CD, Yang L, et al. Hedgehog-mediated epithelial-to-mesenchymal transition and fibrogenic repair in nonalcoholic fatty liver disease. Gastroenterology 2009;137(4):1478–88.

[142] Seki E, de Minicis S, Inokuchi S, Taura K, Miyai K, van Rooijen N, et al. CCR2 promotes hepatic fibrosis in mice. Hepatology 2009;50:185–97.

[143] Omenetti A, Yang L, Li YX, McCall SJ, Jung Y, Sicklick JK, et al. Hedgehog-mediated mesenchymal–epithelial interactions modulate hepatic response to bile duct ligation. Lab Invest 2007;87:499–514.

[144] Chu AS, Diaz R, Hui JJ, Yanger K, Zong Y, Alpini G, Stanger BZ, et al. Lineage tracing demonstrates no evidence of cholangiocyte epithelial-to-mesenchymal transition in murine models of hepatic fibrosis. Hepatology 2011;53:1685–95.

CHAPTER 7

Matrix Metalloproteinases and Their Inhibitors

Lara Campana and John Iredale
MRC/UoE Centre for Inflammation Research Queen's Medical Research Institute, Edinburgh, UK

7.1 INTRODUCTION

7.1.1 Chronic liver diseases

Liver fibrosis represents the wound-healing response of the liver to diverse chronic hepatic insults such as viral infections, autoimmune diseases, genetic diseases, and toxic damage (e.g., alcohol). Because of the high worldwide prevalence of these insults, liver fibrosis is common and associated with high morbidity and mortality. The liver fibrosis process recapitulates the scarring process in other anatomical sites and can be considered a paradigm for the damage and healing process in the body [1]. The result of end-stage fibrosis is cirrhosis [2], a pathological state characterized by architectural disruption with bands of mature fibrosis linking vascular structures and bounding regenerative nodules. Cirrhosis is associated with severe and life-threatening complications including portal hypertension and hepatocellular cancer. The available therapeutic repertoire directed at fibrosis is currently limited. Indeed, our two main approaches to fibrosis entail: (i) removal of underlying injurious stimulus and (ii) liver transplantation. For severe end-stage disease (in effect advanced cirrhosis) transplantation is the only option, although the application of this is limited by several factors. Indeed there is a shortage of available liver donors, recipients are committed to lifelong treatment with anti-rejection immunosuppressant drugs, and finally, exclusion criteria affect transplant eligibility, such as significant medical or psychiatric co-morbidity [3]. Therefore, there remains a significant therapeutic imperative to undertake detailed studies of the cellular and molecular mechanisms of liver fibrosis and cirrhosis in order to define potential therapeutic targets. For the last 20 years our laboratory has focused on the processes defining spontaneous resolution of liver fibrosis (a process critically dependent on matrix-degrading metalloproteinases) in order to define the attributes of an effective anti-fibrotic, or more accurately, pro-resolution therapy.

7.1.2 Hepatic stellate cells and the fibrotic response

During liver injury the cellular and molecular signals regulating the fibrotic response converge on an effector cell, the hepatic stellate cell (HSC). As a result of an insult,

Stellate Cells in Health and Disease
DOI: http://dx.doi.org/10.1016/B978-0-12-800134-9.00007-5

HSCs undergo phenotypic and functional changes to a myofibroblast-like phenotype in which they produce extracellular matrix (ECM) components such as the collagen I and collagen III [4] that constitute the major cellular components of the fibrotic response. In the steady-state healthy condition, HSCs are perisinusoidal vitamin–A-storing cells of the liver. During liver injury HSCs are activated from this quiescent to the so-called activated phenotype. Activated HSCs are characterized by proliferative activity, contractility, increased synthesis, and secretion of ECM and altered matrix-degrading activity (described in detail in Sections 7.2 and 7.3) [5,6]. Activated HSCs, which are indistinguishable from myofibroblasts, are also able to secrete soluble mediators typical of pro-fibrogenic and pro-inflammatory cells (e.g., interleukin-6, IL-6). Moreover, they play a role in the re-vascularization of the damaged organ by secreting pro-angiogenic factors (e.g., vascular endothelial growth factor, VEGF) [7]. If the injurious stimulus is not withdrawn, a vicious cycle of damage and remodeling with increasing dominance of fibrillar collagens is established. Ultimately the repetitive deposition of fibrillar matrix leads to fibrosis within the parenchyma. Fibrotic septa isolate dividing hepatocytes within bundles of ECM components, resulting in the formation of nodules—irregular areas of regeneration that lead to disruption of the regular histological appearance of the liver and consequently to advanced fibrosis and cirrhosis [8–10].

7.1.3 ECM changes in liver fibrosis

The ECM in chronic liver fibrosis is different, both quantitatively and qualitatively, from the one present in health. The physiological ECM is mainly composed of collagens I and III, present in equivalent proportions in the capsule and the perivascular regions (the major blood vessels and the portal tract). Within the parenchyma and away from the vascular tracts the perisinusoidal ECM is non-electron dense and rich in type IV collagen and laminin. During the development of fibrosis, in a whole organ, the expression of type III collagen initially dominates but ultimately type 1, and secondarily type III collagens increase with collagen being the dominant fibrillar matrix component. Other collagens, including collagen V, become expressed. As the matrix load of the liver increases, other matrix components that form a rich substrate for intermolecular crosslinking, including elastin, also become laid down [4,6,11].

Given the position of the HSCs in the space of Disse, it is perhaps not surprising that the fibrillar matrix is initially laid down within the space of Disse and in perisinusoidal distribution. The presence of fibrillar collagen between the sinusoidal endothelium and the hepatocytes, together with products expressed by the incoming extra-vasated inflammatory cells and activated stellate cells, leads to a change in cell matrix interactions with the loss of endothelial fenestrae—a process termed "sinusoidal capillarization"; and in turn a loss of the normal hepatocyte brush border occurs [4,6,12]. Cell culture experiments demonstrate that whereas hepatocytes cultured on a

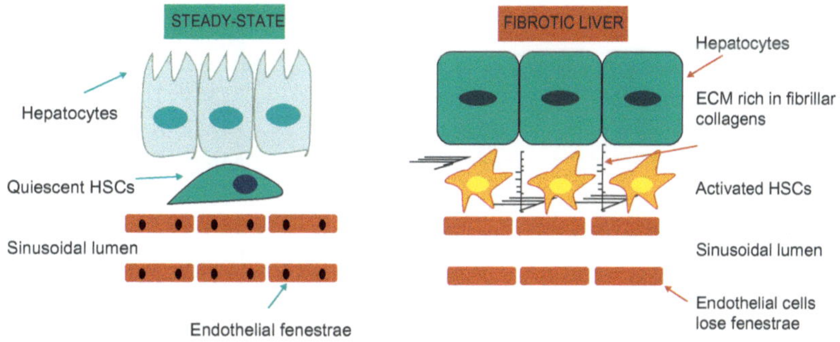

Figure 7.1 In steady state conditions HSCs located in the space of Disse lie in close association with an ECM consisting of collagen IV and other components. The quiescent HSC is filled with fat droplets and retinyl esters. The hepatocytes have a normal morphology with a brush border and the endothelial cells have fenestrae that allow the exchange of nutrients and metabolites. In the fibrotic liver HSCs become activated to a myofibroblast-like phenotype. They proliferate and synthesize the interstitial collagens and other ECM molecules that accumulate in the space of Disse. Hepatocytes lose their brush border and endothelial cells lose their fenestrae (sinusoidal capillarization). *Adapted from Ref. [15].*

matrix rich in type IV collagen retain their gene specific function and structure, those on type I collagen rapidly lose defining features of hepatocyte function including albumin secretion [13,14]. This process of sinusoidal capillarization impairs the ability of the liver to exchange substances with the general circulation and diminishes metabolic function of both the diseased and remaining healthy hepatocytes [1]. General concepts about HSC activation are summarized in Figure 7.1. Ultimately the secreted ECM will bridge vascular structures including the portal and central veins leading to the architectural disruption that culminates in cirrhosis as described earlier [3,16].

7.2 THE METALLOPROTEASES (MMPs)

ECM degradation is mediated by an extensive family of calcium- and zinc-containing endopeptidases that degrade panoply of ECM proteins. MMP nomenclature is complex; MMPs are numbered (largely in order of discovery) but are separately grouped functionally according to substrate specificity (a more useful classification) [15]. Metalloproteinases of specific relevance to the development of liver disease include those expressed early in injury, associated with inflammation and likely to be crucial to degrading the normal hepatic ECM, consisting of type IV collagen, laminin (e.g., the gelatinases, MMP-2 and 9) stromelysins and matrilysins (e.g., MMP-3 and 7), and those with specific activity against components of the mature matrix of fibrosis and cirrhosis, which are necessary for the degradation of fibrillar fibrotic tissue. These include the true interstitial collagenases (e.g., MMPs 1, 8, and 13), some of the

Table 7.1 Classification of MMPs based on target protein

Class	MMP number
Collagenases	MMP-1
	MMP-8
	MMP-13
	MMP-18
Gelatinases	MMP-2
	MMP-9
Elastase	MMP-12
Stromelysins	MMP-3
	MMP-10
	MMP-11
	MMP-27
Matrylisins	MMP-7
	MMP-26
Membrane type (MT-MMPs)	MMP-14
	MMP-15
	MMP-16
	MMP-24

membrane type metalloproteinases, and with respect to other mature matrix components, the elastases (e.g., MMP-12). Representative examples of this enzyme family are listed in Table 7.1 [15,17–19].

7.2.1 Matrix-degrading metalloproteinases structure biochemistry and function

7.2.1.1 Substrate profile

Collagenases are central to the process of remodeling of fibrotic tissue as they cleave to the native helix of fibrillar collagen I, II, and III to render it degradable by other MMPs (see Section 7.2.1.2). Because of the predominance of collagen type I and III in fibrotic liver their expression is necessary to start the matrix degradation process. The stromelysins have a promiscuous substrate profile; they digest collagens II, IV, IX, X, XI, denaturated collagens (gelatin), laminin, and fibronectin. They also have a physiological function in activating pro-MMPs into fully functional MMPs. The gelatinases are the third group of MMPs. MMP-2 (or gelatinase A) targets collagens IV, V, VII, X, XI, and elastin; MMP-9 (or gelatinase B) shares the same substrate profile but has not been conclusively demonstrated to be functional against interstitial collagens. However, gelatinase B is able to form synergistic dimers with other interstitial MMPs. Metalloelastases are active against elastin and encompass MMP-12 or macrophage metalloelastase. The most recent added group is the membrane-type MMPs (MT-MMPs) functional in activating gelatinases [15,17–19].

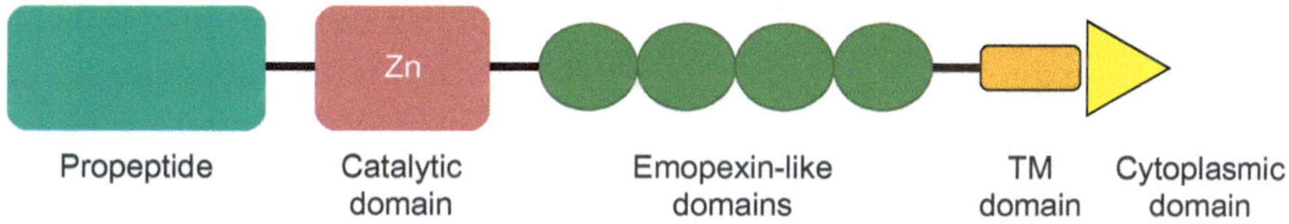

Figure 7.2 MMPs general structure. MMPs are composed by a propeptide that is eventually cleaved to enable the activation of the enzyme. The catalytic Zn-containing domain is shielded in the inactive form of the enzyme. Both the catalytic domain and the hemopexin-like domain interact with specific domains in the target protein of the enzyme. The trans-membrane (TM) and the cytoplasmic domain are present only in membrane-type MMPs. *Adapted from Refs. [17,18].*

7.2.1.2 MMP structure

The basic structure of MMPs is reported in Figure 7.2. In sequence from the N terminus, MMPs consist of a pro-domain, cleaved to give full functionality to the enzyme; there is then a catalytic domain containing zinc [20] and the sequence HEXGHXXGXXHS. The catalytic domain is rendered invisible to potential MMPs substrates by the pro-domain. Most MMPs show several hemopexin-like domains adjacent to the catalytic one. Those domains are of particular importance because they allow each MMP to interact with specific tissue inhibitors (tissue inhibitors of metalloproteinases, TIMPs) described in detail later. TIMPs recognize their specific MMP through a double interaction, one established with the catalytic domain and one with the hemopexin-like modular structure. A specific sub-family of MMPs is composed of MT-MMPs. MT-MMPs possess a trans-membrane and a cytoplasmic domain at their C-terminal end [18,21].

7.2.1.3 Potential cellular sources of MMPs in the liver

There are subtle differences in the expression of MMPs comparing rodents (in which the majority of models have been undertaken) and human disease examples. Most importantly, rodents express MMP13 as their primary interstitial collagenase whereas in humans this is MMP-1 [15,21]. Candidate collagenases within the liver include neutrophil collagenase, MMP-8, which is expressed by both neutrophils and macrophages. These populations are well represented during hepatic inflammation, particularly early in injury or as a consequence of iterative chronic injury [2,16,22,23]. Interstitial collagenase of MMP-1 has been described in inflammatory cells in human liver and we and others have demonstrated expression of MMP-13 by inflammatory cells (particularly macrophages) and also transiently in HSCs in rodent models. Within HSCs, MMP-13 expression is a feature of early activation but downregulation in MMP-13 expression occurs with full activation at a time when cells are expressing interstitial collagens and TIMP-1 particularly [24,25].

Macrophage expression of MMP-13 appears relatively constant and unlike gelatinase and elastases expression (described in detail in Section 7.3) does not seem to be altered by macrophage phenotype [26,27]. Stromelysins are expressed by HSCs transiently during the early phases of activation and have been shown to be expressed by inflammatory cells [28–30]. Both stromelysins and gelatinases A and B (MMP-2 and 9, respectively) [31] have promiscuous activity with respect to collagens, but are potent gelatinases and would therefore be expected to degrade collagens partially denatured by the action of interstitial collagenases. Gelatinase A (MMP-2) has been demonstrated to be expressed by HSCs and gelatinase B (MMP-9) by Kupffer cells and inflammatory macrophages. Both gelatinases have been described to act as shedases and to regulate cytokine and soluble signal activation and deactivation suggesting a further possible role in the early stages of liver injury [27,32–36].

7.2.1.4 Activation and regulation

Perhaps not surprisingly for an enzyme family with such degradative potential against structural proteins, MMP activity is tightly controlled at three levels. These are as follows: transcriptional activation; activation of the latent pro-enzyme by cleavage of the protease; and, the focus of this chapter, extracellular inhibition by specific TIMPs (or other more general protease inhibitors such as α2 macroglobulin) [15,17].

7.2.1.4.1 Transcriptional regulation

All MMPs are regulated to a greater or lesser extent at the level of the gene. This results from stimulation by a range of soluble signals encompassing growth factors and cytokines present in the inflamed liver including, tumor necrosis factor-alpha (TNFα), epidermal growth factor (EGF), platelet derived growth factor (PDGF), basic fibroblast growth factor (BFGF), and IL-1. Cytokines such as TGFβ1 found in high concentration in liver injury and fibrosis may differentially affect MMP expression downregulating interstitial collagenase expression while upregulating the expression of gelatinase A, TIMP-I, and collagen I [15,18,37].

7.2.1.4.2 Pro-enzyme activation

Metalloproteinases are secreted as inactive zymogens with a structural propiece; the activation of these zymogens requires double proteolytic cleavage of the pro-domain at the N-terminal of the MMP. Initial MMP cleavage, for example, by plasmin or other serine proteases, releases a small peptide and unveils a new sequence for the second cleavage to occur. A second small peptide is subsequently removed and the enzyme is fully activated with the catalytic site free to interact with the targets. Discussed in the next section, MMP activity is regulated by interactions with members of the TIMP family. Remarkably, TIMPs may also play a role in the activation process of MMPs by binding to the hemopexin-like domains of adjacent MMPs thus favoring the reciprocal activation. This interaction is of particular importance in the case of a soluble MMP activated by a MT-MMP. In this case, the soluble MMP is brought on the membrane of the cell and this localization can favor the cell migration through the newly degraded ECM [15,18,21,37].

7.2.2 TIMPs, structure, regulation, and function

In addition to overexuberant synthesis of fibrillar (collagens I and III) matrix and proliferation, HSCs have also been demonstrated to express the potent inhibitors of matrix-degrading metalloproteinases: the TIMPs 1 and 2. Early work studying the expression and secretion of TIMP-1 by HSCs demonstrated a powerful MMP inhibitory activity and led directly to the hypothesis that the matrix accumulation that accompanies fibrosis and cirrhosis is not simply the result of excessive matrix secretion, but a failure of matrix degradation mediated by TIMP 1 [24]. This hypothesis has withstood a series of mechanistic experiments, described later in the chapter. Moreover, the model that matrix accumulates in part as the result of a failure of matrix degradation would predict that liver fibrosis is reversible, given current and previous evidence that HSCs and other incoming inflammatory cells express matrix-degrading metalloproteinases [2,3,15,38]. These concepts are discussed in greater detail in Section 7.3.

7.2.2.1 Classification and target definition

Thus far, four TIMPs have been described (TIMPs 1–4). Of these there is strong evidence for expression of TIMP-1, 2, and 3 in the liver. TIMP 1 and 2 are expressed by activated HSCs and TIMP-2 is expressed by both activated HSCs, and to a lesser extent by quiescent HSCs within the normal liver. TIMP-3 is associated with ECM, and is frequently matrix-bound [39,40].

All known MMPs can be inhibited by at least one of the four known TIMPs [21]. Each TIMP is a separate gene product although they share common structural features. TIMP-1 and TIMP-2 display a 40% amino acid sequence homology and both show a three-looped structure stabilized by six disulfide bonds. TIMP-1 and TIMP-2 are both secreted in the intracellular milieu, while TIMP-3 has been associated with the ECM. The binding of TIMPs to MMPs is stoichiometric and effectively irreversible under physiological conditions with the TIMP physically blocking the active site of the MMP. The binding is, however, non-covalent and can be disrupted, with the TIMP retaining its inhibitory activity. The N-terminal domain is sufficient and necessary to inhibition of MMPs, while the C-term fraction of the TIMPs is necessary to bind the pro-MMP [15,41–43].

7.2.2.2 Activation and regulation

Transcriptional regulation of the TIMPs is mediated by cytokines and growth factors such as TGFβ (see Section 7.2.1.4.1). In contrast to the differential regulation of

TIMP-1 and MMPs mediated by TGFβ, TNFα coordinately upregulates TIMP-1 and interstitial collagenases [42,44]. The TIMP promoter region contains regulatory motifs common among the distinct TIMPs and shared also by some MMPs. However, in each case the peculiar assembly of regulatory factors differs. For example, the AP-1 binding site is present in virtually all TIMPs and some MMPs, but it is combined differently with binding site for other transcription factors and it has a distinct configuration between humans and rodents. The combination of regulatory elements specific to each TIMP may explain the different pattern of expression observed in response to individual cytokines and cell messengers [42,45,46].

The TIMP protein family is characterized by redundancy, so one TIMP can inhibit more than one MMP. Within the extracellular milieu the resulting balance between TIMPs and MMPs is a major factor determining the net accumulation or degradation of ECM components [42,47,48].

7.3 RELEVANCE OF MMP ACTIVITY AND MATRIX DEGRADATION IN CHRONIC LIVER DISEASE

In the early stages of fibrotic injury, soluble factors produced by incoming inflammatory cells and damaged resident cells promote the transformation of HSCs to an activated fibrogenic phenotype. There is evidence that early activation of HSCs is associated with the expression of MMPs with degradative activity against the normal basement membrane-like matrix (rich in type IV collagen and laminin). These enzymes include gelatinase A (MMP-2) and stromelysin (MMP-3). Even after activation HSCs have been shown to secrete gelatinase A (MMP-2) and this activity has been linked to HSC proliferation [30,49,50].

Interestingly, disruption of the normal basement membrane itself and the physical effects of inflammation may in turn promote HSC activation. Early studies by Bissell, Arthur, and Friedman demonstrated that contact with a model basement membrane matrix in the form of Matrigel powerfully elicits a quiescent stellate cell phenotype [14,49–51]. It is therefore logical to suppose that degradation of the signals acquired from such a basement membrane-like matrix might themselves contribute to activation. Additionally, using models in which matrix stiffness can be varied, the Wells group has demonstrated that physical factors contribute to the activated HSC phenotype [52,53]. As inflammation is accompanied by edema, this in turn contributes to a rise in tissue turgor and stiffness. This compelling cell culture data suggest a model in which an increasingly stiff physical environment, even prior to the production of fibrillar collagens, might favor HSC activation.

7.3.1 Evidence for changes in the pattern of matrix degradation during liver fibrosis

The presence of inflammatory cells including macrophages and activated HSCs within the inflamed and fibrotic liver means that sources for a range of metalloproteinases, including collagenases and elastases, are present [1,23,38]. Clearly this is compatible with the hypothesis that in progressive fibrosis, matrix degradation is inhibited by co-expressed TIMP-1 (and 2). Thus one might predict that manipulation of TIMP-1 would influence the development or regression of fibrosis. In fact, there is a series of compelling experimental data supporting this assertion.

Early work in the 1970s identified that collagenase activity was detectable in fibrotic rat liver and decreased with duration of insult (carbon tetrachloride, CCl_4) [54,55]. It is entirely plausible that this observation was the result of enhanced levels of TIMP-1, which we and others have demonstrated is expressed by activated HSCs with increasing duration of experimental injury [1,15]. Moreover, TIMP-1 levels are increased in experimental fibrotic injury regardless of the insult used. Clear evidence for the potential of matrix degradation in even comparatively advanced fibrosis comes from experimental studies of rodent models and the deployment of CCl_4 or bile duct ligation. Using CCl_4 induced injury, a period of fibrosis induction can be followed by spontaneous recovery. A parallel model can be established in bile duct ligation where after bile duct ligation induced fibrosis bilio-duodenal anastomosis is undertaken. In both of these models, despite the presence of bridging fibrosis linking vascular structures, it can be readily demonstrated that cessation of injury will result in dramatic remodeling with matrix degradation and a return of near normal architecture [15,56]. Detailed studies of the tissue during this process indicate that with cessation of injury, HSC apoptosis or HSC phenotypic reversion to quiescence occurs. Both are associated with a reduction in hepatic TIMP levels. During this period of time expression of collagenase (MMP-13), gelatinase A and B (MMP-2 and 9), and elastase (MMP-12) continues and there is a net increase in hepatic MMP activity that coincides with the remodeling of the matrix [12,57].

Significantly, a series of mechanistic studies supports the crucial role for expressed TIMPs 1 and 2 in defining the reversibility of fibrosis. Through a variety of elegant approaches, several research groups have shown that by deregulating the TIMP/MMP balance in fibrosis, net matrix degradation can be achieved. These approaches have included the deployment of an enzymatically inactive decoy to act as a sump for TIMP [58], the deployment of excess collagenase via an adenoviral vector delivering neutrophil collagenase (MMP 8) [59], and the inhibition of TIMP 1 using an antibody mediated approach [60]. Most crucially, the studies described by Yoshiji et al. using TIMP-1 overexpressing animals have clearly defined the role of matrix degradation in both progressive fibrosis and spontaneous resolution of fibrosis. Yoshiji et al's approach demonstrated that a transgenic TIMP-1 overexpressing mouse would accumulate matrix rather faster than its wild-type littermate in response to CCl_4 (demonstrating that some fibrillar matrix turnover occurs even in progressive fibrosis) but importantly, TIMP1 overexpressing animals fail to show any evidence of matrix remodeling during spontaneous resolution of fibrosis. These data underscore the crucial role played by TIMP in preventing MMP mediated matrix turnover [25,61]. A further interesting observation in the Yoshiji et al. study was that persistent fibrosis was associated with a persistence of activated myofibroblast-like

HSCs within the hepatic scar [25]. Clearly this might be related to crucial cell matrix interactions including the influence of a matrix with enhanced stiffness [52]. The link between persistently activated HSCs and collagen 1 was highlighted by studies of the rr collagen mouse, which expresses a mutated collagen 1 not susceptible to the key cleavage required to initiate degradation of fibrillar collagens [62]. Like the Yoshiji experiment, this study demonstrated that persistent collagen-I was associated with a failure of HSC apoptosis and HSC persistence in association with ECM. Discussion of the influence of matrix turnover on hepatocyte behavior is beyond the scope of this review, but it is important to note that in a wild-type mouse or rat, restitution of the normal hepatocyte mass rapidly follows withdrawal of injury and fibrillar matrix remodeling. In the rr collagen mouse both the hepatocellular proliferative and hepatic progenitor cell responses to withdrawal of injury are blunted and incomplete compared to those observed in wild-type livers [62]. A view of the dynamics of the HSC life cycle during liver fibrosis establishment and resolution is reported in Figure 7.3.

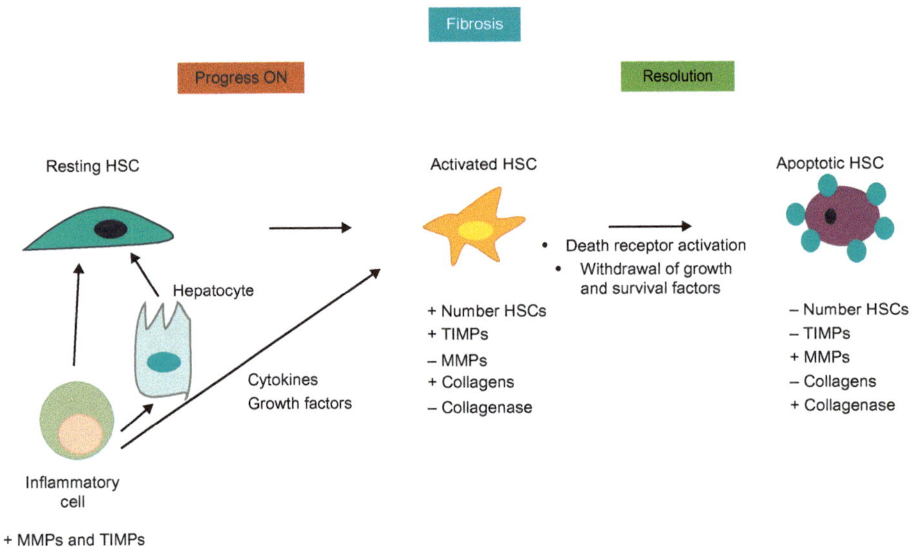

Figure 7.3 Hepatocytes and resident inflammatory cells (Kupffer cells) interact with resting HSCs. Upon damage, HSCs are activated by cytokines and growth factors produced by incoming inflammatory cells. Cytokines and growth factors cause HSC proliferation and survival. Activated HSCs produce TIMPs and ECM components therefore leading to fibrosis progression. Inflammatory cells at this stage produce mainly TIMPs, further favoring ECM deposition. If the injurious stimulus is removed inflammatory cells switch their phenotype and they mainly produce MMPs that help the degradation of ECM and favor fibrosis resolution and tissue healing. Withdrawal of pro-survival factors and engagement of death receptors on the surface of activated HSCs lead them to apoptosis finally leading to the re-establishment of steady-state conditions (adapted from Ref. [16]). Very recent studies provide evidence in animal models that phenotypic reversion to quiescence of HSCs may also play a role [63].

7.3.2 Evidence for MMP mediated matrix remodeling in human disease

The expression of MMPs and TIMPs 1 and 2 has been demonstrated in explanted human tissue and both MMPs and TIMPs have been identified as potential serum markers of fibrotic liver disease; indeed the measurement of TIMP-1 in the serum is a component of at least two marketed serum markers of liver disease [16,64–66]. Evidence that spontaneous resolution of fibrosis is possible in humans is most effectively demonstrated in the large-scale trials of antiviral treatments for chronic hepatitis B and C. In these trials, biopsies were taken and provided clear-cut evidence for fibrosis remodeling with, in some cases, resolution of fibrosis and a return to a more normal hepatic architecture [67–70]. Other studies including those by Wanless et al. have shown evidence for matrix remodeling and a change in fibrosis distribution even in advanced cases of cirrhosis at the time of transplantation [71]. Other disorders in which reversal of fibrosis has been convincingly demonstrated include biliary obstruction, hemochromatosis, and immune mediated liver disease [72–74].

The development of cirrhosis, which represents not simply advanced fibrosis, but a gross architectural disruption with the formation of nodules, vascular changes and angiogenesis likely prevents complete resolution of fibrosis even after withdrawal of the injurious stimulus. Even in rodent models, which are rightly criticized for being more plastic in terms of matrix remodeling than their human counterparts, rats will develop cirrhosis after prolonged CCl_4 exposure [3]. Withdrawal of CCl_4 in this context is associated with some remodeling but a failure to progress beyond a macronodular stage suggesting that the most mature areas of fibrosis are indeed irreversible. Our studies of this rat model suggest that these areas are rich in elastin in addition to collagens I and III and relatively pauci cellular [1,3,24]. Additionally, angiogenesis with the development of new vessels particularly within these persistent septa characterize irreversibility [75,76].

Initial interest in the role of tissue transglutaminase mediated crosslinking in such mature septa has not been borne out by studies using transglutaminase knockouts, which suggest that this form of crosslinking alone is insufficient to render fibrosis irreversible [31,77]. However, other forms of crosslinking including that mediated by lysyl oxidase may have an important effect in determining matrix stability and enzymatic resistance and there is further evidence that lysyl oxidase itself may have a proinflammatory and profibrotic effect [78–80]. Such crosslinking in addition to rendering the matrix not susceptible to degradation is also likely to enhance the stiffness leading to an enhanced profibrotic physical environment as well as potentially hiding epitopes, which may mediate a more beneficial cell matrix interaction with respect to both HSCs and hepatocytes [63,81]. Elastin is a product of HSCs and can be demonstrated to be secreted and laid down in the hepatic scar in CCl_4 and thioacetamide-induced liver injury. In common with the earlier studies of CCl_4 mediated injury examining the balance between specific MMPs and TIMPs in MMP-13 knockout animals, the induction of hepatic fibrosis in MMP 12 (metalloelastase) knockout mice is associated with enhanced accumulation of substrate (elastin) in the hepatic scar and a failure to degrade not only elastin, but other collagenous matrix proteins, once again reinforcing the importance of the TIMP/MMP balance in determining the accumulation of fibrotic tissue [82].

7.3.3 Key cellular origins of MMPs during the resolution of fibrosis

During spontaneous resolution of fibrosis apoptosis or reversion to quiescence of activated HSCs is associated with a drop in whole organ TIMP-1 levels. As described earlier, this is associated with a net increase in MMP activity and matrix degradation [5,16,21]. However, given the loss of activated HSCs, these cells are unlikely to be the major source of MMPs mediating fibrosis resolution. Indeed it would be somewhat counterintuitive to propose that HSCs were the major mediator of matrix degradation in this context. Studies of whole liver, in either experimental animal models or human models, can detect the MMP expression pattern during fibrosis remodeling, but such studies are limited in identifying the cellular source. Moreover, using strategies that delete or induce apoptosis in HSCs in fibrotic injury, the Mann group has demonstrated enhanced fibrosis remodeling suggesting directly that HSC are not necessary for resolution [83,84]. For these reasons, interest has recently focused on macrophages and dendritic cells as potential sources of the MMPs crucial to fibrosis resolution.

Using a model in which macrophages can be specifically deleted at the onset of fibrosis resolution in a CCl_4 mediated liver injury model, we have shown that not only are macrophages crucial to resolution and matrix remodeling (their deletion is associated with a failure of resolution), but the macrophage products MMP-13 (collagenase 3) and MMP-12 (metalloelastase), respectively, are crucial for degradation of collagens 1, 3, and elastin [22,27]. Using MMP-12 and MMP-13 knockout mice we have confirmed that both enzymes (predominantly expresses in macrophages) are required for fibrillar and elastin-rich matrix degradation in the fibrotic liver [82]. A similar manipulation of dendritic cells described in detail by Aloman demonstrates a very similar phenotype with respect to a failure of matrix degradation in spontaneous resolution that appears to be mediated via gelatinase B (MMP-9) [85]. We have further demonstrated that the adoptive transfer of macrophages is sufficient to enhance resolution of fibrosis even during continuing injury and that this effect may be mediated not only by macrophage derived MMPs but through the recruitment of other inflammatory cells producing and releasing MMPs including neutrophils [23,27,86].

We have developed a model in which specific macrophage populations can be identified during spontaneous resolution of liver fibrosis following isolation from the liver. By using the Ly6C marker we have shown that the macrophages mediating matrix degradation in resolution of liver fibrosis are derived from incoming inflammatory macrophages that undergo a phenotypic switch *in situ* during resolution. This phenotypic change is characterized by macrophages switching from $Ly6C^{hi}$ to $Ly6C^{lo}$. Moreover, $Ly6C^{lo}$ cells go on accumulating within the liver even after maximal hepatic damage (defined by transaminase release) has occurred, and the cells are most numerous at the onset of matrix degradation. These $Ly6C^{lo}$ cells are quantitatively the major source of MMPs during spontaneous resolution from CCl_4 induced liver fibrosis and can be demonstrated to express key enzymes including the interstitial collagenase MMP-13, the elastase MMP-12, and the gelatinase MMP-9. A further defining feature of these cells is that they demonstrate ingestion of apoptotic debris. By recapitulating this process in tissue culture, we have demonstrated that the ingestion of apoptotic debris (thymocytes or hepatocyte derived) is associated with a similar phenotypic switch and importantly an upregulation of MMPs 9 and 12. Finally, we have shown that matrix degradation may be enhanced in experimental liver fibrosis resolution simply by administering liposomes in order to deliver an apoptotic debris "meal" to the liver macrophages *in situ* [27].

As noted earlier, liver fibrosis resolution is associated with the restitution of the normal hepatocellular mass. Intriguingly parallel studies undertaken by the Forbes group have demonstrated that macrophages post-ingestion of debris may both stimulate the hepatic progenitor cell response via the expression of tumor necrosis factor-like weak inducer of apoptosis (TWEAK) and define the lineage of the bi-potential hepatic progenitor cell specifying a hepatocyte fate through Wnt expression. A further defining feature of the $Ly6c^{lo}$ macrophage crucial for resolution of fibrosis is the secretion of insulin-like growth factors that in turn have been implicated in hepatocyte proliferation and survival, representing a further mechanism whereby the normal liver mass can be reconstituted following macrophage mediated matrix degradation [86,87]. A schematic view of the relationship between distinct macrophage phenotypes, phases of liver fibrosis process, and MMPs metabolism is found in Figure 7.4.

Recently, the factors mediating macrophage extravasation to contribute to resolution have become better clarified. Work from the Tacke group has demonstrated that CxCl9/CxCr3 axis in macrophage mediates matrix remodeling during fibrosis resolution [88], while the Shah group has implicated a crucial role for VEGF signaling in mediating the extravasation of macrophages equipped to mediate matrix degradation [89]. These observations are important in that both the CxCl9/CxCr3 and VEGF axes have been implicated as potential targets in progressive fibrosis. Our greater understanding of the roles of macrophages and their recruitment axes during resolution indicates that soluble factors active and fibrogenic in one context may be important to enhancing resolution and matrix remodeling in another. A comprehensive view of the role of macrophage during fibrosis establishment and resolution can be gained in Figure 7.5.

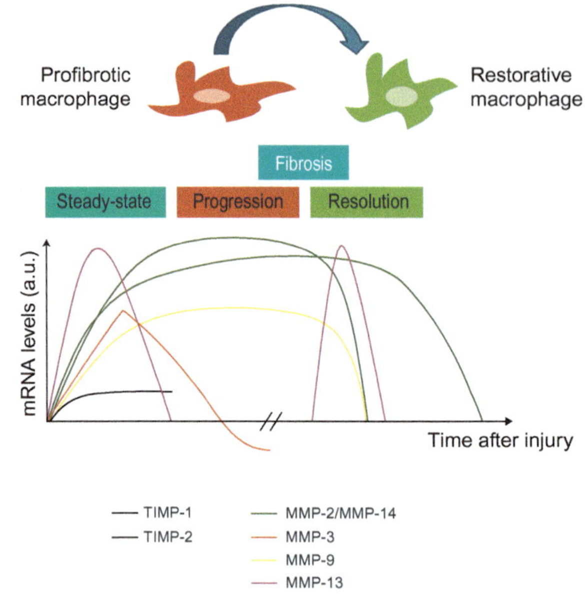

Figure 7.4 mRNA levels are reported in arbitrary unit (a.u.) on the *y*-axis. Data used to depict the graph have been obtained in total liver of rodent models of chronic toxic liver injury (e.g., CCl_4 intoxication). The curves represent general trend of expression and do not refer to any absolute quantification. The graphic illustrates how a balance between MMPs and TIMPs activity is present at the steady state, with characteristic changes of the expression pattern in liver following injury or during resolution of fibrosis that alter that balance. Macrophage phenotype appears to be instrumental for liver repair and fibrosis resolution. *Adapted from Refs. [21,38].*

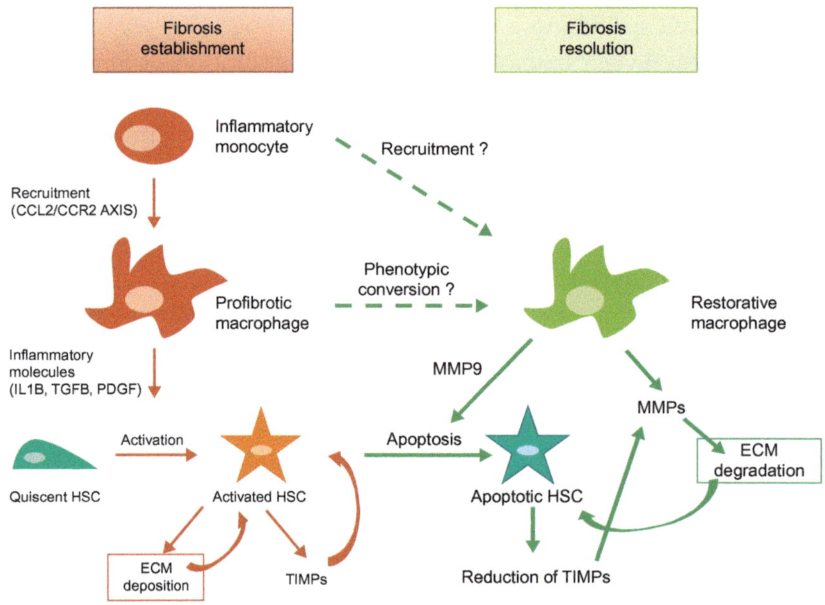

Figure 7.5 Relationship between macrophages and ECM biology during fibrosis establishment and fibrosis resolution. During fibrogenesis, inflammatory monocytes are recruited to the inflamed liver, giving rise to the pro-fibrotic macrophage population. Pro-fibrotic macrophages express inflammatory mediators such as IL-1β, TGF-β, and PDGF, which promote activation of HSCs. Activated HSCs synthesize the ECM and TIMP-1. Both TIMP-1 and ECM interactions promote persistence of the activated myofibroblast phenotype. During fibrosis resolution, there is a change in macrophage phenotype.

7.4 CONCLUSIONS

In human and animal models, liver fibrosis has been demonstrated to be in part a reversible process. Gaining insights in the molecular details of the fibrosis establishment and resolution process will help in the design of targeted therapies. In this context studying and understanding the dynamic of the MMPs/TIMPs balance is of pivotal importance.

Further studies involving the MMPs/TIMPs ratio will allow us to design therapies to regulate the axis between ECM deposition and degradation in favor of ECM degradation in a time- and cell-specific manner and thereby target fibrosis.

REFERENCES

[1] Iredale JP, Thompson A, Henderson NC. Extracellular matrix degradation in liver fibrosis: biochemistry and regulation. Biochim Biophys Acta 2013;1832(7):876–83.
[2] Pellicoro A, Ramachandran P, Iredale JP. Reversibility of liver fibrosis. Fibrogenesis Tissue Repair 2012;5(Suppl. 1 Proceedings of Fibroproliferative disorders: from biochemical analysis to targeted therapiesPetro E Petrides and David Brenner):S26.
[3] Iredale JP. Models of liver fibrosis: exploring the dynamic nature of inflammation and repair in a solid organ. J Clin Invest 2007;117(3):539–48.
[4] Friedman SL. Seminars in medicine of the Beth Israel Hospital, Boston. The cellular basis of hepatic fibrosis. Mechanisms and treatment strategies. N Engl J Med 1993;328(25):1828–35.
[5] Friedman SL. Hepatic stellate cells: protean, multifunctional, and enigmatic cells of the liver. Physiol Rev 2008;88(1):125–72.
[6] Friedman SL, et al. Hepatic lipocytes: the principal collagen-producing cells of normal rat liver. Proc Natl Acad Sci USA 1985;82(24):8681–5.
[7] Carloni V, Luong TV, Rombouts K. Hepatic stellate cells and extracellular matrix in hepatocellular carcinoma: more complicated than ever. Liver Int 2014;34(6):834–43.
[8] Forbes SJ, Parola M. Liver fibrogenic cells. Best Pract Res Clin Gastroenterol 2011;25(2):207–17.
[9] Lemoinne S, et al. Origins and functions of liver myofibroblasts. Biochim Biophys Acta 2013;1832(7):948–54.
[10] Rappaport AM, et al. The scarring of the liver acini (Cirrhosis). Tridimensional and microcirculatory considerations. Virchows Arch A Pathol Anat Histopathol 1983;402(2):107–37.
[11] Maher JJ, et al. Immunolocalization of laminin in normal rat liver and biosynthesis of laminin by hepatic lipocytes in primary culture. Gastroenterology 1988;94(4):1053–62.
[12] Issa R, et al. Spontaneous recovery from micronodular cirrhosis: evidence for incomplete resolution associated with matrix cross-linking. Gastroenterology 2004;126(7):1795–808.
[13] Du WD, et al. Dynamic changes of type I, III and IV collagen synthesis and distribution of collagen-producing cells in carbon tetrachloride-induced rat liver fibrosis. World J Gastroenterol 1999;5(5):397–403.
[14] Friedman SL, et al. Maintenance of differentiated phenotype of cultured rat hepatic lipocytes by basement membrane matrix. J Biol Chem 1989;264(18):10756–62.
[15] Iredale JP, et al. Tissue inhibitor of metalloproteinase-1 messenger RNA expression is enhanced relative to interstitial collagenase messenger RNA in experimental liver injury and fibrosis. Hepatology 1996;24(1):176–84.
[16] Henderson NC, Iredale JP. Liver fibrosis: cellular mechanisms of progression and resolution. Clin Sci (Lond) 2007;112(5):265–80.
[17] Clark IM, et al. The regulation of matrix metalloproteinases and their inhibitors. Int J Biochem Cell Biol 2008;40(6–7):1362–78.
[18] Consolo M, et al. Matrix metalloproteinases and their inhibitors as markers of inflammation and fibrosis in chronic liver disease (Review). Int J Mol Med 2009;24(2):143–52.

[19] Fanjul-Fernandez M, et al. Matrix metalloproteinases: evolution, gene regulation and functional analysis in mouse models. Biochim Biophys Acta 2010;1803(1):3–19.
[20] Wager TT, et al. Discovery of two clinical histamine H(3) receptor antagonists: trans-N-ethyl-3-fluoro-3-[3-fluoro-4-(pyrrolidinylmethyl)phenyl]cyclobutanecarbox amide (PF-03654746) and trans-3-fluoro-3-[3-fluoro-4-(pyrrolidin-1-ylmethyl)phenyl]-N-(2-methylpropyl)cyc lobutanecarboxamide (PF-03654764). J Med Chem 2011;54(21):7602–20.
[21] Hemmann S, et al. Expression of MMPs and TIMPs in liver fibrosis—a systematic review with special emphasis on anti-fibrotic strategies. J Hepatol 2007;46(5):955–75.
[22] Duffield JS, et al. Selective depletion of macrophages reveals distinct, opposing roles during liver injury and repair. J Clin Invest 2005;115(1):56–65.
[23] Pellicoro A, et al. Liver fibrosis and repair: immune regulation of wound healing in a solid organ. Nat Rev Immunol 2014;14(3):181–94.
[24] Iredale JP, et al. Mechanisms of spontaneous resolution of rat liver fibrosis. Hepatic stellate cell apoptosis and reduced hepatic expression of metalloproteinase inhibitors. J Clin Invest 1998;102(3):538–49.
[25] Yoshiji H, et al. Tissue inhibitor of metalloproteinases-1 attenuates spontaneous liver fibrosis resolution in the transgenic mouse. Hepatology 2002;36(4 Pt 1):850–60.
[26] Fallowfield JA, et al. Scar-associated macrophages are a major source of hepatic matrix metalloproteinase-13 and facilitate the resolution of murine hepatic fibrosis. J Immunol 2007;178(8):5288–95.
[27] Ramachandran P, et al. Differential Ly-6C expression identifies the recruited macrophage phenotype, which orchestrates the regression of murine liver fibrosis. Proc Natl Acad Sci USA 2012;109(46):E3186–95.
[28] Herbst H, et al. Temporal and spatial patterns of transin/stromelysin RNA expression following toxic injury in rat liver. Virchows Arch B Cell Pathol Incl Mol Pathol 1991;60(5):295–300.
[29] Knittel T, et al. Expression of matrix metalloproteinases and their inhibitors during hepatic tissue repair in the rat. Histochem Cell Biol 2000;113(6):443–53.
[30] Vyas SK, et al. Rat hepatic lipocytes synthesize and secrete transin (stromelysin) in early primary culture. Gastroenterology 1995;109(3):889–98.
[31] Popov Y, et al. Tissue transglutaminase does not affect fibrotic matrix stability or regression of liver fibrosis in mice. Gastroenterology 2011;140(5):1642–52.
[32] Ohashi N, et al. Matrix metalloproteinase-9 in the initial injury after hepatectomy in mice. World J Gastroenterol 2013;19(20):3027–42.
[33] Theret N, et al. MMP2 activation by collagen I and concanavalin A in cultured human hepatic stellate cells. Hepatology 1999;30(2):462–8.
[34] Theret N, et al. Differential expression and origin of membrane-type 1 and 2 matrix metalloproteinases (MT-MMPs) in association with MMP2 activation in injured human livers. Am J Pathol 1998;153(3):945–54.
[35] Gadd VL, et al. Portal, but not lobular, macrophages express matrix metalloproteinase-9: association with the ductular reaction and fibrosis in chronic hepatitis C. Liver Int 2013;33(4):569–79.
[36] Ehling J, et al. CCL2-dependent infiltrating macrophages promote angiogenesis in progressive liver fibrosis. Gut 2014;63(12):1960–71.
[37] Malemud CJ. Matrix metalloproteinases (MMPs) in health and disease: an overview. Front Biosci 2006;11:1696–701.
[38] Ramachandran P, Iredale JP. Liver fibrosis: a bidirectional model of fibrogenesis and resolution. QJM 2012;105(9):813–7.
[39] Kishnani NS, et al. Identification and characterization of human tissue inhibitor of metalloproteinase-3 and detection of three additional metalloproteinase inhibitor activities in extracellular matrix. Matrix Biol 1995;14(6):479–88.
[40] Staskus PW, et al. The 21-kDa protein is a transformation-sensitive metalloproteinase inhibitor of chicken fibroblasts. J Biol Chem 1991;266(1):449–54.
[41] Howard EW, Banda MJ. Binding of tissue inhibitor of metalloproteinases 2 to two distinct sites on human 72-kDa gelatinase. Identification of a stabilization site. J Biol Chem 1991;266(27):17972–7.

[42] Iredale JP. Tissue inhibitors of metalloproteinases in liver fibrosis. Int J Biochem Cell Biol 1997;29(1):43–54.
[43] Ward RV, et al. Tissue inhibitor of metalloproteinases-2 inhibits the activation of 72 kDa progelatinase by fibroblast membranes. Biochim Biophys Acta 1991;1079(2):242–6.
[44] Edwards DR, et al. Transforming growth factor beta modulates the expression of collagenase and metalloproteinase inhibitor. EMBO J 1987;6(7):1899–904.
[45] Edwards DR, et al. Involvement of AP1 and PEA3 binding sites in the regulation of murine tissue inhibitor of metalloproteinases-1 (TIMP-1) transcription. Biochim Biophys Acta 1992;1171(1):41–55.
[46] Wick M, et al. Structure of the human TIMP-3 gene and its cell cycle-regulated promoter. Biochem J 1995;311(Pt 2):549–54.
[47] Iyer S, et al. Crystal structure of the catalytic domain of matrix metalloproteinase-1 in complex with the inhibitory domain of tissue inhibitor of metalloproteinase-1. J Biol Chem 2007;282(1):364–71.
[48] Will H, et al. The soluble catalytic domain of membrane type 1 matrix metalloproteinase cleaves the propeptide of progelatinase A and initiates autoproteolytic activation. Regulation by TIMP-2 and TIMP-3. J Biol Chem 1996;271(29):17119–23.
[49] Benyon RC, et al. Progelatinase A is produced and activated by rat hepatic stellate cells and promotes their proliferation. Hepatology 1999;30(4):977–86.
[50] Zhou X, et al. Expression of matrix metalloproteinase-2 and -14 persists during early resolution of experimental liver fibrosis and might contribute to fibrolysis. Liver Int 2004;24(5):492–501.
[51] Olaso E, et al. DDR2 receptor promotes MMP-2-mediated proliferation and invasion by hepatic stellate cells. J Clin Invest 2001;108(9):1369–78.
[52] Wells RG. The role of matrix stiffness in hepatic stellate cell activation and liver fibrosis. J Clin Gastroenterol 2005;39(4 Suppl. 2):S158–61.
[53] Olsen AL, et al. Hepatic stellate cells require a stiff environment for myofibroblastic differentiation. Am J Physiol Gastrointest Liver Physiol 2011;301(1):G110–8.
[54] Okazaki I, Maruyama K. Collagenase activity in experimental hepatic fibrosis. Nature 1974;252(5478):49–50.
[55] Montfort I, Perez-Tamayo R. Collagenase in experimental carbon tetrachloride cirrhosis of the liver. Am J Pathol 1978;92(2):411–20.
[56] Paquet KJ, Kamphausen U. The carbon-tetrachloride-hepatotoxicity as a model of liver damage. First report: long-time biochemical changes. Acta Hepatogastroenterol (Stuttg) 1975;22(2):84–8.
[57] Murphy FR, et al. Inhibition of apoptosis of activated hepatic stellate cells by tissue inhibitor of metalloproteinase-1 is mediated via effects on matrix metalloproteinase inhibition: implications for reversibility of liver fibrosis. J Biol Chem 2002;277(13):11069–76.
[58] Roderfeld M, et al. Inhibition of hepatic fibrogenesis by matrix metalloproteinase-9 mutants in mice. FASEB J 2006;20(3):444–54.
[59] Garcia-Banuelos J, et al. Cirrhotic rat livers with extensive fibrosis can be safely transduced with clinical-grade adenoviral vectors. Evidence of cirrhosis reversion. Gene Ther 2002;9(2):127–34.
[60] Parsons CJ, et al. Antifibrotic effects of a tissue inhibitor of metalloproteinase-1 antibody on established liver fibrosis in rats. Hepatology 2004;40(5):1106–15.
[61] Yoshiji H, et al. Tissue inhibitor of metalloproteinases-1 promotes liver fibrosis development in a transgenic mouse model. Hepatology 2000;32(6):1248–54.
[62] Issa R, et al. Mutation in collagen-1 that confers resistance to the action of collagenase results in failure of recovery from CCl_4-induced liver fibrosis, persistence of activated hepatic stellate cells, and diminished hepatocyte regeneration. FASEB J 2003;17(1):47–9.
[63] Perepelyuk M, et al. Hepatic stellate cells and portal fibroblasts are the major cellular sources of collagens and lysyl oxidases in normal liver and early after injury. Am J Physiol Gastrointest Liver Physiol 2013;304(6):G605–14.
[64] Rosenberg WM, et al. Serum markers detect the presence of liver fibrosis: a cohort study. Gastroenterology 2004;127(6):1704–13.
[65] Lieber CS, Weiss DG, Paronetto F. Value of fibrosis markers for staging liver fibrosis in patients with precirrhotic alcoholic liver disease. Alcohol Clin Exp Res 2008;32(6):1031–9.

[66] Fontana RJ, et al. Serum fibrosis markers are associated with liver disease progression in non-responder patients with chronic hepatitis C. Gut 2010;59(10):1401–9.
[67] Tsubota A, et al. Time course of histological changes in patients with a sustained biochemical and virological response to interferon-alpha therapy for chronic hepatitis C virus infection. J Hepatol 1997;27(1):49–55.
[68] de Alava E, et al. Histological outcome of chronic hepatitis C treated with a 12-month course of lymphoblastoid alfa interferon. Liver 1993;13(2):73–9.
[69] Arase Y, et al. Time course of histological changes in patients with a sustained biochemical and virological response to corticosteroid withdrawal therapy for chronic hepatitis B. Am J Gastroenterol 1999;94(11):3304–9.
[70] Lee SS. Histological response to interferon alfa-based therapies in hepatitis C. Semin Liver Dis 2004;24(Suppl. 2):55–60.
[71] Wanless IR, Nakashima E, Sherman M. Regression of human cirrhosis. Morphologic features and the genesis of incomplete septal cirrhosis. Arch Pathol Lab Med 2000;124(11):1599–607.
[72] Falize L, et al. Reversibility of hepatic fibrosis in treated genetic hemochromatosis: a study of 36 cases. Hepatology 2006;44(2):472–7.
[73] Amital H, et al. Fibrosis regression induced by intravenous gammaglobulin treatment. Ann Rheum Dis 2003;62(2):175–7.
[74] Hammel P, et al. Regression of liver fibrosis after biliary drainage in patients with chronic pancreatitis and stenosis of the common bile duct. N Engl J Med 2001;344(6):418–23.
[75] Zimmermann A, Zhao D, Reichen J. Myofibroblasts in the cirrhotic rat liver reflect hepatic remodeling and correlate with fibrosis and sinusoidal capillarization. J Hepatol 1999;30(4):646–52.
[76] Sherman IA, Pappas SC, Fisher MM. Hepatic microvascular changes associated with development of liver fibrosis and cirrhosis. Am J Physiol 1990;258(2 Pt 2):H460–5.
[77] Grenard P, et al. Transglutaminase-mediated cross-linking is involved in the stabilization of extracellular matrix in human liver fibrosis. J Hepatol 2001;35(3):367–75.
[78] Ricard-Blum S, et al. Mechanism of collagen network stabilization in human irreversible granulomatous liver fibrosis. Gastroenterology 1996;111(1):172–82.
[79] Carter EA, et al. Lysyl oxidase and collagenase in experimental acute and chronic liver injury. Gastroenterology 1982;82(3):526–34.
[80] Siegel RC, et al. Biochemical and immunochemical study of lysyl oxidase in experimental hepatic fibrosis in the rat. Proc Natl Acad Sci USA 1978;75(6):2945–9.
[81] Georges PC, et al. Increased stiffness of the rat liver precedes matrix deposition: implications for fibrosis. Am J Physiol Gastrointest Liver Physiol 2007;293(6):G1147–54.
[82] Pellicoro A, et al. Elastin accumulation is regulated at the level of degradation by macrophage metalloelastase (MMP-12) during experimental liver fibrosis. Hepatology 2012;55(6):1965–75.
[83] Elsharkawy AM, Oakley F, Mann DA. The role and regulation of hepatic stellate cell apoptosis in reversal of liver fibrosis. Apoptosis 2005;10(5):927–39.
[84] Oakley F, et al. Inhibition of inhibitor of kappaB kinases stimulates hepatic stellate cell apoptosis and accelerated recovery from rat liver fibrosis. Gastroenterology 2005;128(1):108–20.
[85] Jiao J, et al. Dendritic cell regulation of carbon tetrachloride-induced murine liver fibrosis regression. Hepatology 2012;55(1):244–55.
[86] Bird TG, et al. Bone marrow injection stimulates hepatic ductular reactions in the absence of injury via macrophage-mediated TWEAK signaling. Proc Natl Acad Sci USA 2013;110(16):6542–7.
[87] Boulter L, et al. Macrophage-derived Wnt opposes Notch signaling to specify hepatic progenitor cell fate in chronic liver disease. Nat Med 2012;18(4):572–9.
[88] Wasmuth HE, et al. Antifibrotic effects of CXCL9 and its receptor CXCR3 in livers of mice and humans. Gastroenterology 2009;137(1):309–19. 319 e1–3.
[89] Yang L, et al. Vascular endothelial growth factor promotes fibrosis resolution and repair in mice. Gastroenterology 2014;146(5):1339–50. e1.

CHAPTER 8

Stellate Cells and Portal Hypertension

Don C. Rockey
Department of Internal Medicine, Medical University of South Carolina, Charleston, SC, USA

ABBREVIATIONS

ET endothelin
NO nitric oxide
MLCK myosin light chain kinase
MYPT-1 MLC phosphatase targeting subunit 1
CPI-17 17-kDa PKC-potentiated protein phosphatase 1 inhibitor protein
NOS nitric oxide synthase
ecNOS endothelial cell NO synthase
iNOS inducible NO synthase
ECE-1 endothelin-converting enzyme-1
TGF-β transforming growth factor-beta
GPCR G-protein coupled receptor
ACE angiotensin converting enzyme
RAAS renin–angiotensin–aldosterone
TX thromboxane
PG prostaglandin
TP thromboxane-A2/prostaglandin-endoperoxide
VEGF vascular endothelial growth factor
Tie-2 tyrosine kinase with immunoglobulin G-like and endothelial growth factor-like domains 2
CO carbon monoxide
HO heme oxygenase
PLC-β phospholipase C-β

8.1 OVERVIEW

The pathogenesis of portal hypertension is complex, and involves factors both within the liver (intrahepatic) and outside of the liver (extrahepatic). The variables important in intrahepatic portal hypertension surround largely stellate and endothelial cells

Stellate Cells in Health and Disease
DOI: http://dx.doi.org/10.1016/B978-0-12-800134-9.00008-7

as well as vasoactive mediators that have prominent effects on these cells. Stellate cells have gained attention because of their pericyte-like and contractile properties. Extensive work by our and other laboratories over the last decade has demonstrated that abnormalities in endothelial cells leads to an endothelialopathy and dysregulated production of nitric oxide (NO), which contributes to the vasoconstricted phenotype within the liver typical in liver injury [1–4]. Abnormalities in stellate cells also appear to play an important role in intrahepatic portal hypertension.

Stellate cells (also lipocytes, Ito, perisinusoidal cells) are perisinusoidal, pericyte-like cells found in the space of Disse (Figure 8.1). A major function of stellate cells in the normal liver appears to be as a reservoir for retinoids and stellate cells store a large proportion of the body's vitamin A [5,6]. Given their pericyte-like morphologic appearance, it has also been proposed that they may also function as regulators of blood flow [5] (Figure 8.1). After liver injury, stellate cells undergo a remarkable transformation in which they convert from a "quiescent" (normal) to an "activated" (injured liver) state. Importantly, many cell culture conditions have recapitulated this phenomenon and the activation of stellate cells has been appreciated as a critical element of the liver wound-healing process.

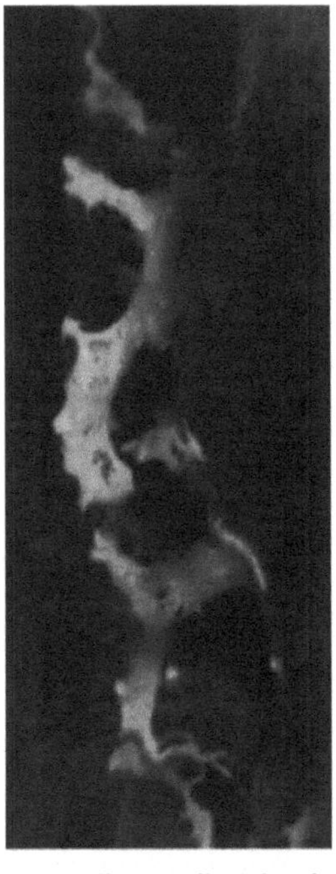

Figure 8.1 ***Sinusoidal anatomy.*** Hepatic stellate cells (also known as Ito cells, perisinusoidal cells, lipocytes) reside in the sinusoidal space of Disse, juxtaposed to sinusoidal endothelial cells. This histological arrangement is similar to that of pericytes found in the capillary system. *From Cell Tissue Research Vol. 273, p. 227–37, figure 10.*

Stellate cell activation is multifaceted; features associated with activation include the loss of vitamin A, acquisition of stress bundles, development of prominent rough endoplasmic reticulum, production of cytokines, chemokines, release of matrix degrading enzymes and their inhibitors, and others [7]. One of the most prominent features of activation is the striking increase in expression of smooth muscle proteins, including upregulation of the smooth muscle specific isoform of actin (smooth muscle α actin), first described by our laboratory [8]. Importantly, this feature has been associated with increased contractility, and suggests a prominent role for stellate cells in the constriction of hepatic sinusoids [9,10]. The increased expression of smooth muscle α actin [8] also identifies stellate cells as liver-specific myofibroblasts, a cell type typical of wound healing in essentially all organs [11,12].

A variety of smooth muscle proteins, including smooth muscle α actin, smooth muscle myosin heavy chain, the smooth muscle isoform of caldesmon and others are upregulated in stellate cells during activation and injury and these proteins appear to be important in mediating the contractile phenotype. The molecular mechanism of upregulation of these smooth muscle contractile proteins appears to be via a robust smooth muscle transcriptional program, including upregulation of a family of transcription factors important in smooth muscle cell biology [13]. Although controversy exists regarding the magnitude of contraction of normal stellate cells, available data emphasize that stellate cells in the injured liver exhibit enhanced contractility and moreover, the degree of contraction appears to be proportional to the degree of liver injury [14,15].

The evidence that stellate cell contraction is a determinant of early increases in intrahepatic resistance associated with liver injury includes data from both cell culture and *in vivo* studies (see Sections 8.2 and 8.3). Further, not only has it been proposed that activated stellate cells restrict hepatic blood flow by constricting individual sinusoids (Figure 8.1), but it has also been suggested that activated and contractile myofibroblasts may physically contract the cirrhotic liver [16], because the collagenous bands typical of end-stage cirrhosis contain large numbers of activated stellate cells (see Ref. [17] for review and also Chapter 4).

8.2 CELL CULTURE BASED STUDIES

Stellate cell contractility has been studied extensively in isolated cell systems [9,10,18]; several points are noteworthy. First, the intensity of contraction appears to be directly proportional to the degree of activation, which is directly related to the extensiveness of upregulation of smooth muscle contractile proteins. For example, stellate cells isolated from normal rat livers that are not activated exhibit minimal if any contractility [15]. In contrast, cells isolated from cirrhotic liver exhibit remarkable contractility.

The mechanism for this enhanced contractility is coupled to enhanced expression of smooth muscle proteins and activation dependent enhanced signaling pathways

[19,20]. Evidence for the importance of the absolute quantity of smooth muscle contractile proteins comes from data demonstrating that when smooth muscle α actin is genetically expressed in stellate cells, contractility is proportional to the amount of smooth muscle α actin expressed [21].

Signaling pathways in stellate cell contractility appear to parallel many of those found in smooth muscle cells, and include calcium-dependent and calcium-independent pathways [19,22–24] (Figure 8.2). The rho-kinase pathway, a calcium-independent contraction mechanism, appears to be prominent. And myosin light chain kinase (MLCK), 17-kDa PKC-potentiated protein phosphatase 1 inhibitor protein (CPI-17), and MLC phosphatase targeting subunit 1 (MYPT1) all also appear to be involved. The complete signaling cascade is not currently fully elucidated, but likely varies depending on the state of stellate cell activation, and the precise agonist stimulating stellate cell contraction. Mechanisms that induce relaxation appear to be less complicated, and involve primarily stimulation of cGMP and myosin dephosphorylation.

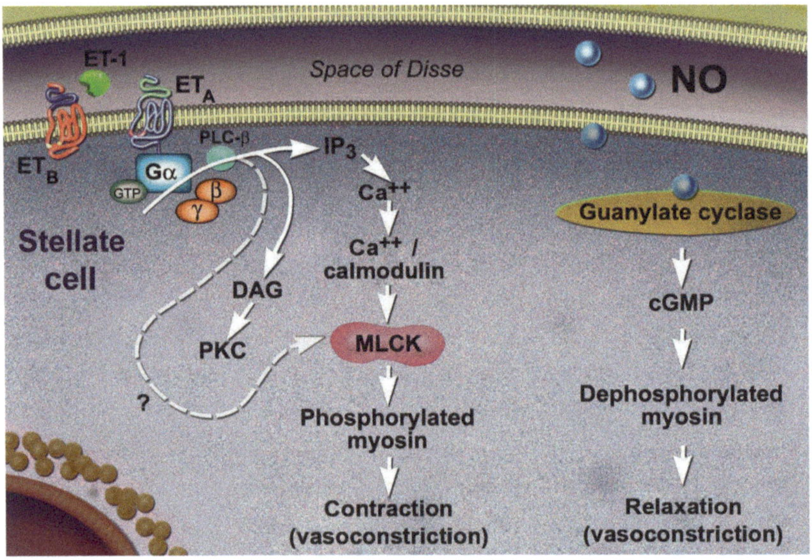

Figure 8.2 ***Stellate cell contractile signaling.*** ET-1 binding to its cognate receptors leads to classic G protein-coupled receptor-mediated activation of phospholipase C-β (PLC-β), cleavage of phosphatidylinositol 4,5-bisphosphate (PIP_2), and subsequent release of the second messengers inositol 1,4,5-trisphosphate (IP_3) and diacylglycerol (DAG). In the canonical signaling pathway, IP_3 mediated stimulation of Ca^{2+} release from endoplasmic reticulum stores leads to activation of Ca^{2+}/calmodulin dependent MLCK, which in turn leads to activation of myosin MgATPase and activation of the acto-myosin medicated contraction. Non-Ca^{2+} dependent pathways, including the RhoA—rho-kinase and as yet unknown pathways also appear to be active in stellate cell contractility. NO, one of the most prominent mediators of stellate cell relaxation binds to guanylate cyclase and leads to production of cGMP, which in turn leads to dephosphorylation of myosin and cellular relaxation. Both endothelin and NO signaling pathways are simplified. *Reproduced from Hepatology Vol. 37, No. 1, p. 4–12, figure 2.*

8.3 *IN VIVO* STUDIES

Stellate cells within the sinusoid possess extensive long and branching cytoplasmic processes that are located in both perisinusoidal and intersinusoidal locations [25] and have been shown to contract *in situ* in the normal sinusoid [26–28]. It should also be pointed out that stellate cells exist in close physical proximity to sinusoidal endothelial cells and that these two cell types have close cell–cell contacts and interactions; their interplay appears to be critical in the physiology of the hepatic microcirculation [29].

In vivo data suggest that contraction of stellate cells in the normal liver is modest, but contraction of activated stellate cells is robust [30–32]. Activated stellate cells, by virtue of their enhanced contractile phenotype, appear to restrict hepatic blood flow by both constricting sinusoids. Additionally, it has been suggested that activated and contractile myofibroblasts may physically contract the cirrhotic liver, because the collagenous bands typical of end-stage cirrhosis contain large numbers of activated stellate cells [16,33] (see also Ref. [17] for review); such contraction also likely disrupts blood flow and increases intrahepatic resistance.

8.4 VASOACTIVE MEDIATORS AND STELLATE CELLS

A number of vasoactive substances have effects on stellate cells. The most prominent agents include the family of endothelins, angiotensin II, arginine vasopressin, adrenomedullin, eicosanoids, and others [17]. Multiple inhibitors of stellate cell contractility have also been identified. While NO appears to be the major inhibitor of stellate cell contractility [34], other compounds have also been identified [17].

8.4.1 Endothelin

The endothelins are made up of a family of three unique 21 amino acid peptides termed ET-1, ET-2, and ET-3 [35]. These potent vasoconstrictors bind to two known G-protein coupled receptors (GPCRs), endothelin A (ET_A) and endothelin B (ET_B) [36]. ET_A receptors are found predominantly on vascular smooth muscle cells while ET_B receptors are classically found on endothelial cells. ET_B receptors appear to have a somewhat broader distribution than ET_A receptors, and have often been associated with the injury state. Affinities of peptides for the ET_A receptor are ET-1 > ET-2>>> ET-3 with the affinity of ET-1 for the ET_A receptor being more than 100-fold that of ET-3 [36]. ET_B receptors have equal affinity for endothelin subtypes [36]. Conventional dogma suggests that ET_A receptors on smooth muscle cells mediate vasoconstriction. Endothelin binding to the ET_B receptor on endothelial cells mediates vasodilation through stimulation of endothelial cell NO synthase (eNOS) and NO release. Typically, endothelins are produced by endothelial cells and exert paracrine effects on adjacent smooth muscle cells and autocrine effects on endothelial cells

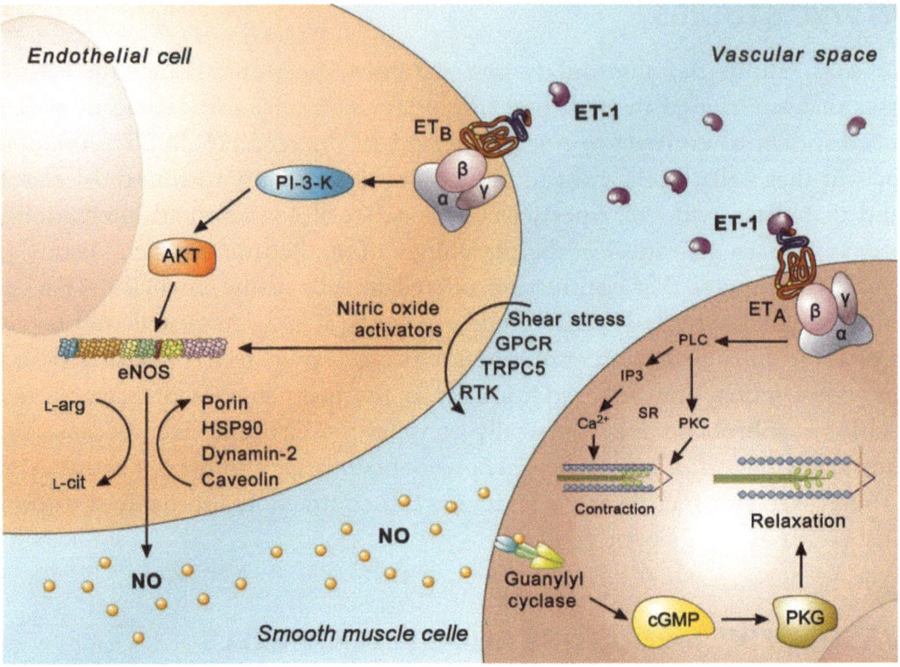

Figure 8.3 ***Endothelial and smooth muscle cell signaling.*** ET-1 stimulates NO production in endothelial cells by activation of eNOS. This occurs via ET-1's activation of the ET-B receptor and the PI3-K/Akt pathway, which in turn stimulates phosphorylation of eNOS, with subsequent conversion of L-arginine to L-citrulline and at the same time, generating NO. NO diffuses to smooth muscle cells, where it directly activates the heme moiety of soluble guanylate cyclase, leading to the production of cGMP. Intracellular cGMP leads to activation of protein kinase G (PKG) resulting in activation of myosin phosphatase, and subsequent dephosphorylation of myosin light chain, which causes the cell to relax. *From Cell Signal Vol. 22, p. 1615–25, figure 4.*

themselves (Figure 8.3). Notably, the sinusoidal endothelial cell and stellate cell system appears to closely mimic this more general vascular biologic system.

Regulation of endothelin synthesis is complex, and may be controlled at the transcriptional or posttranscriptional level (Figure 8.4). ET-1 is derived from precursor preproET-1 (approximately 200 amino acid residues); preproET-1 is cleaved by furin-like enzymes to intermediates termed big endothelin (38–41 amino acid residues). Big endothelins, which have little or no biologic activity, are cleaved at Trp-21-Val/Ile-22 by endothelin converting enzyme (ECE), a neutral membrane-bound metalloprotease, to yield the biologically active, mature 21 amino acid peptide. Two major isoforms of ECE, termed ECE-1 and ECE-2, have been identified [37,38]; in addition, ECE-1 undergoes extensive alternative splicing [38,39]. PreproET-1 (precursor ET-1) mRNA expression may be stimulated by many different factors, including vasopressor hormones such as epinephrine, angiotensin II, and vasopressin, by shear stress, or by

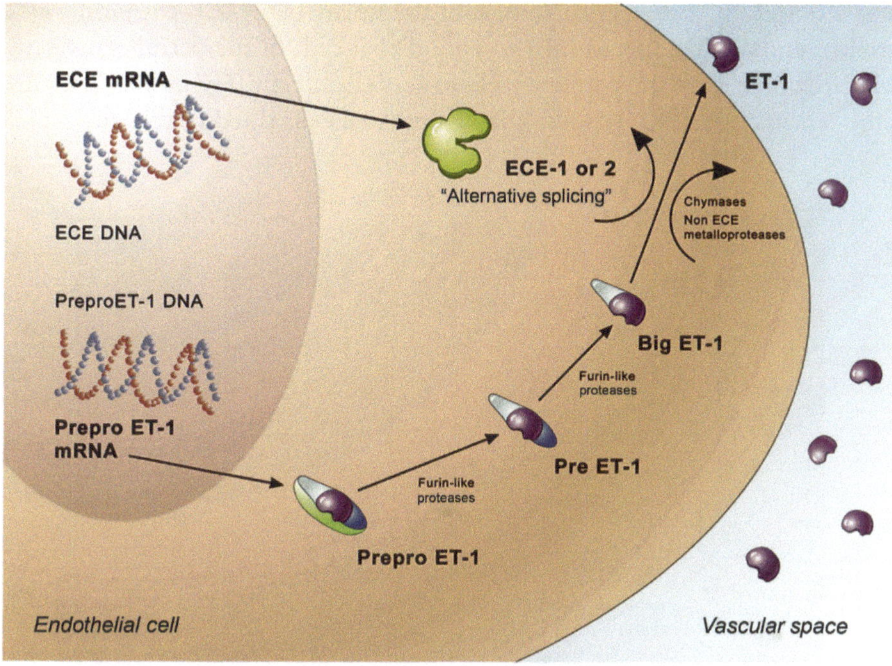

Figure 8.4 ***Endothelin biosynthetic pathway.*** Preproendothelin mRNA is synthesized via transcriptional activation of the preproendothelin gene. The translational product is a 203-amino acid peptide known as preproendothelin, which is cleaved at dibasic sites by furin-like endopeptidases to form big endothelins. These biologically inactive, 37- to 41-amino acid intermediates, are cleaved at Trp21-Val 22 by a family of ECE to produce mature ET-1. The pathway for endothelin-2 and -3 is presumed to be similar. *From Cell Signal Vol. 22, p. 1615–25, figure 2.*

cytokines such as interleukin-1 [40–42]. Thus, ET-1 levels are controlled at the transcriptional level as well as by ECE.

Binding of mature ET-1 to ET_A or ET_B receptors leads to classic G protein-coupled receptor-mediated signaling (Figure 8.3). Not only are the endothelins critical in the regulation of vascular tone, but abundant evidence indicates that they are important in multiple other processes [43,44]. For example, the endothelins play critical roles in cellular growth [45] and even cancer [46,47], neural crest development [48,49], and wound healing [50–53]. ET-1 in particular appears to be important to many different fibrotic tissues (see Sections 8.4.6.2 and 8.4.6.3) [54–56].

Nearly 20 years ago, the family of endothelins emerged as potentially important in liver disease. Initial reports demonstrated that circulating ET-1/ET-3 levels were elevated in patients with cirrhosis [57–59]. Mechanistic work since then has demonstrated that the source of endothelin in patients with cirrhosis is the injured liver itself [60–62]. The mechanism underlying the increased production of ET-1 in the injured liver is related to a transition at its cellular source. In the normal liver, ET-1 is produced largely by sinusoidal endothelial cells (similar to endothelial cells in the peripheral vasculature). The cellular source shifts after injury to hepatic stellate cells (Figure 8.5). Further, after injury, the synthesis of ET-1 by sinusoidal endothelial cells is reduced [63].

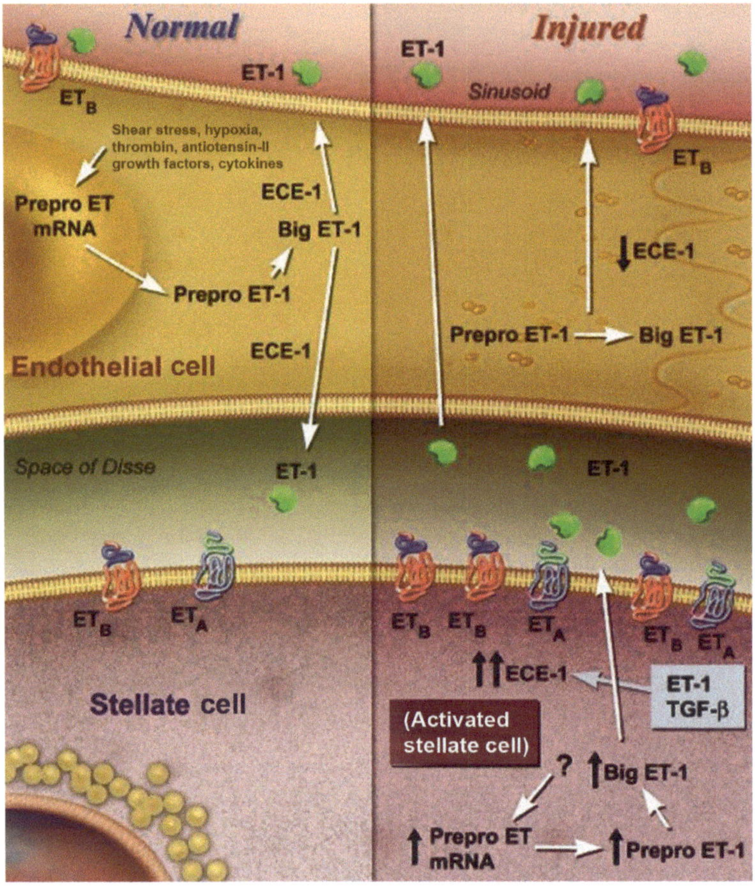

Figure 8.5 ***Endothelin synthesis in normal and injured liver.*** Under normal conditions, control of endothelin synthesis in the sinusoid mirrors that in the systemic vasculature. Hormones, other vascular mediators, and flow conditions appear to modulate precursor ET-1 synthesis in endothelial cells. In this state, proteolytic processing of precursor endothelins leads to production of mature ET-1. ET-1 then has paracrine physiologic effects on neighboring stellate (or smooth muscle) cells. After liver injury, stellate cell undergo "activation" (see Section 8.1) and synthesis of ET-1 shifts dramatically to activated stellate cells. The mechanism underlying enhanced ET-1 synthesis appears to largely involve upregulation of ECE-1, the enzyme responsible for conversion of big ET-1 to the mature peptide. In the injured liver, a host of factors, including components in the wounding milieu such as TGF-β, ET-1 itself, and other elements are likely to modulate ET-1 synthesis. ET-1 in turn has prominent effects on key cellular effectors (see Section 8.4.1), both locally and perhaps in distant vascular beds such as the lungs. Regulation of endothelin synthesis is described as an example of a canonical vascular mediator produced by stellate cells, and with autocrine effects on stellate cells. However, other similar processes likely exist for vasoactive mediator pathways. *From Hepatology Vol. 37, No. 1, p. 4–12, figure 1.*

The molecular mechanisms underlying enhanced production of ET-1 in stellate cells after injury are complex and multifactorial (Figure 8.5). For example, preproET-1 synthesis by stellate cells can be stimulated by extracellular matrix, in particular fibronectin species [64], by cytokines such as TNF-α [65], by ET-1 itself [66], and may be regulated by other factors typical of the wounding milieu [67]. Additionally, after stellate cell activation, ECE-1 expression is upregulated; the mechanism of its upregulation is notably complex [63,68]. For example, TGF-β causes production of at least two proteins that bind to the ECE-1 3′ untranslated region (UTR) to stabilize it and cause protein levels to increase [68].

The effect of endothelins in the liver disease appears to be related to their cellular target. Endothelin receptors in the liver are found predominantly on stellate cells (ET_A or ET_B receptors) and endothelial cells (ET_B receptors only) [69]. Hepatocytes and Kupffer cells express smaller numbers of endothelin receptors [69]. Localization of labeled ET-1 in the liver is consistent with binding to stellate cells [9,70]. It is also notable that endothelin receptors are upregulated in the injured liver [61,71] particularly on stellate cells [20] (Figure 8.5). Together, the increased production of endothelin as well as enhanced endothelin receptor expression on stellate cells in the injured liver contributes to enhanced contractility.

8.4.2 Angiotensin II

Similar to ET-1, angiotensin II has diverse biologic cellular and molecular effects [72]. It is a classic stimulus for smooth muscle cell contraction, mediated via GPCR signaling (also similar to ET-1) occurring as a result of binding to GPCRs, known as angiotensin-1 (AT1) and angiotensin-2 (AT2) receptors (which are distinct and share only 32% homology). Known pharmacological effects of angiotensin II appear to be mediated by the AT1 receptor.

In stellate cells, the biology of angiotensin II appears to be analogous to that of ET-1. That is to say that angiotensin II expression appears to be upregulated during liver injury and stellate cell activation [73–77]. Thus, analogous to ET-1, the major cellular source of angiotensin II in the liver is the stellate cell, also via complex molecular events (i.e., synthesis of angiotensinogen and expression of angiotensin converting enzyme [ACE]) [75]. Angiotensin II appears to have autocrine effects on stellate cells, via AT1 receptor such that angiotensin II binding to stellate cells induces cellular contraction and proliferation [78]. Notably, angiotensin II synthesis is upregulated after liver injury due to pronounced upregulation of ACE [75].

Finally, a physiologic role for angiotensin on stellate cells and in wound healing appears to parallel other examples of tissue fibrosis in that antagonism of the angiotensin II system may abrogate fibrogenesis [79–81]. Of note, while the biology of angiotensin II points to it as an attractive therapeutic target, angiotensin II antagonism in human clinical trials in patients with liver fibrosis has thus far been disappointing [82].

8.4.3 Nitric oxide

A prominent role for the potent vasodilator, NO, has been well established both in the liver as well as in stellate cells. NO has potent effects in the liver; in whole liver studies, NO clearly modulates intrahepatic resistance [27,83–85] and see Ref. [86] for review. Perfusion of the liver with NO donors causes a reduction in portal pressure and dilation of sinusoids, consistent with an effect on stellate cells [87] and likely also on intrahepatic vascular smooth muscle cells. In isolated stellate cells, NO is one of the most potent inhibitors of stellate cell contractility identified [34]. The mechanism of action of NO on stellate cell contractility has not been fully elucidated, but appears to be related to the canonical cGMP pathway. Of note, NO appears to have effects on stellate cells beyond that on contractility. For example, differentiated sinusoidal endothelial cells have been reported to prevent stellate cell activation and promote their quiescence through vascular endothelial growth factor (VEGF)-stimulated NO production; additionally, the capillarization phenotype of sinusoidal endothelial cells may be important in this process [88].

The biology of NO synthesis in the liver is complex. NO is produced from L-arginine by one of the three NO synthase (NOS) isoforms [89], encoded by at least three different genes [90–92] that produce three separate NOS enzymes. Endothelial cells (ecNOS) and neurons (nNOS) each contain what has been coined a "constitutive" NOS, while essentially all cells are capable of expressing an inducible NOS isoform (iNOS) [89]. Regulation of NOS expression is complex and involves both pre- and posttranslational elements. iNOS is typically transcriptionally stimulated by cytokines or lipopolysaccharide [89]. Although nNOS and ecNOS have been thought of as "constitutive," in fact, they are quite dynamic and their expression and function are each regulated on many different levels. ecNOS mRNA is increased by shear stress, and putative shear stress response elements have been identified in this genes' promoter [93]. ecNOS regulation is further complicated by the fact that this enzyme can be extensively posttranslationally modified [94–97] (Figure 8.6).

Current evidence indicates that after liver injury, an endothelialopathy ensues in which ecNOS derived NO production is reduced (see Ref. [98] for review). Sinusoidal endothelial cells produce NO basally [1], and increase its production in response to flow [84]. Interestingly, ecNOS mRNA and protein levels are unaltered after liver injury, yet ecNOS derived NO production is impaired after injury [1]. Current evidence indicates that after liver injury, multiple posttranslational abnormalities occur. For example, caveolin-1 is a protein that binds to ecNOS and inhibits ecNOS phosphorylation, and its function has been shown to be upregulated and to impair NO synthesis [99]. More importantly, a set of complex signaling processes that involve the canonical GPCR signaling pathway appear to be dysregulated after injury to sinusoidal endothelial cells [2–4,100] (see also Ref. [98] for review) (Figure 8.6).

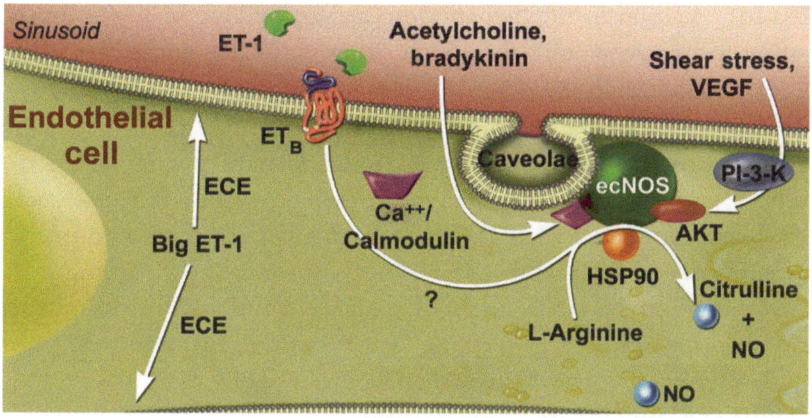

Figure 8.6 ***NO signaling in sinusoidal endothelial cells.*** NO is synthesized by ecNOS; although its production has classically been thought to be triggered by changes in intracellular Ca^{2+} and activation of Ca^{2+}/calmodulin following stimulation by agonists. Alternatively, NO synthesis can be triggered via a pathway involving phosphatidylinositol-3-OH kinase (PI-3-K) and the serine-threonine kinase, AKT. AKT phosphorylation, stimulated by factors such as shear stress and growth factors, leads to ecNOS phosphorylation and NO production. Additionally, eNOS is subject to multiple post-translational effects. *Reproduced from Hepatology Vol. 37, No. 1., p. 4–12, figure 2.*

8.4.4 Carbon monoxide

Carbon monoxide (CO) has been shown to stimulate quanylyl cyclase and cGMP production (similar to NO), and cause smooth muscle relaxation. Thus, it has been postulated that CO, like NO, may have effects on stellate cells and may play a role in regulation of intrahepatic blood flow. Although controversy exists as to the cellular source of CO in the liver [101], endogenously produced CO causes vasorelaxation of the hepatic sinusoid [27]. Notably, the major sites of action of CO in the sinusoid appear to co-localize with stellate cells (see Section 8.4.6.1) and overexpression of CO in the liver reduces intrahepatic vascular resistance [102]. At this point, mechanistic insight as to CO's effects of stellate cells does not exist beyond simple stimulation of cGMP.

8.4.5 Prostaglandins

Prostaglandins, which are derived from arachadonic acid by the action of cyclo-oxygenase (COX), are well known to have pleotropic biologic and physiological effects. Compounds such as thromboxane (Tx) A_2, prostacylcin (PGI_2), PGE_2, $PGF_{2\alpha}$, and PGD_2 have been reported to have effects on stellate cells [103–105], and further appear to have effects on the liver. For example, terutroban, a specific thromboxane-A2/prostaglandin-endoperoxide (TP) receptor antagonist reduced portal pressure in rats with cirrhosis due to carbon tetrachloride or bile duct ligation [106]. The mechanism appeared to be related to reduction in intrahepatic resistance, possibly via effects on stellate cells. Overall, further investigation as to the effects of prostaglandin derivatives is needed.

8.4.6 Others

Multiple other vasoactive compounds are known to have potent effects on multiple circulatory beds, and it is likely that many of these compounds have effects in the liver and perhaps on stellate cells [17,107]. The role of many of these compounds, such as catecholamines, purinergic agents (adenosine, serotonin, etc.), arginine vasopressin, cannabinoids, urotensin-II, atrial naturetic peptide, somatostatin, relaxin, and as yet unidentified endothelium derived relaxing factors requires further investigation.

8.4.6.1 Stellate cells and portal hypertension

Portal hypertension develops early after chronic liver injury and likely proceeds along a continuum through the development of cirrhosis. Portal pressure is a result of the relationship depicted in the hydraulic equivalent of Ohm's law: $\Delta P = Q \times R$ where ΔP is the change in pressure along a vessel, Q is the flow in the vessel, and R is the resistance to that flow. It is well accepted that from a mechanistic standpoint, portal hypertension is caused by both an increase in intrahepatic resistance and an increase in flow through the splanchnic system via a hyperdynamic circulation.

As emphasized earlier, increased intrahepatic resistance appears to be caused by both fixed (i.e., extracellular matrix, regenerative nodules, vascular thrombosis) and modulable elements (i.e., endothelial cells, stellate cells) [32,108]. Although the contribution that each of the modulable and fixed elements account for is difficult to precisely quantitate, it has been shown that portal pressure can be reduced by 20–30% with pharmacologic agents that reduce intrahepatic resistance [107]. Stellate cell contractility appears to be an important component of this modulable cellular contribution to intrahepatic resistance through contracting and relaxing of the sinusoid. As highlighted earlier, while many vasoactive compounds have the potential to regulate intrahepatic resistance, the bulk of the currently available data suggests that ET-1 and NO not only have the most prominent effects on isolated stellate cells, but also that they have the most important effects on intrahepatic resistance and blood flow in the liver *in vivo* [14,15,18,87,109,110].

The evolving paradigm is that in the normal liver, ET-1 (a major sinusoidal vasoconstrictor) and NO (a major vasorelaxing agent) synthesis is balanced, leading to low resistance and low portal pressure. In contrast, during liver injury, there is a shift in the balance of vasoactive substances (Figure 8.7). In the context of exaggerated stellate cell contractility (due to increased expression of contractile proteins, and a hyper-responsive contractile signaling cascade—see Ref. [111] for review of cellular aspects of stellate cell contractility) intrahepatic (sinusoidal) resistance becomes increased [1,85,99].

It should be recognized that although the endothelialopathy typical of liver injury may be in large part related to dysregulated ET-1 and NO biology, another vasoactive

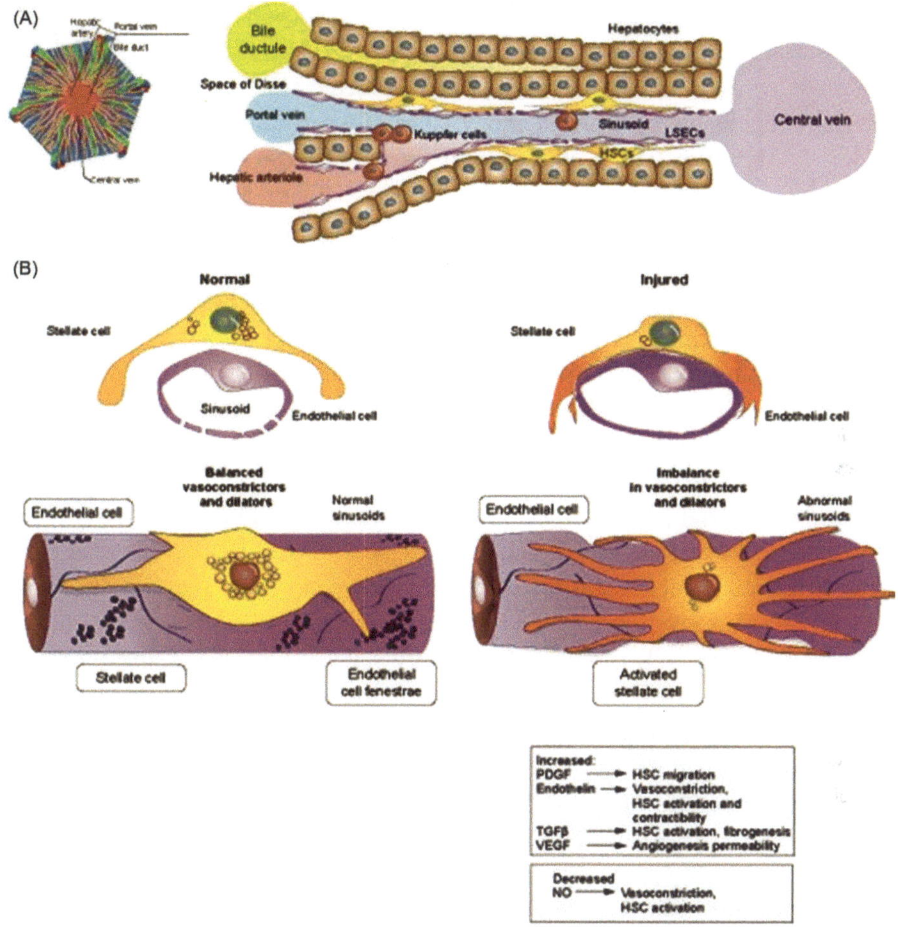

Figure 8.7 ***Changes in the hepatic sinusoid in response to injury***. (A) A simplified version of a portal tract (left), central vein (right), and hepatic sinusoidal morphology with associated cell types within the sinusoid. (B) The sinusoid in response to liver injury; the upper panel depicts a cross-sectional image, while the lower panel shows a three-dimensional view. In the normal state (left), sinusoidal endothelial cells produce NO supporting a low resistance state. In response to liver injury, extensive changes occur in the sinusoid. Liver injury (right) leads to both morphological as well as molecular abnormalities in sinusoidal endothelial cells and stellate cells. For example, endothelial cells become defenestrated, develop a defect in NO production, and concurrently increase production of other proteins such as endothelin and fibronectin that can contribute to stellate cell activation. Upon their activation, stellate cells develop an enhanced contractile phenotype and produce increased extracellular matrix. A number of paracrine and autocrine interactions occur between sinusoidal endothelial cells and stellate cells with some of the putative molecules and their associated functions as shown (gray box). *Abbreviations*: HSC, hepatic stellate cells; LSEC, liver sinusoidal endothelial cell. *From Journal of Hepatology Vol. 61, p. 912–24, figure 2.*

agent may also play a role. For example, prostanoids have effects on cultured stellate cells [103], and studies in whole animals emphasize their potential contribution. Further, angiogenesis may also be important in portal hypertension, though this effect may be indirect. Angiogenesis has been implicated in a number of different forms of tissue injury and fibrosis. In the liver, VEGF is increased during hepatic fibrogenesis and correlates with micro-vessel density. Further, angiopoietin-1, the ligand for the tyrosine kinase with immunoglobulin G-like and endothelial growth factor-like domains 2 (Tie2) receptor, which has angiogenic properties, is produced by stellate cells [112]. Additionally, blocking Tie2 signaling inhibited angiogenesis and experimental hepatic fibrosis [112] and bevacizumab, an anti-angiogenic agent, attenuated angiogenesis and hepatic fibrosis [113]. These data suggest that stellate cells may stimulate angiogenesis. Angiogenesis is in turn associated with enhanced fibrosis, perhaps via effects on stellate cells.

8.4.6.2 Fibrosis and portal hypertension

Liver fibrosis and portal hypertension appear to be linked. The evidence for this stems from both experimental evidence and human data. Consistently, the more extensive the fibrosis, the more severe the portal hypertension. From a mechanistic standpoint, this likely stems from several factors. First, during fibrogenesis, enhanced activation and contractility of stellate cells is typical (the more severe the fibrosis, the more activated are stellate cells [15]), and particularly early in the injury process leads to increased intrahepatic resistance as highlighted earlier. Secondly, simple physical obstruction to blood flow caused by fibrotic tissue causes an increase in intrahepatic resistance and increased pressure. Additionally, as fibrosis develops, as yet unknown signals stimulate increased mesenteric blood flow, which also contributes to portal hypertension.

8.4.6.3 Vascular mediators and stellate cell fibrogenesis

It also appears that several vasoactive mediators have effects on stellate cells, which primarily stimulate fibrogenesis. ET-1 and angiotensin II appear to be the most prominent; as highlighted before, stellate cells possess receptors for each of these ligands and respond to their binding by amplification of typical signaling cascades [69,75]. Exposure of isolated cultured stellate cells to ET-1 led to stimulation of smooth muscle α actin expression, proliferation, and extracellular matrix synthesis [60,62]. Endothelin antagonism during *in vivo* injury leads to reduced stellate cell activation and fibrogenesis (and a reduction in portal pressure) [110,114–116].

Angiotensin II, like ET-1, stimulated stellate cell proliferation and smooth muscle α actin expression, important features of activation [78]. Additionally, like the endothelin system in which there appears to be a substantial autocrine loop for ET-1 (Figure 8.4), angiotensin II is produced by activated stellate cells after liver injury, and can exert effects on stellate cells in an autocrine fashion [75].

It should also be emphasized that for both ET-1 and angiotensin II, these agents are important not only in liver fibrosis, but also appear to be important in many other forms of parenchymal wound healing, including in the skin, heart, lungs, kidneys, pancreas, and others [44,50–53,74,117].

8.5 THERAPEUTIC IMPLICATIONS

Stellate cells and their contractile phenotype represent an attractive therapeutic target. This is particularly true given the reasonably well understood mechanisms of enhanced stellate cell contractility after injury and activation. In addition, the dysregulated vascular biologic system in the injured liver has also pointed to specific potential points of therapeutic intervention. The endothelin and NO systems represent the most attractive targets. There is now abundant experimental evidence suggesting that antagonism of the endothelin system, with either ET_A or mixed $ET_{A/B}$ receptor antagonists reduces portal pressure. This effect is probably associated with inhibition of stellate cell contractility as well as of stellate cell fibrogenesis. Similarly, delivery of NO to the liver [110,118,119] would be attractive because this reduces portal pressure. Better yet, given the potent effects of NO on isolated stellate cells, delivery to stellate cells may also be attractive [34]. Regardless of the approach, given the broad systemic effects of vasoactive compounds, and their potential for off target effects, approaches that target stellate cells and the hepatic sinusoid specifically would be expected to have the best benefit to risk profile.

ACKNOWLEDGMENT

This work was supported by the NIH (R01 DK 57830 and R01 DK 098819 to DCR).

REFERENCES

[1] Rockey DC, Chung JJ. Reduced nitric oxide production by endothelial cells in cirrhotic rat liver: endothelial dysfunction in portal hypertension. Gastroenterology 1998;114:344–51.

[2] Liu S, Premont RT, Kontos CD, Zhu S, Rockey DC. A crucial role for GRK2 in regulation of endothelial cell nitric oxide synthase function in portal hypertension. Nat Med 2005;11:952–8.

[3] Liu S, Premont RT, Rockey DC. G-protein-coupled receptor kinase interactor-1 (GIT1) is a new endothelial nitric-oxide synthase (eNOS) interactor with functional effects on vascular homeostasis. J Biol Chem 2012;287:12309–20.

[4] Liu S, Premont RT, Rockey DC. eNOS is activated through GIT1 tyrosine phosphorylation and Src. J Biol Chem 2014;289:26.

[5] Wake K. Perisinusoidal stellate cells (fat-storing cells, interstitial cells, lipocytes), their related structure in and around the liver sinusoids, and vitamin A-storing cells in extrahepatic organs. Int Rev Cytol 1980;66:303–53.

[6] Blomhoff R, Wake K. Perisinusoidal stellate cells of the liver: important roles in retinol metabolism and fibrosis. FASEB J 1991;5:271–7.

[7] Friedman SL. Hepatic stellate cells: protean, multifunctional, and enigmatic cells of the liver. Physiol Rev 2008;88:125–72.
[8] Rockey DC, Boyles JK, Gabbiani G, Friedman SL. Rat hepatic lipocytes express smooth muscle actin upon activation *in vivo* and in culture. J Submicrosc Cytol Pathol 1992;24:193–203.
[9] Rockey DC, Housset CN, Friedman SL. Activation-dependent contractility of rat hepatic lipocytes in culture and *in vivo*. J Clin Invest 1993;92:1795–804.
[10] Pinzani M, Failli P, Ruocco C, et al. Fat-storing cells as liver-specific pericytes. Spatial dynamics of agonist-stimulated intracellular calcium transients. J Clin Invest 1992;90:642–6.
[11] Hinz B. Masters and servants of the force: the role of matrix adhesions in myofibroblast force perception and transmission. Eur J Cell Biol 2006;85:175–81.
[12] Gabbiani G. The myofibroblast in wound healing and fibrocontractive diseases. J Pathol 2003;200:500–3.
[13] Shi Z, Rockey DC. Interferon-gamma-mediated inhibition of serum response factor-dependent smooth muscle-specific gene expression. J Biol Chem 2010;285:32415–24.
[14] Bauer M, Paquette NC, Zhang JX, et al. Chronic ethanol consumption increases hepatic sinusoidal contractile response to endothelin-1 in the rat. Hepatology 1995;22:1565–76.
[15] Rockey DC, Weisiger RA. Endothelin induced contractility of stellate cells from normal and cirrhotic rat liver: implications for regulation of portal pressure and resistance. Hepatology 1996;24:233–40.
[16] Irle C, Kocher O, Gabbiani G. Contractility of myofibroblasts during experimental liver cirrhosis. J Submicrosc Cytol 1980;12:209–17.
[17] Rockey DC. Vascular mediators in the injured liver. Hepatology 2003;37:4–12.
[18] Kawada N, Tran-Thi TA, Klein H, Decker K. The contraction of hepatic stellate (Ito) cells stimulated with vasoactive substances. Possible involvement of endothelin 1 and nitric oxide in the regulation of the sinusoidal tonus. Eur J Biochem 1993;213:815–23.
[19] Reinehr RM, Kubitz R, Peters-Regehr T, Bode JG, Haussinger D. Activation of rat hepatic stellate cells in culture is associated with increased sensitivity to endothelin 1. Hepatology 1998;28:1566–77.
[20] Yoo DS, Rockey DC. Endothelin receptor subtypes in liver injury: upregulation and differential signaling in myofibroblast contractility. Hepatology 2000;32:328A.
[21] Rockey DC, Weymouth N, Shi Z. Smooth muscle alpha actin (Acta2) and myofibroblast function during hepatic wound healing. PLoS One 2013;8:e77166.
[22] Melton AC, Datta A, Yee Jr. HF. $[Ca^{2+}]_i$-independent contractile force generation by rat hepatic stellate cells in response to endothelin-1. Am J Physiol Gastrointest Liver Physiol 2006;290:G7–G13.
[23] Kruglov EA, Correa PR, Arora G, Yu J, Nathanson MH, Dranoff JA. Molecular basis for calcium signaling in hepatic stellate cells. Am J Physiol Gastrointest Liver Physiol 2007;292:G975–82.
[24] Iizuka M, Murata T, Hori M, Ozaki H. Increased contractility of hepatic stellate cells in cirrhosis is mediated by enhanced Ca^{2+}-dependent and Ca^{2+}-sensitization pathways. Am J Physiol Gastrointest Liver Physiol 2011;300:G1010–21.
[25] Ekataksin W, Kaneda K. Liver microvascular architecture: an insight into the pathophysiology of portal hypertension. Semin Liver Dis 1999;19:359–82.
[26] Zhang JX, Pegoli WJ, Clemens MG. Endothelin-1 induces direct constriction of hepatic sinusoids. Am J Physiol 1994;266:G624–32.
[27] Suematsu M, Goda N, Sano T, et al. Carbon monoxide: an endogenous modulator of sinusoidal tone in the perfused rat liver [see comments]. J Clin Invest 1995;96:2431–7.
[28] Okumura S, Takei Y, Kawano S, et al. Vasoactive effect of endothelin-1 on rat liver *in vivo*. Hepatology 1994;19:155–61.
[29] Watanabe N, Takashimizu S, Nishizaki Y, Kojima S, Kagawa T, Matsuzaki S. An endothelin A receptor antagonist induces dilatation of sinusoidal endothelial fenestrae: implications for endothelin-1 in hepatic microcirculation. J Gastroenterol 2007;42:775–82.
[30] McCuskey RS. A dynamic and static study of hepatic arterioles and hepatic sphincters. Am J Anat 1966;119:455–78.
[31] Clemens MG, Zhang JX. Regulation of sinusoidal perfusion: *in vivo* methodology and control by endothelins. Semin Liver Dis 1999;19:383–96.
[32] McCuskey RS. Morphological mechanisms for regulating blood flow through hepatic sinusoids. Liver 2000;20:3–7.

[33] Olsen AL, Bloomer SA, Chan EP, et al. Hepatic stellate cells require a stiff environment for myofibroblastic differentiation. Am J Physiol Gastrointest Liver Physiol 2011;301:G110–8.
[34] Rockey DC, Chung JJ. Inducible nitric oxide synthase in rat hepatic lipocytes and the effect of nitric oxide on lipocyte contractility. J Clin Invest 1995;95:1199–206.
[35] Yanagisawa M, Inoue A, Ishikawa T, et al. Primary structure, synthesis, and biological activity of rat endothelin, an endothelium-derived vasoconstrictor peptide. Proc Natl Acad Sci USA 1988;85:6964–7.
[36] Sakurai T, Yanagisawa M, Masaki T. Molecular characterization of endothelin receptors. Trends Pharmacol Sci 1992;13:103–8.
[37] Xu D, Emoto N, Giaid A, et al. ECE-1: a membrane-bound metalloprotease that catalyzes the proteolytic activation of big endothelin-1. Cell 1994;78:473–85.
[38] Emoto N, Yanagisawa M. Endothelin-converting enzyme-2 is a membrane-bound, phosphoramidon-sensitive metalloprotease with acidic pH optimum. J Biol Chem 1995;270:15262–68.
[39] Valdenaire O, Rohrbacher E, Mattei MG. Organization of the gene encoding the human endothelin-converting enzyme (ECE-1). J Biol Chem 1995;270:29794–98.
[40] Lee ME, Dhadly MS, Temizer DH, Clifford JA, Yoshizumi M, Quertermous T. Regulation of endothelin-1 gene expression by Fos and Jun. J Biol Chem 1991;266:19034–39.
[41] Marsden PA, Brenner BM. Transcriptional regulation of the endothelin-1 gene by TNF-alpha. Am J Physiol 1992;262:C854–61.
[42] Malek AM, Greene AL, Izumo S. Regulation of endothelin 1 gene by fluid shear stress is transcriptionally mediated and independent of protein kinase C and cAMP. Proc Natl Acad Sci USA 1993;90:5999–6003.
[43] Hocher B, Thone-Reineke C, Bauer C, Raschack M, Neumayer HH. The paracrine endothelin system: pathophysiology and implications in clinical medicine [see comments]. Eur J Clin Chem Clin Biochem 1997;35:175–89.
[44] Khimji AK, Rockey DC. Endothelin—Biology and disease. Cell Signal 2010;22:1615–25.
[45] Battistini B, Chailler P, D'Orleans-Juste P, Briere N, Sirois P. Growth regulatory properties of endothelins. Peptides 1993;14:385–99.
[46] Nelson JB, Hedican SP, George DJ, et al. Identification of endothelin-1 in the pathophysiology of metastatic adenocarcinoma of the prostate. Nat Med 1995;1:944–9.
[47] Nelson J, Bagnato A, Battistini B, Nisen P. The endothelin axis: emerging role in cancer. Nat Rev Cancer 2003;3:110–6.
[48] Kurihara Y, Kurihara H, Suzuki H, et al. Elevated blood pressure and craniofacial abnormalities in mice deficient in endothelin-1 [see comments]. Nature 1994;368:703–10.
[49] Baynash AG, Hosoda K, Giaid A, et al. Interaction of endothelin-3 with endothelin-B receptor is essential for development of epidermal melanocytes and enteric neurons. Cell 1994;79:1277–85.
[50] Appleton I, Tomlinson A, Chander CL, Willoughby DA. Effect of endothelin-1 on croton oil-induced granulation tissue in the rat. A pharmacologic and immunohistochemical study [see comments]. Lab Invest 1992;67:703–10.
[51] Nakamura T, Ebihara I, Fukui M, Tomino Y, Koide H. Effect of a specific endothelin receptor A antagonist on mRNA levels for extracellular matrix components and growth factors in diabetic glomeruli. Diabetes 1995;44:895–9.
[52] Kakugawa Y, Giaid A, Yanagisawa M, et al. Expression of endothelin-1 in pancreatic tissue of patients with chronic pancreatitis. J Pathol 1996;178:78–83.
[53] Saleh D, Furukawa K, Tsao MS, et al. Elevated expression of endothelin-1 and endothelin-converting enzyme-1 in idiopathic pulmonary fibrosis: possible involvement of proinflammatory cytokines. Am J Respir Cell Mol Biol 1997;16:187–93.
[54] Boffa JJ, Tharaux PL, Dussaule JC, Chatziantoniou C. Regression of renal vascular fibrosis by endothelin receptor antagonism. Hypertension 2001;37:490–6.
[55] Seccia TM, Belloni AS, Kreutz R, et al. Cardiac fibrosis occurs early and involves endothelin and AT-1 receptors in hypertension due to endogenous angiotensin II. J Am Coll Cardiol 2003;41:666–73.
[56] Shi-Wen X, Rodriguez-Pascual F, Lamas S, et al. Constitutive ALK5-independent c-Jun N-terminal kinase activation contributes to endothelin-1 overexpression in pulmonary fibrosis: evidence of an autocrine endothelin loop operating through the endothelin A and B receptors. Mol Cell Biol 2006;26:5518–27.

[57] Moore K, Wendon J, Frazer M, Karani J, Williams R, Badr K. Plasma endothelin immunoreactivity in liver disease and the hepatorenal syndrome [see comments]. N Engl J Med 1992;327:1774–8.

[58] Moller S, Emmeluth C, Henriksen JH. Elevated circulating plasma endothelin-1 concentrations in cirrhosis. J Hepatol 1993;19:285–90.

[59] Trevisani F, Colantoni A, Gerbes AL, et al. Daily profile of plasma endothelin-1 and -3 in pre-ascitic cirrhosis: relationships with the arterial pressure and renal function. J Hepatol 1997;26:808–15.

[60] Pinzani M, Milani S, De Franco R, et al. Endothelin 1 is overexpressed in human cirrhotic liver and exerts multiple effects on activated hepatic stellate cells. Gastroenterology 1996;110:534–48.

[61] Gandhi CR, Sproat LA, Subbotin VM. Increased hepatic endothelin-1 levels and endothelin receptor density in cirrhotic rats. Life Sci 1996;58:55–62.

[62] Rockey DC, Fouassier L, Chung JJ, et al. Cellular localization of endothelin-1 and increased production in liver injury in the rat: potential for autocrine and paracrine effects on stellate cells. Hepatology 1998;27:472–80.

[63] Shao R, Yan W, Rockey DC. Regulation of endothelin-1 synthesis by endothelin-converting enzyme-1 during wound healing. J Biol Chem 1999;274:3228–34.

[64] Zhan S, Chan CC, Serdar B, Rockey DC. Fibronectin stimulates endothelin-1 synthesis in rat hepatic myofibroblasts via a Src/ERK-regulated signaling pathway. Gastroenterology 2009;136:2345–55. e1–e4.

[65] Zhan S, Rockey DC. Tumor necrosis factor alpha stimulates endothelin-1 synthesis in rat hepatic stellate cells in hepatic wound healing through a novel IKK/JNK pathway. Exp Cell Res 2011;317:1040–8.

[66] Shao R, Rockey DC. Effects of endothelins on hepatic stellate cell synthesis of endothelin-1 during hepatic wound healing. J Cell Physiol 2002;191:342–50.

[67] Li T, Shi Z, Rockey DC. Preproendothelin-1 expression is negatively regulated by IFNgamma during hepatic stellate cell activation. Am J Physiol Gastrointest Liver Physiol 2012;302:G948–57.

[68] Shao R, Shi Z, Gotwals PJ, Koteliansky VE, George J, Rockey DC. Cell and molecular regulation of endothelin-1 production during hepatic wound healing. Mol Biol Cell 2003;14:2327–41.

[69] Housset C, Rockey DC, Bissell DM. Endothelin receptors in rat liver: lipocytes as a contractile target for endothelin 1. Proc Natl Acad Sci USA 1993;90:9266–70.

[70] Gondo K, Ueno T, Sakamoto M, Sakisaka S, Sata M, Tanikawa K. The endothelin-1 binding site in rat liver tissue: light- and electron-microscopic autoradiographic studies. Gastroenterology 1993;104:1745–9.

[71] Leivas A, Jimenez W, Bruix J, et al. Gene expression of endothelin-1 and ET(A) and ET(B) receptors in human cirrhosis: relationship with hepatic hemodynamics. J Vasc Res 1998;35:186–93.

[72] Cody RJ. The integrated effects of angiotensin II. Am J Cardiol 1997;79:9–11.

[73] Paizis G, Gilbert RE, Cooper ME, et al. Effect of angiotensin II type 1 receptor blockade on experimental hepatic fibrogenesis. J Hepatol 2001;35:376–85.

[74] Yoshiji H, Kuriyama S, Yoshii J, et al. Angiotensin-II type 1 receptor interaction is a major regulator for liver fibrosis development in rats. Hepatology 2001;34:745–50.

[75] Bataller R, Gines P, Lora J, et al. Evidence for a local renin-angiotensin system in human liver: expression in activated hepatic stellate cells. Hepatology 2001;34:399A.

[76] Bataller R, Schwabe RF, Choi YH, et al. NADPH oxidase signal transduces angiotensin II in hepatic stellate cells and is critical in hepatic fibrosis. J Clin Invest 2003;112:1383–94.

[77] Bataller R, Gabele E, Parsons CJ, et al. Systemic infusion of angiotensin II exacerbates liver fibrosis in bile duct-ligated rats. Hepatology 2005;41:1046–55.

[78] Bataller R, Gines P, Nicolas JM, et al. Angiotensin II induces contraction and proliferation of human hepatic stellate cells. Gastroenterology 2000;118:1149–56.

[79] Mezzano SA, Ruiz-Ortega M, Egido J. Angiotensin II and renal fibrosis. Hypertension 2001;38:635–8.

[80] Podowski M, Calvi C, Metzger S, et al. Angiotensin receptor blockade attenuates cigarette smoke-induced lung injury and rescues lung architecture in mice. J Clin Invest 2012;122:229–40.

[81] Roubille F, Busseuil D, Merlet N, Kritikou EA, Rheaume E, Tardif JC. Investigational drugs targeting cardiac fibrosis. Expert Rev Cardiovasc Ther 2014;12:111–25.

[82] Rockey DC. Translating an understanding of the pathogenesis of hepatic fibrosis to novel therapies. Clin Gastroenterol Hepatol 2013;11:224–31. e1–e5.
[83] Mittal MK, Gupta TK, Lee FY, Sieber CC, Groszmann RJ. Nitric oxide modulates hepatic vascular tone in normal rat liver. Am J Physiol 1994;267:G416–22.
[84] Shah V, Haddad FG, Garcia-Cardena G, et al. Liver sinusoidal endothelial cells are responsible for nitric oxide modulation of resistance in the hepatic sinusoids. J Clin Invest 1997;100:2923–30.
[85] Gupta TK, Toruner M, Chung MK, Groszmann RJ. Endothelial dysfunction and decreased production of nitric oxide in the intrahepatic microcirculation of cirrhotic rats. Hepatology 1998;28:926–31.
[86] Wiest R, Groszmann RJ. The paradox of nitric oxide in cirrhosis and portal hypertension: too much, not enough. Hepatology 2002;35:478–91.
[87] Edwards C, Feng HQ, Reynolds C, Mao L, Rockey DC. Effect of the nitric oxide donor V-PYRRO/NO on portal pressure and sinusoidal dynamics in normal and cirrhotic mice. Am J Physiol Gastrointest Liver Physiol 2008;294:G1311–17.
[88] Deleve LD, Wang X, Guo Y. Sinusoidal endothelial cells prevent rat stellate cell activation and promote reversion to quiescence. Hepatology 2008;48:920–30.
[89] Moncada S, Higgs A. The L-arginine-nitric oxide pathway. N Engl J Med 1993;329:2002–12.
[90] Bredt DS, Hwang PM, Glatt CE, Lowenstein C, Reed RR, Snyder SH. Cloned and expressed nitric oxide synthase structurally resembles cytochrome P-450 reductase. Nature 1991;351:714–8.
[91] Lowenstein CJ, Glatt CS, Bredt DS, Snyder SH. Cloned and expressed macrophage nitric oxide synthase contrasts with the brain enzyme. Proc Natl Acad Sci USA 1992;89:6711–5.
[92] Lamas S, Marsden PA, Li GK, Tempst P, Michel T. Endothelial nitric oxide synthase: molecular cloning and characterization of a distinct constitutive enzyme isoform. Proc Natl Acad Sci USA 1992;89:6348–52.
[93] Ranjan V, Xiao Z, Diamond SL. Constitutive NOS expression in cultured endothelial cells is elevated by fluid shear stress. Am J Physiol 1995;269:H550–5.
[94] Garcia-Cardena G, Martasek P, Masters BS, et al. Dissecting the interaction between nitric oxide synthase (NOS) and caveolin. Functional significance of the NOS caveolin binding domain *in vivo*. J Biol Chem 1997;272:25437–40.
[95] Feron O, Saldana F, Michel JB, Michel T. The endothelial nitric-oxide synthase-caveolin regulatory cycle. J Biol Chem 1998;273:3125–8.
[96] Feron O, Dessy C, Moniotte S, Desager JP, Balligand JL. Hypercholesterolemia decreases nitric oxide production by promoting the interaction of caveolin and endothelial nitric oxide synthase. J Clin Invest 1999;103:897–905.
[97] Fulton D, Gratton JP, McCabe TJ, et al. Regulation of endothelium-derived nitric oxide production by the protein kinase Akt. Nature 1999;399:597–601.
[98] Iwakiri Y, Shah V, Rockey DC. Vascular pathobiology in chronic liver disease and cirrhosis—current status and future directions. J Hepatol 2014;61:912–24.
[99] Shah V, Toruner M, Haddad F, et al. Impaired endothelial nitric oxide synthase activity associated with enhanced caveolin binding in experimental cirrhosis in the rat. Gastroenterology 1999;117:1222–8.
[100] Liu S, Premont RT, Kontos CD, Huang J, Rockey DC. Endothelin-1 activates endothelial cell nitric-oxide synthase via heterotrimeric G-protein betagamma subunit signaling to protein kinase B/Akt. J Biol Chem 2003;278:49929–35.
[101] Goda N, Suzuki K, Naito M, et al. Distribution of heme oxygenase isoforms in rat liver. Topographic basis for carbon monoxide-mediated microvascular relaxation. J Clin Invest 1998;101:604–12.
[102] Wakabayashi Y, Takamiya R, Mizuki A, et al. Carbon monoxide overproduced by heme oxygenase-1 causes a reduction of vascular resistance in perfused rat liver. Am J Physiol 1999;277:G1088–96.
[103] Kawada N, Klein H, Decker K. Eicosanoid-mediated contractility of hepatic stellate cells. Biochem J 1992;285:367–71.
[104] Gardi C, Arezzini B, Monaco B, De Montis MG, Vecchio D, Comporti M. F2-isoprostane receptors on hepatic stellate cells. Lab Invest 2008;88:124–31.
[105] Paik YH, Kim JK, Lee JI, et al. Celecoxib induces hepatic stellate cell apoptosis through inhibition of Akt activation and suppresses hepatic fibrosis in rats. Gut 2009;58:1517–27.

[106] Rosado E, Rodriguez-Vilarrupla A, Gracia-Sancho J, et al. Terutroban, a TP-receptor antagonist, reduces portal pressure in cirrhotic rats. Hepatology 2013;58:1424–35.
[107] Bhathal PS, Grossman HJ. Reduction of the increased portal vascular resistance of the isolated perfused cirrhotic rat liver by vasodilators. J Hepatol 1985;1:325–37.
[108] Lautt WW, Greenway CV, Legare DJ, Weisman H. Localization of intrahepatic portal vascular resistance. Am J Physiol 1986;251:G375–81.
[109] Gandhi CR, Stephenson K, Olson MS. Endothelin, a potent peptide agonist in the liver. J Biol Chem 1990;265:17432–35.
[110] Feng HQ, Weymouth ND, Rockey DC. Endothelin antagonism in portal hypertensive mice: implications for endothelin receptor-specific signaling in liver disease. Am J Physiol Gastrointest Liver Physiol 2009;297:G27–33.
[111] Reynaert H, Thompson MG, Thomas T, Geerts A. Hepatic stellate cells: role in microcirculation and pathophysiology of portal hypertension. Gut 2002;50:571–81.
[112] Taura K, De Minicis S, Seki E, et al. Hepatic stellate cells secrete angiopoietin 1 that induces angiogenesis in liver fibrosis. Gastroenterology 2008;135:1729–38.
[113] Huang Y, Feng H, Kan T, et al. Bevacizumab attenuates hepatic fibrosis in rats by inhibiting activation of hepatic stellate cells. PLoS One 2013;8:e73492.
[114] Rockey DC, Chung JJ. Endothelin antagonism in experimental hepatic fibrosis. Implications for endothelin in the pathogenesis of wound healing. J Clin Invest 1996;98:1381–8.
[115] Cho JJ, Hocher B, Herbst H, et al. An oral endothelin a receptor antagonist blocks collagen synthesis and depostion in advanced rat liver fibrosis. Gastroenterology 2000;118:1169–78.
[116] Thirunavukkarasu C, Yang Y, Subbotin VM, Harvey SAK, Fung J, Gandhi CR. Endothelin receptor antagonist TAK-044 arrests and reverses the development of carbon tetrachloride-induced cirrhosis in rats. Gut 2004;53:1010–9.
[117] Weber KT. Fibrosis, a common pathway to organ failure: angiotensin II and tissue repair. Semin Nephrol 1997;17:467–91.
[118] Yu Q, Shao R, Qian HS, George SE, Rockey DC. Gene transfer of the neuronal NO synthase isoform to cirrhotic rat liver ameliorates portal hypertension. J Clin Invest 2000;105:741–8.
[119] Fiorucci S, Antonelli E, Morelli O, et al. NCX-1000, a NO-releasing derivative of ursodeoxycholic acid, selectively delivers NO to the liver and protects against development of portal hypertension. Proc Natl Acad Sci USA 2001;98:8897–902.

CHAPTER 9

Hepatic Stellate Cells and Liver Cancer

Dianne H. Dapito[1] and Robert F. Schwabe[1,2]
[1]Institute of Human Nutrition, Columbia University, New York, NY, USA
[2]Department of Medicine, Columbia University, New York, NY, USA

9.1 FIBROSIS AND HCC

Fibrosis is the hepatic response to chronic injury, common to all chronic liver diseases with hepatocellular injury [1]. Most complications of chronic liver disease—including the development of portal hypertension, gastric and esophageal varices, and subsequent variceal bleeding, ascites, decreased hepatic function, and detoxification—are directly or indirectly related to the development of hepatic fibrosis. With improved therapeutic options and prolonged survival of patients experiencing these grave consequences of chronic liver disease, the development of hepatocellular carcinoma (HCC) has become a leading cause of mortality in patients with compensated liver cirrhosis. Worldwide, HCC causes almost 700,000 deaths annually [2]. Importantly, HCC is strongly associated with the presence of hepatic fibrosis with about 80–90% of HCCs developing in fibrotic or cirrhotic livers [2]. Moreover, recurrence of HCC following resection is strongly linked to gene expression patterns of surrounding nontumor tissue than the tumor itself, further suggesting a role of the environment in the process of carcinogenesis [3]. Based on the highly increased risk of patients with cirrhosis to develop HCC [2], it is widely believed that myofibroblasts and/or extracellular matrix (ECM) represent a critical factor in the chronically injured liver that promotes the development of HCC. Despite the association of fibrosis and HCC, evidence proving a functional link between these two is limited. Moreover, several other factors including chronic inflammation, chronic hepatic regeneration, and angiogenesis represent additional potential cancer-promoting factors within the chronically injured liver [4]. Here, we will first review key components of the hepatic premaligant microenvironment (PME) and the tumor microenvironment (TME) in the liver. Next, we will discuss evidence for a tumor-promoting role of hepatic stellate cells (HSCs) in HCC, the underlying mechanisms, and opportunities for therapeutic targeting.

9.2 THE PREMALIGNANT MICROENVIRONMENT

HCC typically develops in severely perturbed livers, which display profound changes of the hepatic microarchitecture as well as functional alterations of virtually every hepatic cell type. Besides HSC activation and liver fibrosis, chronic liver disease leads to

Stellate Cells in Health and Disease
DOI: http://dx.doi.org/10.1016/B978-0-12-800134-9.00009-9

inflammatory cell recruitment, hepatocyte proliferation, progenitor cell accumulation, angiogenesis, and altered endothelial cell biology such as the loss of fenestration. As such, inflammation has been shown to promote HCC development in many murine HCC models [5]. Inflammation generates a tumor-prone environment through multiple pathways and cellular targets [6]. The tumor-promoting effects of many inflammatory pathways have been linked to their ability to activate cell survival pathways leading to resistance to cell death, one of the hallmarks of cancer [7]. Moreover, recruitment of inflammatory cells leads to increased generation of reactive oxygen species (ROS), which, in turn, can exert genotoxic effects. Inflammatory cells are also a rich source of cytokines and growth factors, which may (i) promote signals in epithelial and tumor cells, (ii) proliferation and activation of HSCs, or (iii) induce angiogenesis. Chronically injured livers also display enhanced angiogenesis with increases in CD31-positive vessels [8,9], which may also favor carcinogenesis. One of the most characteristic features is the accumulation of α-smooth muscle actin (α-sma)-positive myofibroblasts along with the presence of liver fibrosis. In chronic liver injury, HSCs represent the dominant cellular source of ECM-producing myofibroblasts [10]. ECM not only alters mechanical properties of the liver, but also the behavior of epithelial cells via ECM-associated growth factors and ECM-induced integrin signals as discussed in subsequent sections of this chapter. Finally, it has been suggested that normal cells present in healthy organs provide signals that suppress the growth of epithelial cells. Among these are normal fibroblasts, that is, quiescent HSCs in the liver, and epithelia themselves, which suppress growth via contact inhibition [7,11]. Hence, not only the increase in myofibroblastic/activated HSCs but also the loss of quiescent HSCs (or other quiescent hepatic cell types) may contribute to a cancer-promoting microenvironment.

9.3 THE TUMOR MICROENVIRONMENT

In addition to the hepatic PME providing a fertile soil for malignant transformation, established tumors also create a niche that may favor their growth, called the tumor microenvironment (TME). The TME is composed of multiple cell types including cancer-associated fibroblasts (CAFs), tumor-associated macrophages (TAMs), B and T cells, neutrophils, mast cells, platelets, endothelial cells, and pericytes [7]. It is believed that there is a constant and close interaction between tumor and TME, and that they co-evolve, ultimately resulting in a symbiotic relationship in which the tumor supports the TME and vice versa. Importantly, it has been suggested that the inefficiency of most of today's pharmacologic therapies may in part be related to the fact that these therapies focus exclusively on targeting tumor cells, and that they largely ignore nontumor cells, which may not only form a large proportion of the tumor, but also provide important functions.

The importance of an altered microenvironment, and fibroblasts in particular, has been demonstrated in the prostate, where genetic alterations of fibroblasts or co-culture with CAFs were sufficient to initiate malignant transformation of epithelial cells [12,13]. These findings clearly demonstrate that the presence of an altered microenvironment plays a key role in controlling epithelial behavior, thereby contributing to cancer development.

9.4 HSCs AS POTENTIAL CONTRIBUTORS TO CANCER PROMOTION

HSCs have long been considered the most relevant fibrogenic cell population of the liver. However, additional cellular sources of hepatic myofibroblasts exist and include portal fibroblasts, as well as bone marrow-derived fibroblasts [14]. Recent fate tracing studies using lecithin retinol acyltransferase (LRAT)-driven Cre that labels up to 99% of HSC have confirmed HSCs as the principal source of myofibroblasts in toxic, cholestatic, and fatty liver diseases, contributing 82–96% of myofibroblasts [10]. Hence, myofibroblasts that may promote carcinogenesis in the setting of chronic injury are most likely HSC-derived. Although current data do not exclude the possibility that fibrosis *per se* may be sufficient to trigger HCC, the current concept is that myofibroblasts represent one of the factors that promote HCC in chronically injured livers.

9.4.1 Evidence for HSCs exerting a tumor-promoting role in the PME

Although the role of HSCs in promoting liver fibrosis is well-established, direct evidence for their role in tumor promotion in the PME is lacking to date. As most HCCs arise in a fibrotic and cirrhotic setting [2], it is important to establish animal models that can recapitulate the course of human disease that promotes HCC, namely chronic injury, inflammation, and fibrosis. Therefore, emphasis needs to be placed in generating experimental mouse models that can mimic the injury-inflammation-fibrosis preceding HCC (reviewed in Refs. [15,16]). Genetic models of HCC such as overexpression of transforming growth factor (TGF)-α, c-Myc and hepatitis viral protein HBx, or single injection of the carcinogen diethylynitrosamine (DEN) result in the development of tumors in the absence of substantial fibrosis (reviewed in Ref. [16]), thus rendering them unsuitable to examine the interaction between HSCs and transformed hepatocytes in the fibrotic PME. As the great majority of hepatocarcinogenesis studies rely on either the use of very potent carcinogens or genetic HCC drivers that reach a penetrance of nearly 100%, the more subtle and long-term effects of the PME may therefore not be revealed in many of these potent models. As such, HCC development has 5-year cumulative incidence of 4–30% in patients with cirrhosis [2], which is often preceded by several decades of chronic injury to develop cirrhosis in the first place. Hence, studying the role of HSCs and fibrosis in the PME requires the use of appropriate models and needs to reflect principles of human disease. Mice with hepatocyte-specific deletion of Nemo or TAK1 (Transforming growth factor β activated kinase 1) display spontaneous and chronic liver injury that, similar to human liver disease, progresses from inflammation to severe fibrosis and HCC [17–19]. However, multiple mechanisms including chronic liver regeneration and inflammation may be responsible for the development of HCC, and the role of HSCs and liver fibrosis in these models has not been sorted out.

Transgenic mice with hepatocyte-specific overexpression of human platelet-derived growth factor C (PDGF-CC mouse) show increased HSC activation, collagen deposition, steatosis, and hepatomegaly, followed by the development of HCC [20]. The strong expression of PDGF receptors PDGFRα and PDGFRβ in HSCs in preneoplastic PDGF-CC mice suggest that PDGF signaling in HSCs alters the liver microenvironment resulting in HCC development [21]. Accordingly, hepatocyte growth factor (HGF) and heparin-binding epidermal growth factor (EGF)-like growth factor (HB-EGF) as well as EGFR ligands Egf and epiregulin are increased within stromal cells [21] suggesting that HSCs, activated by PDGF-CC, can promote tumor growth through secretion of these mitogenic factors. Furthermore, treatment of PDGF-CC transgenic mice with peritinoin, an acyclic retinoid that inhibits the transdifferentiation of HSCs into myofibroblasts, repressed the development of hepatic fibrosis and tumors in a PDGFR-dependent manner [22]. Although this transgenic mouse model has elucidated a role for HSC/stroma in tumor promotion, one important caveat of this model is that overexpressed PDGF-CC and/or peritinoin might act on epithelial cells, and alter their proliferation and malignant transformation, as suggested in other models in which PDGF-B overexpression in combination with p53 deletion was sufficient to trigger cancer development [23]. Hence, further studies are required to prove the contribution of HSCs to hepatocarcinogenesis by more specific methods.

Given that activated HSCs may contribute to a permissive tumorigenic microenvironment, modulating their activation may seem a viable approach to reduce HCC incidence. Indeed, Fuchs et al. used the EGFR inhibitor erlotinib, which inhibited EGFR phosphorylation in HSCs, reduced the total number of HSCs, fibrosis, and progression to cirrhosis [24]. At the same time, effects on the epithelial compartment, including decreased hepatocyte proliferation and decreased liver injury were observed in three rodent models of progressive cirrhosis: repeated injections of low-dose DEN, chronic injections of CCl_4, and bile duct ligation [24]. Importantly, EGFR inhibition prevented the development of HCC compared to untreated control rats [24]. Importantly, they observed the effect of erlotinib specifically in nontumor tissue and not in the HCC lesions, suggesting that altering the PME, rather than direct effects on the tumor, explains the effects of erlotinib. However, as one cannot completely exclude the possibility that some effects of erlotinib are mediated through effects on epithelial/tumor cells, more specific methods need to demonstrate the role of HSCs (e.g., using HSC-specific ablation of EGFR during hepatocarcinogenesis).

9.4.2 HSCs as component of the TME and contributors to HCC progression

The presence of α-sma-positive stromal cells has been demonstrated in both human and murine HCC: liver biopsies of patients with HBV-related HCC displayed desmin-positive and α-sma-positive cells in the perisinusoidal space between tumor cells [25]. The α-sma-positive myofibroblasts have been identified between endothelial cells and cancer cell trabeculae of patients with HCC [26]. Studies have shown that α-sma-positive myofibroblasts in the stroma were responsible for collagen synthesis around tumors [27,28]. Ju et al. evaluated the prognostic potential of α-sma-expressing myofibroblasts in 130 cases of HCC and observed that increasing peritumoral infiltration of myofibroblasts was correlated to poorer prognosis and increased recurrence of HCC after liver resection [29]. Similarly, patients with high expression of α-sma from resected liver biopsies exhibited worse outcomes in a cohort of patients with intrahepatic cholangiocarcinoma [30]. Clinical observations correlating abundance of myofibroblasts with poorer prognosis coupled with murine studies of metastatic HCC highlight that activated HSCs and myofibroblasts are highly evident within the tumor stroma and may participate in generating a permissive microenvironment for cancer cell seeding and growth.

Despite previous studies showing that HSCs are the predominant source of myofibroblasts in the fibrotic liver [10], it remains unknown whether HSCs also constitute the cellular source of myofibroblasts in HCC. The fact that myofibroblasts in the TME express α-sma does not mean that they are HSC-derived. Bone marrow-derived myofibroblasts have been shown in a number of solid tumors [31–34], suggesting that bone marrow also could be a source for myofibroblasts in HCC. Future fate tracing studies are required to resolve this point and determine whether bone marrow-derived or liver resident myofibroblast precursors, in particular HSCs, are the main source of myofibroblasts in the TME.

To address the contribution of HSCs/myofibroblasts in the TME, one would ideally manipulate HSCs/myofibroblasts in the TME in HCC models that mimic human hepatocarcinogenesis. Up to now, virtually all studies have relied on *in vitro* culture systems or tumor transplantation models. Although these studies may provide valuable insights to the interaction between HSCs and tumor cells, they lack the natural co-evolution of tumor and TME. Moreover, co-culturing or co-injecting tumor cells with HSCs or other myofibroblasts may put these cells into an anatomical contact that does not necessarily exist in endogenously arising tumors. Finally, injecting fully transformed HCC cells into the liver or other anatomical sites could be viewed as an intra- or extrahepatic metastasis model rather than the natural co-evolution of tumor and TME. Despite these limitations, *in vitro* co-culture studies and *in vivo* co-injection studies of HSCs and HCC cells have given valuable insight into mechanisms by which HSCs may affect HCC, and provide several lines of evidence for the hypothesis that HSC-derived myofibroblasts promote the growth of established HCC.

Numerous *in vitro* studies suggest that HSCs can directly or indirectly promote growth of HCC cell lines: conditioned media from culture-activated primary human HSCs significantly enhanced the mitogenic capacity and migratory potential of three human HCC lines: PLC, HepG2, and Hep3B [35]. Incubation of McA-RH777 rat HCC cells with conditioned media of culture-activated rat HSCs significantly increased the proliferation, migration, and the number of invasive HCC cells [36]. Interactions between activated HSCs and cancer cells have also been demonstrated through the use of co-culture systems, revealing a bidirectional crosstalk between the hepatoma cells and HSCs with increased expression of proinflammatory cytokines, proliferation and migration in hepatoma cells, while increasing the expression of proangiogenic mediators such as vascular endothelial growth factor A (VEGFA) and matrix metalloproteinase 9 (MMP9) in HSCs [37].

Several transplant studies have explored the role of activated HSCs on tumor cells *in vivo* through xenograft models. In a subcutaneous co-transplantation model, the presence of an HSC cell line increased tumor formation by neoplastic MIMR hepatocytes two- to threefold [38]. Enhanced proliferation of the implanted tumor cells resulted from HSC-induced autocrine TGF-β signaling and hyperactivation of nuclear Wnt/β-catenin in the neoplastic hepatocytes [38]. Subcutaneous implantation of HepG2 cells and the HSC cell line LX2 in nude mice resulted in more rapid tumor growth and increased tumor size and weight relative to control animals who received HepG2 alone [39]. The tumors of co-transplanted mice exhibited increased fibrotic bands containing the LX2 cells that promoted increased angiogenesis and proliferation and reduced apoptosis of HepG2 cells [39]. These findings were corroborated by another study where *in vivo* implantation of human HepG2 HCC cells and primary human HSCs subcutaneously in nude mice promoted tumor growth, invasiveness, and inhibited necrosis [35]. Orthotopic injection of H22 HCC cells with activated T6 HSCs into the hepatic lobes of nude mice resulted in increased tumor number and tumor volume 7 weeks after transplant, whereas injection of H22 cells alone failed to form tumors [40]. Increased tumorigenicity was attributed to increased angiogenesis within the tumor. Moreover, co-injection of the H22 with activated HSCs resulted in increased invasiveness of these cells as evidenced by metastatic nodules in the lungs and pancreas of tumor-bearing mice [14]. In addition to promoting angiogenesis, proliferation, and tumor cell motility, activated HSCs have been shown to play a role in immunosuppression allowing for increased tumor growth (discussed later).

9.5 MECHANISMS BY WHICH HSCs PROMOTE HCC

9.5.1 HSCs generate quantitatively and qualitatively altered ECM in the PME and TME

Besides providing mechanical strength, the ECM regulates multiple signaling cascades through its ability to bind specific receptors such as integrins, and to bind to growth factors and regulate their distribution, activation, and presentation [41]. Hence, ECM proteins play a key role in regulating growth in many settings including development, stem cell niches, and cancer [41]. In the fibrotic liver, HSCs are the predominant ECM-producing myofibroblast and hence the most important cell type that modulate the hepatic PME, and most likely also the TME, via secretion of ECM proteins and affect the overall behavior of transformed hepatocytes as well as other stromal cells [42].

ECM components such as collagens, laminins, fibronectin, proteoglycans, and glycosaminoglycans interact directly or indirectly with HCC cells and other stromal cells resulting in phenotypic and physiological changes within these cells [43]. Activated HSCs secrete laminin-5, which can stimulate HCC cell migration [44]. Changes within the ECM that result in mechanical stress are conveyed to surrounding cells by transmembrane receptors such as integrins that regulate specific signaling pathways within the cell in response to external stimuli. Several studies using mouse HCC models [45] and human HCC samples [46] demonstrate a positive correlation between collagen and integrin expression and tumorigenicity. When integrins are organized into focal adhesions on the surface of the cell, they can activate mitogenic signaling cascades such as phosphoinositide 3 kinase (PI3K) and mitogen-activated protein kinase (MAPK) that can therefore promote cell growth and survival [47]. Therefore increased ECM may stimulate integrin signaling in transformed hepatocytes, promoting their growth and survival. Integrin signaling has also been attributed to promoting cancer cell migration and invasion as well as enhancing tumor cell survival through anti-apoptotic signaling (reviewed in Ref. [48]).

Enhanced production and reorganization of the ECM leads to a mechanically stiff microenvironment, which has been shown to promote tumor cell proliferation and invasion and also changes in gene expression and behavior of stromal cells, thereby enhancing tumor progression [43,49,50]. Hepatoma cells grown on increasingly stiff polyacrylamide gels exhibited increased proliferation and chemotherapeutic resistance, which was mediated by Fak, Erk, Pkb/Akt, and Stat3 signaling pathways downstream of integrin β1 activation [51]. These findings can further be extended to patients, as elevated liver stiffness is associated with enhanced risk for HCC development [52,53]. Interestingly, one of the most frequent alterations seen in liver carcinogenesis is increased expression of integrin α6β1 (β1) as HCC patients with multiple tumors, vascular invasion, and the absence of tumor cell encapsulation exhibit an upregulation of its expression, whereas its expression is absent in the normal liver [54]. Increased integrin β1 expression, enhanced adhesion capacity, and upregulated integrin-associated focal adhesions were evident in human HCC tumor tissue [55]. This study also found a positive correlation between ECM mechanical stiffness and increased integrin β1 expression with the Edmonson pathologic grade, tumor encapsulation, and metastasis [55], further emphasizing the relevance of a stiff ECM in HCC.

ECM deposition and degradation are highly dynamic processes within the injured liver. Beyond simple removal of ECM, degradation may also modulate the release of sequestered growth factors, generate bioactive cleavage products, or cleave cell surface receptors [48]. In this regard, MMPs and their inhibitors (TIMPs) are considered key regulators of ECM degradation and regulate cell proliferation, increase hepatocyte growth, and promote tumor progression (reviewed in Ref. [48]). Moreover, MMPs are required to promote tumor spread within the often highly fibrotic livers in which HCCs arise as they help break down tumor boundaries and facilitate rearrangement of the liver tissue [4]. Dysregulated MMP expression can compromise the basement membrane and facilitate invasion by cancer cells [4]. HCC is associated with increased MMP2 expression and higher proteolytic activity and an imbalance between MMP2/TIMP2 can be used as a prognostic marker as low expression of TIMP2 in serum and tissues of HCC patients correlates with the occurrence of metastasis [56]. Well differentiated hepatoma cells express MT1-MMP, MMP2, and MMP9, which promote stromal cell invasion [57]. HSC-derived epimorphin, an ECM protein that is highly expressed in HSCs in tumor stroma, mediates the invasive potential of cancer cells by upregulating MMP9 expression [58].

9.5.2 HSCs may decrease immune surveillance

In line with altered ECM production and desmoplasia, activated HSCs can modulate the activity of immune cells by generating a physical barrier separating tumor cells from inflammatory cells [59]. Additionally, several lines of evidence suggest that HSCs reduce tumor immunosurveillance through paracrine signals to surrounding immune cells: (i) Zhao et al. showed in co-transplantation studies of H22 tumor cells and HSCs that HSCs inhibited lymphocyte infiltration in tumors and the spleens of tumor-bearing mice, induced apoptosis of infiltrating mononuclear cells, and increased the number of infiltrating FoxP3-positive immunosuppressive regulatory T cells (Treg cells), resulting in reduced immune surveillance [60]. (ii) *In vitro* studies examining the immunomodulatory role of activated HSCs demonstrated that HSCs induced T-cell hyporesponsiveness, inhibited T-cell-mediated cytotoxicity, increased the number of Treg cells, and enhanced cytokine production of numerous cytokines with proposed functions to modulate cytotoxic activity of T cells [61]. Further *in vivo* analysis using co-injection of activated HSCs and Hepa1–6 HCC cells showed increased tumor growth rate and volume relative to mice injected with tumor cells alone, which can be attributed to an increase in the Treg cells around the tumors [61]. The immunosuppressive properties of activated HSCs were further demonstrated in an orthotopic model of HCC wherein intrahepatic co-injection of HSCs and H22 cells resulted in a threefold increase in tumor burden relative to H22 cells alone owing this response to increased angiogenesis, lymphangiogenesis, and increased numbers of Tregs and myeloid-derived suppressor cells in the spleen, bone marrow, and tumor tissues of tumor-bearing mice [52]. These studies indicate that activated HSCs facilitate immunologic escape of transformed hepatocytes thereby promoting HCC progression. Table 9.1 summarizes the current evidence for the functional role of HSCs in promoting the development of HCC.

Table 9.1 Functional Evidence for HSCs Promoting Hepatocarcinogenesis

	Experimental approach	Model	Effect of HSC on hepatocarcinogenesis	Refs.
PME	Transgenic PDGF-CC Mouse Model	Ectopic overexpression of PDGF-CC in hepatocytes	Mice have increased HSC activation, develop progressive fibrosis, steatosis and hepatomegaly, significant liver injury then HCC; HSCs secrete mitogenic factors in this model	[20,21]
	Chronic liver injury	Repeated low-dose DEN injections in rats	Rats treated with EGFR inhibitor erlotinib displayed reduction in HSCs, prevented cirrhosis and HCC development	[24]
	Determination of HSC SASP effects on hepatocarcinogenesis using GFAPCre-mediated po53 deletion	DEN + 12x CCl_4 hepatocarcinogenesis; co-culture of macrophages and hepatoblasts in HSC medium	SASP in HSCs reduces HCC formation; SASP shifts macrophage polarization to anti-tumorigenic M2 phenotype, reflected by increased HCC formation in mice with GFAPCre-induced p53 deletion	[83]
	Determination of HSC SASP effects in the setting of obesity-associated hepatocarcinogensis	DMBA+obesity-induced HCC, using gut sterilization with antibiotics, IL-1β global knockout, and liposome-mediated delivery of HSP47 siRNA	SASP in HSCs promotes HCC formation; high fat diet or genetic obesity induces SASP which in turn promotes HCC proliferation	[84]
TME	Conditioned medium from activated HSCs	HepG2, PLC, Hep3B	Increased proliferation and migratory potential of all HCC cell lines	[35]
		McA-RH7777	Increased proliferation, migration and invasive potential	[36]
		HepaRG	Increased migratory potential	[37]
		Hep3B, Alexander, HLE, HLF	Increased HCC cell adhesion on plates coated with ECM substrates and migration resulting from HSC-secreted laminin 5 activation of ERK	[44]
		HepG2, MHCC97H, MHCC97L	Increased proliferation, migration and invasion capacity of HCC cells when cultured in activated LX-2 conditioned medium	[58]

(Continued)

Table 9.1 (Continued)

	Experimental approach	Model	Effect of HSC on hepatocarcinogenesis	Refs.
	Co-culture	Hep3B	Accelerated tumor cell/HSC spheroid formation in 3D culture system	[35]
		McA-RH7777	Increased secretion of HGF, IL-6, MMP2, and MMP9 from HCC cells when co-cultured with rat HSCs	[36]
		HepaRG, HuGB, Huh7, HepG2	Increased expression of inflammatory mediators (IL1β, IL6, IL8, CCL2); increased proliferation and chemotactic and migratory ability	[37]
	Subcutaneous co-transplantation	McA-RH7777	Increased tumor weight and volume in HCC cells co-injected with culture-activated rat HSCs	[36]
		MIMR	Twofold increase in tumor formation resulting from co-transplantation with M1-4HSCs	[38]
		Hepa1-6	Increased tumor proliferation and volume due to increased Treg accumulation	[61]
		H22	Increased tumor volume resulting from decreased lymphocyte infiltration, increased apoptosis of mononuclear cells, and increased Treg infiltration in tumor	[60]
		MHCC97H	Accelerated tumor growth when co-injected with LX-2 cells	[58]
	Orthotopic co-transplantation	H22	Threefold increase in tumor burden resulting from increased angiogenesis, lymphangiogenesis, and increased infiltration of Tregs and myeloid-derived suppressor cells	[52]
		MHCC97L	Decreased metastatic ability of HCC cells when co-injected with epimorphin-null LX-2 cells	[58]

9.5.3 HSCs may promote tumor angiogenesis

Accelerated tumor growth is accompanied by a hypoxic microenvironment necessitating neovascularization within the tumor. Angiogenesis is a hypoxia-stimulated and growth-factor mediated process in which new vessels are formed from the existing vasculature [62]. Activated HSCs produce numerous angiogenic factors such as VEGF [40,63] and angiopoietin (Ang) 1 or 2 [8,64,65] which activate their respective receptors on endothelial cells to enhance their functions [59] and promote tumor vascularization and growth. This has been shown in numerous *in vitro* and *in vivo* studies using HSC/endothelial cell co-culture systems and in tumor cell/HSC co-implantation models [39,40,66,67]. VEGF is a potent survival factor for endothelial cells that promotes their proliferation and vascular permeability thus supporting the formation of new vasculature. Angiopoietins, Ang-1 and Ang-2, play a role in vessel stability wherein binding of Ang-1 to its receptor Tie-2 promotes interactions between endothelial cells and pericytes resulting in increased vascular stability. Ang-2, in the presence of VEGF, initiates and augments angiogenesis to promote new vascular growth (reviewed in detail in Ref. [68]). Overexpression of Ang-1 and Ang-2 was positively correlated with tumor de-differentiation in HCC [65] and the level of Ang-2 was found to be associated with clinicopathological parameters of HCC patients [69]. Moreover, histological staining of Ang-1 and Ang-2 in HCC tissues displayed co-localization with α-sma-positive HSCs suggesting that these cells actively produce proangiogenic mediators [65].

In addition to active secretion of proangiogenic factors, HSCs are also considered pericytes, and as such may be actively involved in vascular remodeling [59,70]. Accordingly, Wirz et al. observed that a functional three-dimensional spheroid co-culture of HSCs with endothelial cells resulted in spontaneous differentiation and organization into a structure with a core of HSCs and a surface of endothelial cells resembling the physiological assembly of blood vessels [70]. Additionally, they observed that co-culture of activated HSCs and liver sinusoidal endothelial cells formed capillary-like sprouts within gels in angiogenesis assays thus suggesting that HSCs can induce formation of tubular sprouts with liver sinusoidal endothelial cells (LSECs). As tumor cells can induce HSC activation, this suggests that HSCs may partake in tumor-induced neovascularization [70].

9.5.4 HSC-secreted cytokines may contribute to tumor growth, survival, and invasion

Activated HSCs secrete cytokines and growth factors that can stimulate the growth, proliferation, and migration of cancer cells as well as other nontumor cells within the stroma. As such, HSC culture supernatant contains PDGF-AB, HGF, and TGF-β, which, in turn, enhance the proliferation and migration of tumor cells *in vitro* [35,38,71].

TGF-β is perhaps the best-characterized cytokine in cancer as it acts in an autocrine and paracrine manner on tumor cells and surrounding HSCs. TGF-β is a pleiotropic cytokine with known roles in liver fibrosis and cirrhosis and therefore is key contributor to shaping the hepatic PME. Moreover, TGF-β also contributes to several aspects of liver carcinogenesis leading to tumor progression, and metastasis via epithelial cell autonomous signals and interdependent stromal–epithelial interactions [72,73]. HSC-derived TGF-β stimulated autocrine TGF-β, signaling in premalignant hepatocytes and accumulation of β-catenin within the nucleus of transformed hepatocytes *in vivo* [38]. Additionally, TGF-β is a well-known inducer of the epithelial-to-mesenchymal transition that can contribute to the increased motility of cancer cells.

In addition to TGF-β, activated HSCs secrete potent mitogenic factors such as HGF and epiregulin that promote tumor cell proliferation and growth [35,74]. HGF, present in infiltrating mesenchymal cells and the cytoplasm of cancer cells, has been detected in cirrhotic liver tissue, adenomas, and in 80% of HCC cases [75]. Neutralization of HSC-secreted HGF inhibits proliferation and invasion capacity of HepG2, Huh7, Hep3b, and PLC cells *in vitro* [35,76]. Tumor promotion by HGF is attributed to increased activation of MAPK/ERK, which promotes proliferation and survival of HCC cells. Epiregulin is a member of the EGF family of peptide growth hormones that play a role in both normal and pathological conditions. Epiregulin stimulates the proliferation of epithelial cells [77,78], including HCC cell lines [79], as well as inflammation [80–82]. Epiregulin is used as a prognostic marker in colorectal cancer [77]. Neufert et al. found that epiregulin is primarily secreted by tumor fibroblasts in colitis-associated neoplasms and that it promotes proliferation of intestinal epithelial cells as well as tumor growth *in vivo* [78]. Similarly, in a model of chronic liver injury preceding cancer, activated HSCs were a main source of epiregulin and epiregulin-deficient mice exhibited less tumor burden than wild type controls when subjected to a model of DEN-induced carcinogenesis [74]. These data suggest that HSC secretion of growth factors into the TME is another tumor-promoting mechanism. Figure 9.1 provides an overview of the currently known and putative pathways by which HSCs may regulate tumor growth in HCC.

9.5.5 Role of HSC senescence in hepatocarcinogenesis

Two recent studies have implicated HSC senescence in the promotion of hepatocarcinogenesis. In one study, suppression of HSC senescence (achieved by p53 deletion) resulted in M2 macrophage polarization and proliferation of premalignant hepatocytes [83], implying that the senescence-associated secretory phenotype (SASP) of HSCs suppresses HCC. In contrast, a second study found that SASP promoted HCC development in the setting of obesity [84]. Although the GFAP promoter has been employed to target HSCs, it is neither specific (also deleting in cholangiocytes) nor efficient. With confirmation lacking that p53 was indeed efficiently deleted in HSCs

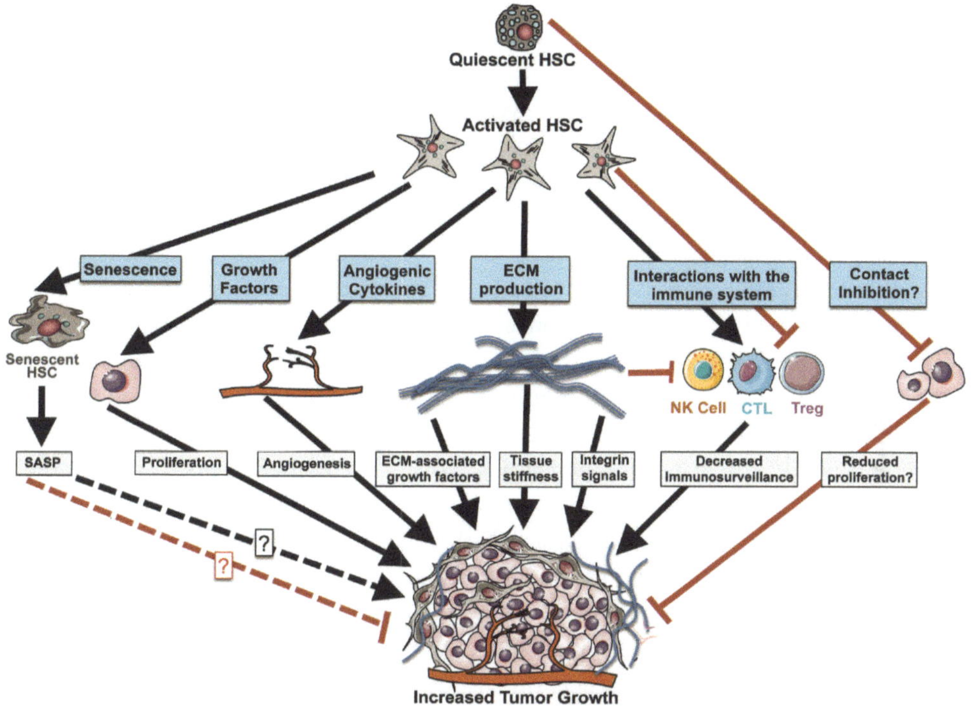

Figure 9.1 ***Proposed mechanisms by which activated HSCs promote HCC.*** Activated HSCs in the PME or TME can potentially support hepatocarcinogenesis by secreting growth factors and cytokines (promoting proliferation and preventing cell death of hepatocytes and tumor cells), proangiogenic cytokines (promoting angiogenesis and tumor blood supply), and ECM components (providing physical support, increasing stiffness, activating integrin signaling, and increasing ECM-associated growth factors). Changes in ECM composition can promote both tumor growth as well as inhibit NK cell function, which decreases tumor immunosurveillance. Although it is known that HSCs undergo senescence, it remains unclear whether the senescence-associated secretory phenotype (SASP) promotes or suppresses HCC. Additionally, activated HSCs can stimulate the proliferation of Treg cells and inhibit the activity of cytotoxic T cells (CTL). Loss of quiescent stellate cells and normal epithelia may promote growth of premalignant hepatocytes by loss of contact inhibition.

and not deleted in other hepatic cell types [83], it is possible that the phenotype in this study may have resulted from combined p53 deletion in multiple or non-HSC cell types. Likewise, it is not clear how the high fat diet employed by Yoshimoto *et al.* induced HSC activation and HSC senescence as high fat diet regimens do not cause HSC activation in genetically unaltered mice unless they are combined with high levels of dietary fructose and/or cholesterol. Further investigations are required to address these open questions to reconcile these opposite conclusions on the role of HSC SASP in hepatocarcinogenesis.

9.6 CONCLUDING REMARKS

The profound changes in the hepatic microenvironment as well as the strong clinical association between fibrosis and HCC development strongly support the hypothesis that HSCs and liver fibrosis exert relevant roles in hepatocarcinogenesis. Moreover, a large number of co-culture and co-transplantation experiments have supplied ample evidence that direct interaction between HSCs and tumor cells promotes tumor growth via multiple mechanisms. Consequently, targeting of HSCs to abrogate their tumor-supporting role presents a putative therapeutic strategy to combat HCC development. However, the hypothesis that HSCs and liver fibrosis promote HCC development remains to be proven *in vivo*. This has been hampered by the lack of suitable animal models to study the direct contribution of HSCs to progression of liver cancer. Animal models allowing to ablate HSCs or maintain their quiescence *in vivo* during hepatocarcinogenesis development would elucidate whether HSCs actively participate in HCC progression.

In vivo evidence that activated HSCs promote the growth of tumor cells is scarce and warrants the development of animal models that can dissect the role of activated HSCs in the TME. To fully characterize the role of HSCs in hepatocarcinogenesis, their specific ablation would serve as a highly useful tool, particularly if depletion can be temporally controlled. In this regard, the role of HSCs in early versus late carcinogenesis will allow for teasing out protumorigenic effects during different stages of the disease progression (i.e., ECM deposition and fibrosis in early HCC vs. secretion of angiogenic and proliferative cytokines in established HCCs). Several chemical agents such as gliotoxin [85] or L-cysteine [86] can act as inhibitors of HSC activation in rodent models. However, it is difficult to exclude that these drugs can possibly affect other hepatic cells [87]. Direct targeting and delivery of chemical agents to inhibit HSC activation or selectively kill HSCs using chimeric molecules that home to activated HSCs [88] can prove to be a more useful tool to study HSCs in hepatocarcinogenesis.

Currently there are very few options for patients affected by HCC including liver transplantation, resection of the affected liver lobes, and a number of chemotherapeutic agents that act on the EGF pathway or block angiogenesis. Expanding this repertoire to include therapies that target HSCs/myofibroblasts in the PME or TME could be important for both HCC prevention and therapy. Sorafenib, a small molecule multikinase inhibitor with an established role in HCC therapy [89], also reduces fibrosis, inflammation, and angiogenesis [90,91]. Hence, it is conceivable that multikinase such as sorafenib have a role in primary or secondary HCC prevention. It is also conceivable that the combination of chemotherapeutic drugs with drugs that can alter the behavior of activated HSCs/myofibroblasts in the TME may exert additive effects, for example, blocking survival signals and rendering tumor cells more sensitive to cytostatic therapies. In summary, the hepatic PME and TME and the HSC/myofibroblast compartment may present additional and yet unexplored strategies for medical HCC treatment, and need to be studied in more detail.

REFERENCES

[1] Luedde T, Kaplowitz N, Schwabe RF. Cell death and cell death responses in liver disease: mechanisms and clinical relevance. Gastroenterology 2014;147(4):765–83. e4.

[2] Fattovich G, et al. Hepatocellular carcinoma in cirrhosis: incidence and risk factors. Gastroenterology 2004;127(5 Suppl. 1):S35–50.

[3] Hoshida Y, et al. Gene expression in fixed tissues and outcome in hepatocellular carcinoma. N Engl J Med 2008;359(19):1995–2004.

[4] Hernandez-Gea V, et al. Role of the microenvironment in the pathogenesis and treatment of hepatocellular carcinoma. Gastroenterology 2013;144(3):512–27.

[5] Luedde T, Schwabe RF. NF-kappaB in the liver-linking injury, fibrosis and hepatocellular carcinoma. Nat Rev Gastroenterol Hepatol 2011;8(2):108–18.

[6] Grivennikov SI, Greten FR, Karin M. Immunity, inflammation, and cancer. Cell 2010;140(6):883–99.

[7] Hanahan D, Weinberg RA. Hallmarks of cancer: the next generation. Cell 2011;144(5):646–74.

[8] Taura K, et al. Hepatic stellate cells secrete angiopoietin 1 that induces angiogenesis in liver fibrosis. Gastroenterology 2008;135(5):1729–38.

[9] Ehling J, et al. CCL2-dependent infiltrating macrophages promote angiogenesis in progressive liver fibrosis. Gut 2014.

[10] Mederacke I, et al. Fate tracing reveals hepatic stellate cells as dominant contributors to liver fibrosis independent of its aetiology. Nat Commun 2013;4:2823.

[11] Bissell MJ, Hines WC. Why don't we get more cancer? A proposed role of the microenvironment in restraining cancer progression. Nat Med 2011;17(3):320–9.

[12] Olumi AF, et al. Carcinoma-associated fibroblasts direct tumor progression of initiated human prostatic epithelium. Cancer Res 1999;59(19):5002–11.

[13] Bhowmick NA, et al. TGF-beta signaling in fibroblasts modulates the oncogenic potential of adjacent epithelia. Science 2004;303(5659):848–51.

[14] Xu J, et al. The types of hepatic myofibroblasts contributing to liver fibrosis of different etiologies. Front Pharmacol 2014;5:167.

[15] Bakiri L, Wagner EF. Mouse models for liver cancer. Mol Oncol 2013;7(2):206–23.

[16] Caviglia JM, Schwabe RF. Experimental hepatocarcinogenesis. McManus LM, Mitchell RN, editors. Pathobiology of Human Disease. San Diego: Elsevier; 2014. p. 1866–80.

[17] Bettermann K, et al. TAK1 suppresses a NEMO-dependent but NF-kappaB-independent pathway to liver cancer. Cancer Cell 2010;17(5):481–96.

[18] Inokuchi S, et al. Disruption of TAK1 in hepatocytes causes hepatic injury, inflammation, fibrosis, and carcinogenesis. Proc Natl Acad Sci USA 2010;107(2):844–9.

[19] Luedde T, et al. Deletion of NEMO/IKKgamma in liver parenchymal cells causes steatohepatitis and hepatocellular carcinoma. Cancer Cell 2007;11(2):119–32.

[20] Campbell JS, et al. Platelet-derived growth factor C induces liver fibrosis, steatosis, and hepatocellular carcinoma. Proc Natl Acad Sci USA 2005;102(9):3389–94.

[21] Wright JH, et al. Paracrine activation of hepatic stellate cells in platelet-derived growth factor C transgenic mice: evidence for stromal induction of hepatocellular carcinoma. Int J Cancer 2014;134(4):778–88.

[22] Okada H, et al. Acyclic retinoid targets platelet-derived growth factor signaling in the prevention of hepatic fibrosis and hepatocellular carcinoma development. Cancer Res 2012;72(17):4459–71.

[23] Hede SM, et al. GFAP promoter driven transgenic expression of PDGFB in the mouse brain leads to glioblastoma in a Trp53 null background. Glia 2009;57(11):1143–53.

[24] Fuchs BC, et al. Epidermal growth factor receptor inhibition attenuates liver fibrosis and development of hepatocellular carcinoma. Hepatology 2014;59(4):1577–90.

[25] Chau KY, et al. Myofibroblasts in hepatitis B related cirrhosis and hepatocellular carcinoma. J Clin Pathol 1992;45(5):446–8.

[26] Enzan H, et al. Alpha-smooth muscle actin-positive perisinusoidal stromal cells in human hepatocellular carcinoma. Hepatology 1994;19(4):895–903.

[27] Faouzi S, et al. Myofibroblasts are responsible for collagen synthesis in the stroma of human hepatocellular carcinoma: an *in vivo* and *in vitro* study. J Hepatol 1999;30(2):275–84.

[28] Ooi LPJ, et al. Evidence that "myofibroblast-like" cells are the cellular source of capsular collage in hepatocellular carcinoma. J Hepatol 1997;26(4):798–807.

[29] Ju MJ, et al. Peritumoral activated hepatic stellate cells predict poor clinical outcome in hepatocellular carcinoma after curative resection. Am J Clin Pathol 2009;131(4):498–510.
[30] Okabe H, et al. Hepatic stellate cells may relate to progression of intrahepatic cholangiocarcinoma. Ann Surg Oncol 2009;16(9):2555–64.
[31] Worthley DL, et al. Bone marrow cells as precursors of the tumor stroma. Exp Cell Res 2013.
[32] Barcellos-de-Souza P, et al. Tumor microenvironment: bone marrow-mesenchymal stem cells as key players. Biochim Biophys Acta 2013;1836(2):321–35.
[33] Mishra PJ, et al. Carcinoma-associated fibroblast-like differentiation of human mesenchymal stem cells. Cancer Res 2008;68(11):4331–9.
[34] Quante M, et al. Bone marrow-derived myofibroblasts contribute to the mesenchymal stem cell niche and promote tumor growth. Cancer Cell 2011;19(2):257–72.
[35] Amann T, et al. Activated hepatic stellate cells promote tumorigenicity of hepatocellular carcinoma. Cancer Sci 2009;100(4):646–53.
[36] Wang ZM, et al. Rat hepatic stellate cells alter the gene expression profile and promote the growth, migration and invasion of hepatocellular carcinoma cells. Mol Med Rep 2014.
[37] Coulouarn C, et al. Hepatocyte-stellate cell cross-talk in the liver engenders a permissive inflammatory microenvironment that drives progression in hepatocellular carcinoma. Cancer Res 2012;72(10):2533–42.
[38] Mikula M, et al. Activated hepatic stellate cells induce tumor progression of neoplastic hepatocytes in a TGF-beta dependent fashion. J Cell Physiol 2006;209(2):560–7.
[39] Geng ZM, et al. Activated human hepatic stellate cells promote growth of human hepatocellular carcinoma in a subcutaneous xenograft nude mouse model. Cell Biochem Biophys 2014;70(1):337–47.
[40] Lin N, et al. Role of activated hepatic stellate cells in proliferation and metastasis of hepatocellular carcinoma. Hepatol Res 2014.
[41] Hynes RO. The extracellular matrix: not just pretty fibrils. Science 2009;326(5957):1216–9.
[42] Rombouts K, Carloni V. The fibrotic microenvironment as a heterogeneity facet of hepatocellular carcinoma. Fibrogenesis Tissue Repair 2013;6(1):17.
[43] Carloni V, Luong TV, Rombouts K. Hepatic stellate cells and extracellular matrix in hepatocellular carcinoma: more complicated than ever. Liver Int 2014;34(6):834–43.
[44] Santamato A, et al. Hepatic stellate cells stimulate HCC cell migration via laminin-5 production. Clin Sci (Lond) 2011;121(4):159–68.
[45] Lai KK, et al. Extracellular matrix dynamics in hepatocarcinogenesis: a comparative proteomics study of PDGFC transgenic and Pten null mouse models. PLoS Genet 2011;7(6):e1002147.
[46] Lee SK, et al. Integrin alpha V polymorphisms and haplotypes in a Korean population are associated with susceptibility to chronic hepatitis and hepatocellular carcinoma. Liver Int 2009;29(2):187–95.
[47] Cox D, Brennan M, Moran N. Integrins as therapeutic targets: lessons and opportunities. Nat Rev Drug Discov 2010;9(10):804–20.
[48] Zhang DY, Friedman SL. Fibrosis-dependent mechanisms of hepatocarcinogenesis. Hepatology 2012;56(2):769–75.
[49] Levental KR, et al. Matrix crosslinking forces tumor progression by enhancing integrin signaling. Cell 2009;139(5):891–906.
[50] Lu P, Weaver VM, Werb Z. The extracellular matrix: a dynamic niche in cancer progression. J Cell Biol 2012;196(4):395–406.
[51] Schrader J, et al. Matrix stiffness modulates proliferation, chemotherapeutic response, and dormancy in hepatocellular carcinoma cells. Hepatology 2011;53(4):1192–205.
[52] Zhao W, et al. Hepatic stellate cells promote tumor progression by enhancement of immunosuppressive cells in an orthotopic liver tumor mouse model. Lab Invest 2014;94(2):182–91.
[53] Masuzaki R, et al. Risk assessment of hepatocellular carcinoma in chronic hepatitis C patients by transient elastography. J Clin Gastroenterol 2008;42(7):839–43.
[54] Begum NA, et al. Differential display and integrin alpha 6 messenger RNA overexpression in hepatocellular carcinoma. Hepatology 1995;22(5):1447–55.
[55] Zhao G, et al. Mechanical stiffness of liver tissues in relation to integrin beta1 expression may influence the development of hepatic cirrhosis and hepatocellular carcinoma. J Surg Oncol 2010;102(5):482–9.
[56] Giannelli G, et al. Altered expression of integrins and basement membrane proteins in malignant and pre-malignant lesions of oral mucosa. J Biol Regul Homeost Agents 2001;15(4):375–80.

[57] Okazaki I. Novel cancer-targeting agents/application strategies developed from MMP science. Anticancer Agents Med Chem 2012.
[58] Jia YL, et al. Epimorphin promotes human hepatocellular carcinoma invasion and metastasis through activation of focal adhesion kinase/extracellular signal-regulated kinase/matrix metalloproteinase-9 axis. Hepatology 2011;54(5):1808–18.
[59] Kang N, Gores GJ, Shah VH. Hepatic stellate cells: partners in crime for liver metastases? Hepatology 2011;54(2):707–13.
[60] Zhao W, et al. Activated hepatic stellate cells promote hepatocellular carcinoma development in immunocompetent mice. Int J Cancer 2011;129(11):2651–61.
[61] Zhao W, et al. The role of hepatic stellate cells in the regulation of T-cell function and the promotion of hepatocellular carcinoma. Int J Oncol 2012;41(2):457–64.
[62] Rosmorduc O, Housset C. Hypoxia: a link between fibrogenesis, angiogenesis, and carcinogenesis in liver disease. Semin Liver Dis 2010;30(03):258–70.
[63] Ankoma-Sey V, Wang Y, Dai Z. Hypoxic stimulation of vascular endothelial growth factor expression in activated rat hepatic stellate cells. Hepatology 2000;31(1):141–8.
[64] Sanz-Cameno P, et al. Hepatitis B virus promotes angiopoietin-2 expression in liver tissue: role of HBV x protein. Am J Pathol 2006;169(4):1215–22.
[65] Torimura T, et al. Overexpression of angiopoietin-1 and angiopoietin-2 in hepatocellular carcinoma. J Hepatol 2004;40(5):799–807.
[66] Olaso E, et al. Proangiogenic role of tumor-activated hepatic stellate cells in experimental melanoma metastasis. Hepatology 2003;37(3):674–85.
[67] Kang N, et al. Focal adhesion assembly in myofibroblasts fosters a microenvironment that promotes tumor growth. Am J Pathol 2010;177(4):1888–900.
[68] Chen JA, et al. Angiogenesis: multiple masks in hepatocellular carcinoma and liver regeneration. Hepatol Int 2010;4(3):537–47.
[69] Zhang ZL, Liu ZS, Sun Q. Expression of angiopoietins, Tie2 and vascular endothelial growth factor in angiogenesis and progression of hepatocellular carcinoma. World J Gastroenterol 2006;12(26): 4241–5.
[70] Wirz W, et al. Hepatic stellate cells display a functional vascular smooth muscle cell phenotype in a three-dimensional co-culture model with endothelial cells. Differentiation 2008;76(7):784–94.
[71] Shimizu S, et al. *In vivo* and *in vitro* interactions between human colon carcinoma cells and hepatic stellate cells. Jpn J Cancer Res 2000;91(12):1285–95.
[72] Meindl-Beinker NM, Matsuzaki K, Dooley S. TGF-beta signaling in onset and progression of hepatocellular carcinoma. Dig Dis 2012;30(5):514–23.
[73] Stover DG, Bierie B, Moses HL. A delicate balance: TGF-beta and the tumor microenvironment. J Cell Biochem 2007;101(4):851–61.
[74] Dapito DH, et al. Promotion of hepatocellular carcinoma by the intestinal microbiota and TLR4. Cancer Cell 2012;21(4):504–16.
[75] Ljubimova JY, et al. Expression of HGF, its receptor c-met, c-myc, and albumin in cirrhotic and neoplastic human liver tissue. J Histochem Cytochem 1997;45(1):79–87.
[76] Neaud V, et al. Human hepatic myofibroblasts increase invasiveness of hepatocellular carcinoma cells: evidence for a role of hepatocyte growth factor. Hepatology 1997;26(6):1458–66.
[77] Riese Ii DJ, Cullum RL. Epiregulin: roles in normal physiology and cancer. Semin Cell Dev Biol 2014;28:49–56.
[78] Neufert C, et al. Tumor fibroblast-derived epiregulin promotes growth of colitis-associated neoplasms through ERK. J Clin Invest 2013;123(4):1428–43.
[79] Zhao M, et al. Dual knockdown of N-ras and epiregulin synergistically suppressed the growth of human hepatoma cells. Biochem Biophys Res Commun 2009;387(2):239–44.
[80] Brandl K, et al. MyD88 signaling in nonhematopoietic cells protects mice against induced colitis by regulating specific EGF receptor ligands. Proc Natl Acad Sci USA 2010;107(46):19967–72.
[81] Shirasawa S, et al. Dermatitis due to epiregulin deficiency and a critical role of epiregulin in immune-related responses of keratinocyte and macrophage. Proc Natl Acad Sci USA 2004;101(38):13921–6.
[82] Zhang Y, et al. Important role of epiregulin in inflammatory responses during corneal epithelial wound healing. Invest Ophthalmol Vis Sci 2012;53(4):2414–23.

[83] Lujambio A, et al. Non-cell-autonomous tumor suppression by p53. Cell 2013;153(2):449–60.
[84] Yoshimoto S, et al. Obesity-induced gut microbial metabolite promotes liver cancer through senescence secretome. Nature 2013;499(7456):97–101.
[85] Wright MC, et al. Gliotoxin stimulates the apoptosis of human and rat hepatic stellate cells and enhances the resolution of liver fibrosis in rats. Gastroenterology 2001;121(3):685–98.
[86] Horie T, et al. L-cysteine administration prevents liver fibrosis by suppressing hepatic stellate cell proliferation and activation. Biochem Biophys Res Commun 2003;305(1):94–100.
[87] Yin C, et al. Hepatic stellate cells in liver development, regeneration, and cancer. J Clin Invest 2013;123(5):1902–10.
[88] Bansal R, et al. Targeted recombinant fusion proteins of IFNgamma and mimetic IFNgamma with PDGFbetaR bicyclic peptide inhibits liver fibrogenesis in vivo. PLoS One 2014;9(2):e89878.
[89] Llovet JM, et al. Sorafenib in advanced hepatocellular carcinoma. N Engl J Med 2008;359(4):378–90.
[90] Mejias M, et al. Beneficial effects of sorafenib on splanchnic, intrahepatic, and portocollateral circulations in portal hypertensive and cirrhotic rats. Hepatology 2009;49(4):1245–56.
[91] Hennenberg M, et al. Hepatic and HSC-specific sorafenib effects in rats with established secondary biliary cirrhosis. Lab Invest 2011;91(2):241–51.

CHAPTER 10

Stellate Cells in Alcoholic Hepatitis

Palash Mandal[1] and Laura E. Nagy[2]

[1]Department of Biological Sciences, BITS PILANI Hyderabad Campus, Hyderabad, India
[2]Center for Liver Disease Research, Department of Pathobiology, Cleveland Clinic, Cleveland, OH, USA, Departments of Molecular Medicine and Nutrition, Case Western Reserve University, Cleveland, OH, USA

10.1 INTRODUCTION

Alcoholic liver disease (ALD) is a complex multifactorial disease caused by chronic alcohol consumption and is a major cause of morbidity and mortality worldwide. According to the 2014 WHO report on global status report on alcohol and health, alcohol consumption is responsible for 3.3 million deaths and 139 million disability-adjusted life years (DALYs) in the world in 2012. The chronic form of ALD progresses in stages of steatosis, steatohepatitis, fibrosis, and cirrhosis, which is the end stage of liver disease; heavy drinkers also have increased incidence of hepatocellular carcinoma. Chronic ALD typically develops over decades of alcohol abuse. Acute alcoholic steatohepatitis (ASH), a severe acute inflammatory disease, can occur at any stage of ALD and is associated with a very high short-term mortality rate [1]. This high short-term mortality is due to liver failure, portal hypertension, and bacterial infections [2]. Interestingly, many patients with ASH exhibit stage IV fibrosis, suggesting an increased susceptibility to ASH in patients with more advanced fibrosis [1]. While it has long been appreciated that early stages of ALD, including steatosis and steatohepatitis, are reversible, recent evidence indicates that early stages of fibrosis may also be reversible, suggesting that specific therapeutic interventions targeting fibrosis may improve the long-term outcome in patients with ALD.

10.2 PATHOPHYSIOLOGY OF FIBROSIS IN ALD

The liver is mainly comprised of five cell types: hepatocytes or parenchymal cells, resident macrophages (Kupffer cells), sinusoidal endothelial cells, hepatic stellate cells (HSCs), the pericytes of the liver (also known as Ito cells, fat storing cells, lipocytes, or interstitial cells), and liver specific natural killer (NK) cells (also known as pit cells). The progression of fibrosis in ALD involves complex interactions between resident liver parenchymal and non-parenchymal cells, as well as recruitment of other immune cells to the liver, particularly targeted to the areas of damage and inflammation within the liver parenchyma [3,4].

Stellate Cells in Health and Disease
DOI: http://dx.doi.org/10.1016/B978-0-12-800134-9.00010-5

10.2.1 Hepatic stellate cells

HSCs play a particularly significant role in the progression of fibrosis during the pathogenesis of ALD. HSC were first identified by Carl von Kupffer in 1876 as dark intralobular star-shaped cells via a selective gold chloride staining method [5]. Approximately 10% of the total resident liver cells are represented by HSC in rats and in a healthy liver; they play a critical role in the uptake and storage of vitamin A [6]. HSCs are localized to the sub-endothelial space of Disse; their star-shaped branches encompass the sinusoidal endothelial tube [7]. HSCs are characterized by the presence of rough endoplasmic reticulum, cytoplasmic droplets rich with triacylglycerol, small numbers of mitochondria, presence of collagen bundles in proximity to the cell body, and relative absence of cytosolic vesicles and vacuoles [5]. Markers of both mesenchymal cells, such as vimentin, desmin, and collagen type I, as well as glial or neuronal cells, like nestin, neural cell adhesion molecule, glial fibrillary acidic protein, p75Ntr, and synaptophysin, are expressed by HSCs [8]. While still not well understood, recent data suggest that the origin of HSCs is from mesoderm via septum transversum and mesothelium [9].

10.2.2 Function of HSCs

The most important function of HSCs in the normal liver is the storage and transport of vitamin A (retinol); up to 80% of the total body retinol, in the form of retinyl esters, is stored as cytoplasmic lipid droplets in HSCs [10]. HSCs also contribute to the homeostatic maintenance of the three-dimensional architecture of liver by regulating the turnover of extracellular matrix (ECM) in the healthy liver [11]. HSCs are also involved in the regulation of liver regeneration through cell-to-cell interactions and cytokine networks [12], modulation of sinusoidal blood flow [13], and contribute to hepatic immune tolerance via their immunological properties [14] (see Chapter 14). HSCs have a complex secretome, including growth factors, prostaglandins, cytokines, and apolipoproteins; all these factors contribute to the homeostasis of ECM [5,15].

10.2.3 Activation and function of HSCs

In the healthy liver, HSCs are quiescent, but, in response to specific activation signals, they undergo a transition into a proliferative, contractile, and fibrogenic phenotype characteristic of myofibroblasts. Signals are generated in response to liver injury "prime" quiescent HSCs; subsequent signals then induce their transdifferentiation from a "quiescent" to an "active" phenotype. Pro-fibrogenic signals, such as platelet derived growth factor (PDGF), stimulate HSC proliferation and migration to areas of injury. Additional autocrine and paracrine signals, including transforming growth factor-β (TGFβ) and connective tissue growth factor (CTGF), stimulate HSC activation and fibrogenesis. HSCs are maintained in their active state both by the sustained activity of initiation signals, along with a number of additional autocrine and paracrine mediators produced by HSC and other cell types within the liver.

Activated HSCs are morphologically characterized by enlargement of the rough endoplasmic reticulum (ER), disheveled nuclear membrane, loss of vitamin A droplets, and the appearance of contractile filaments. Alpha-smooth muscle actin (α-sma) expression is increased after HSC activation, increasing their contractile potential; α-sma is considered a marker for activated HSCs. Importantly, activated HSCs also synthesize and secrete ECM components, primarily type I collagen, contributing to the accumulation of a fibrotic matrix in the injured areas of the liver [16] (see Chapter 4).

Activated HSC exhibit many features of innate immune cells and play an important role in regulation of the immune environment of the liver. The activation of HSC leads to changes in both the profile and quantity of inflammatory cytokines, chemokines, and growth factors that are expressed and secreted [17,18]. For example, activated HSC contribute to the regulation of the hepatic immune system by secreting cytokines and chemokines, including IL-8, MCP-1, TNF-α, MIP-2, and IL-6 [17–23]; this results in increased inflammation in the liver and infiltration of leukocytes. Factors secreted by HSC are important mediators of crosstalk with hepatic macrophages, promoting their differentiation to both profibrotic and proinflammatory phenotypes [24]. Activated HSCs also contribute to hepatic inflammation by secreting C4, a protein in the complement system [25], and can stimulate proliferation or apoptosis of lymphocytes by functioning as antigen-presenting cells [26]. They also alter hepatic immune responses by expressing co-stimulatory molecules like programmed death ligand I, which regulates apoptosis of T cells [27]. Finally, HSCs also express Toll-like receptors (TLR) 4 and 9 on their surface, which can interact with bacterial products and danger signals [20,28].

10.3 HSC AND ASH

While it is clear that chronic, heavy alcohol consumption results in the activation of HSC, accumulation of extracellular matrix, and the development of fibrosis, we know very little about the mechanisms by which alcohol activates HSC. It is likely that alcohol has both direct and indirect effects on HSC activation and function. Direct effects of ethanol and its metabolites have been studied in primary cultures of HSC, while indirect effects of ethanol have been studied both in co-culture systems and *in vivo*.

Progress in the field has been hampered by the lack of suitable animal models to study. Hepatic fibrosis does not develop in rodent models in response to ethanol exposure alone, so investigators have developed models of fibrosis that combine alcohol with other hepatotoxins, such as carbon tetrachloride (CCl_4), or other injurious conditions, such as overfeeding high fat diets [29]. The hepatotoxicity of CCl_4 is limited by its activation via cytochrome P450 2E1 (CYP2E1) [30]. Because heavy ethanol exposure induces CYP2E1, alternative strategies need to be taken to normalize the activation of CCl_4 between control and ethanol-exposed mice. Gao et al. equalized the exposure active CCl_4 in ethanol-fed mice (assessed by serum ALT concentrations) by decreasing the dose of CCl_4 given to ethanol-fed mice [31], while more recent studies from our group have used concentrations of ethanol (2%, v/v, or 11% of calories) in the diet that do not induce CYP2E1 [32–34]. The availability of these different models to study the acceleration of fibrosis by ethanol now makes it possible to study the impact of ethanol on the progression of fibrosis in rodent models.

10.4 DIRECT EFFECTS OF ETHANOL AND ITS METABOLITES ON HSC

10.4.1 Pathways of ethanol metabolism

Ethanol is metabolized mainly in the liver and involves three major steps: (i) reversible oxidative breakdown of alcohol to toxic acetaldehyde; (ii) nonreversible conversion of toxic acetaldehyde to acetate via acetaldehyde dehydrogenase (ALDH); and (iii) acetate breakdown to water and carbon dioxide. Metabolism of alcohol to acetaldehyde is effectuated by three enzymes: alcohol dehydrogenase (ADH), microsomal ethanol oxidizing system (MEOS) catalyzed by CYP2E1, and catalase. Ethanol is mainly metabolized by ADH, but chronic, heavy ethanol consumption increases the activity of CYP2E1 [4,35].

10.4.2 Interaction of ethanol metabolism with HSCs

10.4.2.1 Acetaldehyde and oxidative stress: regulation of collagen synthesis

When CYP2E1 is induced after chronic, heavy ethanol consumption, there is an increase in reactive oxygen species (ROS) generation leading to oxidative stress [36,37]. Acetaldehyde formed during the metabolism of alcohol also enhances ROS formation [38]. Increased ROS production mediated by CYP2E1 generates highly reactive free radicals such as superoxide anion, and hydroxyethyl radical that form adducts with proteins, DNA, and lipids [39,40]. Acetaldehyde also forms protein and DNA adducts thereby impairing gene expression and cellular function [39,41–43]. Acetaldehyde can upregulate collagen I expression directly or indirectly via increased expression of TGFβ. Studies have also found that acetaldehyde sensitizes HSC to stimulation by TGFβ [44]. Ethanol-induced oxidative stress and adduct formation can also contribute to the activation of HSC resulting in secretion of cytokines, chemokines, and ECM proteins [45].

Ethanol-induced ROS production also promotes formation of lipid peroxidation products such as malondialdehyde (MDA) and 4-hydroxynonenal that bind to proteins to form adducts [41]. MDA and acetaldehyde synergistically react together with proteins to form malondialdehyde–acetaldehyde (MAA) protein adducts [46]; adduct accumulation correlates with disease severity in patients with ALD [45]. Further, cell culture studies suggest that MAA adducts activate HSC and stimulate secretion of chemokines, thus enhancing the infiltration of monocytes and neutrophils to the liver parenchyma and exacerbating inflammatory and fibrogenic response [21].

10.4.2.2 ADH in retinoid homeostasis in HSCs

Retinoid homeostasis influences the activity of HSC. Alcohol dehydrogenase III (ADH III) is the major enzyme responsible for conversion of retinol to retinaldehyde in HSC [47]. Retinaldehyde is further metabolized to retinoic acid by retinaldehyde dehydrogenase enzyme. The regulation of retinoid homeostasis by alcohol is not well understood. Several reports suggest that retinol metabolites can activate HSC via activation of latent TGF-β, as well as increase expression of suppressor of cytokine signaling 1 (SOCS 1) and pro-collagen 1α proteins [48]. There is also likely interaction between retinoic acid and suppression of NK cell activity; decreased NK cell activity is associated with decreased apoptosis of activated HSC [47]. While these studies clearly demonstrate a role for the enzymatic machinery of ethanol metabolism in regulation of HSC activation and function, very little is known about the influence of ethanol exposure on these pathways. Future studies investigating the interaction among ethanol, ADH III, retinoid homeostasis, and fibrosis should provide this important information.

10.4.2.3 Discoidin domain receptor 2

Discoidin domain receptor 2 (DDR2) is a tyrosine kinase receptor-like protein expressed in activated HSC [49–51]. Prolonged activation of DDR2 upregulates MMP2, contributing to further activation and proliferation of HSC during ethanol-induced liver injury [52–54]. *In vitro* and *in vivo* studies demonstrate that alcohol-induced HSC activation and proliferation is inhibited by DDR2 gene silencing using shRNA (short hairpin RNA). DDR2 knock-down also resulted in HSC apoptosis, reduced collagen deposition, and improved chronic ethanol-induced liver injury [55].

10.4.2.4 Alcohol metabolism: hypoxia, hypoxia inducible factors, and adenosine

Ethanol metabolism leads to a transient and localized hypoxia in the liver [32,56] resulting in the stabilization of an oxygen-sensitive transcription factor known as the hypoxia-inducible factor 1α (HIF1α). Activation and stabilization of HIF1α can contribute via multiple mechanisms to increase fibrosis. HIF1α promotes the expression of pro-inflammatory and pro-fibrogenic genes, including VEGF, iNOS, and COX-2 in murine models of fibrosis [57]. HIF1α also contributes to ethanol-induced apoptosis of hepatocytes [33]; apoptosis of hepatocytes is an important activator of HSC [58] and contributes to the acceleration of fibrosis by ethanol [59]. Hepatocyte apoptosis in response to challenge with ethanol and CCl_4 is associated with HIF1α-dependent activation of p53, as well as the downstream effectors of p53, including p53 upregulated modulator of apoptosis (PUMA) [33].

Ethanol-induced hypoxia also results in the release of adenosine and activates adenosine receptor [32]. Adenosine signals through G-protein coupled adenosine receptors—A1, A2A, A2B, and A3 [60]. Ethanol consumption sensitizes HSC to activation, a process known as priming of HSC. This is due to the establishment of hypoxia conditions in the liver as a consequence of ethanol metabolism resulting in adenosine generation [61] and expression of angiogenic and chemotactic factors [62]. Hypoxia-induced angiogenic factors like ANGP-1 and VEGF also increase HSC activation and migration, thus contributing to both the initiation and progression of fibrosis [63,64].

10.5 INTERACTIONS OF HSCs WITH INNATE IMMUNE SYSTEMS IN ASH

HSC exhibit a number of innate immune functions, expressing toll-like receptors (TLRs), components of the inflammasome and receptors for complement. The progression of ASH is known to activate multiple innate immune pathways, likely contributing to the activation of HSC and progression of fibrosis.

10.5.1 Role of lipopolysaccharide and TLRs

Chronic and excessive alcohol consumption increases intestinal permeability resulting in the translocation of bacterially derived products, such as lipopolysaccharide, into the portal circulation. Bacterial products can then activate TLR signaling in the liver, leading to the production of cytokines and chemokines. Chronic ethanol also increases the expression of TLRs in the liver and sensitizes Kupffer cells, the resident macrophage in the liver, to stimulation by TLR4 ligands [4]. Activation of Kupffer cells in response to chronic alcohol can activate HSC both by direct paracrine effects due to secreted cytokines, such as TGFβ, or via indirect injury to hepatocytes by ROS and TNF-α [44].

In addition to the impact of TLR4 on Kupffer cells to the activation of HSC, TLR4 is also expressed by HSC [17]. In HSC, TLR4 activation activates IKK/NF-κB signaling pathway [20], contributing to a downregulation of BAMBI, a transmembrane inhibitor of TGF-β. Downregulation of BAMBI thus enhances TGF-β signaling, exacerbating liver fibrosis [65]. TLR4 signaling in HSC also increases the expression of chemokines, including MCP-1, MIP-1α, MIP-1β, and regulated on activation, normal T cell expressed and secreted (RANTES) [66,67]. LPS also upregulates CCL20 chemokine production from HSC. *In vivo* silencing of CCL20 gene is shown to reduce hepatic pro-inflammatory and pro-fibrogenic gene levels, suggesting that CCL20 plays a role in hepatic inflammation, injury, and fibrosis [68]. Taken together, these data illustrate the critical crosstalk between the gut and liver, as well as the important TLR4-dependent cell–cell communication between HSC and Kupffer cells, in the progression of ALD [69].

10.5.2 Complement

Complement is a critical component of the innate immune system that provides links to adaptive immunity [70]. The complex system is comprised of both plasma proteins and membrane-bound receptors and regulatory proteins. C5 is identified as a quantitative trait gene in mouse and human that contributes to hepatic fibrosis across species [71]. The precise role of complement in fibrosis is not yet well understood; however, it is likely that complement contributes to multiple steps in the response to injurious signals to the liver. For example, complement is activated in response to both chronic ethanol feeding [72,73] as well as in response to CCl_4-induced liver injury [74]. Complement activation products generated in response to fibrotic insults are involved in both the priming of hepatocytes for regeneration and the clearance of cellular debris. Further, recent reports suggest that activated HSCs express C5a receptors and that C5a can stimulate migration of activated HSCs [75].

10.5.3 Inflammasomes

HSCs also express components of the inflammasome. Activation of the inflammasome appears to regulate multiple functions of HSCs and both $ASC^{-/-}$ and $NLRP3^{-/-}$ mice are protected from CCl_4-induced fibrosis [76]. Importantly, chronic ethanol exposure also activates inflammasomes [77], suggesting that this pathway may also contribute to the stimulation of HSCs and fibrosis by ethanol. However, future studies using cell-specific knock-outs of the inflammasome in HSCs are required to determine if ethanol-mediated activation of inflammasomes specifically occurs in HSCs.

10.6 INTERACTIONS OF HSCs WITH ADAPTIVE IMMUNITY IN ASH

Adaptive T-cell responses contribute to hepatic inflammation and fibrosis; however, little is known about the interaction of adaptive T cells and ethanol in the regulation of HSC.

The available data suggest that $CD4^+CD25^+Foxp3^+$ regulatory T cells (T_{regs}) can reduce hepatic fibrosis. T_{regs} are critical to maintenance of immune homeostasis; failure to induce T_{regs} could thus contribute to dysregulation of immune responses during wound healing, ultimately leading to unchecked fibrotic responses. Interestingly, patients with alcoholic hepatitis have decreased circulating numbers of T_{regs} [78].

Effector T cells are also involved in the fibrotic response. Studies on human liver biopsies and human HSC *in vitro* have shown that IL-17 receptor is expressed on HSCs. Furthermore, Th17 cells secrete IL-17 after ethanol exposure [79]. IL-17 binds to IL-17 receptor on HSCs resulting in increased expression of IL-6, adhesion molecules, growth factors, such as granulocyte-macrophage colony-stimulating factor and granulocyte-colony stimulating factor, as well as CXC chemokines, including CXCL6,

CXCL1, and CXCL8. Thus, IL-17 can stimulate both neutrophil infiltration and activation, contributing to the pathogenesis of ALD [80]. IL-17 receptor ligation also activates HSCs via STAT3 and stimulates expression of collagen-1α [81].

10.7 SUMMARY: POTENTIAL THERAPEUTIC STRATEGIES FOR REDUCING HSC ACTIVATION IN ASH

As more mechanistic information on the interactions between ethanol and HSC activation becomes available from cellular and animal-based studies, there is a growing potential for the development of rationally designed therapeutic interventions in the prevention and treatment of the fibrosis associated with ASH. While general anti-fibrotic therapies will likely be useful in treating fibrosis associated with ASH, it is also possible that there may be specific targets related to alcohol and alcohol metabolism. For example, based on mechanistic studies revealing that ethanol metabolism is associated with increased local adenosine concentrations in the liver, further *in vivo* studies established that administration of adenosine 2A receptor (A2AR) antagonists, like KW-6002, to ethanol-fed/CCl_4 exposed mice attenuated HSC activation. Treatment with the A2AR antagonist not only prevented fibrosis progression but also reversed fibrosis [32]. Another therapeutic approach is to target IL22 because HSC express high levels of IL22 receptor R1 (IL22R1). *In vitro* or *in vivo* treatment with IL22 ameliorates ethanol-induced injury by activating STAT3 in HSC and inducing SOCS3 to stimulate the senescence of HSCs [82,83]. As a final example, activated NK cells are an important cellular pathway for killing activated HSCs via the release of TRAIL [1]. However, because alcohol consumption suppresses the anti-fibrotic effects of NK cells [1], strategies to promote NK cell activity could be an important approach to limiting fibrosis in ASH.

In summary, HSCs play a key role in the development of fibrosis in ASH and in the progression of ALD. Though much progress has been made in understanding interactions between ethanol and the biology and function of HSC, further research is needed to better understand the impact of alcohol and ASH on the complex cell–cell interactions in the liver, as well as the role of both innate and adaptive immunity in regulating HSCs in ASH.

REFERENCES

[1] Gao B, Bataller R. Alcoholic liver disease: pathogenesis and new therapeutic targets. Gastroenterology 2011;141:1572–85.
[2] Boetticher NC, Peine CJ, Kwo P, Abrams GA, Patel T, Aqel B, et al. A randomized, double-blinded, placebo-controlled multicenter trial of etanercept in the treatment of alcoholic hepatitis. Gastroenterology 2008;135:1953–60.
[3] Gressner AM, Bachem MG. Molecular mechanisms of liver fibrogenesis—a homage to the role of activated fat-storing cells. Digestion 1995;56:335–46.

[4] Nagy LE. Molecular aspects of alcohol metabolism: transcription factors involved in early ethanol-induced liver injury. Annu Rev Nutr 2004;24:55–78.
[5] Geerts A. History, heterogeneity, developmental biology, and function of quiescent hepatic stellate cells. Semin Liver Dis 2001;21:311–35.
[6] Hendriks HF, Verhoofstad WA, Brouwer A, de Leeuw AM, Knook DL. Perisinusoidal fat-storing cells are the main vitamin storage sites in the rat liver. Exp Cell Res 1985;160:138–49.
[7] Ballardini G, Groff P, Badiali de Giorgi L, Schuppan D, Bianchi FB. Ito cell heterogeneity: desmin-negative Ito cells in normal rat liver. Hepatology 1994;19:440–6.
[8] Friedman SL. Hepatic stellate cells: protean, multifunctional, and enigmatic cells of the liver. Physiol Rev 2008;1:125–72.
[9] Asahina K, Tsai SY, Li P, Ishii M, Maxson Jr. RE, Sucov HM, et al. Mesenchymal origin of hepatic stellate cells, submesothelial cells, and perivascular mesenchymal cells during mouse liver development. Hepatology 2009;3:998–1011.
[10] Blomhoff R, Blomhoff HK. Overview of retinoid metabolism and function. J Neurobiol 2006;7:606–30.
[11] Thabut D, Shah V. Intrahepatic angiogenesis and sinusoidal remodeling in chronic liver disease: new targets for the treatment of portal hypertension? J Hepatol 2010;5:976–80.
[12] Chen L, Zhang W, Zhou QD, Yang HQ, Liang HF, Zhang BX, et al. HSCs play a distinct role in different phases of oval cell-mediated liver regeneration. Cell Biochem Funct 2012;7:588–96.
[13] Reynaert H, Urbain D, Geerts A. Regulation of sinusoidal perfusion in portal hypertension. Anat Rec (Hoboken) 2008;6:693–8.
[14] Ichikawa S, Mucida D, Tyznik AJ, Kronenberg M, Cheroutre H. Hepatic stellate cells function as regulatory bystanders. J Immunol 2011;10:5549–55.
[15] Li D, Friedman SL. Liver fibrogenesis and the role of hepatic stellate cells: new insights and prospects for therapy. J Gastroenterol Hepatol 1999;14:618–33.
[16] Mormone E, George J, Nieto N. Molecular pathogenesis of hepatic fibrosis and current therapeutic approaches. Chem Biol Interact 2011;3:225–31.
[17] Bataller R, Brenner DA. Liver fibrosis. J Clin Invest 2005;115:209–18.
[18] Tacke F, Weiskirchen R. Update on hepatic stellate cells: pathogenic role in liver fibrosis and novel isolation techniques. Expert Rev Gastroenterol Hepatol 2012;1:67–80.
[19] Marra F. Chemokines in liver inflammation and fibrosis. Front Biosci 2002;7:1899–914.
[20] Paik YH, Schwabe RF, Bataller R, Russo MP, Jobin C, Brenner DA. Toll-like receptor 4 mediates inflammatory signalling by bacterial lipopolysaccharide in human hepatic stellate cells. Hepatology 2003;5:1043–55.
[21] Kharbanda KK, Todero SL, Shubert KA, Sorrell MF, Tuma DJ. Malondialdehyde–acetaldehyde–protein adducts increase secretion of chemokines by rat hepatic stellate cells. Alcohol 2001;25:123–8.
[22] Fujimiya T, Liu J, Kojima H, Shirafuji S, Kimura H, Fujimiya M. Pathological roles of bone marrow-derived stellate cells in a mouse model of alcohol-induced fatty liver. Am J Physiol Gastrointest Liver Physiol 2009;297:G451–60.
[23] Quiroz SC, Bucio L, Souza V, Hernández E, González E, Gómez-Quiroz L, et al. Effect of endotoxin pretreatment on hepatic stellate cell response to ethanol and acetaldehyde. J Gastroenterol Hepatol 2001;16:1267–73.
[24] Chang J, Hisamatsu T, Shimamura K, Yoneno K, Adachi M, Naruse H, et al. Activated hepatic stellate cells mediate the differentiation of macrophages. Hepatol Res 2012. http://dx.doi.org/10.1111/j.1872-034X.2012.01111.x.
[25] Fimmel CJ, Brown KE, O'Neill R, Kladney RD. Complement C4 protein expression by rat hepatic stellate cells. J Immunol 1996;6:2601–9.
[26] Ueno T, Bioulac-Sage P, Balabaud C, Rosenbaum J. Innervation of the sinusoidal wall: regulation of the sinusoidal diameter. Anat Rec A Discov Mol Cell Evol Biol 2004;1:868–73.
[27] Muhlbauer M, Fleck M, Schutz C, Weiss T, Froh M, Blank C, et al. PD-L1 is induced in hepatocytes by viral infection and by interferon-alpha and -gamma and mediates T cell apoptosis. J Hepatol 2006;4:520–8.
[28] Gabele E, Muhlbauer M, Dorn C, Weiss TS, Froh M, Schnabl B, et al. Role of TLR9 in hepatic stellate cells and experimental liver fibrosis. Biochem Biophys Res Commun 2008;2:271–6.

[29] Xu J, Lai KK, Verlinsky A, Lugea A, French SW, Cooper MP, et al. Synergistic steatohepatitis by moderate obesity and alcohol in mice despite increased adiponectin and p-AMPK. J Hepatol 2011;55(3):673–82.
[30] Wong FW, Chan WY, Lee SS. Resistance to carbon tetrachloride-induced hepatotoxicity in mice which lack CYP2E1 expression. Toxicol Appl Pharmacol 1998;153:109–18.
[31] Jeong WI, Park O, Gao B. Abrogation of the antifibrotic effects of natural killer cells/interferon-gamma contributes to alcohol acceleration of liver fibrosis. Gastroenterology 2008;134:248–58.
[32] Chiang DJ, Roychowdhury S, Bush K, McMullen MR, Pisano S, Niese K, et al. Adenosine 2A receptor antagonist prevented and reversed liver fibrosis in a mouse model of ethanol exacerbated liver fibrosis. PLoS One 2013;8(7):e69114. 1–11.
[33] Roychowdhury S, Chiang D, McMullen MR, Nagy LE. Moderate, chronic ethanol feeding exacerbates carbon-tetrachloride-induced hepatic fibrosis via hepatocyte-specific hypoxia inducible factor 1α. Pharmcol Res Perspect 2014;2(5):e00061.
[34] Roychowdhury S, McMullen MR, Pisano SG, Liu X, Nagy LE. Absence of receptor interacting protein kinase 3 prevents ethanol-induced liver injury. Hepatology 2013;57:1773–83.
[35] Setshedi M, Wands JR, de la Monte SM. Acetaldehyde adducts in alcoholic liver disease. Oxid Med Cell Longev 2010;3(3):178–85.
[36] Lu Y, Cederbaum AI. CYP2E1 and oxidative liver injury by alcohol. Free Radic Biol Med 2008;44:723–38.
[37] Brooks PJ. DNA damage, DNA repair, and alcohol toxicity—a review. Alcohol Clin Exp Res 1997;21:1073–82.
[38] Adhikari KB, Tirosh O. Alcoholic liver disease and immune response. Open Gastroenterol J 2012;6:8–15.
[39] Cederbaum AI. CYP2E1—Biochemical and toxicological aspects and role in alcohol-induced liver injury. Mt Sinai J Med 2006;73:657–72.
[40] Brooks PJ, Theruvathu JA. DNA adducts from acetaldehyde: implications for alcohol-related carcinogenesis. Alcohol 2005;35:187–93.
[41] Tuma DJ, Casey CA. Dangerous byproducts of alcohol breakdown—focus on adducts. Alcohol Res Health 2003;27:285–90.
[42] Biewald J, Nilius R, Langner J. Occurrence of acetaldehyde protein adducts formed in various organs of chronically ethanol fed rats: an immunohistochemical study. Int J Mol Med 1998;2:389–96.
[43] Seitz HK, Meier P. The role of acetaldehyde in upper digestive tract cancer in alcoholics. Transl Res 2007;149:293–7.
[44] Purohit V, Brenner DA. Mechanisms of alcohol-induced hepatic fibrosis: a summary of the Ron Thurman symposium. Hepatology 2006;43:872–8.
[45] Mello T, Ceni E, Surrenti C, Galli A. Alcohol induced hepatic fibrosis: role of acetaldehyde. Mol Aspects Med 2008;29:17–21.
[46] Tuma DJ, Theile GM, Xu D, Klassen LW, Sorrell MF. Acetaldehyde and malondialdehyde react together to generate distinct protein adducts in the liver during long-term ethanol administration. Hepatology 1996;23:872–80.
[47] Yi H-S, Lee Y-S, Byun J-S, Seo W, Jeong J-M, Park O, et al. Alcohol dehydrogenase III exacerbates liver fibrosis by enhancing stellate cell activation and suppressing natural killer cells in mice. Hepatology 2014:1–10.
[48] Radaeva S, Wang L, Radaev S, Jeong WI, Park O, Gao B. Retinoic acid signaling sensitizes hepatic stellate cells to NK cell killing via upregulation of NK cell activating ligand RAE1. Am J Physiol Gastrointest Liver Physiol 2007;293:G809–16.
[49] Kazuo I, Friedman SL. Liver fibrosis and DDR2 signalling (Special issue adhesion signals). Connective Tissue 2003;35:165–8.
[50] Zhang X-H, Yan M, Liu L, Wu T-J, Ma L-L, et al. Expression of Discoidin Domain Receptors (DDR2) in alcoholic liver fibrosis in rats. Arch Med Res 2010;41:586–92.
[51] Olaso E, Ikeda K, Eng FJ, Xu L, Wang LH, et al. DDR2 receptor promotes MMP-2-mediated proliferation and invasion by hepatic stellate cells. J Clin Invest 2001;108:1369–78.
[52] Olaso E, Labrador JP, Wang L, Ikeda K, Eng FJ, et al. Discoidin domain receptor 2 regulates fibroblast proliferation and migration through the extracellular matrix in association with transcriptional activation of matrix metalloproteinase-2. J Biol Chem 2002;277:3606–13.

[53] Friedman SL. Seminars in medicine of the Beth Israel Hospital, Boston. The cellular basis of hepatic fibrosis. Mechanisms and treatment strategies. N Engl J Med 1993;328:1828–35.
[54] Gukovsky I, Lugea A, Shahsahebi M, Cheng JH, Hong PP, et al. A rat model reproducing key pathological responses of alcoholic chronic pancreatitis. Am J Physiol Gastrointest Liver Physiol 2008;294:G68–79.
[55] Luo Z, Liu H, Sun X, Guo R, Cui R, Ma X, et al. RNA interference against discoidin domain receptor 2 ameliorates alcoholic liver disease in rats. PLoS One 2013;8(2):e55860. 1–14.
[56] Nishiyama Y, Goda N, Kanai M, Niwa D, Osanai K, Yamamoto Y, et al. HIF-1alpha induction suppresses excessive lipid accumulation in alcoholic fatty liver in mice. J Hepatol 2012;56:441–7.
[57] Hirota SA, Beck PL, MacDonald JA. Targeting hypoxia-inducible factor-1 (HIF-1) signaling in therapeutics: implications for the treatment of inflammatory bowel disease. Recent Pat Inflamm Allergy Drug Discov 2009;3:1–16.
[58] Canbay A, Friedman S, Gores GJ. Apoptosis: the nexus of liver injury and fibrosis. Hepatology 2004;39:273–8.
[59] Roychowdhury S, Chiang DJ, Mandal P, McMullen MR, Liu X, Cohen JI, et al. Inhibition of apoptosis protects mice from ethanol-mediated acceleration of early markers of CCl(4)-induced fibrosis but not steatosis or inflammation. Alcohol Clin Exp Res 2012;36:1139–47.
[60] Burnstock G. Purine and pyrimidine receptors. Cell Mol Life Sci 2007;64:1471–83.
[61] Hashmi AZ, Hakim W, Kruglov EA, Watanabe A, Watkins W, et al. Adenosine inhibits cytosolic calcium signals and chemotaxis in hepatic stellate cells. Am J Physiol Gastrointest Liver Physiol 2007;292:G395–401.
[62] Novo E, Cannito S, Zamara E, Valfre di Bonzo L, Caligiuri A, et al. Proangiogenic cytokines as hypoxia-dependent factors stimulating migration of human hepatic stellate cells. Am J Pathol 2007;170:1942–53.
[63] Pinhal-Enfield G, Ramanathan M, Hasko G, Vogel SN, Salzman AL, et al. An angiogenic switch in macrophages involving synergy between Toll-like receptors 2, 4, 7, and 9 and adenosine A(2A) receptors. Am J Pathol 2003;163:711–21.
[64] Yoshiji H, Kuriyama S, Yoshii J, Ikenaka Y, Noguchi R, et al. Vascular endothelial growth factor and receptor interaction is a prerequisite for murine hepatic fibrogenesis. Gut 2003;52:1347–54.
[65] Seki E, De Minicis S, Osterreicher CH, Kluwe J, Osawa Y, Brenner DA, et al. TLR4 enhances TGF-beta signaling and hepatic fibrosis. Nat Med 2007;13(11):1324–32.
[66] Seki E, De Minicis S, Gwak GY, Kluwe J, Inokuchi S, Bursill CA, et al. CCR1 and CCR5 promote hepatic fibrosis in mice. J Clin Invest 2009;119(7):1858–70.
[67] Seki E, de Minicis S, Inokuchi S, Taura K, Miyai K, van Rooijen N, et al. CCR2 promotes hepatic fibrosis in mice. Hepatology 2009;50(1):185–97.
[68] Affò S, Morales-Ibanez O, Rodrigo-Torres D, Altamirano J, Blaya D, Dapito DH, et al. CCL20 mediates lipopolysaccharide induced liver injury and is a potential driver of inflammation and fibrosis in alcoholic hepatitis. Gut 2014;63:1782–92.
[69] Inokuchi S, Tsukamoto H, Park E, Liu Z-X, Brenner DA, Seki E. Toll-like receptor 4 mediates alcohol-induced steatohepatitis through bone marrow-derived and endogenous liver cells in mice. Alcohol Clin Exp Res 2011;35(8):1509–18.
[70] Carroll MC. The complement system in regulation of adaptive immunity. Nat Immunol 2004;5:981–6.
[71] Hillebrandt S, Wasmuth HE, Weiskirchen R, Hellerbrand C, Keppeler H, Werth A, et al. Complement factor 5 is a quantitative trait gene that modifies liver fibrogenesis in mice and humans. Nat Genet 2005;37:835–43.
[72] Pritchard MT, McMullen MR, Stavitsky AB, Cohen JI, Lin F, Medof ME, et al. Differential contributions of C3, C5 and decay accelerating factor to ethanol-induced fatty liver in mice. Gastroenterology 2007;132(3):1117–26.
[73] Roychowdhury S, McMullen MR, Pritchard MT, Hise AG, van Rooijen N, Medof ME, et al. An early complement-dependent and TLR-4-independent phase in the pathogenesis of ethanol-induced liver injury in mice. Hepatology 2009;49:1326–34.
[74] Xu R, Lin F, He J, Jin L, Zhang JY, Fu J, et al. Complement 5a stimulates hepatic stellate cells *in vitro*, and is increased in the plasma of patients with chronic hepatitis B. Immunology 2013;138:228–34.
[75] Das D, Barnes MA, Nagy LE. Anaphylatoxin C5a modulates hepatic stellate cell migration. Fibrogenesis Tissue Repair 2014;7:9.

[76] Watanabe A, Sohail MA, Gomes DA, Hashmi A, Nagata J, Sutterwala FS, et al. Inflammasome-mediated regulation of hepatic stellate cells. Am J Physiol Gastrointest Liver Physiol 2009;296(6):G1248–57.

[77] Petrasek J, Bala S, Csak T, Lippai D, Kodys K, Menashy V, et al. IL-1 receptor antagonist ameliorates inflammasome-dependent alcoholic steatohepatitis in mice. J Clin Invest 2012;122(10):3476–89.

[78] Almeida J, Polvorosa MA, Gonzalez-Quintela A, Marcos M, Pastor I, Hernandez Cerceño ML, et al. Decreased peripheral blood $CD4^+/CD25^+$ regulatory T cells in patients with alcoholic hepatitis. Alcohol Clin Exp Res 2013;37(8):1361–9.

[79] Lemmers A, Moreno C, Gustot T, Maréchal R, Degré D, Demetter P, et al. The interleukin-17 pathway is involved in human alcoholic liver disease. Hepatology 2009;49:2.

[80] Weaver CT, Hatton RD, Mangan PR, Harrington LE. IL-17 family cytokines and the expanding diversity of effector T cell lineages. Annu Rev Immunol 2007;25:821–52.

[81] Meng F, Wang K, Aoyama T, Grivennikov SI, Paik Y, Scholten D, et al. IL-17 signaling in inflammatory cells, kupffer cells and hepatic stellate cells exacerbates liver fibrosis. Gastroenterology 2012;143(3):765–76. e1–e3.

[82] Kong X, Feng D, Mathews S, Gao B. Hepatoprotective and anti-fibrotic functions of interleukin-22: therapeutic potential for the treatment of alcoholic liver disease. J Gastroenterol Hepatol 2013;28(Suppl. 1):56–60.

[83] Kong X, Feng D, Wang H, Hong F, Bertola A, Wang F-S, et al. Interleukin-22 induces hepatic stellate cell senescence and restricts liver fibrosis in mice. Hepatology 2012;56:1150–9.

CHAPTER 11

Hepatic Stellate Cells as Target for Reversal of Fibrosis/Cirrhosis

Sophie Lotersztajn[1,2,3] and Ariane Mallat[4,5,6]
[1]INSERM U1149, Center for Research on Inflammation, Paris, France
[2]Université Paris Diderot, Paris, France
[3]Laboratory of Excellence Inflammex, Paris, France
[4]INSERM U955, Créteil, France
[5]Université Paris-Est, Créteil, France
[6]AP-HP, Hôpital Henri Mondor-Albert Chenevier, Service d'Hépatologie, Créteil, France

11.1 INTRODUCTION

Liver fibrosis, a common pathological consequence of all chronic liver diseases, results from the progressive accumulation of a qualitatively altered extracellular matrix highly enriched in type I and III fibrillar collagens. Scar deposition arises from an altered wound-healing response to prolonged liver injury that is characterized by increased production of matrix proteins and decreased matrix remodeling (for reviews, see Refs. [1–3]) (see Chapter 4). Early stages of fibrosis have no consequences on the symptomatology of chronic liver diseases. However, cirrhosis, which occurs in a number of patients following iterative parenchymal insult, is a major factor responsible for life-threatening complications. Cirrhosis is a process that involves several alterations besides the presence of fibrotic septa surrounding nodules. Thus, pathological changes also include regenerating hepatocytes, and marked alterations of hepatic vascularization including intrahepatic shunts due to angiogenesis, capillarization of liver sinusoids, and obliterative lesions in portal tracts and hepatic venules. Overall, these changes expose cirrhotic patients to life-threatening complications of liver failure and portal hypertension, including ascites, variceal rupture, hepatic encephalopathy, bacterial infections, coagulation disorders, and hepatorenal syndrome. Moreover, cirrhosis is a precancerous state that requires systematic screening for hepatocellular carcinoma.

The fibrogenic process is driven by activated hepatic myofibroblasts that are predominantly derived from hepatic stellate cells (HSCs). Hepatic myofibroblasts constitute a major source of both fibrillar extracellular matrix and tissue inhibitors of metalloproteinases (TIMPs) that inactivate metalloproteinases produced by neighboring non-parenchymal cells, thereby inhibiting degradation of matrix molecules [1,2] (see Chapter 7).

Stellate Cells in Health and Disease
DOI: http://dx.doi.org/10.1016/B978-0-12-800134-9.00011-7

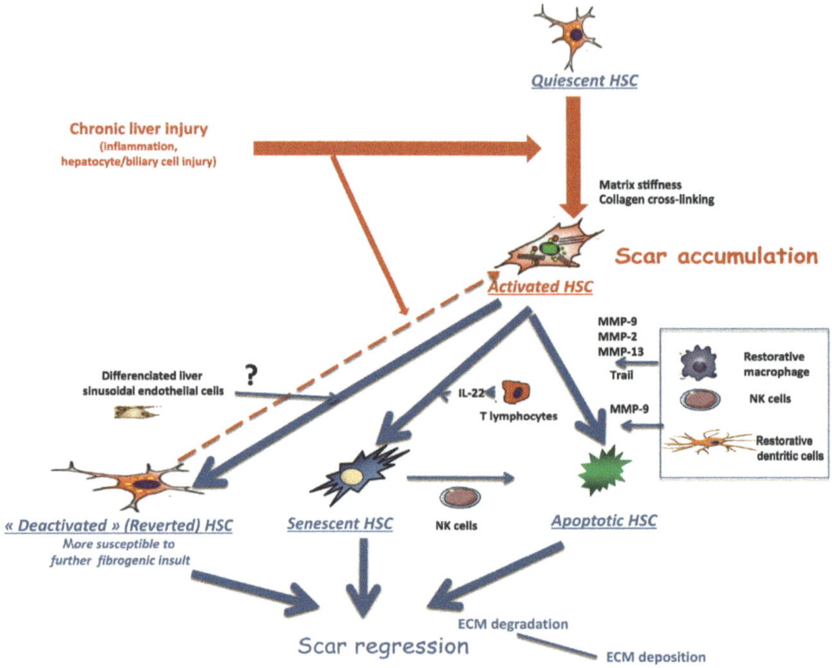

Figure 11.1 ***Mechanisms of fibrosis resolution: fate of HSCs.*** Upon liver injury, HSCs undergo transdifferentiation from a quiescent to a myofibroblastic phenotype. Activated HSCs secrete excessive extracellular matrix and cause liver fibrosis. Clearance of activated HSCs is a key step in fibrosis regression, and may result (i) from apoptosis, promoted by NK cells, dendritic cells, and proresolution macrophages, (ii) senescence that exposes senescent HSC to NK-mediated HSC killing, or (iii) from reversion to a deactivated phenotype more susceptible to an additional fibrogenic insult.

The long-standing dogma of the irreversibility of liver fibrosis and cirrhosis was first challenged at the end of the 1970s with the report of three cases of regression of advanced fibrosis [1]. Subsequent reports have largely documented the potential for histological regression of fibrosis and early-stage cirrhosis in patients with various liver diseases including chronic viral hepatitis, autoimmune hepatitis, hemochromatosis, and alcoholic hepatitis, provided that liver insult is controlled or eradicated [3–7]. In addition, regression of cirrhosis has been linked to improved clinical outcome. These findings have generated a major research effort dedicated to the characterization of mechanisms underlying the fibrolytic capacity of the liver, in the hope of identifying antifibrotic therapeutic approaches (Figure 11.1).

11.2 FIBROSIS REVERSIBILITY: GENERAL OVERVIEW

Mechanisms underlying fibrosis regression have been delineated following withdrawal of liver insult, in rodents with liver fibrosis induced by diverse types of injury, including carbon tetrachloride intoxication [8], bile duct ligation [9], or methionine- and choline-deficient diet [10]. These models identified HSCs as key determinants of fibrosis regression, together with other non-parenchymal cells and infiltrating immune cells.

Complete regression of fibrosis is observed in animals with fibrosis or early-stage cirrhosis within weeks upon termination of the toxic insult. Sequential analysis of events has shown a very early clearance of hepatic myofibroblasts leading to a drop in hepatic TIMP-1 and followed by the upregulation and activation of hepatic matrix metalloproteinases (MMPs) [8] predominantly produced by macrophages/Kupffer cells associated with fibrotic septa [11].

In contrast, experiments in animals with advanced cirrhosis due to prolonged injury have shown that withdrawal of the liver insult is associated with partial remodeling of cirrhosis from a micronodular to a macronodular pattern, even after a prolonged period of observation [12], indicating that fibrosis regression is hampered at advanced stages in keeping with clinical data [13]. Mechanisms underlying limited remodeling of the scar matrix in models of advanced fibrosis remain partially understood. Available data suggest that increased matrix stiffness and extensive collagen crosslinking by tissue transglutaminase and lysyl oxidase in older septae induces resistance to the collagenolytic effects of MMPs [12–14]. Enzymes implicated in collagen crosslinking comprise tissue transglutaminase and lysyl-oxidase, which stabilize extracellular matrix [13,15]. Other factors incriminated in irreversibility of fibrolysis include the paucity of fibrolytic cells in densely crosslinked septa [12] and advanced architectural distortion including extensive neoangiogenesis [16].

Follow up of patients included in large phase III trials of potent antivirals has allowed to convincingly demonstrate the regression of fibrosis and early-stage cirrhosis in a significant proportion of patients. However, the resolution process requires prolonged periods extending from several months to years, as illustrated in the recent phase III tenofovir trial in patients with chronic hepatitis B [17]. Among 94 patients with early-stage cirrhosis prior to treatment, an increasing proportion exhibited non-cirrhotic liver biopsies over time, achieving 74% after 5 years [17]. Importantly, compared to rodents, regression of fibrosis in humans is both slower and less extensive, probably as the consequence of a highly crosslinked fibrosis and greater vascular alterations after several years to decades of fibrogenesis in patients.

11.3 MECHANISMS OF HSC CLEARANCE DURING FIBROSIS RESOLUTION

Seminal experimental studies of fibrosis recovery have identified apoptosis of hepatic myofibroblasts as a major pathway responsible for HSC clearance and initiation of fibrolysis. More recently, senescence and reversion into a partially quiescent phenotype have also been incriminated. However, the respective contribution of the three mechanisms to fibrosis resolution remains to be determined and their relevance in a clinical setting is still unclear.

11.3.1 HSC apoptosis

Spontaneous recovery of hepatic fibrosis is associated with an early decrease in the density of activated HSCs due to apoptosis [8]. Experiments in TIMP-1 transgenic mice and with TIMP-1 scavengers demonstrated the causal relationship between hepatic TIMP-1 expression, failure of fibrolysis, and increased HSC survival [13]. Several apoptotic pathways have been identified in HSCs, including the death receptor pathway, mitochondria-related pathways, or the signaling pathways stimulated following activation of the p75 receptor for nerve growth factor (NGF). In addition, survival of activated HSCs is tightly controlled by activation of nuclear factor kappa-light-chain-enhancer of activated B cells (NF-κB), resulting in upregulation of anti-apoptotic genes. Accordingly, inhibitors of NF-κB signaling induce apoptosis of activated HSCs and reversal of fibrosis [18]. Factors that increase NF-κB and survival in HSC are largely derived from Kupffer cells, including IL-1β and TNF-α, but also include lipopolysaccharide (LPS)and its receptor TLR4 [2]. Type I collagen has also been identified as a survival signal via $\alpha_v\beta_3$ integrins [19,20].

11.3.2 HSC senescence

Liver fibrogenesis is associated with progressive accumulation of senescent hepatic myofibroblasts, a process that plays a key role in limiting the fibrogenic response [21]. Senescent HSCs are characterized by cell cycle arrest, upregulation of metalloproteinases and cytokines, and downregulation of extracellular matrix gene expression, including MMP-9 and TIMP-1 and 2 [21]. Interestingly, compared to control mice, double p53/INK4/ARF knock-out mice showed deficient HSC senescence and delayed recovery following carbon tetrachloride intoxication, indicating that HSC senescence facilitates fibrosis resolution [21]. Interleukin-22 (IL-22) has been identified as a key cytokine in the induction of HSC senescence, via a STAT3/SOCS53/p53-dependent signaling pathway [22]. Mechanisms underlying the impact of senescence on fibrogenesis and recovery appear multifactorial, including downregulation of TIMPs and collagen expression, upregulation of MMP expression, and activation of natural killer (NK) cells resulting in clearance of senescent HSCs following NK cell-induced apoptosis (see Section 11.4.1.1.3) [21].

11.3.3 Reversion of HSC phenotype to an inactivated state

The generation of mouse models allowing specific HSC fate tracking studies have recently allowed to demonstrate that activated HSCs also undergo deactivation to a quiescent phenotype in models of fibrosis recovery, in keeping with earlier observations in culture [23–25]. Permanent stamping with a fluorescent label of HSCs was performed at the onset of fibrosis recovery under the control of HSC specific promoters of type 1 collagen and vimentin, in two independent mouse models [23,24]. The recovery phase was studied following carbon tetrachloride administration [23,24], or chronic alcohol feeding [23]. Follow up of animals during the recovery phase showed the presence of fluorescent labeling in cells that expressed vitamin A, arguing for the reversion of activated myofibroblastic HSCs to a quiescent state. In addition, fluorescence-activated cell sorting (FACS) and single-cell polymerase chain reaction (PCR) analysis of labeled cells isolated at various times after cessation of injury indicated that the expression of characteristic markers (Collagen I, TIMP1, α-SMA, TGF-βR) of activated HSCs gradually decreases during the recovery phase.

Intriguingly, reverted HSCs in the recovering liver reacquire some but not all features of quiescent HSCs that had never been exposed to a fibrogenic stimulus. Indeed, reverted HSCs in culture were more susceptible to subsequent activation by TGF-β1 or mitogens [23,24]. In keeping with this observation, administration of a second round of carbon tetrachloride after a 6-month period of recovery induces more severe fibrosis, compared to mice that received one round of the hepatotoxin [23]. Comparative gene analysis of reverted HSCs and quiescent HSCs showed differences in a number of genes, including the absence, in reverted cells, of glial fibrillary acidic protein (GFAP) or adipose tissue-related genes such as Adfp or AdipoR1, which are characteristic of quiescent HSCs.

Mechanisms underlying HSC phenotype reversion remain to be defined. A recent study suggests that paracrine interactions with liver sinusoidal endothelial cells are implicated in this process (see Section 11.4.2) [26].

11.4 CONTROL OF HSC FATE AND MATRIX DEGRADATION BY NEIGHBORING LIVER CELLS DURING FIBROSIS RESOLUTION

Clearance of activated HSCs and matrix degradation is controlled by complex interactions with non-parenchymal cells, including endothelial and immune cells.

11.4.1 Immune cells

Sustained hepatic inflammation resulting from parenchymal injury is a driving force of fibrosis progression in the liver, as described in other organs. Selective manipulation of individual inflammatory cells, either pharmacologically or genetically, has recently allowed to characterize the impact of resident (Kupffer and dendritic cells) and infiltrating immune cells (macrophages, dendritic cells, and NK cells) on HSCs and resolution of fibrosis.

11.4.1.1 Innate immune cells

11.4.1.1.1 Monocyte/Macrophages

Kupffer cells and infiltrating monocyte/macrophages have been identified as key actors in the fibrogenic and fibrolytic phases. Liver metabolic or toxic insult induces massive infiltration of monocyte-derived macrophages exhibiting a $GR1^{hi}$ ($Ly6C^{hi}$) proinflammatory phenotype characterized by the expression of high levels of CD11b and F4/80 [27,28]. These proinflammatory monocytes enhance liver fibrogenesis as shown by targeted pharmacological or conditional genetic ablation of these cells [29,30]. These findings have been corroborated by *in vitro* studies showing that Kupffer cells promote both activation of the myofibroblastic phenotype and survival of activated HSCs [31].

In sharp contrast, experiments during the recovery phase have shown that macrophages harboring a distinct phenotype (restorative macrophages) accumulate following cessation of liver insult, induce apoptosis of HSCs, and produce active metalloproteinases such as MMP12 and MMP13 involved in the fibrolytic process [30,32]. The crucial role of "restorative" macrophages has now been demonstrated in several experimental models of liver fibrosis, including chronic administration of carbon tetrachloride, methionine-choline-deficient diet, and bile duct ligation [33]. "Restorative macrophages" show no expression of TGF-β, low levels of TNF-α mRNA, and may promote HSC apoptosis through expression of HSC death ligands such as tumor necrosis factor-related apoptosis-inducing ligand (TRAIL) and MMP-9 [32]. Their signature is characterized by a mixed expression of M1 and M2 markers that can be distinguished from their profibrogenic counterparts by a low versus high Gr1 (Ly6c) expression, a high phagocytic and matrix degrading activity, including high expression of MMP13 and MMP 12 [31,34].

Overall, these data indicate that in experimental models, distinct macrophage subsets or the same macrophage population undergoing a phenotypic switch regulate fibrosis progression and resolution by modulating HSC survival and matrix degradation.

11.4.1.1.2 Dendritic cells

Dendritic cells are $CD11c^+$ cells that link the innate and adaptive immune responses, via recruitment of immune effectors cells such as NK, NKT (natural killer T) cells, and neutrophils. Similarly to monocyte/macrophages, dendritic cells have been shown to orchestrate the inflammatory response during chronic liver injury [35–37], although a direct contribution to the fibrogenic process is still a matter of debate [31,37]. Current evidences rather indicate a role of dendritic cells in the fibrolytic process as shown in models of recovery following cessation of carbon tetrachloride intoxication, or interruption of the administration of a methionine-choline-deficient diet [36] or nonalcoholic steatohepatitis [37]. Thus, conditional ablation of $CD11c^+$ cells was associated with delayed regression of fibrosis and reduced clearance of activated HSCs, whereas pharmacological or exogenous amplification of the dendritic cell pool accelerated fibrosis recovery. The fibrolytic effect of dendritic cell was correlated in part to the release of MMP-9, a gelatinase that cleaves collagen IV and elastin [36]. Detailed comparison of the respective contribution of dendritic cells and macrophages to fibrolysis will require better phenotypic discrimination between the two cell types.

11.4.1.1.3 NK and NKT cells

NK and NKT cells are lymphocytes with antiviral and antitumorigenic properties that are enriched in the liver. Cytotoxic effects of NK cells are governed by the balance between inhibitory and stimulatory NK cell receptors. In the context of liver fibrosis, compelling evidence indicates that NK cells reduce fibrogenesis and promote fibrosis recovery by inducing apoptosis of early activated and senescent HSCs [21,38,39]. Early activated HSCS acquire susceptibility to NK-mediated killing via TRAIL following combined induction of retinoic acid early inducible 1 (RAE1), a ligand for the NK cell-activating receptor NKG2, and downregulation of the NK cell-inhibitory ligand MHC-1 [38,39]. Interestingly, fully activated HSCs lose their capacity to produce RAE1, which may contribute to their survival in the chronically injured liver [38,40]. Of note, it has been shown that excessive alcohol intake downregulates TRAIL and IFNγ expression in NK cells, a mechanism that contributes to accelerated fibrogenesis in this setting [41]. Senescent HSCs become sensitive to NK induced killing following upregulation of NK-cell activating ligands such as MICA and ULBP2 [21]. Contact with activated NK cells triggers apoptosis of senescent HSCs by the granule exocytosis pathway, whereas the TRAIL pathway is inactivated by the decoy receptor Dcr2, which is upregulated in these cells [42]. Beyond their direct impact on HSC survival, activated NK cells also limit HSC accumulation by releasing IFN gamma, thereby inducing growth inhibition and apoptosis of HSCs.

In contrast to NK cells, anti-fibrogenic or profibrogenic properties of NKT cells have been described, depending on the model and of the extent of liver injury [40]. Profibrogenic effects rely on Th2 and Th17 mediators produced by NKT cells, including IL4, IL13, or IL17, whereas NKT cell-mediated apoptosis of HSCs has been ascribed to the release of IFNγ [40]. The impact of NKT cells on fibrosis recovery remains unknown. Further studies are required to understand whether the heterogeneous nature of NKT cells may account for these dual effects.

11.4.1.2 Adaptive immune cells

The role of adaptive immune cells in fibrosis progression has been well characterized, with the reported contribution of $CD4^+$ T lymphocytes (Th2 and Th17 cells), B cells, and CD8 lymphocytes [2], but their impact on fibrosis regression is much less characterized. It has been shown that IL-22, a cytokine produced in particular by Th17 cells and γδ T cells accelerates fibrosis regression by inducing HSC senescence; however, the origin of the T lymphocyte releasing IL-22 in this setting remains to be determined [22]. A recent study also showed that a subset of γδ T cells expressing CCR6, IL-17 and IL-22 induces apoptosis of quiescent and activated HSC via Fas/Fas-ligand interactions thereby reducing fibrosis progression independently of IL-17 and IL-22 [43]. Whether these adaptive immune cells modulate fibrosis recovery has not been investigated heretofore.

11.4.2 Liver sinusoidal endothelial cells

In the normal liver, sinusoidal endothelial cells (LSEC) are characterized by specific features including the presence of fenestrae and a high endocytic activity [44]. Vascular endothelial growth factor (VEGF) released by hepatocytes and HSCs plays a key role in the maintenance of this differentiated phenotype via nitric oxide (NO)-dependent and independent pathways [33]. Chronic liver injury is associated with an early loss of cell fenestration (i.e., capillarization of sinusoids) that precedes the onset of fibrosis. Activated endothelial cells produce TGF-β, PDGF-BB, and collagen I, thereby contributing to HSC activation. Conversely, a recent study has shown that restoration of LSEC differentiation may accelerate fibrosis regression by promoting HSC quiescence [26]. Indeed, *in vitro* experiments indicate that differentiated LSEC promote reversal of activated HSCs into a quiescent phenotype via paracrine interactions, as shown in co-culture studies [26]. The *in vivo* relevance of these findings has been explored in rats with thioacetamide-induced cirrhosis. In this model early restoration of the normal LSEC phenotype is induced by a 7-day course with a soluble guanylate cyclase activator (sGC) without altering the density of activated phenotype and the level of fibrosis. Following discontinuation of sGC, differentiated LSEC promote subsequent reversion of activated HSCs to quiescence and accelerated fibrosis recovery [26]. Whether HSCs revert to a so-called reverted phenotype or a physiological (i.e., quiescent) phenotype is currently unknown. Interestingly, a recent study identifies an additional impact of VEGF on fibrosis regression. Indeed, in a BDL recovery model, VEGF promotes monocyte-endothelial cell adhesion and restores sinusoidal permeability, thereby allowing monocyte infiltration, accumulation of "restorative macrophages" in scar tissue, and MMP13 upregulation to remodel scar and promote fibrosis resolution [33]. Taken together, these data indicate that differentiated LSEC play a gatekeeper role in the regulation of both fibrogenesis and fibrolysis.

11.5 CONCLUSION

Advances in the understanding of HSC biology have uncovered a pivotal role in fibrosis resolution together with other non-parenchymal cells within a complex network. Increasing evidences indicate that clearance of activated HSCs occur through various mechanisms including apoptosis, senescence, and reversion to a so-called quiescent phenotype and *in vitro* findings have progressively been confirmed in experimental models owing to the increasing availability of models allowing cell-targeted gene deletion or expression. Future studies should more precisely delineate the mechanisms underlying HSC fate. The respective roles of apoptosis, senescence or reversion to physiologic phenotype of activated HSCs in fibrosis reversal also deserve further clarification. Finally, the issue of the relevance of experimental findings to human liver diseases warrants additional investigations inasmuch as previous studies have shown that human activated HSCs are more resistant to apoptosis than their rodent counterparts.

ACKNOWLEDGMENTS

This work was supported by the INSERM, the Université Paris-Diderot, the Université Paris-Est, the Université Paris-Descartes, the Investissement d'Avenir program ANR-11-Idex-005-02, Sorbonne Paris Cité Laboratoire d'Excellence Inflammex (SL).

REFERENCES

[1] Perez-Tamayo R. Cirrhosis of the liver: a reversible disease? Pathol Annu 1979;14(Pt 2)):183–213.
[2] Mallat A, Lotersztajn S. Cellular mechanisms of tissue fibrosis. 5. Novel insights into liver fibrosis. Am J Physiol Cell Physiol 2013;305(8):C789–99.
[3] Chang TT, et al. Long-term entecavir therapy results in the reversal of fibrosis/cirrhosis and continued histological improvement in patients with chronic hepatitis B. Hepatology 2010;52(3):886–93.
[4] Falize L, et al. Reversibility of hepatic fibrosis in treated genetic hemochromatosis: a study of 36 cases. Hepatology 2006;44(2):472–7.
[5] Lotersztajn S, et al. Hepatic fibrosis: molecular mechanisms and drug targets. Ann Rev Pharmacol Toxicol 2005;45:605–28.
[6] Hernandez-Gea V, Friedman SL. Pathogenesis of liver fibrosis. Annu Rev Pathol 2010;6:425–56.
[7] Schuppan D, Kim YO. Evolving therapies for liver fibrosis. J Clin Invest 2013;123(5):1887–901.
[8] Iredale J, et al. Mechanisms of spontaneous resolution of rat liver fibrosis. Hepatic stellate cell apoptosis and reduced hepatic expression of metalloproteinase inhibitors. J Clin Invest 1998;102:538–49.
[9] Issa R, et al. Apoptosis of hepatic stellate cells: involvement in resolution of biliary fibrosis and regulation by soluble growth factors. Gut 2001;48(4):548–57.
[10] Baeck C, et al. Pharmacological inhibition of the chemokine C–C motif chemokine ligand 2 (monocyte chemoattractant protein 1) accelerates liver fibrosis regression by suppressing Ly-6C(+) macrophage infiltration in mice. Hepatology 2014;59(3):1060–72.
[11] Ramachandran P, Iredale JP. Macrophages: central regulators of hepatic fibrogenesis and fibrosis resolution. J Hepatol 2012;56(6):1417–9.
[12] Issa R, et al. Spontaneous recovery from micronodular cirrhosis: evidence for incomplete resolution associated with matrix cross-linking. Gastroenterology 2004;126(7):1795–808.
[13] Iredale JP, et al. Extracellular matrix degradation in liver fibrosis: biochemistry and regulation. Biochim Biophys Acta 2013;1832(7):876–83.
[14] Olsen AL, et al. Hepatic stellate cells require a stiff environment for myofibroblastic differentiation. Am J Physiol Gastrointest Liver Physiol 2011;301(1):G110–8.
[15] Popov Y, et al. Tissue transglutaminase does not affect fibrotic matrix stability or regression of liver fibrosis in mice. Gastroenterology 2011;140(5):1642–52.
[16] Fernandez M, et al. Angiogenesis in liver disease. J Hepatol 2009;50(3):604–20.
[17] Marcellin P, et al. Regression of cirrhosis during treatment with tenofovir disoproxil fumarate for chronic hepatitis B: a 5-year open-label follow-up study. Lancet 2013;381(9865):468–75.
[18] Oakley F, et al. Angiotensin II activates I kappaB kinase phosphorylation of RelA at Ser 536 to promote myofibroblast survival and liver fibrosis. Gastroenterology 2009;136(7):2334–44. e2331.
[19] Murphy FR, et al. Inhibition of apoptosis of activated hepatic stellate cells by TIMP-1 is mediated via effects on MMP inhibition: implications for reversibility of liver fibrosis. J Biol Chem 2002;277(13):11069–76.
[20] Zhou X, et al. Engagement of alpha Vbeta 3 integrin regulates proliferation and apoptosis of hepatic stellate cells. J Biol Chem 2004;24:24.
[21] Krizhanovsky V, et al. Senescence of activated stellate cells limits liver fibrosis. Cell 2008;134(4):657–67.
[22] Kong X, et al. Interleukin-22 induces hepatic stellate cell senescence and restricts liver fibrosis in mice. Hepatology 2012;56(3):1150–9.
[23] Kisseleva T, et al. Myofibroblasts revert to an inactive phenotype during regression of liver fibrosis. Proc Natl Acad Sci USA 2012;109(24):9448–53.
[24] Troeger JS, et al. Deactivation of hepatic stellate cells during liver fibrosis resolution in mice. Gastroenterology 2012;143(4):1073–83. e1022.

[25] Mallat A, Lotersztajn S. Reversion of hepatic stellate cell to a quiescent phenotype: from myth to reality? J Hepatol 2013;59(2):383–6.
[26] Xie G, et al. Role of differentiation of liver sinusoidal endothelial cells in progression and regression of hepatic fibrosis in rats. Gastroenterology 2012;142(4):918–27. e916.
[27] Karlmark KR, et al. Hepatic recruitment of the inflammatory Gr1+ monocyte subset upon liver injury promotes hepatic fibrosis. Hepatology 2009;50(1):261–74.
[28] Liaskou E, et al. Monocyte subsets in human liver disease show distinct phenotypic and functional characteristics. Hepatology 2013;57(1):385–98.
[29] Imamura M, et al. Suppression of macrophage infiltration inhibits activation of hepatic stellate cells and liver fibrogenesis in rats. Gastroenterology 2005;128(1):138–46.
[30] Duffield JS, et al. Selective depletion of macrophages reveals distinct, opposing roles during liver injury and repair. J Clin Invest 2005;115(1):56–65.
[31] Pradere JP, et al. Hepatic macrophages but not dendritic cells contribute to liver fibrosis by promoting the survival of activated hepatic stellate cells. Hepatology 2013;58(4):1461–73.
[32] Fischer R, et al. Caspase 9-dependent killing of hepatic stellate cells by activated Kupffer cells. Gastroenterology 2002;123(3):845–61.
[33] Yang L, et al. Vascular endothelial growth factor promotes fibrosis resolution and repair in mice. Gastroenterology 2014;146(5):1339–50. e1331.
[34] Ramachandran P, et al. Differential Ly-6C expression identifies the recruited macrophage phenotype, which orchestrates the regression of murine liver fibrosis. Proc Natl Acad Sci USA 2012;109(46):E3186–95.
[35] Connolly MK, et al. In liver fibrosis, dendritic cells govern hepatic inflammation in mice via TNF-alpha. J Clin Invest 2009;119(11):3213–25.
[36] Jiao J, et al. Dendritic cell regulation of carbon tetrachloride-induced murine liver fibrosis regression. Hepatology 2012;55(1):244–55.
[37] Henning JR, et al. Dendritic cells limit fibroinflammatory injury in nonalcoholic steatohepatitis in mice. Hepatology 2013;58(2):589–602.
[38] Radaeva S, et al. Natural killer cells ameliorate liver fibrosis by killing activated stellate cells in NKG2D-dependent and tumor necrosis factor-related apoptosis-inducing ligand-dependent manners. Gastroenterology 2006;130(2):435–52.
[39] Melhem A, et al. Anti-fibrotic activity of NK cells in experimental liver injury through killing of activated HSC. J Hepatol 2006;45(1):60–71.
[40] Gao B, Radaeva S. Natural killer and natural killer T cells in liver fibrosis. Biochim Biophys Acta 2013;1832(7):1061–9.
[41] Jeong WI, et al. STAT1 inhibits liver fibrosis in mice by inhibiting stellate cell proliferation and stimulating NK cell cytotoxicity. Hepatology 2006;44(6):1441–51.
[42] Sagiv A, et al. Granule exocytosis mediates immune surveillance of senescent cells. Oncogene 2013;32(15):1971–7.
[43] Hammerich L, et al. Chemokine receptor CCR6-dependent accumulation of gammadelta T cells in injured liver restricts hepatic inflammation and fibrosis. Hepatology 2014;59(2):630–42.
[44] DeLeve LD. Liver sinusoidal endothelial cells in hepatic fibrosis. Hepatology 2014.http://dx.doi.org/10.1002/hep.27376

CHAPTER 12

Interactions of Stellate Cells with Other Non-Parenchymal Cells

Norifumi Kawada[1] and Maurizio Parola[2]

[1]Department of Hepatology, Graduate School of Medicine, Osaka City University, Osaka, Japan
[2]Department of Clinical and Biological Sciences, Unit of Experimental Medicine and Clinical Pathology, School of Medicine, University of Torino, Torino, Italy

12.1 HEPATIC STELLATE CELLS: INTRODUCTORY REMARKS

Hepatic stellate cells (HSCs), formerly known as Ito cells, fat-storing cells, or lipocytes, reside in the space of Disse in the hepatic sinusoid, encapsulate sinusoidal endothelial cells with their well-branching dendritic processes on one side, and face hepatocytes on the other side [1]. In physiological conditions, the principal function of quiescent HSCs is storing vitamin A in their cytoplasm; 50–80% of vitamin A in the body is accumulated in the liver and 90% is stored in HSCs. Quiescent HSCs express neural crest markers, such as glial fibrillary acidic protein (GFAP) and neurotrophins and their receptor (p75), and an intermediate filament, desmin. Quiescent HSCs secrete extracellular matrix materials (ECMs) including laminin, proteoglycan, and type IV collagen, which form basement membrane-like structure [2–4]. Because HSCs function also as liver-specific pericytes, their contractility in response to endothelin-1 (ET-1), angiotensin-II, and relaxation by nitric oxide (NO) controls the diameter of the sinusoidal lumen and regulates the local microcirculation [5].

Following liver injury caused by hepatitis virus B or C (HBV or HCV) infection, alcohol abuse, drug toxicity, autoimmunity, or steatosis, HSCs undergo activation and transdifferentiate to myofibroblast (MF)-like cells [8]. These MF-like cells are characterized by a loss of vitamin A droplets, increased expression of α-smooth muscle actin (α-sma) and growth factor receptors, augmented contractile activity, and increased generation of multiple ECMs, largely types I and III collagens [2–4]. Activation of HSCs is triggered by paracrine stimulation by hepatic constituent cells, including sinusoidal endothelial cells, Kupffer cells (liver macrophages), hepatocytes, and cholangiocytes as well as platelets and following interactions with cells of the immune system. Activated HSCs secrete profibrogenic transforming growth factor β (TGF-β) as their autocrine stimulant and liver sinusoidal endothelial cells (LSECs) participate in conversion of TGF-β from its latency associated peptide-binding form to active form [6]. Cholangiocytes and platelets are also a source of platelet-derived growth factor

Stellate Cells in Health and Disease
DOI: http://dx.doi.org/10.1016/B978-0-12-800134-9.00012-9

(PDGF), TGF-β, and epidermal growth factor (EGF). Hepatocytes constitute another source of fibrogenic lipid peroxides and their apoptosis initiates the activation of HSCs via a process mediated by Fas and tumor-necrosis factor (TNF)-related apoptosis-inducing ligand (TRAIL). Apoptotic fragments derived from hepatocytes stimulate HSC activation in culture, and phagocytosis of apoptotic hepatocytes by MFs stimulates their fibrogenic activity via reduced nicotinamide adenine dinucleotide phosphate (NADPH) oxidase 2 and the janus kinase/signal transducer and activator of transcription (STAT) and phosphoinositide 3-kinase/Akt pathways [7,8].

Activation of HSC is controlled by transcription factors, such as activated protein-1, Jun D, Sp1, Kruppel-like factor 6, and nuclear factor kappa B (NF-κB), leading to transcriptional upregulation of latent TGF-β [9]. In this process, intracellular signaling molecules, such as Smad, Ras, Raf-1, and mitogen-activated protein (MAP) kinase, play important roles [10]. In addition, augmented production of the tissue inhibitor of matrix metalloproteinases (TIMPs) hampers the degradation of ECMs and conversely stimulates their accumulation in the inflamed liver [11]. Involvement of leptin and other adipocytokines in the HSC activation process is also notable [12]. Activated HSCs are also characterized by an increased expression of receptors for several polypeptides, including PDGF, TGF-β, vascular endothelial growth factor (VEGF), angiotensin-II, and ET-1 [13].

Along these lines, chronic activation of wound-healing reaction and of hepatic MFs is then essentially sustained by several growth factors, cytokines, and chemokines, as well as several additional mediators and environmental conditions (i.e., hypoxia). Thus, the development of progressive CLD may be envisaged to rely on a long-standing postulate of crosstalk/interactions between quiescent or activated HSCs and other populations of liver cells, with non-parenchymal cells playing a fundamental role.

12.2 THE CROSSTALK OF HSCs WITH MACROPHAGES

12.2.1 Liver macrophages (Kupffer cells)

Kupffer cells are resident liver macrophages adherent to sinusoidal endothelial cells inside the sinusoid and are the largest population of innate immune cells in the liver [14]. The principal function of Kupffer cells is to perform scavenger and phagocytic functions to remove protein complexes, small particles, senescent red blood cells, and cell debris from portal blood flow through pattern recognition receptors (PRRs). Kupffer cells also perform primary immune surveillance against gut-derived toxic materials, including endotoxin lipopolysaccharide (LPS) and pathogens from the intestinal flora. Thus, Kupffer cells, the most abundant pool of macrophages in the body, participate in the homeostasis by protecting the host and are able to trigger both immunogenic and tolerogenic immune responses [15,16].

Although it has been long considered that circulating monocytes contribute to the Kupffer cell pool, Naito et al. already showed in 1997 that Kupffer cells originate from fetal yolk sac precursor and undergo self-renewal throughout adult life [17]. More recent fate-mapping studies by Yona et al., using a mouse expressing constitutive and conditional CX3CR1 promoter-driven Cre recombinase, demonstrated that Kupffer cells are established prior to birth, maintained in adult steady state via self-renewal, and are independent of the input of bone marrow-derived monocytes [18]. Thus, Kupffer cells are able to renew themselves under stimulation with granulocyte/macrophage-colony stimulating factor (GM-CSF) and M-CSF. Although a specific marker to identify hepatic Kupffer cells is absent, human Kupffer cells are identified by their expression of CD14, CD16, and CD68, rat Kupffer cells by CD68 or CD163 (ED1 and ED2, respectively), and mouse Kupffer cells by F4/80.

12.2.2 Kupffer cells in hepatic inflammation and fibrogenesis

Upon inflammatory stimuli, Kupffer cells become activated and initiate their biological response. Endotoxin LPS derived from intestinal flora is a well-known stimulus of Kupffer cell activation. LPS binds to the toll-like receptor 4 (TLR4) with co-receptor CD14 and MD-2. TLR4 activates MyD88-dependent pathway to induce NF-κB and p38/c-jun N-terminal kinase (JNK) activation. When activated, Kupffer cells produce multiple bioactive substances, including cytokines such as TNF-α, interleukin 1 (IL-1), IL-6, IL-10, and IL-18, chemokines such as macrophage inflammatory protein 2 (MIP-2, CXCL2), macrophage chemotactic protein-1 (MCP-1, CCL2), RANTES (CCL5), MIP-1α (CCL3), MIP-1β (CCL4) and osteopontin, and reactive oxygen species (ROS), which lead to the infiltration of inflammatory cells into the liver [19]. In addition, activated Kupffer cells generate PDGF and TGF-β1, which in turn induce HSC activation (Figure 12.1).

In vivo macrophage depletion or blockade strategies revealed a major role for macrophages in liver fibrosis. For instance, Ide et al. showed that, in an animal model, depletion of Kupffer cells using the administration of gadolinium chloride ($GdCl_3$) suppresses α-sma–positive MF activation and ameliorates hepatic fibrosis in response to thioacetamide administration [20]. Rivera et al. also demonstrated that destruction of Kupffer cells with $GdCl_3$ or their inactivation with glycine ameliorates liver fibrosis induced by carbon tetrachloride (CCl_4) accompanied with downregulation of α-sma, α1(I) collagen mRNA and TGF-β1 protein expressions in the liver [21]. Aoyama et al. showed that CX3CR1 expression in Kupffer cells regulates HSC activation via the binding of HSC-derived CX3CL1 in CCl_4-induced mouse liver fibrosis [22]. Pradere et al. also demonstrated that hepatic macrophages enhance the survival of activated HSCs in a NF-κB-dependent manner and thereby promote liver fibrosis induced by bile duct ligation (BDL) in mice [23]. Taken together, liver macrophages

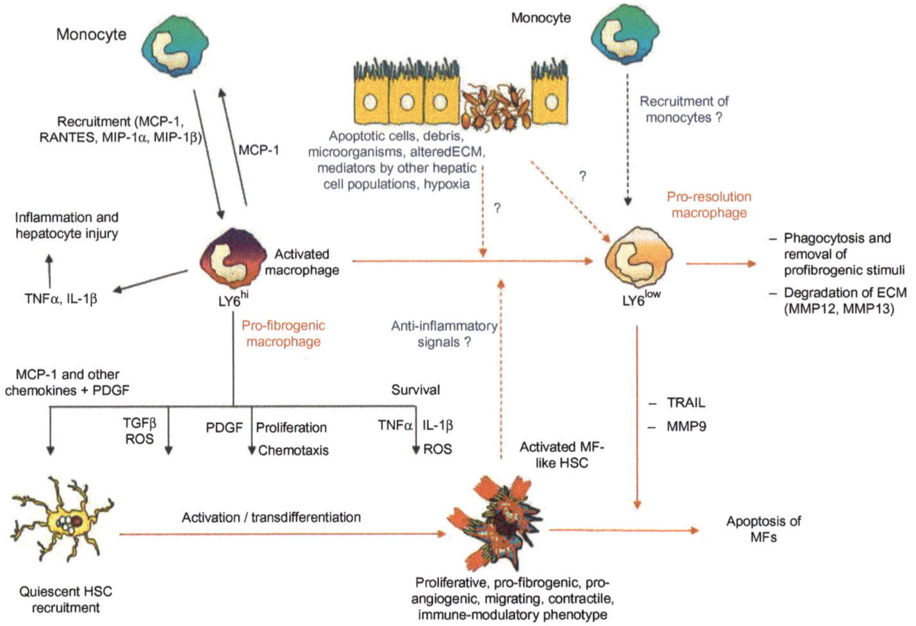

Figure 12.1 ***Simplified scheme of interactions between macrophages and HSCs in CLD.*** During the course of fibrogenic progression of a CLD, inflammatory monocytes are recruited to the site of injury/inflammation by different chemoattractants, with a major role described for MCP-1 (CCL2), leading to a population of proinflammatory and profibrogenic macrophages ($LY6^{hi}$). These macrophages, in addition to sustained inflammation and further hepatocyte injury by releasing TNF and IL-1β, can release mediators able to sustain fibrogenesis by favoring the process of activation/transdifferentiation of HSC to MF-like, to regulate the related classic phenotypic responses of activated cells and enhance their survival. Pro-resolution macrophages ($LY6^{low}$) originate as a consequence of signals/mediators coming either from activated MFs (possibly CX3CL1) or from other hepatic cell populations in the scenario of chronic live injury. Pro-resolution macrophages remove cell debris and potential profibrogenic signals and express TRAIL and MMP9 that can promote apoptosis of MFs. In addition, these macrophages can release matrix metalloproteases such as MMP12 and MMP13 that can degrade/remodel extracellular matrix. More details in the text.

generally participate in the augmentation of liver fibrosis through the activation of HSCs primarily via TGF-β1 cascade.

12.2.3 Macrophage heterogeneity

Macrophage heterogeneity has been described by a diversity in cytokine production, cell surface markers, and transcriptional profiles. Macrophages have been classified either into M1 or M2 macrophages; M1 macrophages are linked to Th1 primed CD4 T cells and induced by IL-12, interferon (IFN)-γ, and LPS, whereas M2 macrophages associate with Th2 CD4 T cells and are controlled by IL-4, IL-13, and GM-CSF. It is generally accepted that M1 macrophages are activated immediately as a defense against bacteria and viruses and release pro-inflammatory cytokines, such as TNF, IL-1β, IL-12, and ROS. M2 macrophages are generally thought to promote tissue-remodeling and secrete immune-modulatory mediators such as IL-4 and IL-13 via IL-4 receptor and IL-13 receptor α1, and characterized by unique signal transducers such as STAT6, arginase, or scavenger receptor (CD206) [24,25]. However, liver macrophages are shown to express markers of M1 and M2 differentiation simultaneously, indicating their ability to change phenotype markedly in response to injury.

As mentioned earlier, resident Kupffer cells play a role in early response of liver injury and secrete CCL2 and CCL5 chemokines and multiple cytokines. HSCs, upon activation by TLR4 ligands, are also a source of CCL2 and initiate monocyte recruitment. While the number of Kupffer cells decreases during inflammation and fibrogenesis, the number of monocyte-derived macrophages increases in inflamed liver. By using a model of conditionally depleted macrophages ($CD11b^+F4/80^+$) in CD-11b-diphtheria toxin receptor transgenic mice, Duffield et al. showed that macrophage depletion resulted in marked reduction of ECM deposition and of the number of α-sma–positive MFs in CCl_4 intoxication for 12 weeks, indicating that infiltrating monocytes exert a pro-fibrogenic role [26]. Karmark et al. tracked LY-6C, a cell surface glycoprotein that is widely used to identify functionally discrete mouse circulating monocyte population, and found that $Gr1^+$ ($Ly6C^{hi}$) inflammatory monocytes (their human counterpart is $CD14^+CD16^-$ monocytes expressing CCR2, CD64, and CD62L) are recruited into the liver in a CCR2-dependent manner during CCl_4-induced chronic liver injury in mice [27]. This interaction caused increase in the inducible NO synthase (iNOS)-positive $CD11b^+F4/80^+$ intrahepatic macrophages, which promote profibrogenic actions through direct activation of HSCs [28]. Thus, in the context of liver fibrogenesis in mice, infiltrating $Ly6C^{hi}CD11b^+F4/80^+$ macrophages play a central role in wound-healing response compared to $Ly6C^{low}CD11b^{low}F4/80^{hi}$ matured monocyte-derived and resident Kupffer cells.

12.2.4 Role of macrophages in fibrosis regression

It is generally accepted that regression of liver fibrosis occurs clinically in patients who achieve a sustained viral response (SVR) after the eradication of HCV by antiviral therapy or whose HBV viral level is well controlled by using nucleot(s)ide analogs, such as entecavir or tenofovir. In addition, substantial clinical data indicate that these treatments can not only regress cirrhosis but also promote recovery of hepatic functions of the liver and improve the prognosis of patients. Thus, the development of common anti-fibrotic therapy is anticipated regardless of the etiology of liver disease [6,29]. Regression of liver fibrosis is mechanistically explained by the following four aspects: (i) regeneration of hepatocytes, (ii) reversal of activated and MF-like HSCs to vitamin A-storing quiescent phenotype, (iii) removal of MFs by apoptosis, and (iv) lysis of ECMs. Advanced liver fibrosis is far less reversible in humans while rodent models exhibit induction and spontaneous resolution of fibrosis.

Macrophages are crucial in the resolution process of liver fibrosis (Figure 12.1). They are a source of fibrolytic MMPs, including MMP-12 and MMP-13, and also express TRAIL that promotes apoptosis of activated MFs [30]. Recent understanding of macrophage heterogeneity, as shown by Ramachandran et al., has illustrated that $Ly6C^{low}CD11b^{hi}F4/80^{int}$ subset of macrophages is most abundant in the liver in a resolution stage and represents the principle MMP-expressing subset. It is noteworthy that $Ly6C^{low}$ macrophages are derived from a phenotypic transition of the profibrogenic $Ly6C^{hi}$ macrophages and characterized by evidence of prior phagocytosis of dying cells [31].

Regression of liver fibrosis is also characterized by the removal of activated HSCs and MFs from the scar. Two potential pathways, either reversal to a quiescent phenotype or clearance through apoptosis, are considered. While the former can be completely accomplished in culture model [32], recent studies by Troeger et al. using genetic tracking techniques have revealed that reversal of activated HSCs toward a more quiescent phenotype occurs in a mouse model of liver injury; approximately 50% of HSC-derived MFs adopted an intermediate phenotype and reacquired features of quiescence [33]. On the other hand, activated HSCs express CD95, equivalent to Fas, TNF receptor 1, p75 and TRAIL receptors, resulting in the apoptosis after the binding of respective ligands. In fact, gliotoxin, an inducer of apoptosis of activated HSCs *in vivo*, enhances the resolution of liver fibrosis [34,35].

12.3 HSC/MFs AND INTERACTIONS WITH OTHER CELLS OF INNATE AND ADAPTIVE IMMUNITY

12.3.1 HSCs in the scenario of an immunological organ like the liver

As mentioned in the previous section, activated HSCs closely interact with liver macrophages and this interaction is believed to be critical for fibrogenic progression of CLD, regardless of the etiology. In this scenario, several laboratories have provided compelling evidence that HSCs, in addition to the widely accepted conventional "profibrogenic" role, can serve as immune-regulatory cells because of their ability to secrete a number of mediators and polypeptides, including critical chemokines like MCP-1 or CCL2, regulated and normal T cell expressed and secreted (RANTES or CCL5) as well as several isoforms of MIPs. Moreover, it should be considered that HSCs express several TLRs and chemokine receptors including CCR5, CCR7, CXCR3, and CXCR7, thus also becoming a "target cell" for several pro-inflammatory mediators. Finally, HSCs have also been reported to function as putative antigen presenting cells (APCs) [2,36]. For these reasons, HSCs may be considered innate immune cells [3].

The interactions between HSCs and cells of either innate or adaptive immunity (Figure 12.2) are relevant for two main reasons: (i) they occur in the scenario of an organ like the liver, which is considered to be an "immunological" or "lymphoid" organ and (ii) these interactions as well as the activation of specific immune signaling pathways within HSC function together to promote or modulate liver fibrogenesis [3,37–39].

The definition of the liver as an "immunological" organ relies on its unique structural and anatomical organization, with the organ receiving blood flow mostly from the gastrointestinal (GI) tract through the portal vein. This antigen-rich blood from the GI tract is then forced to pass through liver sinusoids where it is continuously monitored by resident APCs (including Kupffer cells, LSECs, and dendritic cells but also even HSCs) [3,36–39] as well as by a heterogeneous lymphocyte population that is enriched particularly in natural killer (NK) and natural killer T (NKT) cells, and B lymphocytes as well as "conventional" T lymphocytes ($CD8^+$ and $CD4^+$ T cells). The definition of "conventional" T lymphocytes is useful in order to distinguish them from

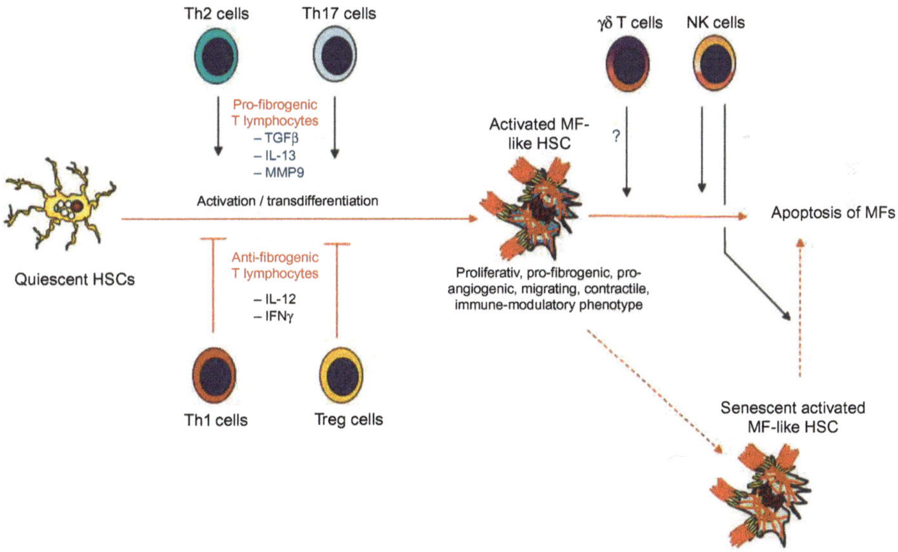

Figure 12.2 ***A simplified scheme of interactions between immune cells and HSC in a CLD.*** Immune cells can differently affect fibrogenesis by modulating the process of activation/transdifferentiation of HSCs to activated and MF-like cells. In synthesis, Th2 and Th17 lymphocytes, directly by releasing mediators like IL-13 and MMP9 or indirectly by stimulating the expression of TGFβ, can sustain the process of activation. By contrast, Th1 and Treg cells are believed to slow down or inhibit the same process through the release of IFNγ and IL-12 and then by inhibiting/limiting Th2 response. The atypical γδ T cells, together with ROS and other cytokines and death ligands, can induce apoptosis of activated and MF-like HSC. NK cells can kill early activated HSC and senescent activated HSC but not fully activated and MF-like HSC.

those defined as "unconventional" T cells. The latter comprise various cell types that can be categorized into two major populations: (i) cells that express NK cell markers, also defined as classical NKT cells (either CD4-positive or CD4/CD8 double negative cells), which are abundant in the liver, and their migration or expansion in the liver being controlled by NK cells and (ii) cells that do not express NK cell markers or TCR γδ T cells (i.e., nonclassical NKT) a population representing 15–25% of all intrahepatic T cells; the liver is one of the richest sources of TCR γδ T cells, which are able to recognize a limited range of antigens such as stress proteins and non-protein antigens [37,40].

In general terms, NK and NKT cells play critical roles as first-line cellular immune defense against invading pathogens and they also play an important role in the modulation of liver injury and in the recruitment of circulating lymphocytes [3,37–39]. Indeed, circulating lymphocytes make contact with liver resident APC displaying antigens and can also contact hepatocytes directly, due to fenestrated sinusoidal endothelium that lacks a basement membrane. This overall "hepatic" scenario is believed to facilitate direct or indirect priming of lymphocytes as well as to modulate the immune response to hepatotrophic pathogens. Such interactions contribute to some of the unique immunological properties of the liver, including its capacity to induce antigen-specific tolerance [3,37–39].

12.3.2 NK and NKT cells: Their role in modulating HSC activity and fibrogenesis

Interest in the role of NK and NKT cells in modulating liver fibrogenesis has recently emerged on the basis of the fact that these cells are enriched among liver lymphocytes and are markedly altered in CLDs of different etiology, with NK cells proposed to prevent fibrogenic progression of CLDs and NKT cells to act by either inhibiting or favoring liver fibrosis [41].

NK cells, which belong to the innate immune system, are able to recognize and kill target cells by employing several cell surface receptors. NK cells detect changes in the expression of host cell surface molecules that usually appear on either viral-infected, transformed, or injured cells [42]. This results in NK cell activation followed by the killing of target cell mediated by either exocytosis of perforin and granzyme granules or through FAS ligand, TNF-α, and TRAIL [41,42]. NK cells are somewhat strategically located in the hepatic sinusoids close to liver non-parenchymal cells and represent a rather unique organ-associated NK cell population. They are characterized by a rapid turnover, with continuous substitution by bone marrow-derived cells. NK cells, which normally account for approximately 30–50% of human liver lymphocytes, increase dramatically in number in pathological conditions such as those related to viral infection as well as acute and chronic inflammation [41]. The most important message from studies in this field (reviewed in Ref. [41]) is that liver NK cells can exert their anti-fibrogenic activity by selectively killing early activated HSCs but, remarkably, not quiescent or fully activated and MF-like cells. This phenomenon was first demonstrated by several laboratories in co-culture experiments (i.e., co-culture of HSCs and NK cells), then confirmed in rodent models of CLDs and, even more relevant, in human patients affected by CLD of different etiology [41]. Bing Gao et al. provided evidence for this peculiar and selective killing by showing that early during the process of activation, HSCs produce retinoic acid that, in turn, upregulates the NK cell activating ligand retinoic acid (RA) inducible gene 1 (RAE1) expression on HSCs. RAE1 binds NKG2D on NK cells and subsequently activates NK cells to kill the early activated HSCs through TRAIL- and NKG2D-dependent mechanisms [43,44]. This does not occur in chronically activated HSCs or MFs because these fully activated cells lose cytoplasmic stores of retinol/retinoids and do not produce RA and RAE1, thereby surviving the lethal NK cell attack. This hypothesis is confirmed by studies in NK cells from HCV-infected patients that can induce the apoptosis of activated HSCs through TRAIL-, Fas L-, and NKG2D-dependent mechanisms [45]. The NK cell-induced death of activated HSCs is facilitated by three additional critical features: (i) after activation, HSCs upregulate their expression of TRAIL receptor, which increases their sensitivity to the action of NK cells [46]; (ii) killing of either human and murine HSCs critically depends not only on NKG2D but also on the expression of the NK cell activating receptor NKp46 or of its murine ortholog NCR1, a receptor that recognizes viral hemagglutinins and other unknown ligands [47]; (iii) activated HSCs have been reported to lose the ability to express MHC-1 antigen, an NK cell inhibitory ligand that can suppress NK cell function by binding the inhibitory killer-cell immunoglobulin-like receptors (iKIRs) [48,49]. On the other hand, NK cells have been shown to kill senescent activated HSCs (which accumulate in the fibrotic septa following chronic liver injury) by a different mechanism, involving the granule exocytosis pathway. This represents the consequence of the fact that senescent activated HSCs express elevated levels of NK cell-activating ligands, thus becoming sensitive to NK cell killing [50]. Finally, activated NK cells are known to produce large amount of IFN-γ that has been suggested to contribute to anti-fibrotic effect by directly inducing cell cycle arrest and apoptosis of HSCs [48,51] as well as by enhancing NK cell killing of activated HSCs [43].

The scenario is obviously much more complex and intriguing for the role of NKT cells in relation to HSCs and liver fibrogenesis primarily due to the fact that these cells represent a heterogenous population, with types I and II of NKT cells playing different if not opposite functions in the liver parenchyma [41]. The fact that the mechanism(s) by which NKT cells are activated by endogenous ligands and cytokines is largely unknown adds further variables. Once activated, NKT cells become tolerant and nonresponsive to subsequent stimulation and are able to release high amount of either anti-fibrotic (e.g., IFN-γ) and profibrotic (e.g., IL-4, IL-13,

Hedgehog ligands, and osteopontin) cytokines, together with other cytokines, chemokines, and mediators that can differentially regulate liver fibrogenesis [41]. This is particularly evident by analyzing data from human patients and it has been suggested that this scenario may be significantly affected by the specific etiology. The postulate here is that NKT cells can promote liver fibrogenesis in HCV patients [52] and possibly also in nonalcoholic steatohepatitis (NASH) patients [53]. In the case of chronic viral infections, significant increases in intrahepatic iNKT cells and hepatic expression of CD1d have been reported; in addition, the production of profibrogenic cytokines, IL-4 and IL-13, by iNKT cells was significantly increased in HCV patients with cirrhosis. However, this may not apply to all conditions with NKT cells at times being stimulated to release IFN-γ, perforin or FasL enabling them to induce HSC cell cycle arrest and/or apoptosis, thus potentially inhibiting liver fibrogenesis. According to the hypothesis of Gao and Radaeva, the net effect of NKT cells in progressive CLD is likely regulated by a balance between the anti- and profibrotic effects of NKT cells. Literature data, as reviewed by Gao and Radaeva, also imply that NKT cells may play different roles in the control of liver fibrogenesis in different stages and various types of liver diseases [41].

12.3.3 Cells of the adaptive immunity: Their role in modulating fibrogenesis

As mentioned in the previous section, conventional T lymphocytes are believed to contribute to modulation of liver fibrogenesis in the overall interactions between profibrogenic cells and other cells of innate and adaptive immunity. As for many other tissues and organs, the balance between Th1 and Th2 lymphocytes is likely to affect fibrogenic progression of CLDs also, as indirectly suggested by data from two different strains of mice, with C57BL/6 mice (in which Th1 response predominates) displaying a weaker fibrotic reaction than BALB/c mice (in which Th2 response predominates) [54]. The reference paradigm here is that Th1 cells and of course their cytokines (mainly IL-12 and IFN-γ) may act by limiting/inhibiting fibrogenesis [3,55]. For example, administration of IL-12 has been shown to decrease granuloma formation and fibrosis induced by *Schistosoma* spp. [56]. Evidence in the literature also indicates that IFN-γ can suppress collagen deposition in fibrogenic cells by regulating the balance of MMP and TIMP expression. Moreover, IFN-γ or IL-12 are postulated to also negatively affect the expression and release of profibrogenic cytokines by Th2 cells [57]. In this regard, Th2 cells are believed to exert a predominant profibrogenic action that, apart from the established role in the experimental models by *Schistosoma* spp. [56], mainly relies on the ability of Th2 cells to release IL-13 [3,57]. Indeed, IL-13 availability is critical for at least two reasons: (i) first, this cytokine can stimulate TGF-β1 synthesis and upregulate MMP9 expression, which in turn is known to activate TGF-β1; (ii) second, IL-13 can promote fibrogenesis independently of TGF-β1, through a mechanism that is controlled by the relative expression of IL-13 receptor-α1 (IL-13Rα1; the signaling receptor) versus IL-13Rα2 (the decoy receptor) on hepatic MFs [57]. The critical role of IL-13 is proven by improvement of liver fibrosis by antagonizing IL-13 signaling [58].

Our understanding of the relevance of the Th1/Th2 paradigm for liver fibrogenesis has been modified by a number of studies outlining the role of Th17 cells as well as regulatory T (Treg) cells. Treg cells, a subset of $CD4^+$ T helper cells, are characterized by the expression of CD25 and by the release of the potent antiinflammatory and immunosuppressive cytokine IL-10. They are able to regulate other immune cells in a dominant-negative manner. The number of Treg cells strongly increase in the liver of patients affected by chronic HCV infection as well as autoimmune liver diseases and primary biliary cirrhosis (PBC) [3]. Data from experimental models suggest that Treg cells possess anti-fibrogenic property. Depletion of Treg cells has been reported to exacerbate fibrosis in the rat BDL model [59], and the anti-fibrogenic function of Treg cells has been attributed to their ability to suppress Th2 cells in *Schistosoma* spp.-induced fibrosis [60] However, it should be noted that $Foxp3^+CD4^+$ Treg cells can release IL-8 in patients affected by chronic HCV infection. Of interest, these IL-8 positive Tregs were found in close proximity of fibrotic areas and their number was found to correlate with the stage of fibrosis *in vivo*. Moreover, in primary human HSCs, these Tregs were found to upregulate the expression of classic genes related to activation, including TGF-β1, α-sma, procollagen, CCL2, TIMP1, and MMP2 in an IL-8-dependent manner [61].

On the other hand, Th17 cells, which are characterized by expression and release of IL-17 and IL-22, have emerged as putative profibrogenic cells and are usually increased in number (serum and liver tissue) in patients affected by several forms of either acute or chronic liver injury. The role of IL-17, in particular, is believed to be relevant because its receptor is expressed by HSCs, cholangiocytes as well as monocytes, and Kupffer cells. HSCs respond to IL-17 by upregulating collagen type I through activation of the STAT3 signaling pathway. More generally, IL-17 upregulates expression of mediators like IL-1β, IL-6, TNF-α and TGF-β in target cells [62].

Here a final mention should be dedicated to cytotoxic $CD8^+$ T cells by recalling data from an elegant adoptive transfer study in which these cells have been shown to be profibrogenic, a property that was negatively affected by IL-10 [63]. Other studies, performed in mice deficient for either $CD4^+$ or $CD8^+$ T cells, did not apparently confirm such a profibrogenic role of these cells [64] and it has been hypothesized that the role of $CD8^+$ T cells may in the end depend on the underlying etiology that drives the fibrotic process [3]. However, one has to recall a very recent study that has shown that fusion of microparticles released by T cells, which circulate in the blood of hepatitis patients, can result in fibrolytic activation of HSCs that leads to the downregulation of pro-collagen α1 and TGF-β1 by these cells [64].

12.4 INTERACTIONS OF HSCs AND HSC/MFs WITH LSECs: FROM LIVER SPECIFIC PERICYTES TO PRO-ANGIOGENIC CELLS

12.4.1 HSCs as liver specific pericytes in normal and pathophysiological conditions

Under physiological conditions, peri-sinusoidal HSCs reside in the space of Disse and are able to interact through their cytosolic processes (hepatocyte-contacting, inter-sinusoidal or inter-hepatocellular, peri-sinusoidal, or sub-endothelial) with hepatocytes as well as LSECs [1–5,65]. In addition to their significant contribution to vitamin A metabolism and deposition/remodeling of the extracellular matrix components of the space of Disse [1–5,65–67] quiescent HSCs from their strategic position are believed to act as liver specific pericytes and regulate sinusoidal blood flow [66,67]. Indeed, under physiological conditions quiescent HSCs in the space of Disse can maintain contact with sinusoidal endothelial cells through their peri-sinusoidal or sub-endothelial processes that run along one or more sinusoids. Moreover, their secondary processes appear to encircle the sinusoids [65,68] and HSCs are also reached by axonal processes of autonomic nerve containing several vasoactive peptides [65]. Although cultured human HSCs were found to contract in response to vasoconstrictors like thrombin, angiotensin-II, and ET-1 and to relax in response to NO and NO donors, prostaglandin E2, somatostatin, and adrenomedullin (reviewed in Refs. [65,67]), morphological and cell culture data were not sufficient to prove that quiescent HSCs may support or modulate sinusoidal diameter. More compelling data in favor of quiescent HSC contraction were produced from *in vivo* microscopy study [69]. On the other hand, overwhelming literature supports the paradigm that liver injury, particularly when chronic, can lead to vascular disorder in which ET-1 is overproduced by HSCs whereas NO release by LSECs is reduced [2,5,65,67,70]. Along these lines, although ET-1 was originally identified as a potent vasoconstrictor produced mainly by endothelial cells [71], HSCs were recognized by different laboratories as a major source of ET-1 as well as a target for this vasoactive molecule during liver injury [72–75]. ET-1 has a prominent contractile effect on HSCs and MFs, which has been proposed to contribute to portal hypertension in the cirrhotic liver [2,5,65,67,70,72–75] (see Chapter 8).

In a recent review by Iwakiri et al. [70], interactions between LSECs and activated HSCs are believed to be critical in intrahepatic vascular pathophysiology. A relevant point is represented by the regulation of the LSEC phenotype, which is the net result of the action of soluble factors (mainly VEGF, angiopoietins, ephrins, and fibroblast growth factors (FGFs)) as well as mechanical forces like those due to shear stress that, in turn, modulate endothelial NO synthase (eNOS) activity in LSECs, thereby regulating flow and vascular tone in the sinusoids [76]. Mediators released by LSECs in a paracrine manner are also involved in maintaining the "physiological" LSEC phenotype and can affect the early response of HSCs. In this regard, in response to injury, LSECs produce the cellular isoform of fibronectin that can stimulate in a paracrine way early HSC activation and, more specifically, HSC synthesis of ET-1 [77]. ET-1, in turn, can promote the proliferation of early-cultured HSCs, but inhibits fully activated HSCs (i.e., cultured for more than 1 week) [78]. As mentioned, the LSECs phenotype changes dramatically during chronic liver injury undergoing the so-called "capillarization" in association with a reduction in eNOS activity and NO synthesis after injury, likely due to extensive post-translational dysregulation of eNOS [70]. A reduction in NO release by LSECs may play a role in the fibrogenic progression of CLDs because NO has been shown to maintain quiescence of HSCs and reduced exposure of HSCs to NO may facilitate their activation [79,80]. The very complex scenario of interactions between LSECs and HSCs is critical in the maintenance/alteration of microcirculation during liver injury because HSCs also produce NO (i.e., a physiological antagonist to ET-1) [81], likely as a consequence of the activity of iNOS in response to pro-inflammatory cytokines or endotoxemia (reviewed in Ref. [70]). The overall message here is that all these changes are deemed to facilitate remodeling and constriction of the sinusoidal vasculature, which in turn can increase hepatic vascular resistance and is believed to represent an early feature of intrahepatic portal hypertension (also see Chapter 8).

12.4.2 HSCs and their proangiogenic role in CLD

During chronic liver injury, the sinusoids undergo significant vascular remodeling with changes in the LSEC phenotype and their interactions with HSCs. The most evident change is represented by capillarization of sinusoids, which is characterized by the loss of fenestrae in LSECs and by abnormal deposition of a basement membrane matrix in the space of Disse. In addition to the previously mentioned changes in NO and ET-1 synthesis, a number of paracrine and autocrine interactions have been described to occur between LSECs and HSCs. If one refers to polypeptide mediators produced and released by activated LSECs, it is worth mentioning that they produce PDGF, the most potent mitogenic and chemotactic stimulus for HSCs (both quiescent and activated) and TGF-β1, although the release of the latter cytokine by LSECs is likely less relevant for the progression of a CLD than that released by activated macrophages or by activated HSCs in a paracrine/autocrine loop [1–9,70]. In this scenario, one has to recall that VEGF also has a role in modulating LSEC phenotype. Indeed, under physiological conditions VEGF is the most critical molecule to modulate the size and number of LSEC fenestrae [82–85]. This has been shown *in vitro* by disappearance/restoration of fenestrae upon removing the supply of VEGF [83] as well as *in vivo* by disruption of VEGF signaling in mice carrying conditional deletion for VEGFR1 [85]. In conditions of chronic liver injury interactions between LSECs and HSCs become relevant in relation to the emerging evidence proposing a close link among hypoxia, angiogenesis, and fibrogenesis in the progression of CLDs [8,70,86]. Hypoxic areas within the liver parenchyma are common stimulus in the development of CLD and hypoxia inducible factors or HIFs have been shown to switch on the transcription of proangiogenic genes [78,79]. The involvement of angiogenesis in the progression of CLDs, is intrinsically related to chronic activation of wound healing and histopathological changes in the liver tissue, with increased deposition of ECMs and formation of fibrotic septa (paralleled by vascular changes) in turn leading to an impairment of oxygen diffusion [8,70,86]. In fact, both sprouting and intussusceptive angiogenesis are likely relevant in normal liver physiology and, particularly, in pathophysiologic conditions (i.e., CLDs but also tumor angiogenesis), with a role also in the genesis of portal hypertension in both intra- and extrahepatic circulation [70]. In this context, data obtained in conditional Notch1 knockout mice have suggested that Notch1 in LSECs may be required for retaining/maintaining fenestration of LSECs and that the loss of Notch1 can result in pathological intussusceptive angiogenesis and the development of nodular regenerative hyperplasia and portal hypertension in intrahepatic circulation [44]. In any case, histopathological analyses of cirrhotic livers indicate an increased number of new vessels in the fibrotic septa and surrounding regenerative nodules [87].

Evidence linking hypoxia and angiogenesis to liver fibrogenesis is now overwhelming based on the fact that angiogenesis and fibrogenesis develop in parallel in human patients and experimental model of CLD. More importantly, experimental antiangiogenic therapy was found to be effective in reducing fibrogenic progression, inflammatory infiltrate, the number of α-sma positive MFs as well as the increase in portal pressure [8,13,70,86–89]. In any event, what is relevant for this section is that VEGF expression, detected in hypoxic areas of chronically injured liver and in proximity of fibrotic septa, is mostly limited, in addition to LSECs, to hepatocytes as well as to activated HSC/MFs [8,12,70,86–88]. HSC/MFs or MFs are likely to be then affected in their behavior by proangiogenic cytokines released by hypoxic hepatocytes and LSECs but at the same time these profibrogenic cells are, in response to hypoxia, an additional source of VEGF-A and angiopoietin 1 and also express related receptors, such as VEGFR2 and Tie-2, in CLDs [12,88,90]. Literature indicates that hepatic MFs, in particular HSC/MFs, also represent a target for VEGF and angiopoietin I, with VEGF being able to stimulate their proliferation and increased deposition of extracellular matrix as well as increased migration and chemotaxis [13,88], the latter response being also significantly elicited in cells just exposed to hypoxia [13]. Oriented migration of liver MFs in response to either hypoxia or VEGF and other chemoattractants relies on a biphasic mechanism: (i) an early phase that is switched on by ROS released by either mitochondria or through ligand-receptor related activation of NADPH-oxidase, then resulting in redox-dependent activation of Ras/ERK and JNKs; (ii) a late and delayed phase of migration depending on HIF-1α-mediated, ROS-stabilized, upregulation of

VEGF expression, resulting in the subsequent chemotactic action of extracellularly released VEGF [13]. In addition to these *in vitro* data, immunohistochemical analysis for HIFs and hypoxia gene targets (VEGF, VEGFR2, angiopoietin I, Tie2) performed on human and rodent fibrotic/cirrhotic liver has led to the hypothesis that hypoxia, through a HIF-mediated pathway, may affect both the migration of MF-like cells and their proangiogenic behavior, leading these cells to align with developing septa and then drive both fibrogenesis and angiogenic response [13,86,88].

12.5 INTERACTIONS OF HSCs AND PORTAL FIBROBLASTS WITH CHOLANGIOCYTES

12.5.1 Cholangiocytes

The intrahepatic bile ducts construct a complicated network of conduits composed of the cholangiocytes. Besides their important roles in bile modification and transport, cholangiocytes actively respond to the inflammatory reactions associated with (chronic) liver injury. They are also sensitive to bioactive molecules in bile, including endogenous bile acids, lipids, and nucleotides as well as exogenous microbially derived and xenobiotic molecules that regulate cholangiocyte function and phenotype. Cholangiocytes are known to express multiple TLRs and associated molecular machinery and express receptors for cytokines, such as TNF-α, IL-6, IL-8, growth factors, such as PDGF, VEGF, and TGF-β1, and morphogens, such as Hedgehog (Hh) ligands and Notch. Thus, cholangiocytes affect repair and remodeling of the biliary epithelium [91]. Cholangiocytes are the target of various diseases termed cholangiopathies, including (i) inherited disorders, such as Alagille syndrome and cystic fibrosis, (ii) autoimmune disorders, such as primary sclerosing cholangitis (PSC), PBC, autoimmune cholangitis, and graft-versus-host disease, (iii) infection, such as cholangitis due to bacteria, fungi, parasite, or viruses, (iv) drug-induced injury, and (v) ischemic injury [92].

During organogenesis, cholangiocytes are derived from liver stem/progenitor cells (LPCs), which express cell surface markers including epithelial cell adhesion molecule (EpCAM), CD13, and CD133 in adult liver. Cholangiocytes vary in size, shape, and function along the biliary tree [93]. Large cholangiocytes are located at the level of interlobular and major bile ducts and maintain normal biliary homeostasis by expressing different ion channels and transporters at the basolateral or apical domain, and undergo proliferation in acute injury. On the other hand, small cholangiocytes at smaller bile duct branches, including terminal cholangioles and canals of Hering, behave as LPCs, playing a role in initiation and progression of biliary fibrosis. Ductular reaction initiate the differentiation of resting cholangiocytes to reactive cholangiocytes, which secrete multiple cytokines, chemokines, and growth factors recruiting inflammatory and mesenchymal cells [94].

12.5.2 Portal fibroblasts

Portal fibroblasts (PFs) are the resident fibroblasts of the portal tract, which exist in the mesenchyme surrounding the bile ducts. PFs were first described more than 50 years ago in rat liver and are considered to produce ECMs specifically in biliary fibrosis in addition to HSC-derived MFs [95]. Although PFs as well as HSCs are thought to derive from the septum transversum-derived mesothelial cells, lack of reliable markers to distinguish PFs from HSCs has hampered the definition of the roles of PFs in normal and injured liver. PFs are generally labeled with fibulin 2, elastin, IL-6, cofilin, and the ectonucleotidase NTPDase 2. Recently Mederacke et al. used a mouse model carrying a bacterial artificial chromosome (BAC) with a Cre-driven lecithin-retinol acyltransferase and found that 82–96% of fibrogenic MFs were derived from HSCs in seven models of fibrosis although the possibility that PFs are necessary for, in particular, biliary fibrosis could not be ruled out. In particular, bridging fibrosis expanded from portal and central vein area where PFs and desmin-negative cells exist, respectively [96].

One of the characteristics of PFs, distinct from HSCs-MFs, is their production of elastin. As mentioned earlier, PFs were first defined by their expression of fibulin 2 [97], which is a linker protein occupying the interface between microfibrils and their elastin core. Fibulin 2 sequestrates latent TGF-β binding proteins and might thereby mediate the progression of fibrosis. PFs also produce lysyl oxidase (LOX) and the related protein LOX-like 1, which crosslink elastin. Thus, the function of PFs should be considered with regard to elastin metabolism because elastin fibers may provide mechanical stability to the bile ducts and the portal vasculature [98].

With regard to the interaction of PFs with cholangiocytes, neurotrophin receptor p75-positive progenitors of PFs, and cholangiocytes express the laminin α1 and α5 subunits, respectively, which interact with the β1 integrin subunit to maintain the polarity and lumen of the bile ducts [99]. Jhandier et al. reported an interesting observation; PF-derived MFs fail to express ectonucleotidase NTPDase2 (also known as ecto-ATPase or CD39L1), which metabolizes extracellular nucleotides, resulting in the initiation of cholangiocyte proliferation through P2Y receptor activation in chronic cholangiopathies [100].

12.5.3 Interaction between cholangiocytes and mesenchymal cells in cholangiopathies

In the presence of chronic inflammation, the interactions between reactive cholangiocytes, mesenchymal cells, including HSCs, PFs, and MFs, and the inflammatory infiltrates contribute to the formation of biliary fibrosis [101].

Reactive cholangiocytes are a source of proinflammatory and chemotactic cytokines such as TNF-α, IL-1, IL-6, IL-8, MCP-1, and IFN-γ. Among them, MCP-1 initiates proliferation of PFs, their differentiation to PF-derived MFs and their collagen mRNA expression. Stromal cell derived factor-1 (SDF-1) is a cytokine with chemoattractive properties for inflammatory cells. SDF-1 is selectively upregulated in cholangiocytes in PBC and PSC and recruits CXCR4, SDF-1 receptor-positive infiltrating T lymphocytes [102].

TGF-β is produced by reactive cholangiocytes and is known to stimulate PF differentiation to PF-derived MFs. Reactive cholangiocytes also produce PDGF and stimulate HSC chemotaxis toward bile ducts and conversion of PFs into portal MF. In addition, in BDL mice, both VEGF and VEGF receptors are upregulated in cholangiocytes and stimulate their proliferation. Reactive cholangiocytes are the main source of connective tissue growth factor (CTGF) in experimental BDL animal models. CTGF induces extensive fibrosis in biliary atresia and desmoplastic reactions in cholangiocarcinoma [103].

Soluble, lipid-modified morphogens interact with patched (Ptc), a membrane-spanning receptor on the surface of Hh-responsive cells. Such signaling by Hh ligands controls tissue construction and remodeling. Hh ligands released by MFs activate Hh signaling in reactive cholangiocytes. In PBC patients, Hh ligands and Hh target genes are present in bile ductules and stromal cells. Hh may promote the acquisition of features of epithelial–mesenchymal transition (EMT) by reactive cholangiocytes. It has been reported that Wnt/β-catenin pathways and Notch signaling pathways are also involved in the proliferation and differentiation of cholangiocytes and in the formation of cholangiopathies [104].

12.5.4 Involvement of EMT in biliary fibrosis?

EMT is a phenomenon in which epithelial cells acquire structural and functional characteristics of mesenchymal cells, in particular, MFs [8]. Cholangiocytes are considered to participate in the generation of fibrosis around bile ducts through EMT. In the course of activation, cholangiocytes lose the expression of E-cadherin, CK-7, or CK-19 whereas they express fibroblast specific protein-1 (FSP-1) or vimentin and generate ECMs including collagen, fibronectin, elastin, and tenascin [105]. Cholangiocytes in culture undergo EMT and exhibit a MF-like morphology and express α-sma and collagen I. Harada et al. demonstrated that poly (I:C) increased the expression of basic FGF, S100A4, and Snail and decrease in the expression of epithelial marker, such as CK19 and E-cadherin, and bone morphogenetic protein and activin membrane-bound inhibitor (BAMBI), a TGF-β1 pseudoreceptor, in cultured human biliary epithelial cells indicating the occurrence of EMT [106]. In human liver with PBC, EMT of cholangiocytes was demonstrated immunohistochemically by co-localization of CK-19 and vimentin and the increase in the expression of snail and FSP-1.

However, two recent reports showed, by using genetic labeling technology, no evidence of EMT in case of liver fibrosis in mice. Scholten et al. studied EMT with

Cre-loxP system to map cell fate of K19 cholangiocytes in K19^{YFP} or FSP-1^{YFP} mice under BDL or CCl_4 administration. They revealed that K19^{YFP} cholangiocytes failed to express EMT markers, such as α-sma, desmin, or FSP-1 while FSP-1^{YFP}-positive cells had no expression of cholangiocyte marker K19 or E-cadherin [107]. Chu et al. demonstrated by using Alfp-Cre × Rosa26-YFP mice that, although primary cholangiocytes isolated from the mice underwent EMT in culture under the stimulation with TGF-β1 and TNF-α, no occurrence of EMT was detected in fibrotic liver induced by CCl_4 administration for 3 weeks or 3,5-diethoxycarbonyl-1,4-dihydrocollidine (DDC) for 2–3 weeks, which was determined by co-localization of YFP with S100A4, vimentin, α-sma, or pro-collagen 1α2 [108]. Taken together, these two reports denied the occurrence of EMT in case of mouse models of liver fibrosis. However, surprisingly, more recent reports by Michelotti et al. showed the ability of HSC's differentiation into hepatocytes and ductular cells by using α-sma-Cre-ER^{T2} or GFAP-Cre-ERTM mice crossed with Rosa-Stop-flox-YFP mice [109]. On the contrary, Lua et al. demonstrated by using Mesoderm posterior 1 (MesP1)^{Cre} and Rosa26lacZ^{flox} mice that mesodermal liver mesenchymal cells, including HSCs and PFs, undergo transformation into MFs, but not hepatocytes, cholangiocytes, and oval cells [110]. There has been no fixed scenario to finalize the EMT and mesenchymal–epithelial transition (MET) battle in liver fibrosis. Detailed analyses of the occurrence of EMT in human liver are also anticipated.

12.6 CONCLUDING REMARKS

Quiescent HSCs as well as hepatic MFs play a central role in modulating/regulating either physiological and metabolic responses in normal liver as well as the most critical events (i.e., inflammatory response, innate and adaptive immunity, fibrogenesis, and angiogenesis) that are determinant for the progression of a CLD, respectively. Experimental and clinical studies performed in the last two decades have been fundamental in order to progressively outline the fascinating major features of the very complex scenario of close interactions and molecular crosstalk among HSCs and MFs and all the other liver resident or nonresident cells. Our increased knowledge of the intimate mechanisms involved in these interactions and of their consequences, especially in conditions of disease, has emerged as crucial in order to identify novel diagnostic and prognostic biomarkers as well as to design novel and more selective therapeutic strategies to affect CLD progression.

REFERENCES

[1] Wake K. Perisinusoidal stellate cells (fat-storing cells, interstitial cells, lipocytes), their related structure in and around the liver sinusoids, and vitamin A-storing cells in extrahepatic organs. Int Rev Cytol 1980;66:303–53.

[2] Friedman SL. Hepatic stellate cells: protean, multifunctional, and enigmatic cells of the liver. Physiol Rev 2008;88:125–72.
[3] Pellicoro A, Ramachandran P, Iredale JP, Fallowfield JA. Liver fibrosis and repair: immune regulation of wound healing in a solid organ. Nat Rev Immunol 2014;14:181–94.
[4] Yin C, Evason KJ, Asahina K, Stainier DY. Hepatic stellate cells in liver development, regeneration, and cancer. J Clin Invest 2013;123:1902–10.
[5] Kawada N, Tran-Thi TA, Klein H, Decker K. The contraction of hepatic stellate (Ito) cells stimulated with vasoactive substances. Possible involvement of endothelin 1 and nitric oxide in the regulation of the sinusoidal tonus. Eur J Biochem 1993;213:815–23.
[6] Schuppan D, Kim YO. Evolving therapies for liver fibrosis. J Clin Invest 2013;123:1887–901.
[7] Wree A, Eguchi A, McGeough MD, Pena CA, Johnson CD, Canbay A, et al. NLRP3 inflammasome activation results in hepatocyte pyroptosis, liver inflammation, and fibrosis in mice. Hepatology 2014;59:898–910.
[8] Novo E, Cannito S, Paternostro C, Bocca C, Miglietta A, Parola M. Cellular and molecular mechanisms in liver fibrogenesis. Arch Biochem Biophys 2014;548:20–37.
[9] Mann DA, Smart DE. Transcriptional regulation of hepatic stellate cell activation. Gut 2002;50:891–6.
[10] Lopez-Sanchez I, Dunkel Y, Roh YS, Mittal Y, De Minicis S, Muranyi A, et al. GIV/Girdin is a central hub for profibrogenic signalling networks during liver fibrosis. Nat Commun 2014;5:4451. http://dx.doi.org/10.1038/ncomms5451.
[11] Perugorria MJ, Murphy LB, Fullard N, Chakraborty JB, Vyrla D, Wilson CL, et al. Tumor progression locus 2/Cot is required for activation of extracellular regulated kinase in liver injury and toll-like receptor-induced TIMP-1 gene transcription in hepatic stellate cells in mice. Hepatology 2013;57:1238–49.
[12] Choi SS, Syn WK, Karaca GF, Omenetti A, Moylan CA, Witek RP, et al. Leptin promotes the myofibroblastic phenotype in hepatic stellate cells by activating the Hedgehog pathway. J Biol Chem 2010;285:36551–60.
[13] Novo E, Povero D, Busletta C, Paternostro C, Valfrè di Bonzo L, Cannito S, et al. The biphasic nature of hypoxia-induced directional migration of activated human hepatic stellate cells. J Pathol 2012;226:588–97.
[14] Smith K. Liver disease: Kupffer cells regulate the progression of ALD and NAFLD. Nat Rev Gastroenterol Hepatol 2013;10:503. http://dx.doi.org/10.1038/nrgastro.2013.140.
[15] Petrasek J, Bala S, Csak T, Lippai D, Kodys K, Menashy V, et al. IL-1 receptor antagonist ameliorates inflammasome-dependent alcoholic steatohepatitis in mice. J Clin Invest 2012;122:3476–89.
[16] Olefsky JM, Glass CK. Macrophages, inflammation, and insulin resistance. Annu Rev Physiol 2010;72:219–46.
[17] Naito M, Hasegawa G, Takahashi K. Development, differentiation, and maturation of Kupffer cells. Microsc Res Tech 1997;39:350–64.
[18] Yona S, Kim KW, Wolf Y, Mildner A, Varol D, Breker M, et al. Fate mapping reveals origins and dynamics of monocytes and tissue macrophages under homeostasis. Immunity 2013;38:79–91.
[19] Tsutsui H, Matsui K, Okamura H, Nakanishi K. Pathophysiological roles of interleukin-18 in inflammatory liver diseases. Immunol Rev 2000;174:192–209.
[20] Ide M, Kuwamura M, Kotani T, Sawamoto O, Yamate J. Effects of gadolinium chloride (GdCl(3)) on the appearance of macrophage populations and fibrogenesis in thioacetamide-induced rat hepatic lesions. J Comp Pathol 2005;133:92–102.
[21] Rivera CA, Bradford BU, Hunt KJ, Adachi Y, Schrum LW, Koop DR, et al. Attenuation of CCl(4)-induced hepatic fibrosis by GdCl(3) treatment or dietary glycine. Am J Physiol Gastrointest Liver Physiol 2001;281:G200–7.
[22] Aoyama T, Inokuchi S, Brenner DA, Seki E. CX3CL1–CX3CR1 interaction prevents carbon tetrachloride-induced liver inflammation and fibrosis in mice. Hepatology 2010;52:1390–400.
[23] Pradere JP, Kluwe J, De Minicis S, Jiao JJ, Gwak GY, Dapito DH, et al. Hepatic macrophages but not dendritic cells contribute to liver fibrosis by promoting the survival of activated hepatic stellate cells in mice. Hepatology 2013;58(4):1461–73.
[24] Tacke F, Zimmermann HW. Macrophage heterogeneity in liver injury and fibrosis. J Hepatol 2014;60:1090–6.

[25] Sica A, Invernizzi P, Mantovani A. Macrophage plasticity and polarization in liver homeostasis and pathology. Hepatology 2014;59:2034–42.

[26] Duffield JS, Forbes SJ, Constandinou CM, Clay S, Partolina M, Vuthoori S, et al. Selective depletion of macrophages reveals distinct, opposing roles during liver injury and repair. J Clin Invest 2005;115:56–65.

[27] Karlmark KR, Weiskirchen R, Zimmermann HW, Gassler N, Ginhoux F, Weber C, et al. Hepatic recruitment of the inflammatory Gr1+ monocyte subset upon liver injury promotes hepatic fibrosis. Hepatology 2009;50:261–74.

[28] Rockey DC. Translating an understanding of the pathogenesis of hepatic fibrosis to novel therapies. Clin Gastroenterol Hepatol 2013;11:224–31. e1–5.

[29] Schuppan D, Pinzani M. Anti-fibrotic therapy: lost in translation? J Hepatol 2012;56(Suppl. 1):S66–74.

[30] Fallowfield JA, Mizuno M, Kendall TJ, Constandinou CM, Benyon RC, Duffield JS, et al. Scar-associated macrophages are a major source of hepatic matrix metalloproteinase-13 and facilitate the resolution of murine hepatic fibrosis. J Immunol 2007;178:5288–95.

[31] Ramachandran P, Pellicoro A, Vernon MA, Boulter L, Aucott RL, Ali A, et al. Differential Ly-6C expression identifies the recruited macrophage phenotype, which orchestrates the regression of murine liver fibrosis. Proc Natl Acad Sci USA 2012;109:E3186–95.

[32] Hazra S, Xiong S, Wang J, Rippe RA, Krishna V, Chatterjee K, et al. Peroxisome proliferator-activated receptor gamma induces a phenotypic switch from activated to quiescent hepatic stellate cells. J Biol Chem 2004;279:11392–401.

[33] Troeger JS, Mederacke I, Gwak GY, Dapito DH, Mu X, Hsu CC, et al. Deactivation of hepatic stellate cells during liver fibrosis resolution in mice. Gastroenterology 2012;143:1073–83.

[34] Kendall TJ, Hennedige S, Aucott RL, Hartland SN, Vernon MA, Benyon RC, et al. p75 Neurotrophin receptor signaling regulates hepatic myofibroblast proliferation and apoptosis in recovery from rodent liver fibrosis. Hepatology 2009;49:901–10.

[35] Wright MC, Issa R, Smart DE, Trim N, Murray GI, Primrose JN, et al. Gliotoxin stimulates the apoptosis of human and rat hepatic stellate cells and enhances the resolution of liver fibrosis in rats. Gastroenterology 2001;121:685–98.

[36] Friedman SL. Mechanisms of hepatic fibrogenesis. Gastroenterology 2008;134:1655–69.

[37] Racanelli V, Rehermann B. The liver as an immunological organ. Hepatology 2006;43:S54–62.

[38] Gao B, Jeong WI, Tian Z. Liver: an organ with predominant innate immunity. Hepatology 2008;47:729–36.

[39] Crispe IN. The liver as a lymphoid organ. Annu Rev Immunol 2009;27:147–63.

[40] Welsh RM, Lin MY, Lohman BL, Varga SM, Zarozinski CC, Selin LK. Alpha beta and gamma delta T-cell networks and their roles in natural resistance to viral infections. Immunol Rev 1997;159:79–93.

[41] Gao B, Radaeva S. Natural killer and natural killer T cells in liver fibrosis. Biochim Biophys Acta 2013;1832:1061–9.

[42] Lanier LL. NK cell recognition. Annu Rev Immunol 2005;23:225–74.

[43] Radaeva S, Sun R, Jaruga B, Nguyen VT, Tian Z, Gao B. Natural killer cells ameliorate liver fibrosis by killing activated stellate cells in NKG2D-dependent and tumor necrosis factor-related apoptosis-inducing ligand-dependent manners. Gastroenterology 2006;130:435–52.

[44] Radaeva S, Wang L, Radaev S, Jeong WI, Park O, Gao B. Retinoic acid signalling sensitizes hepatic stellate cells to NK cell killing via upregulation of NK cell activating ligand RAE1. Am J Physiol Gastrointest Liver Physiol 2007;293:G809–16.

[45] Glassner A, Eisenhardt M, Kramer B, Korner C, Coenen M, Sauerbruch T, et al. NK cells from HCV-infected patients effectively induce apoptosis of activated primary human hepatic stellate cells in a TRAIL-, FasL and NKG2D-dependent manner. Lab Invest 2012;92:967–77.

[46] Taimr P, Higuchi H, Kocova E, Rippe RA, Friedman S, Gores GJ. Activated stellate cells express the TRAIL receptor-2/death receptor-5 and undergo TRAIL mediated apoptosis. Hepatology 2003;37:87–95.

[47] Gur C, Doron S, Kfir-Erenfeld S, Horwitz E, Abu-Tair L, Safadi R, et al. NKp46-mediated killing of human and mouse hepatic stellate cells attenuates liver fibrosis. Gut 2012;61:885–93.

[48] Melhem A, Muhanna N, Bishara A, Alvarez CE, Ilan Y, Bishara T, et al. Anti-fibrotic activity of NK cells in experimental liver injury through killing of activated HSC. J Hepatol 2006;45:60–71.
[49] Muhanna N, Abu Tair L, Doron S, Amer J, Azzeh M, Mahamid M, et al. Amelioration of hepatic fibrosis by NK cell activation. Gut 2011;60:90–8.
[50] Krizhanovsky V, Yon M, Dickins RA, Hearn S, Simon J, Miething C, et al. Senescence of activated stellate cells limits liver fibrosis. Cell 2008;134:657–67.
[51] Rockey DC, Maher JJ, Jarnagin WR, Gabbiani G, Friedman SL. Inhibition of rat hepatic lipocyte activation in culture by interferon-gamma. Hepatology 1992;16:776–84.
[52] de Lalla C, Galli G, Aldrighetti L, Romeo R, Mariani M, Monno A, et al. Production of profibrotic cytokines by invariant NKT cells characterizes cirrhosis progression in chronic viral hepatitis. J Immunol 2004;173:1417–25.
[53] Syn WK, Oo YH, Pereira TA, Karaca GF, Jung Y, Omenetti A, et al. Accumulation of natural killer T cells in progressive nonalcoholic fatty liver disease. Hepatology 2010;51:1998–2007.
[54] Shi Z, Wakil AE, Rockey DC. Strain-specific differences in mouse hepatic wound healing are mediated by divergent T helper cytokine responses. Proc Natl Acad Sci USA 1997;94:10663–8.
[55] Muhanna N, Doron S, Wald O, Horani A, Eid A, Pappo O, et al. Activation of hepatic stellate cells after phagocytosis of lymphocytes: a novel pathway of fibrogenesis. Hepatology 2008;48:963–77.
[56] Wynn TA, Cheever AW, Jankovic D, Poindexter RW, Caspar P, Lewis FA, et al. An IL-12-based vaccination method for preventing fibrosis induced by schistosome infection. Nature 1995;376:594–6.
[57] Wynn TA. Fibrotic disease and the Th1/Th2 paradigm. Nat Rev Immunol 2004;4:583–94.
[58] Chiaramonte MG, Donaldson DD, Cheever AW, Wynn TA. An IL-13 inhibitor blocks the development of hepatic fibrosis during a T-helper type 2-dominated inflammatory response. J Clin Invest 1999;104:777–85.
[59] Katz SC, Ryan K, Ahmed N, Plitas G, Chaudhry UI, Kingham TP, et al. Obstructive jaundice expands intrahepatic regulatory T cells, which impair liver T lymphocyte function but modulate liver cholestasis and fibrosis. J Immunol 2011;187:1150–6.
[60] Turner JD, Jenkins GR, Hogg KG, Aynsley SA, Paveley RA, Cook PC, et al. $CD4^+CD25^+$ regulatory cells contribute to the regulation of colonic TH2 granulomatous pathology caused by schistosome infection. PLoS Negl Trop Dis 2011;5:e1269.
[61] Langhans B, Krämer B, Louis M, Nischalke HD, Hüneburg R, Staratschek-Jox A, et al. Intrahepatic IL-8 producing $Foxp3^+CD4^+$ regulatory T cells and fibrogenesis in chronic hepatitis C. J Hepatol 2013;59:229–35.
[62] Meng F, Wang K, Aoyama T, Grivennikov SI, Paik Y, Scholten D, et al. Interleukin-17 signaling in inflammatory, Kupffer cells, and hepatic stellate cells exacerbates liver fibrosis in mice. Gastroenterology 2012;143:765–76. e1–3.
[63] Safadi R, Ohta M, Alvarez CE, Fiel MI, Bansal M, Mehal WZ, et al. Immune stimulation of hepatic fibrogenesis by CD8 cells and attenuation by transgenic interleukin-10 from hepatocytes. Gastroenterology 2004;127:870–82.
[64] Kornek M, Popov Y, Libermann TA, Afdhal NH, Schuppan D. Human T cell microparticles circulate in blood of hepatitis patients and induce fibrolytic activation of hepatic stellate cells. Hepatology 2011;53:230–42.
[65] Geerts A. History, heterogeneity, developmental biology and functions of quiescent hepatic stellate cells. Semin Liver Dis 2001;21:311–35.
[66] Pinzani M, Failli P, Ruocco C, Casini A, Milani S, Baldi E, et al. Fat-storing cells as liver-specific pericytes. Spatial dynamics of agonist-stimulated intracellular calcium transients. J Clin Invest 1992;90:642–6.
[67] Rockey DC. Hepatic blood flow regulation by stellate cells in normal and injured liver. Semin Liver Dis 2001;21:337–49.
[68] Blomhoff R, Wake K. Perisinusoidal stellate cells of the liver: important roles in retinol metabolism and fibrosis. FASEB J 1991;5:271–7.
[69] Clemens MG, Zhang JX. Regulation of sinusoidal perfusion: *in vivo* methodology and control by endothelins. Semin Liver Dis 1999;19:383–93.

[70] Iwakiri Y, Shah V, Rockey DC. Vascular pathobiology in chronic liver disease and cirrhosis—current status and future directions. J Hepatol 2014 pii: S0168-8278(14)00395-X. http://dx.doi.org/10.1016/j.jhep.2014.05.047.
[71] Yanagisawa M, Kurihara H, Kimura S, Tomobe Y, Kobayashi M, Mitsui Y, et al. A novel potent vasoconstrictor peptide produced by vascular endothelial cells. Nature 1988;332:411–5.
[72] Kawada N, Kuroki T, Kobayashi K, Inoue M, Kaneda K, Decker K. Action of endothelins on hepatic stellate cells. J Gastroenterol 1995;30:731–8.
[73] Mallat A, Preaux AM, Serradeil-Le G, Raufaste D, Gallois C, Brenner DA, et al. Growth inhibitory properties of endothelin-A in activated human stellate cells: a cyclic adenosine monophosphate-mediated pathway. J Clin Invest 1996;98:2771–8.
[74] Pinzani M, Milani S, De Franco R, Grappone C, Caligiuri A, Gentilini A, et al. Endothelin 1 is overexpressed in human cirrhotic liver and exerts multiple effects on activated hepatic stellate cells. Gastroenterology 1996;110:534–48.
[75] Rockey DC, Weisiger RA. Endothelin induced contractility of stellate cells from normal and cirrhotic rat liver: implications for regulation of portal pressure and resistance. Hepatology 1996;24:233–40.
[76] Shah V, Haddad FG, Garcia-Cardena G, Frangos JA, Mennone A, Groszmann RJ, et al. Liver sinusoidal endothelial cells are responsible for nitric oxide modulation of resistance in the hepatic sinusoids. J Clin Invest 1997;100:2923–30.
[77] Jarnagin WR, Rockey DC, Koteliansky VE, Wang SS, Bissell DM. Expression of variant fibronectins in wound healing: cellular source and biological activity of the EIIIA segment in rat hepatic fibrogenesis. J Cell Biol 1994;127:2037–48.
[78] Rockey DC, Fouassier L, Chung JJ, Carayon A, Vallee P, Rey C, et al. Cellular localization of endothelin-1 and increased production in liver injury in the rat: potential for autocrine and paracrine effects on stellate cells. Hepatology 1998;27:472–80.
[79] Langer DA, Das A, Semela D, Kang-Decker N, Hendrickson H, Bronk SF, et al. Nitric oxide promotes caspase-independent hepatic stellate cell apoptosis through the generation of reactive oxygen species. Hepatology 2008;47:1983–93.
[80] Deleve LD, Wang X, Guo Y. Sinusoidal endothelial cells prevent rat stellate cell activation and promote reversion to quiescence. Hepatology 2008;48:920–30.
[81] Rockey DC, Chung JJ. Inducible nitric oxide synthase in rat hepatic lipocytes and the effect of nitric oxide on lipocyte contractility. J Clin Invest 1995;95:1199–206.
[82] Shah V, Toruner M, Haddad F, Cadelina G, Papapetropoulos A, Choo K, et al. Impaired endothelial nitric oxide synthase activity associated with enhanced caveolin binding in experimental cirrhosis in the rat. Gastroenterology 1999;117:1222–8.
[83] Funyu J, Mochida S, Inao M, Matsui A, Fujiwara K. VEGF can act as vascular permeability factor in the hepatic sinusoids through upregulation of porosity of endothelial cells. Biochem Biophys Res Commun 2001;280:481–5.
[84] Yokomori H, Oda M, Yoshimura K, Nagai T, Ogi M, Nomura M, et al. Vascular endothelial growth factor increases fenestral permeability in hepatic sinusoidal endothelial cells. Liver Int 2003;23:467–75.
[85] May D, Djonov V, Zamir G, Bala M, Safadi R, Sklair-Levy M, et al. A transgenic model for conditional induction and rescue of portal hypertension reveals a role of VEGF-mediated regulation of sinusoidal fenestrations. PLoS One 2011;6:e21478.
[86] Fernández M, Semela D, Bruix J, Colle I, Pinzani M, Bosch J. Angiogenesis in liver disease. J Hepatol 2009;50:604–20.
[87] Medina J, Arroyo AG, Sánchez-Madrid F, Moreno-Otero R. Angiogenesis in chronic inflammatory liver disease. Hepatology 2004;39:1185–95.
[88] Novo E, Cannito S, Zamara E, Valfre di Bonzo L, Caligiuri A, Cravanzola C, et al. Proangiogenic cytokines as hypoxia-dependent factors stimulating migration of human hepatic stellate cells. Am J Pathol 2007;170:1942–53.
[89] Moon J-K, Welch TP, Gonzalez FJ, Copple BL. Reduced liver fibrosis in hypoxia-inducible factor-1alpha-deficient mice. Am J Physiol Gastrointest Liver Physiol 2009;296:G582–92.
[90] Taura K, De Minicis S, Seki E, Hatano E, Iwaisako K, Osterreicher CH, et al. Hepatic stellate cells secrete angiopoietin 1 that induces angiogenesis in liver fibrosis. Gastroenterology 2008;135:1729–38.

[91] O'Hara SP, Tabibian JH, Splinter PL, LaRusso NF. The dynamic biliary epithelia: molecules, pathways, and disease. J Hepatol 2013;58(3):575–82.

[92] Masyuk T, Masyuk A, LaRusso N. Cholangiociliopathies: genetics, molecular mechanisms and potential therapies. Curr Opin Gastroenterol 2009;25:265–71.

[93] Tanimizu N, Mitaka T. Re-evaluation of liver stem/progenitor cells. Organogenesis 2014;10(2) [Epub ahead of print].

[94] Strazzabosco M, Fabris L. Development of the bile ducts: essentials for the clinical hepatologist. J Hepatol 2012;56:1159–70.

[95] Wells RG. The portal fibroblast: not just a poor man's stellate cell. Gastroenterology 2014;147(1):41–7.

[96] Mederacke I, Hsu CC, Troeger JS, Huebener P, Mu X, Dapito DH, et al. Fate tracing reveals hepatic stellate cells as dominant contributors to liver fibrosis independent of its aetiology. Nat Commun 2013;4:2823. http://dx.doi.org/10.1038/ncomms3823.

[97] Knittel T, Kobold D, Saile B, Grundmann A, Neubauer K, Piscaglia F, et al. Rat liver myofibroblasts and hepatic stellate cells: different cell populations of the fibroblast lineage with fibrogenic potential. Gastroenterology 1999;117:1205–21.

[98] Perepelyuk M, Terajima M, Wang AY, Georges PC, Janmey PA, Yamauchi M, et al. Hepatic stellate cells and portal fibroblasts are the major cellular sources of collagens and lysyl oxidases in normal liver and early after injury. Am J Physiol Gastrointest Liver Physiol 2013;304:G605–14.

[99] Tanimizu N, Kikkawa Y, Mitaka T, Miyajima A. α1- and α5-containing laminins regulate the development of bile ducts via β1 integrin signals. J Biol Chem 2012;287(34):28586–97.

[100] Jhandier MN, Kruglov EA, Lavoie EG, Sévigny J, Dranoff JA. Portal fibroblasts regulate the proliferation of bile duct epithelia via expression of NTPDase2. J Biol Chem 2005;280:22986–92.

[101] Park SM. The crucial role of cholangiocytes in cholangiopathies. Gut Liver 2012;6:295–304.

[102] Gentilini A, Rombouts K, Galastri S, Caligiuri A, Mingarelli E, Mello T, et al. Role of the stromal-derived factor-1 (SDF-1)-CXCR4 axis in the interaction between hepatic stellate cells and cholangiocarcinoma. J Hepatol 2012;57:813–20.

[103] Paradis V, Dargere D, Vidaud M, De Gouville AC, Huet S, Martinez V, et al. Expression of connective tissue growth factor in experimental rat and human liver fibrosis. Hepatology 1999;30:968–76.

[104] Greenbaum LE. Hedgehog signaling in biliary fibrosis. J Clin Invest 2008;118(10):3263–5.

[105] Xie G, Diehl AM. Evidence for and against epithelial-to-mesenchymal transition in the liver. Am J Physiol Gastrointest Liver Physiol 2013;305:G881–90.

[106] Harada K, Sato Y, Ikeda H, Isse K, Ozaki S, Enomae M, et al. Epithelial–mesenchymal transition induced by biliary innate immunity contributes to the sclerosing cholangiopathy of biliary atresia. J Pathol 2009;217:654–64.

[107] Scholten D, Osterreicher CH, Scholten A, Iwaisako K, Gu G, Brenner DA, et al. Genetic labeling does not detect epithelial-to-mesenchymal transition of cholangiocytes in liver fibrosis in mice. Gastroenterology 2010;139:987–98.

[108] Chu AS, Diaz R, Hui JJ, Yanger K, Zong Y, Alpini G, et al. Lineage tracing demonstrates no evidence of cholangiocyte epithelial-to-mesenchymal transition in murine models of hepatic fibrosis. Hepatology 2011;53:1685–95.

[109] Michelotti GA, Xie G, Swiderska M, Choi SS, Karaca G, Krüger L, et al. Smoothened is a master regulator of adult liver repair. J Clin Invest 2013;123:2380–94.

[110] Lua I, James D, Wang J, Wang KS, Asahina K. Mesodermal mesenchymal cells give rise to myofibroblasts, but not epithelial cells, in mouse liver injury. Hepatology 2014;60:311–22.

CHAPTER 13

Hepatic Stellate Cells and Hepatocyte Survival

Chandrashekhar R. Gandhi
Department of Pediatrics, Cincinnati Children's Hospital Medical Center, Department of Surgery, University of Cincinnati and Cincinnati VA Medical Center, Cincinnati, OH, USA

13.1 INTRODUCTION

The liver is exposed to a variety of noxious stimuli such as food and environmental-derived antigens and chemicals, and toxins including gram-negative bacterial endotoxin (lipopolysaccharide, LPS) on a regular basis via the portal vein. Toll-like receptors (TLRs) recognize the gut-derived, pathogen-associated molecular patterns (PAMPs) such as LPS, peptidoglycans, and RNA/DNA; lipoproteins are recognized by TLR1 and TLR2, double-stranded (ds)RNA by TLR3, LPS by TLR4, Flagelin by TLR5, single-stranded RNA by TLR7 and TLR8, and CpG-containing DNA by TLR9. The cell's response to the PAMPs is dependent upon the presence and association of TLRs with their co-receptors and adaptor molecules (e.g., CD14, MD2, and MyD88) [1]. While the gut-derived factors may have a direct effect on hepatocytes, they are generally taken up by the cells of the hepatic innate immune system including the resident macrophages Kupffer cells and eliminated effectively. Kupffer cells phagocytose (or endocytose) and degrade particulate or soluble matter including microbes and viruses and products thereof such as LPS. During this process, Kupffer cells produce cytokines such as tumor necrosis factor-α (TNFα), interleukin-6 (IL6) and IL1β, and reactive oxygen species (ROS) that kill the invading microorganisms [2–5]. However, the same cytokines and free radicals can elicit damaging effect on hepatocytes. Much of this information has been gathered by the *in vitro* experiments and supported by *in vivo* animal experiments in which Kupffer cells are blocked by chemicals such as gadolinium chloride ($GdCl_3$) or eliminated by liposome-encapsulated clodronate [6]. Over the last decade, increasing evidence has accumulated to indicate that hepatic stellate cells (HSCs) play an important role in hepatic defense or injury mechanism.

HSCs, located in the space of Disse, are in direct physical contact with hepatocytes. Similar to Kupffer cells, HSCs also produce numerous cytokines and chemokines and thus have the ability to regulate hepatocyte survival in physiology and pathology. While HSCs can impart protection to hepatocytes, a recently developed model of HSC

Stellate Cells in Health and Disease
DOI: http://dx.doi.org/10.1016/B978-0-12-800134-9.00013-0

depletion [7] has provided evidence that, under certain conditions, HSCs can induce injury to hepatocytes directly and by interacting with other cells (e.g., infiltrating inflammatory cells and Kupffer cells). The characteristics of HSCs to produce mediators of hepatocyte injury and protection are significantly altered during the progression of liver disease as they become activated. However, not much attention has been given to this important aspect of HSC biology. This chapter describes the current understanding of the mechanisms by which HSCs regulate hepatocyte survival.

13.2 HSC PHENOTYPES AND LIVER INJURY

It is important to understand which HSC phenotype is predominantly present during a specific type of liver injury (acute, ongoing, or chronic). Physiologically, the perisinusoidal HSCs are quiescent, store major quantities of body's retinoids including vitamin A, and express desmin or glial fibrillary acidic protein (GFAP). Within a short time after initiation of liver injury (e.g., partial hepatectomy, ischemia/reperfusion (I/R) or toxin-induced), HSCs begin the activation process characterized by progressive loss of retinoids, expression of platelet-derived growth factor (PDGF) receptor, and expression of α-smooth muscle actin (α-sma). During this activation process, HSCs enter cell cycle, start proliferating, and migrate to the site of injury. They also become increasingly contractile, and produce increasing amounts of abnormal extracellular matrix proteins, cytokines, and chemokines (see Chapters 4–8). Thus, it is apparent that the quiescent and very early-activated HSCs influence hepatic pathophysiology at the beginning or early phases of liver injury, transitionally activated HSCs (with moderate retinoids and α-sma expression) during intermediate stages of liver injury, and fully activated HSCs (very low or no retinoids and robust α-sma expression) regulate fibrosis and influence hepatocyte injury or proliferation within the nodules encompassed by fibrous deposits in the chronic phase of liver injury. HSCs at these three steps of activation are shown in Figure 13.1. It should, however, be considered that not all HSCs are at the same stage of activation at any given time during progression of liver injury (and even in primary culture) and also that the cells in zone 1 are quite distinct from those in zone 3, which again may have subtle differences from the cells in zone 2 [8]. Furthermore, the characteristics of *in vivo* and *in vitro* activated HSCs are not identical as indicated by differential expression of many molecules by gene array analysis [9].

Several murine and human stellate cell lines are also available, which have provided an effective tool to understand the mechanisms of fibrosis, but these immortalized cells are derived from a very select (minor) population of HSCs in normal or injured (fibrotic) liver and may not necessarily represent the actual cell population of the normal or diseased liver. Certainly, they should not be used to elucidate mechanisms of the role of HSCs in acute injury to the previously normal liver.

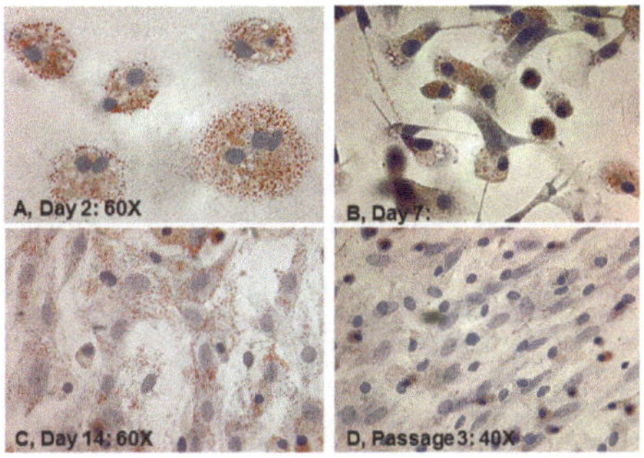

Figure 13.1 Rat stellate cells at indicated times in culture after isolation. The cells were stained with Oil Red O to show lipid stores. With time in culture, HSCs progressively lose their lipid content and become flattened, gaining a myofibroblast-like phenotype.

13.3 HSCs AND LIVER REGENERATION

It was found in the late 1980s and early 1990s that HSCs have a potential to produce cytokines and growth mediators [10–14]. Originally, HSCs were noted to be the major hepatic cell type to produce a potent mitogen hepatocyte growth factor (HGF), but this ability is reduced upon their activation [15,16]. Nevertheless, by producing HGF, HSCs can be an important cell type to promote liver regeneration (LR). Indeed, the significance of HSCs in such regulation is evident from their appearance before other non-parenchymal cells into the hepatocyte clusters during LR following partial hepatectomy [17]. The bidirectional interactions between the two cell types could be an important mechanism of LR as hepatocytes release insulin-like growth factor-1 (IGF1), which stimulates the synthesis of HGF in HSCs [18]. Subsequently in the three-dimensional (3D) co-culture system, hepatocytes demonstrated superior cytochrome P450 function whereas HSCs had increased HGF expression [19]. The 3D co-culture also provided evidence for the role of HSCs in revascularization of the liver; HSCs perform this function by secreting laminin between hepatocytes that is used as a trail by the endothelial cells to form hepatic microvasculature [20]. These observations were fortified by another 3D culture study in which HSCs were shown to mediate hepatocyte-induced morphogenesis of endothelial cell [21]. Furthermore, kinetics of the expression of epimorphin (expressed exclusively in HSCs) and connexin 32 (in hepatocytes) following partial hepatectomy also indicated that HSCs may be critically important in LR in a contact-dependent manner [22]. Indeed, HSCs were found to form tight junctions with hepatocytes during early phases of partial hepatectomy-induced LR [23]. During this time HSCs in contact with hepatocytes were found to be activated, indicating that their transient activation may be an essential component of LR. In this regard, HSC activation promotes engraftment of transplanted hepatocytes in rats [24]. The authors also reported that Kupffer cells were not required for the activation of HSCs, and depletion of HSCs in CCl_4-treated fibrotic liver by gliotoxin reduced engraftment of transplanted hepatocytes [24]. A caveat is that gliotoxin also depletes Kupffer cells from the CCl_4-treated fibrotic liver [25], and the depletion of both cell types in a hostile (CCl_4-injury) environment may have adverse effect on hepatocyte engraftment. Interestingly, fibronectin (of which HSCs are a major source) has been shown to be a survival factor for hepatocytes, but its deficiency leads to hepatocyte regeneration following toxin-induced acute liver injury, mediated by HGF [26]. Thus the previously somewhat overlooked role of HSCs in LR is becoming increasingly apparent and further understanding of the underlying mechanisms will be critical in developing new therapeutic modalities for hepatic injury including I/R and transplantation.

13.4 HSCs IN HEPATOCYTE INJURY DURING ENDOTOXEMIA

Being constantly exposed to low levels of LPS, the liver cells do not respond to its actions in a way that would cause hepatic injury in physiology. Physiologically, up to 1 ng/mL LPS is present in the portal venous blood and is not detected in the systemic circulation [27]. The concentration of LPS increases during acute and chronic liver injury as a consequence of increased intestinal translocation [27,28], and reduced hepatic clearance due to intra- and extrahepatic shunts and impaired activity of the reticuloendothelial system [29]. In such situations, injury to hepatocytes may occur due to LPS's direct effects or via mediators produced by the non-parenchymal cells in response to LPS, Kupffer cells being the cell type studied extensively in this regard. In Kupffer cells and recruited inflammatory and immune cells, LPS stimulates production of cytokines such as IL1β, TNFα, and interferon (IFN)-γ [2,4,30,31] that cause induction of nitric oxide (NO) synthesis in hepatocytes [32,33]; these effects are mediated by TLR4 [1]. These and other cytokines as well as free radicals produced by LPS-stimulated Kupffer cells can be damaging to hepatocytes. Because HSCs can also produce the same mediators, we began to investigate the effects of LPS on quiescent and transitionally activated HSCs in the mid-1990s.

13.4.1 Hepatic vasoconstriction

Endothelin-1 (ET-1) [34,35] and NO [36,37] are potent constrictor and dilator, respectively, of blood vessels including the hepatic vasculature. LPS upregulates the expression of ET-1 and NO by induction of inducible NO synthase (iNOS) in both quiescent and transitionally activated HSCs at a concentration as low as 1 ng/mL [38]. LPS also increases the expression of ET-1 receptors in HSCs, and blockade of NOS inhibits upregulation of ET-1 receptors in activated but not in quiescent HSCs [38]. The effect of NO on activated HSCs is similar to that in vascular smooth muscle cells in which NO was reported to increase ET-1 receptors [39]. Earlier, it was shown that ET-1 can cause contraction of HSCs in the normal rat liver [40], and thus the autocrine effects of ET-1 on the upregulated ET-1 receptors on HSCs and the paracrine actions on the smooth muscle cells of the hepatic veins can be an important mechanism of hepatic vasoconstriction and ischemic damage. Indeed, ET-1 and its receptors are upregulated upon administration of LPS as well as CCl_4 (acute necrotic injury) in rats, and the acute increase in portal resistance is ameliorated by prior treatment with a mixed ET-1 receptor antagonist TAK-044 [41,42]. The question is why hepatic vasoconstriction prevails following endotoxin administration when NO synthesis is also increased contemporaneously. The most plausible explanation is that ET-1–induced hepatic vasoconstriction is extremely powerful and long-lasting [34], and overcomes the counter-regulatory effect of NO. Interactions between ET-1 and activated HSC are also considered to be an important factor in portal hypertension during chronic liver injury, which is addressed in Chapter 8.

13.4.2 Hepatocyte injury

Parallel work in our laboratory (along with LPS-induced upregulation of ET-1 and its receptors [38]) showed that at a very low concentration (1 ng/mL), and in the absence of serum, LPS induced iNOS expression and NO synthesis and increased the expression and synthesis/release of IL1β, IL6, and TNFα from quiescent as well as transitionally activated rat HSCs [43–45]. Such effect of LPS in the absence of serum was not observed in Kupffer cells (unpublished observation). The rat quiescent HSCs were found to express very low (almost undetectable) levels of TLR4 but its expression increases upon activation of HSCs [44]. Furthermore, the effects of LPS occurred in the absence of serum (i.e., in absence of lipopolysaccharide-binding protein, LBP). This observation is interesting as serum is required for providing LBP in the initiation of binding of LPS to its receptor CD14 [46], and HSCs do not produce LBP [44]. These findings provided evidence for an alternate way by which LPS exerts effects on HSCs that could be potentially important in endotoxemic acute liver injury. Brenner's group also showed that quiescent human HSCs express very low levels of TLR4, which increases upon their activation, and that the inflammatory effects of LPS are mediated by TLR4 in activated human HSCs [47]. Subsequently, Schwabe's group reported that quiescent and activated mouse HSCs express TLR4 similarly, and LPS binding to TLR4 downregulates TGFβ pseudoreceptor BAMBI (BMP and the activin membrane bound inhibitor), allowing the effects of Kupffer cell-derived TGFβ on HSCs in fibrogenic response [48]. Such species-specific differences in variable expression of TLR4 should be considered while investigating the effects of LPS on HSCs from rat, mouse, or human.

With our interest in how HSCs might influence hepatocyte survival and function during endotoxemia, we initially examined the effects of LPS-stimulated HSCs on hepatocytes. Culture media from both quiescent and transitionally activated rat HSCs upon stimulation with LPS were found to secrete mediators that strongly inhibited DNA synthesis in hepatocytes [43,44]. Interestingly, medium conditioned in the absence of LPS by activated HSCs, but not quiescent HSCs, stimulated DNA synthesis in hepatocytes, suggesting that activated HSCs release mediators that promote hepatocyte growth. Both phenotypes produced TNFα, IL6, IL1β, and NO upon stimulation with LPS. IL6, which is a priming agent for hepatocyte proliferation during LR [49–51], also protects against I/R injury [52], but can also inhibit cell cycle progression [53]. TNFα is implicated in promotion of liver injury, and IL10 and IL6 in its suppression [54–59]. TNFα causes only minor apoptosis of hepatocytes in culture (unpublished observation). However, under specific conditions TNFα is pro-apoptotic [60–62]. For example, in the absence of NF-κB activation beyond the basal level, TNFα can induce apoptosis/necrosis [63]. IL1β not only has anti-proliferative property [64], but also has been found to exert no effect on hepatocyte proliferation [65], and in yet other studies it was found to be anti-apoptotic [66,67]. While NO was found to inhibit DNA synthesis and growth and to induce apoptosis of various cell types [68–77], it was shown to protect hepatocytes from heat- and TNFα/actinomycin-induced apoptosis [78–80]. Furthermore, lack of iNOS was found to increase hepatocyte apoptosis but decreased fibrosis in mice after chronic CCl_4 administration [81].

Neutralization of IL1β, TNFα, IL6, and TGFβ with specific antibodies and inhibition of NO synthesis (independently and together) had only minor effect on the LPS/HSC-induced inhibition of DNA synthesis [43,44]. Moreover, despite the strong inhibition of DNA synthesis, only a minor population of hepatocytes was found to undergo apoptosis upon incubation with LPS/HSC medium [44]. These findings led to the hypothesis that HSCs may have an ability to produce additional mediators of hepatocyte apoptosis.

Because LPS-stimulated quiescent and transitionally activated HSCs appeared to exert similar effects on hepatocytes (inhibition of DNA synthesis and apoptosis of a minor population), we performed transcripomic analysis of LPS-stimulated transitionally activated rat HSCs at time points up to 24 h [82]. In addition to TNFα, IL6, IL1β, and iNOS, this analysis led to the identification of type I interferons in LPS-stimulated HSCs. Type I interferons are known to contribute to I/R-induced liver injury [83–86]. However, the expression of IFNβ increased only transiently at 3 h and rapidly decreased thereafter, reaching undetectable basal levels by 6–12 h [87]. In rats administered LPS also mRNA expression of IFNβ increased transiently within 1 h and declined at 3 h [87]. Similarly, TNFα expression increased time-dependently in LPS-treated HSCs till 12 h and declined subsequently [45,82]. In contrast, expression of IL6 and IL10 continued to rise till 12 h and remained at this level at 24 h following LPS stimulation; TGFβ1 expression was very low in HSCs and increased modestly by LPS treatment [82]. Because IL6 and IL10 [54–59,88] are hepatoprotective and TNFα is

Table 13.1 Alterations in LPS-induced gene transcripts in HSCs

Cytokines	Maximal increase (time and fold)
IL1a	24 h (60-fold)
IL1b	12 h (12-fold)
IL6	12 h (79-fold)
IL10	24 h (240-fold)
IL17f	24 h (40-fold)
TNF	3 h (420-fold)
TGFβ1	24 (55%)

The data are derived from Ref. [82].
Rat HSCs were incubated in medium containing 100 ng/mL LPS for up to 24 h, and gene expression was determined via microarray analysis. The time and magnitude of maximal increase in the gene transcripts are shown.

injurious to hepatocytes only under conditions such as blockade or inhibition of gene transcription [60–62], these results suggest that LPS-stimulated HSCs may act primarily as a cell type that imparts protection to hepatocytes during endotoxemia. The fact that all of the rats survived the endotoxin challenge at 3–5 mg/kg LPS [41] and even 10 mg/kg [89] suggests that HSCs from their close location to hepatocytes play an important hepatoprotective role. Table 13.1 lists some of the critical cytokines whose expression is increased in HSCs upon stimulation by LPS [82].

Similar to rat HSCs, LPS stimulated the synthesis and secretion of TNFα, IL6, IL10, and IL1β in transitionally activated mouse HSCs [90]. Interestingly, LPS does not stimulate NO synthesis in mouse HSCs (unpublished observation) indicating a major difference in species-specific characteristics of these cells. Both mouse and rat HSCs, however, also produce various chemokines with the ability to recruit immune cells (CD4 and CD8 T cells, dendritic cells, and regulatory T cells), and LPS increases their expression similarly [48,82,90,91]. Recruitment and modulation of immune cells by HSCs can be an important mechanism of liver tolerance, which is described in Chapter 14.

13.4.3 LPS-Stimulated HSCs induce autophagy in hepatocytes

Because all mice and rats survive endotoxemia induced by up to 10 mg/kg LPS, and only a few hepatocytes can be seen undergoing apoptosis *in vivo* [87,89], the data suggested initiation and implementation of a robust survival mechanism. It has been established that endoplasmic reticulum (ER) stress and autophagy are physiological processes that clear undesirable material including abnormal or misfolded proteins and injured/damaged organelles, and thus critical mechanisms of cell survival [92–94].

ER stress triggers activation of the ER membrane-associated double stranded RNA-activated protein kinase-like ER kinase (PERK), activating transcription factor-6 (ATF6) and inositol-requiring enzyme 1 (IRE1), with dissociation of their negative regulator, glucose-regulated protein-78 (GRP78/Bip). Activated PERK phosphorylates the α-subunit of eukaryotic translation initiation factor 2 (eIF2α); P-eIF2α attenuates translation initiation and protein synthesis, and induces translation of a transcription factor, ATF4. ATF6 is cleaved and its cytosolic N-terminal fragment translocates to nucleus. IRE1 is dimerized and autophosphorylated, and induces splicing of X-Box binding protein 1 mRNA (XBP1) that alters carboxy-terminus of the protein creating a potent transcription factor. ATF4, XBP1, and ATF6 initiate expression of genes encoding ER chaperones, enzymes and structural components of the ER, and normalize cell homeostasis. P-eIF2α and ATF6 also induce gene expression and synthesis of c/EBP homologous protein-10 (CHOP/GADD153) that negatively regulates cell growth and promotes apoptosis [92–94].

Ubiquitination and autophagic degradation of the unfolded or abnormal protein aggregates causes attenuation of ER stress and protect cells from death [95]. Although the mechanisms of recognition and removal of misfolded/unfolded proteins by autophagosomes are not completely understood, activated IRE1 and PERK/eIF2α have been implicated in the induction of autophagy [95–99]. During autophagy microtubule-associated protein 1 light chain-3 (LC3 or Atg6) is converted into phosphatidylethanolamine-conjugated membrane-bound LC3-II. Localization and accumulation of LC3-II in autolysosomes is required for the formation and elongation of their double-membrane. However, cells undergo apoptosis when there is overproduction of autophagic vesicles that interferes with normal membrane trafficking, or upon blockade of autophagocytic degradation of abnormal proteins [100,101]. Thus it is critical that a well-orchestrated response is in place to prevent cells from injury/death during increased stress.

Ultrastructural examination of the livers of rats treated with 10 mg/kg LPS revealed increased dilation of ER (with loss of ribosomes) and autophagy (Figure 13.2A). This effect was accompanied by increased phosphorylation of PERK, eIF2α, and increased expression of CHOP [89]. *In vitro*, ultrastructural examination showed that soluble mediators released by LPS-stimulated HSCs induced ER stress and autophagy in hepatocytes (Figure 13.2B). This was accompanied by increased phosphorylation of PERK, eIF2α, and increased expression of CHOP, as well as formation of LC3-II from LC3-I [87,89], an essential requirement for the autophagy to continue and complete. Although a definitive identification of the signaling pathways as well as HSC-derived mediator(s) responsible for these cell survival mechanisms remains undetermined, NO, TNFα, IL6, and IFNβ all are known to exert such effects [102–106], whereas IL10 inhibits autophagy [104]. It is likely that instigation of ER stress and autophagy by LPS-stimulated HSCs in hepatocytes and at the same time limiting this process by producing IL10 can be a cell survival response in endotoxemia.

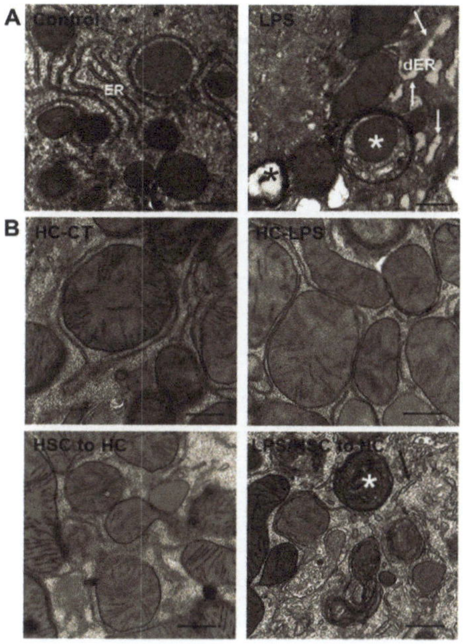

Figure 13.2 Transmission electron micrograph showing LPS/HSC-induced autophagy in hepatocytes. (A) At 6h following administration of PBS or LPS (10mg/kg), the livers were excised, fixed, and prepared for transmission electron microscopy (TEM). TEM shows rough endoplasmic reticulum seen in control liver is dilated significantly (dER) with loss of ribosomes (long arrows) upon LPS treatment. Several cytosolic vacuoles (short arrows) and autophagosomes containing mitochondria and cellular debris (asterisks) are seen in the LPS-treated liver. Bar = 500nm. (B) Hepatocytes were incubated in control medium (HC-CT), medium containing 100ng/mL LPS (HC-LPS) or medium conditioned by HSCs without (HSC to HC) or with LPS (LPS/HSC to HC) for 6h. Hepatocytes incubated with LPS or HSC-conditioned medium are ultrastructurally similar to control cells. Autophagosomes (asterisks) and dilated ER can be seen in LPS/HSC-treated hepatocytes. Bar = 500nm.

13.5 HSC DEPLETION MODEL TO STUDY ACUTE LIVER INJURY

While most of the evidence for the role of HSCs in regulating hepatocyte injury came from *in vitro* experiments with cultured cells, the relevance of these findings to the *in vivo* pathogenesis has been speculative due to lack of HSC-depleted *in vivo* model. To investigate such relevance, we developed a mouse model of HSC depletion. Mouse expressing herpes simplex virus thymidine kinase (HSV-TK) transgene under the GFAP promoter (B6.Cg-Tg(Gfap-Tk)7.1Mvs/J) (GFAP-Tg) (The Jackson Laboratory) was used. Nearly 80% of the murine HSCs express GFAP [8]. HSV-TK phosphorylates nontoxic ganciclovir (GCV) to GCV-monophosphate, which is converted into GCV-triphosphate by cellular guanylate kinase; phosphorylated GCV incorporates into the DNA causing death of replicating cells [107,108]. However, HSCs are physiologically quiescent and for their elimination by the action of GCV-monophosphate, it is important that HSCs are in replication mode. This was achieved by three administrations (2 days apart) of CCl_4, which induced liver injury but stimulated HSCs to enter cell cycle; discontinuation of CCl_4 and administration of GCV for the next 10 days caused effective elimination of a major HSC population (up to 75%) from the liver [7]. At this time the liver injury due to initial CCl_4 administration had resolved and the other organs did not show any obvious effects of CCl_4 followed by the GCV treatment [7].

Administration of LPS (10 mg/kg) to the HSC-sufficient and HSC-depleted mice demonstrated amelioration of the liver injury in the latter. Even the warm I/R injury was strongly ameliorated in HSC-depleted mice [7]. These results were surprising because Kupffer cells were not eliminated and they and the infiltrating monocytes and neutrophils are considered to be the major cell types responsible for acute liver injury of various types including that due to endotoxin and I/R [5,85]. Thus, the overall results of the experiments with HSC-depleted mice were contrary to the hypothesis that HSCs primarily promote cell survival (see Sections 13.4.2 and 13.4.3). In this regard, using a similar model, depletion of HSCs during concomitant treatment with CCl_4 and GCV was found to ameliorate not only fibrosis, but also the ongoing CCl_4-induced hepatic injury in GFAP-Tg mice [109]. The interpretation of these data is that HSCs exert strong influence on the functions of Kupffer cells as well as the recruitment of infiltrating neutrophils and monocytes. HSCs were already known to express monocyte chemoattractant protein-1 (MCP1 or CCL2) [110–113], and the expression of this and other chemoattractants increase upon LPS treatment [82,90]. Thus HSCs are likely to play an important role in hepatic inflammation by recruiting neutrophils and monocytes. Indeed, hepatic infiltration of neutrophils and expression of CXCL1 and TNFα were significantly reduced in HSC-depleted mice after I/R or LPS administration relative to the HSC-sufficient mice [7]. In contrast, IL6 expression was similarly increased in HSC-depleted and HSC-sufficient mice following I/R and LPS treatment [7], and that of IL10 increased to a much greater extent in HSC-depleted mice (unpublished observation). It is important to note that both HSCs and Kupffer cells produce all these mediators. TNFα causes injury to hepatocytes [114,115], while IL6 is protective [116]. Treatment with neutralizing TNFα antibody ameliorated I/R injury [117,118], and carbon monoxide inhalation by recipient animals caused inhibition of hepatic graft TNF-α and IL-6 expression and reduced I/R injury [119]. LPS also stimulates production of TNFα and IL6 by Kupffer cells [2,120–122]. Incomplete abrogation of neutrophil infiltration in HSC-depleted mice might be due to the chemokines released by Kupffer cells [5,123]. Thus, it is clear that in addition to eliciting effects on hepatocytes directly, HSCs also regulate the functions of other cells (e.g., Kupffer cells) during acute liver injury.

The interactions between Kupffer cells and HSCs, however, must be considered to be very complex due to heterogeneity in both cell populations, and their distribution within the liver. M1 Kupffer cells are suggested to play an important role in inflammatory diseases whereas M2 Kupffer cells promote resolution of inflammation and tissue injury [124]. Heterogeneity in Kupffer cells is also apparent from their differential phagocytic activity and expression of mediators in periportal versus pericentral regions [125,126]. Similarly, heterogeneity between periportal and pericentral HSCs has also been demonstrated [8]. Nevertheless, amelioration of I/R and LPS-induced liver injury in HSC-depleted mice [7] indicate that HSCs elicit proinflammatory response in Kupffer cells as evident from reduced expression of TNFα and CXCL1, and increased or unaltered expression of IL6.

13.6 PERSPECTIVE

Despite the discoveries in the 1980s and 1990s that HSCs produce growth mediators and chemokines, their possible role in liver injury (or its prevention) remained poorly investigated. This is not surprising since overwhelming evidence pointed to Kupffer cells as a primary cell type of liver inflammation, and injury of various types. Establishment of procedures to block Kupffer cells or their elimination made it easier to elucidate their role in hepatic physiology and injury. Furthermore, early recognition that HSCs could be the major cell type to cause liver fibrosis led to exponential interest in elucidation of the mechanisms of their transdifferentiation into fibrogenic phenotype and of ECM production/deposition. Even the immune cell interactions with HSCs have been largely focused on fibrogenic function of HSCs. However, work during the past decade led to the realization that HSCs can also be a significant participant in acute liver injury. The development of an HSC-depleted mouse model presents a valuable tool to investigate not only the mechanisms of liver injury by direct actions of HSCs (physical as well as soluble-mediator-induced) on hepatocytes, but also how they might regulate the functions of Kupffer cells as well as resident and infiltrating immune and inflammatory cells in causing liver injury. In the coming years, it is expected that much will be discovered about this aspect of HSC biology, which may be clinically relevant in treating acute liver failure.

REFERENCES

[1] Mencin A, Kluwe J, Schwabe RF. Toll-like receptors as targets in chronic liver diseases. Gut 2009;58:704–20.
[2] Decker K. Biologically active products of stimulated liver macrophages (Kupffer cells). Eur J Biochem 1990;192:245–61.
[3] Nain M, Hinder F, Gong JH, Schmidt A, Bender A, Sprenger H, et al. Tumor necrosis factor-α production of influenza A virus infected macrophages and potentiating effect of lipopolysaccharides. J Immunol 1990;145:1921–8.

[4] Klocker U, Schultz U, Schaller H, Protzer U. Endotoxin stimulates liver macrophages to release mediators that inhibit an early step in hepadnavirus replication. J Virol 2000;74:5525–33.
[5] Jaeschke H. Reactive oxygen and mechanisms of inflammatory liver injury: present concepts. J Gastroenterol Hepatol 2011;26(Suppl. 1):173–9.
[6] Gandhi CR. Kupffer cells Monga SP, editor. Molecular pathology of liver diseases. Springer; 2010. p. 81–96. <http://www.springer.com/978-1-4419-7106-7>.
[7] Stewart R, Dangi A, Huang C, Murase N, Kimura S, Stolz DB, et al. A novel mouse model of depletion of stellate cells clarifies their role in ischemia/reperfusion- and endotoxin-induced acute liver injury. J Hepatol 2014;60:298–305.
[8] Geerts A. History, heterogeneity, developmental biology, and functions of quiescent hepatic stellate cells. Semin Liver Dis 2001;21:311–35.
[9] De Minicis S, Seki E, Uchinami H, Kluwe J, Zhang Y, Brenner DA, et al. Gene expression profiles during hepatic stellate cell activation in culture and *in vivo*. Gastroenterology 2007;132:1937–46.
[10] Ramadori G. The stellate cell (Ito-cell, fat-storing cell, lipocyte, perisinusoidal cell) of the liver. New insights into pathophysiology of an intriguing cell. Virchows Arch B Cell Pathol 1991;61:147–58.
[11] Geerts A, DeBleser P, Hautekeete ML, Niki T, Wisse E. Fat-storing (Ito) cell biology Arias IM, Boyer JL, Fasuto N, Jakoby WB, Schachter DL, Shafritz DA, editors. The liver: biology and pathobiology. New York, NY: Raven Press Ltd; 1994. p. 819–38.
[12] Friedman SL. Molecular regulation of hepatic fibrosis, an integrated cellular response to tissue injury. J Biol Chem 2000;275:2247–50.
[13] Maher JJ. Interactions between hepatic stellate cells and the immune system. Semin Liver Dis 2001;21:417–26.
[14] Pinzani M, Marra F. Cytokine receptors and signaling in hepatic stellate cells. Semin Liver Dis 2001;21:397–416.
[15] Ramadori G, Neubauer K, Odenthal M, Nakamura T, Knittel T, Schwögler S, et al. The gene of hepatocyte growth factor is expressed in fat-storing cells of rat liver and is downregulated during cell growth and by transforming growth factor-beta. Biochem Biophys Res Commun 1992;183:739–42.
[16] Schirmacher P, Geerts A, Pietrangelo A, Dienes HP, Rogler CE. Hepatocyte growth factor/hepatopoietin A is expressed in fat-storing cells from rat liver but not myofibroblast-like cells derived from fat-storing cells. Hepatology 1992;15:5–11.
[17] Martinez-Hernandez A, Amenta PS. The extracellular matrix in hepatic regeneration. FASEB J 1995;9:1401–10.
[18] Skrtic S, Wallenius V, Ekberg S, Brenzel A, Gressner AM, Jansson JO. Hepatocyte-stimulated expression of hepatocyte growth factor (HGF) in cultured rat hepatic stellate cells. J Hepatol 1999;30:115–24.
[19] Thomas RJ, Bhandari R, Barrett DA, Bennett AJ, Fry JR, Powe D, et al. The effect of three-dimensional co-culture of hepatocytes and hepatic stellate cells on key hepatocyte functions *in vitro*. Cells Tissues Organs 2005;181:67–79.
[20] Abu-Absi SF, Hansen LK, Hu WS. Three-dimensional co-culture of hepatocytes and stellate cells. Cytotechnology 2004;45:125–40.
[21] Kasuya J, Sudo R, Mitaka T, Ikeda M, Tanishita K. Hepatic stellate cell-mediated three-dimensional hepatocyte and endothelial cell triculture model. Tissue Eng Part A 2011;17:361–70.
[22] Watanabe S, Hirose M, Wang XE, Ikejima K, Oide H, Kitamura T, et al. A novel hepatic stellate (Ito) cell-derived protein, epimorphin, plays a key role in the late stages of liver regeneration. Biochem Biophys Res Commun 1998;250:486–90.
[23] Mabuchi A, Mullaney I, Sheard P, Hessian P, Zimmermann A, Senoo H, et al. Role of hepatic stellate cells in the early phase of liver regeneration in rat: formation of tight adhesion to parenchymal cells. Comp Hepatol 2004;3(Suppl. 1):S29.
[24] Benten D, Kumaran V, Joseph B, Schattenberg J, Popov Y, Schuppan D, et al. Hepatocyte transplantation activates hepatic stellate cells with beneficial modulation of cell engraftment in the rat. Hepatology 2005;42:1072–81.
[25] Anselmi K, Stolz DB, Nalesnik M, Watkins SC, Kamath R, Gandhi CR. Gliotoxin causes apoptosis and necrosis of rat Kupffer cells *in vitro* and *in vivo* in the absence of oxidative stress: exacerbation by caspase and serine protease inhibition. J Hepatol 2007;47:103–13.

[26] Moriya K, Sakai K, Yan MH, Sakai T. Fibronectin is essential for survival but is dispensable for proliferation of hepatocytes in acute liver injury in mice. Hepatology 2012;56:311–21.
[27] Lumsden AB, Henderson JM, Kutner MH. Endotoxin levels measured by a chromogenic assay in portal, hepatic and peripheral venous blood in patients with cirrhosis. Hepatology 1988;8:232–6.
[28] Triger DR, Boyer TD, Levin J. Portal and systemic bacteraemia and endotoxaemia in liver disease. Gut 1978;19:935–9.
[29] Bomzon A, Blendis LM. The nitric oxide hypothesis and the hyperdynamic circulation in cirrhosis. Hepatology 1994;20:1343–50.
[30] Nathan CF. Interferon-gamma and macrophage activation in cell-mediated immunity Steinman RM, North RJ, editors. Mechanisms of host resistance to infectious agents, tumors, and allografts. New York, NY: The Rockefeller University Press; 1968. p. 165–84.
[31] Adams DO, Hamilton TA. The cell biology of macrophage activation. Annu Rev Immunol 1984;2:283–318.
[32] Curran RD, Billiar TR, Stuehr DJ, Ochoa JB, Harbrecht BG, Flint SG, et al. Multiple cytokines are required to induce hepatocyte nitric oxide production and inhibit total protein synthesis. Ann Surg 1990;212:462–9.
[33] Nussler AK, Di Silvio M, Billiar TR, Hoffman RA, Geller DA, Selby R, et al. Stimulation of nitric oxide synthase pathway in human hepatocytes by cytokines and endotoxin. J Exp Med 1992;176:261–4.
[34] Gandhi CR, Stephenson K, Olson MS. Endothelin: a potent peptide agonist in the liver. J Biol Chem 1990;265:17432–5.
[35] Gandhi CR, Berkowitz DE, Watkins WD. Endothelins. Biochemistry and pathophysiologic actions. Anesthesiology 1994;80:892–905.
[36] Ignarro LJ. Nitric oxide. A novel signal transduction mechanism for transcellular communication. Hypertension 1990;16:477–83.
[37] Feng Q, Hedner T. Endothelium-derived relaxing factor (EDRF) and nitric oxide (NO). II. Physiology, pharmacology and pathophysiological implications. Clin Physiol 1990;10:503–26.
[38] Gandhi CR, Uemura T, Kuddus RH. Endotoxin causes up-regulation of endothelin receptors in cultured hepatic stellate cells via nitric oxide-dependent and -independent mechanisms. Br J Pharmacol 2000;131:319–27.
[39] Redmond EM, Cahill PA, Hodges R, Zhang S, Sitzmann JV. J Cell Physiol 1996;166:469–79.
[40] Zhang JX, Pegoli Jr. W, Clemens MG. Endothelin-1 induces direct constriction of hepatic sinusoids. Am J Physiol 1994;266:G624–32.
[41] Gandhi CR, Kuddus RH, Nemoto EM, Murase N. Endotoxin treatment causes up-regulation of endothelin system in the liver: amelioration of increased portal resistance by endothelin receptor antagonism. J Gastroenterol Hepatol 2001;6:61–9.
[42] Gandhi CR, Watkins S, Nemoto EM, Subbotin VM. An endothelin receptor antagonist TAK-044 ameliorates carbon tetrachloride-induced acute liver injury and portal hypertension in rats. Liver 1998;18:39–48.
[43] Uemura T, Gandhi CR. Inhibition of DNA synthesis in cultured hepatocytes by endotoxin-conditioned medium of activated stellate cells is transforming growth factor-β- and nitric oxide-independent. Br J Pharmacol 2001;133:1125–33.
[44] Thirunavukkarasu C, Uemura T, Wang LF, Watkins S, Gandhi CR. Normal rat hepatic stellate cells respond to endotoxin in LBP-independent manner to produce inhibitor(s) of DNA synthesis in hepatocytes. J Cell Physiol 2005;204:654–65.
[45] Thirunavukkarasu C, Watkins S, Gandhi CR. Mechanisms of endotoxin-induced nitric oxide, interleukin-6 and tumor necrosis factor-α production in activated rat hepatic stellate cells: role of p38MAPK. Hepatology 2006;44:389–98.
[46] Martin TR, Mathison JC, Tobias PS, Letúrcq DJ, Moriarty AM, Maunder RJ, et al. Lipopolysaccharide binding protein enhances the responsiveness of alveolar macrophages to bacterial lipopolysaccharide. Implications for cytokine production in normal and injured lungs. J Clin Invest 1992;90:2209–19.
[47] Paik YH, Schwabe RF, Bataller R, Russo MP, Jobin C, Brenner DA. Tolllike receptor 4 mediates inflammatory signaling by bacterial lipopolysaccharide in human hepatic stellate cells. Hepatology 2003;37:1043–55.

[48] Seki E, De Minicis S, Osterreicher CH, Kluwe J, Osawa Y, et al. TLR4 enhances TGF-beta signaling and hepatic fibrosis. Nat Med 2007;13:1324–32.
[49] Fausto N, Laird AD, Webber EM. Role of growth factors and cytokines in hepatic regeneration. FASEB J 1995;9:1527–36.
[50] Michalopoulos GK, DeFrances MC. Liver regeneration. Science 1997;276:60–6.
[51] Ren X, Hogaboam C, Carpenter A, Colletti L. Stem cell factor restores hepatocyte proliferation in IL-6 knockout mice following 70% hepatectomy. J Clin Invest 2003;112:1407–18.
[52] Camargo Jr. CA, Madden JF, Gao W, Selvan RS, Clavien PA. Interleukin-6 protects liver against warm ischemia/reperfusion injury and promotes hepatocyte proliferation in the rodent. Hepatology 1997;26:1513–20.
[53] Wustefeld T, Rakemann T, Kubicka S, Manns MP, Trautwein C. Hyperstimulation with interleukin 6 inhibits cell cycle progression after hepatectomy in mice. Hepatology 2000;32:514–22.
[54] Mizuhara H, O'Neill E, Seki N, Ogawa T, Kusunoki C, Otsuka K, et al. T cell activation-associated hepatic injury: mediation by tumor necrosis factors and protection by interleukin 6. J Exp Med 1994;179:1529–37.
[55] Harada M, Kawano T, Yamashita M, Shibata Y, Gejyo F, Nakayama T, et al. Augmentation of Valpha14 NKT cell-mediated cytotoxicity by interleukin 4 in an autocrine mechanism resulting in the development of concanavalin A-induced hepatitis. J Exp Med 2000;191:105–14.
[56] Chiu H, Gardner CR, Dambach DM, Durham SK, Brittingham JA, Laskin JD, et al. Role of tumor necrosis factor receptor 1 (p55) in hepatocyte proliferation during acetaminophen-induced toxicity in mice. Toxicol Appl Pharmacol 2003;193:218–27.
[57] Sun R, Tian Z, Kulkarni S, Gao B. IL-6 prevents t cell-mediated hepatitis via inhibition of NKT cells in $CD4^+$ T cell- and STAT3-dependent manners. J Immunol 2004;172:5648–55.
[58] Tiegs G. Cellular and cytokine-mediated mechanisms of inflammation and its modulation in immune-mediated liver injury. Z Gastroenterol 2007;45:63–70.
[59] Chen L, Xie XJ, Ye YF, Zhou L, Xie HY, Xie QF, et al. Kupffer cells contribute to concanavalin A induced hepatic injury through a Th1 but not Th17 type response-dependent pathway in mice. Hepatobiliary Pancreat Dis Int 2011;10:171–8.
[60] Schwabe RF, Brenner DA. Mechanisms of liver injury. I. TNF-alpha-induced liver injury: role of IKK, JNK, and ROS pathways. Am J Physiol Gastrointest Liver Physiol 2006;290:G583–9.
[61] Hatano E. Tumor necrosis factor signaling in hepatocyte apoptosis. J Gastroenterol Hepatol 2007;22(Suppl. 1):S43–4.
[62] Shuh M, Bohorquez H, Loss Jr. GE, Cohen AJ. Tumor necrosis factor-α: life and death of hepatocytes during liver ischemia/reperfusion injury. Ochsner J 2013;13:119–30.
[63] Maeda S, Chang L, Li ZW, Luo JL, Leffert H, Karin M. IKKbeta is required for prevention of apoptosis mediated by cell-bound but not by circulating TNFalpha. Immunity 2003;19:725–37.
[64] Boulton R, Woodman A, Calman D, Selden C, Tam F, Hodgson H. Nonparenchymal cells from regenerating rat liver generate interleukin-1a and -1b: a mechanism of negative regulation of hepatocyte proliferation. Hepatology 1997;26:49–58.
[65] Morita M, Watanabe Y, Akaike T. Protective effect of hepatocyte growth factor on interferon gamma-induced cytotoxicity in mouse hepatocytes. Hepatology 1995;21:1585–93.
[66] Bohlinger I, Leist M, Barsig J, Uhlig S, Tiegs G, Wendel A. Interleukin-1 and nitric oxide protect against tumor necrosis factor a-induced liver injury through distinct pathway. Hepatology 1995;22:1829–37.
[67] Takehara T, Hayashi N, Tatsumi T, Kanto T, Mita E, Sasaki Y, et al. Interleukin 1b protects mice from Fas-mediated hepatocyte apoptosis and death. Gastroenterology 1999;117:661–8.
[68] Helyar L, Bundschuh DS, Laskin JD, Laskin DL. Induction of hepatic Ito cell nitric oxide production after acute endotoxemia. Hepatology 1994;20:1509–15.
[69] Mannick JB, Asano K, Izumi K, Kieff E, Stamler JS. Nitric oxide produced by human B lymphocytes inhibits apoptosis and Epstein-Barr virus reactivation. Cell 1994;79:1137–46.
[70] Fehsel K, Kroncke KD, Meyer KL, Huber H, Wahn V, Kolb-Bachofen V. Nitric oxide induces apoptosis in mouse thymocytes. J Immunol 1995;155:2858–65.
[71] Dimmeler S, Haendeler J, Nehls M, Zeiher AM. Suppression of apoptosis by nitric oxide via inhibition of interleukin-1beta-converting enzyme (ICE)-like and cysteine protease protein (CPP)-32-like proteases. J Exp Med 1997;185:601–7.

[72] Paul A, Bryant C, Lawson MF, Chilvers ER, Plevin R. Dissociation of lipopolysaccharide-mediated nitric oxide synthase and inhibition of DNA synthesis in RAW264.7 macrophages and rat aortic smooth muscle cells. Br J Pharmacol 1997;120:1439–44.

[73] Kawada N, Seki S, Kuroki T, Inoue M. Regulation of stellate cell proliferation by lipopolysaccharide: role of endogenous nitric oxide. J Gastroenterol Hepatol 1998;13:S6–S13.

[74] Calderone A, Thaik CM, Takahashi N, Chang DL, Colucci WS. Nitric oxide, atrial natriuretic peptide, and cyclic GMP inhibit the growth-promoting effects of norepinephrine in cardiac myocytes and fibroblasts. J Clin Invest 1998;101:812–8.

[75] Rupprecht HD, Akagi Y, Keil A, Hofer G. Nitric oxide inhibits growth of glomerular mesangial cells: role of the transcription factor EGR-1. Kidney Int 2000;57:70–82.

[76] Aramaki Y. Liposomes as immunomodulator–inhibitory effect of liposomes on NO production from macrophages. Biol Pharm Bull 2000;23:1267–74.

[77] Iwai-Kanai E, Hasegawa K, Fujita M, Araki M, Yanazume T, Adachi S, et al. Basic fibroblast growth factor protects cardiac myocytes from iNOS-mediated apoptosis. J Cell Physiol 2002;190:54–62.

[78] Kim YM, Talanian RV, Billiar TR. Nitric oxide inhibits apoptosis by preventing increases in caspase-3-like activity via two distinct mechanisms. J Biol Chem 1997;272:31138–48.

[79] Li J, Bombeck CA, Yang S, Kim YM, Billiar TR. Nitric oxide suppresses apoptosis via interrupting caspase activation and mitochondrial dysfunction in cultured hepatocytes. J Biol Chem 1999;274:17325–33.

[80] Kim JS, He L, Qian T, Lemasters JJ. Role of the mitochondrial permeability transition in apoptotic and necrotic death after ischemia/reperfusion injury to hepatocytes. Curr Mol Med 2003;3:527–35.

[81] Aram G, Potter JJ, Liu X, Torbenson MS, Mezey E. Lack of inducible nitric oxide synthase leads to increased hepatic apoptosis and decreased fibrosis in mice after chronic carbon tetrachloride administration. Hepatology 2008;47:2051–8.

[82] Harvey SAK, Dangi A, Gandhi CR. The transcriptomic response of rat hepatic stellate cells to endotoxin: implications for hepatic inflammation and immune regulation. PLoS One 2013;8:e82159.

[83] Geller DA, Nguyen D, Shapiro RA, Nussler A, Di SM, Freeswick P, et al. Cytokine induction of interferon regulatory factor-1 in hepatocytes. Surgery 1993;114:235–42.

[84] Tsung A, Stang MT, Ikeda A, Critchlow ND, Izuishi K, Nakao A, et al. The transcription factor interferon regulatory factor-1 mediates liver damage during ischemia-reperfusion injury. Am J Physiol Gastrointest Liver Physiol 2006;290:G1261–8.

[85] Montalvo-Jave EE, Escalante-Tattersfield T, Ortega-Salgado JA, Piña E, Geller DA. Factors in the pathophysiology of the liver ischemia-reperfusion injury. J Surg Res 2008;147:153–9.

[86] Dhupar R, Klune JR, Evankovich J, Cardinal J, Zhang M, Ross M, et al. Interferon regulatory factor-1 mediates acetylation and release of high mobility group box-1 from hepatocytes during murine liver ischemia-reperfusion injury. Shock 2011;35:293–301.

[87] Gandhi CR, Huang C, Stolz DB. Endotoxin-stimulated hepatic stellate cells promote survival of hepatocytes in endotoxemia by inducing ER stress and autophagy. FASEB J 2012;26: 396.2.

[88] Louis H, Moine O, Peny MO, Quertinmont E, Fokan D, Goldman M, et al. Production and role of interleukin-10 in concanavalin A-induced hepatitis in mice. Hepatology 1997;25:1382–9.

[89] Huang C, Yang S, Geller DA, Stolz D, Gandhi CR. Interferon-β derived from endotoxin-stimulated hepatic stellate cells cause hepatocyte injury via JNK-mediated IRF-1 expression. FASEB J 2011;25 794.11.

[90] Dangi A, Sumpter TL, Kimura S, Stolz DB, Murase N, Raimondi G, et al. Selective expansion of allogeneic regulatory T cells by hepatic stellate cells: role of endotoxin and implications for allograft tolerance. J Immunol 2012;188:3667–77.

[91] Sumpter TL, Dangi A, Matta B, Huang C, Stolz D, Vodovotz Y, et al. Hepatic stellate cells undermine the allostimulatory function of liver myeloid dendritic cells via STAT3-dependent induction of IDO. J Immunol 2012;189:3848–58.

[92] Gotoh T, Mori M. Nitric oxide and endoplasmic reticulum stress. Arterioscler Thromb Vasc Biol 2006;26:1439–46.

[93] Kaufman RJ. Orchestrating the unfolded protein response in health and disease. J Clin Invest 2002;110:1389–98.

[94] Ma Y, Hendershot LM. The mammalian endoplasmic reticulum as a sensor for cellular stress. Cell Stress Chaperones 2002;7:222–9.

[95] Ding WX, Yin XM. Sorting, recognition and activation of the misfolded protein degradation pathways through macroautophagy and the proteasome. Autophagy 2008;4:141–50.
[96] Hoyer-Hansen M, Jaattela M. Connecting endoplasmic reticulum stress to autophagy by unfolded protein response and calcium. Cell Death Differ 2007;14:1576–82.
[97] Ogata M, Hino S, Saito A, Morikawa K, Kondo S, Kanemoto S, et al. Autophagy is activated for cell survival after endoplasmic reticulum stress. Mol Cell Biol 2006;26:9220–31.
[98] Kouroku Y, Fujita E, Tanida I, Ueno T, Isoai A, Kumagai H, et al. ER stress (PERK/eIF2alpha phosphorylation) mediates the polyglutamine induced LC3 conversion, an essential step for autophagy formation. Cell Death Differ 2007;14:230–9.
[99] Ding WX, Ni HM, Gao W, Hou YF, Melan MA, Chen X, et al. Differential effects of endoplasmic reticulum stress-induced autophagy on cell survival. J Biol Chem 2007;282:4702–10.
[100] Klionsky DJ, Abeliovich H, Agostinis P, Agrawal DK, Aliev G, Askew DS, et al. Guidelines for the use and interpretation of assays for monitoring autophagy in higher eukaryotes. Autophagy 2008;4:151–75.
[101] Czaja MJ, Geerts A, Xu J, Schmiedeberg P, Ju Y. Monocyte chemoattractant protein 1 (MCP-1) expression occurs in toxic rat liver injury and human liver disease. J Leukoc Biol 1994;55:120–6.
[102] Xue X, Piao JH, Nakajima A, Sakon-Komazawa S, Kojima Y, Mori K, et al. Tumor necrosis factor alpha (TNFalpha) induces the unfolded protein response (UPR) in a reactive oxygen species (ROS)-dependent fashion, and the UPR counteracts ROS accumulation by TNFalpha. J Biol Chem 2005;280:33917–25.
[103] Maddox JF, Amuzie CJ, Li M, Newport SW, Sparkenbaugh E, Cuff CF, et al. Bacterial- and viral-induced inflammation increases sensitivity to acetaminophen hepatotoxicity. J Toxicol Environ Health A 2010;73:58–73.
[104] Harris J. Autophagy and cytokines. Cytokine 2011;56:140–4.
[105] Schmeisser H, Fey SB, Horowitz J, Fischer ER, Balinsky CA, Miyake K, et al. Type I interferons induce autophagy in certain human cancer cell lines. Autophagy 2013;9:683–96.
[106] Schmeisser H, Bekisz J, Zoon KC. New function of type I IFN: induction of autophagy. J Interferon Cytokine Res 2014;34:71–8.
[107] Miller WH, Miller RL. Phosphorylation of acyclovir (acycloguanosine) monophosphate by GMP kinase. J Biol Chem 1980;255:7204–7.
[108] Bush TG, Savidge TC, Freeman TC, Cox HJ, Campbell EA, Mucke L, et al. Fulminant jejuno-ileitis following ablation of enteric glia in adult transgenic mice. Cell 1998;93:189–201.
[109] Puche JE, Lee YA, Jiao J, Aloman C, Fiel MI, Muñoz U, et al. A novel murine model to deplete hepatic stellate cells uncovers their role in amplifying liver damage in mice. Hepatology 2013;57:339–50.
[110] Marra F, Valente AJ, Pinzani M, Abboud HE. Cultured human liver fat-storing cells produce monocyte chemotactic protein-1. Regulation by proinflammatory cytokines. J Clin Invest 1993;92:1674–80.
[111] Czaja MJ, Ding WX, Donohue Jr. TM, Friedman SL, Kim JS, Komatsu M, et al. Functions of autophagy in normal and diseased liver. Autophagy 2013;9:1131–58.
[112] Mühlbauer M, Bosserhoff AK, Hartmann A, Thasler WE, Weiss TS, Herfarth H, et al. A novel MCP-1 gene polymorphism is associated with hepatic MCP-1 expression and severity of HCV-related liver disease. Gastroenterology 2003;125:1085–93.
[113] Stephenson K, Harvey SA, Mustafa SB, Eakes AT, Olson MS. Endothelin association with the cultured rat Kupffer cell: characterization and regulation. Hepatology 1995;22:896–905.
[114] Teoh NC, Farrell GC. Hepatic ischemia reperfusion injury: pathogenic mechanisms and basis for hepatoprotection. J Gastroenterol Hepatol 2003;18:891–902.
[115] Fondevila C, Busuttil RW, Kupiec-Weglinski JW. Hepatic ischemia/reperfusion injury—a fresh look. Exp Mol Pathol 2003;74:86–93.
[116] Jaruga B, Hong FH, Sun R, Radaeva S, Gao B. Crucial role of IL-4/STAT6 in T cell-mediated hepatitis: up-regulating eotaxins and IL-5 and recruiting leukocytes. J Immunol 2003;171:3233–44.
[117] Shiratori Y, Ohmura K, Hikiba Y, Matsumura M, Nagura T, Okano K, et al. Hepatocyte nitric oxide production is induced by Kupffer cells. Dig Dis Sci 1998;43:1737–46.
[118] Shito M, Wakabayashi G, Ueda M, Shimazu M, Shirasugi N, Endo M, et al. Interleukin 1 receptor blockade reduces tumor necrosis factor production, tissue injury, and mortality after hepatic ischemia-reperfusion in the rat. Transplantation 1997;63:143–8.

[119] Tomiyama K, Ikeda A, Ueki S, Nakao A, Stolz DB, Koike Y, et al. Inhibition of Kupffer cell-mediated early proinflammatory response with carbon monoxide in transplant-induced hepatic ischemia/reperfusion injury in rats. Hepatology 2008;48:1608–20.
[120] Cutrn JC, Perrelli MG, Cavalieri B, Peralta C, Rosell Catafau J, et al. Microvascular dysfunction induced by reperfusion injury and protective effect of ischemic preconditioning. Free Radic Biol Med 2002;33:1200–8.
[121] Su GL. Lipopylysaccharides in liver injury: molecular mechanisms of Kupffer cell activation. Am J Physiol Gastrointest Liver Physiol 2002;283:G256–65.
[122] Tsukamoto H. Redox regulation of cytokine expression in Kupffer cells. Antioxid Redox Signal 2002;4:741–8.
[123] Sadik CD, Kim ND, Luster AD. Neutrophils cascading their way to inflammation. Trends Immunol 2011;32:452–60.Sakamoto M, Ueno T, Kin M, Ohira H, Torimura T, Inuzuka S, et al. Ito cell contraction in response to endothelin-1 and substance P. Hepatology 1993;18:978–83.
[124] Kinoshita M, Uchida T, Sato A, Nakashima M, Nakashima H, Shono S, et al. Characterization of two F4/80-positive Kupffer cell subsets by their function and phenotype in mice. J Hepatol 2010;53:903–10.
[125] Sleyster EC, Knook DL. Relation between localization and function of rat liver Kupffer cells. Lab Invest 1982;47:484–90.
[126] Hoedemakers RM, Morselt HW, Scherphof GL, Daemen T. Heterogeneity in secretory response of rat liver macrophages of different size. Liver 1995;15:313–9.

CHAPTER 14

Stellate Cells in Hepatic Immunological Tolerance

Chandrashekhar R. Gandhi
Department of Pediatrics, Cincinnati Children's Hospital Medical Center, Department of Surgery, University of Cincinnati and Cincinnati VA Medical Center, Cincinnati, OH, USA

14.1 HEPATIC IMMUNE SYSTEM

The liver contains an abundant lymphocyte population distributed throughout the sinusoids and also in the portal tracts. $CD3^-CD56^+$ NK (natural killer) cells (pit cells) occur at greater frequency in the liver (about 50% of all lymphocytes) than in the other tissues. NK cells are activated by interleukin (IL)-2, IL-12, IL-15, IL-15/IL-15RA complexes and IL-18, and produce cytokines such as tumor necrosis factor (TNF)-α, interferon (IFN)-γ, IL-17 and IL-22, as well as the cytotoxic molecules perforin and granzyme, that cause pores in target cells and induce their apoptosis. The liver is also enriched in $CD3^+NK1.1^+$ NKT (natural killer T) cells, and conventional $CD4^+$ and $CD8^+$ T cells. NKT cells are activated by glycolipid antigens in a CD1d-restricted manner. B cells and T-cell receptor (TCR)-γδ cells form only a minor portion of the total hepatic lymphocyte population. Migratory dendritic cells (DCs) in liver sinusoids and periportal areas and endothelium-adherent stationary sinusoidal Kupffer cells (KCs) are professional hepatic antigen-presenting cells (APCs). Sinusoidal endothelial cells (SECs) gain antigen-presenting function primarily upon stimulation with cytokines, and are considered nonprofessional APCs [1–7].

With this repertoire of innate and adaptive immune cells, the liver has an impressive ability to eliminate cytopathic microorganisms. Whereas macrophages (KCs), γδ cells, NK cells, and NKT cells provide the first line of defense, subsequently activated T and B cells control and eliminate cytopathic microorganisms. Cells infected with microorganisms are phagocytosed by macrophages that process and present microbial antigens in an MHC-restricted manner to T cells bearing TCRs. Complete activation of T cells requires a second signal known as co-stimulation, which is achieved by ligation of APC's CD80/CD86 to CD28 receptor on T cells. In the absence of co-stimulation, T cells undergo anergy and apoptosis. Activated T cells produce cytokines that kill and eliminate microorganisms. Antibody production by activated B cells is also a major component of this defense mechanism.

Stellate Cells in Health and Disease
DOI: http://dx.doi.org/10.1016/B978-0-12-800134-9.00014-2

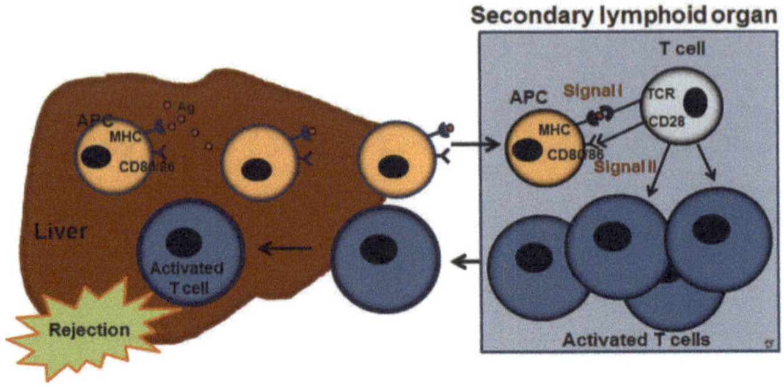

Figure 14.1 ***Immune cell-mediated allograft rejection.*** Donor or recipient APC (DC) pick up antigen (Ag) within the allograft and display it on the MHC peptide. Activated APCs also upregulate co-stimulatory B7 (CD80, CD86) molecules. Ag-loaded APCs then migrate to secondary lymphoid organs (lymph nodes and spleen) where they present the Ag to effector T cells expressing TCR (signal I). Co-stimulation with ligation of CD80/86 and CD28 (signal II) is essential for activation of T cells. Activated T cells expand and migrate to the allograft, and promote organ rejection by releasing cytotoxic mediators.

The physiologic response of elimination of infected cells by the immune system is equivalent to allograft rejection. Following transplantation, graft interstitial DCs undergo maturation by increasing expression of MHC class II and co-stimulatory molecules (e.g., CD80, CD86). Alloantigen-bearing DCs then migrate to secondary lymphoid tissues, and stimulate host alloreactive T cells that proliferate in response to IL2 and other cytokines. These activated T cells migrate to the allograft, where they induce an inflammatory response and, along with activated macrophages, B cells, and plasma cells, inflict organ damage and promote transplant rejection (Figure 14.1) [8]. Control and prevention of allograft rejection has been a topic of intense research since the 1960s. Many anti-rejection immunosuppressive drugs have been developed and used effectively. However, improved understanding of the intrinsic mechanisms of immunoregulation and transplant tolerance is extremely important in order to eliminate or reduce drug-based immunosuppressive therapy because of the morbidity and mortality associated with its use.

14.2 LIVER TOLEROGENICITY

The liver receives ~75% of its blood supply from the portal vein and oxygenated blood (~25%) from the hepatic artery. The portal venous blood delivers almost continuously a variety of food- and environment-derived antigens, drugs, xenobiotics, toxins, and gut-derived microbial products including ribonucleic acid (RNA), deoxyribonucleic acid (DNA), and gram-negative bacterial cell wall product lipopolysaccharide (LPS). Through well-orchestrated mechanisms involving both parenchymal cells (hepatocytes), and nonparenchymal cells (NPCs) (KCs, endothelial cells, hepatic stellate cells (HSCs)), and cells of the hepatic innate immune system, the liver degrades or converts blood-derived elements into innocuous forms and thus prevents their harmful effects on itself and other organs. However, the liver also exhibits immune tolerance, as evidenced by easy occurrence of metastasis of cancer cells from other organs, and its harboring of viruses (e.g., hepatitis B virus (HBV) and hepatitis C virus (HCV)) and other infectious organisms (e.g., malaria parasite). Persistence of such infection can eventually lead to life-threatening fibrosis and cirrhosis (HBV and HCV) or potentially lethal infection with liver cells acting as hosts for multiplication of the organisms such as malaria parasite. Furthermore, oral or portal venous administration of antigens can induce systemic tolerance [9–11].

The liver's tolerogenicity is further exemplified by acceptance of hepatic allografts and maintenance of their function without immunosuppressive therapy in many animal models across the MHC barriers and a proportion (~20%) of stable human liver transplant recipients who discontinue immunosuppressive drugs [12,13]. Hepatic transplant tolerance was first recognized in 1969 in two independent studies in which liver grafts transplanted between outbred pigs were found to survive without a requirement for immunosuppressive therapy [14,15]. These findings were confirmed in dogs [16], rats [17], and mice [18]. In contrast, other organs such as the kidney and heart are readily rejected in the absence of immunosuppression treatment [12,13]. Interestingly, prior or concomitant transplantation of the liver can induce tolerance to other organs from the same donor transplanted in the same host [15,19–23].

With greatly refined technical expertise and more specific/less toxic immunosuppressive drugs, liver transplantation has become the standard-of-care procedure for end-stage chronic liver disease and fulminant liver failure. By current estimates, the 1- and 5-year survival of liver transplant patients is 85% and 75%, respectively. However, the demand for donor organs has increased as the incidence of irreversible chronic liver disease due to viral infection (e.g., HCV and HCB), alcohol abuse, and nonalcoholic steatohepatitis has risen and is expected to increase further in the future. This has led to a shortage of donor organs, with about 16,000 patients on the transplant list at present in the United States alone, and nearly one-third of them dying while awaiting transplantation (http://optn.transplant.hrsa.gov/). The shortage of donor organs can be potentially addressed by transplantation of marginal, steatotic, older, and small-for-size liver grafts, which can increase complications of the procedure or by use of live donors, that carries a potential risk to the donor's life [24,25].

Despite the tolerogenic property of the liver allograft [12,13,26–32], the incidence of rejection is common in human recipients (50–75%). This may be reversed by aggressive or long-term immunosuppression therapy, but its complications (80% developing renal dysfunction, 75% hypertension, 50% hyperlipidemia, and 20–25% new-onset diabetes mellitus) [25] continue to be major clinical challenges. Furthermore, a majority of liver transplant recipients require lifelong immunosuppression [30]. However, the inherent ability of the liver allograft to overcome immune-mediated rejection in animal models and in a minor percentage of patients suggest that the liver environment may modulate immune cell function to favor allograft acceptance rather than rejection [4,5].

14.3 IMPORTANCE OF HSCs IN TRANSPLANTATION

Due to their location in hepatic sinusoids, KCs can intimately interact with trafficking immune cells prompting investigation of their role in liver transplantation. Experimental studies have reported that KCs (i) do not alter the outcome of liver allograft transplantation [33]; (ii) are detrimental for allograft survival [34,35]; and (iii) promote tolerance by suppressing T cells in rat allograft transplantation [36]. Although the precise reasons for these inconsistent findings remain undetermined, it is apparent that other hepatic cell type(s) may influence the immune system to promote acceptance of the liver allograft. In this regard, SECs were found to induce $CD4^+$ T-cell division but inhibit Th1 differentiation [1,37] and are associated with liver's tolerogenicity via PD1/PDL1 (B7-H1)-mediated co-inhibitory mechanism [38,39]. However, sinusoidal endothelium is severely disrupted during transplantation reducing the ability of SECs to interact with the immune cells indicating that cells other than KCs or SECs may be important in regulating immunological reactions following transplantation. In early 1990s, HSCs were reported to produce cytokines and chemokines critical to hepatic inflammation as well as immune regulation [40,41] (see Chapter 5). However, it took nearly a decade before investigations to study interactions of HSCs with the immune cells with respect to liver transplantation were initiated.

HSCs reside in the perisinusoidal space of Disse and form 8–10% of the liver cell population. They are a major storage site of retinoids, including vitamin A, express glial fibrillary acidic protein (GFAP) or desmin, and are physiologically quiescent. During liver injury, HSCs release retinoids and transdifferentiate into myofibroblast-like cells, a process known as activation. Transdifferentiation of HSCs from quiescent (high retinoids and very low or no expression of α-smooth muscle actin (α-sma)) to fully activated phenotype (very low or no retinoids and high expression of α-sma) progresses through a transitional phenotype (moderate retinoids and α-sma expression) (Figure 14.2). The transitionally activated HSCs are relevant to acute liver injury and the early period after liver transplantation. Activated HSCs are highly responsive to cytokines, growth mediators, and contractile agents (e.g., endothelin), and produce increased amounts of extracellular matrix (ECM) as well as numerous cytokines, chemokines, and growth factors. Persistence of liver injury (e.g., due to HCV, HBV, alcohol, toxins) perpetuates activation and proliferation of HSCs, and continued deposition of ECM by these cells leads to fibrosis and eventually cirrhosis (see Chapter 4).

Perturbation in hepatic homeostasis such as fulminant liver failure and *ex vivo* liver perfusion [42,43], partial hepatectomy [44,45], toxin (e.g., CCl_4)-induced liver injury

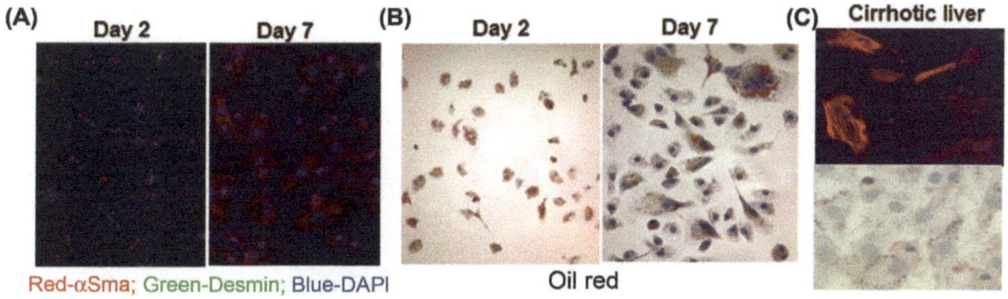

Figure 14.2 ***Activation of HSCs.*** (A) Normal rat HSCs on day 2 of culture do not express α-sma, but are positive on day 7, indicating an activated phenotype. (B) HSCs on day 2 and day 7 of culture show quiescent and transiently activated phenotypes with cytosolic lipid stores (Oil Red staining). (C) Rat HSCs in passage 3, showing a fully activated α-sma-positive myofibroblast-like phenotype (upper panel) and very low or no retinoids (oil red O staining) (lower panel). Magnification 10×.

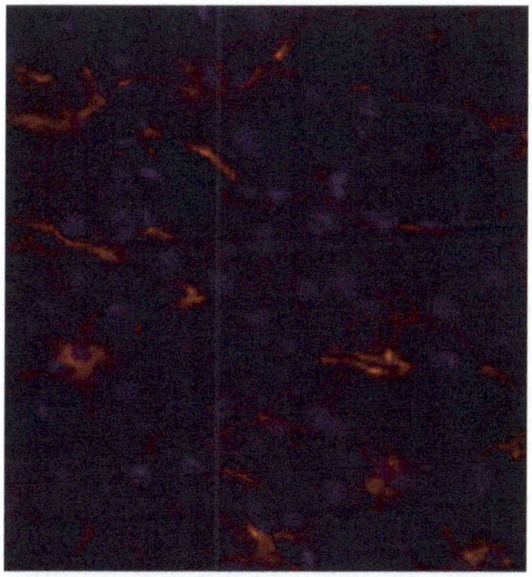

Figure 14.3 ***Transiently activated α-sma⁺ HSCs in the rat liver 48 h after transplantation, with 18 h prior cold preservation.*** Transplanted livers were kindly provided by Dr. Noriko Murase, TE Starzl Transplantation Institute, University of Pittsburgh. Magnification 10×.

[46], and liver transplantation induces rapid activation of HSCs (Figure 14.3). Reactive oxygen species that accumulate during cold-ischemic preservation of the graft and their increased generation upon reperfusion are a primary stimulus of HSC activation. HSCs display a high capacity to phagocytose apoptotic bodies and thus apoptosis of hepatocytes is an inflammatory stimulus that also activates HSCs [47,48]. This response to apoptotic hepatocytes reflects, in part, the interaction of hepatocyte DNA with Toll-like receptor 9 (TLR-9) expressed on HSCs [49]. As LPS is elevated in liver disease and transplantation, and reperfusion causes death of hepatocytes in the transplanted

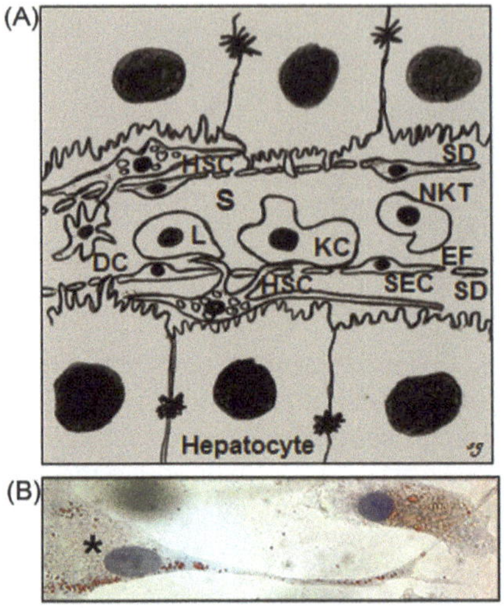

Figure 14.4 (A) In normal liver, HSCs are located in the space of Disse between sinusoidal endothelial cells (EC) and hepatocytes. The cytoplasmic processes of HSCs can span 2–3 hepatocytes. Fenestrations (EF); KC; lymphocytes (L); NK cells. (B) An HSC (*) isolated from rat liver after overnight culture is shown extending cytoplasmic processes to contact a neighboring HSC. Oil red O staining.

allograft with the release of DNA, increased LPS-induced expression of TLR-9 on HSCs [50] appears relevant to HSC-phenotypic change in transplantation.

Sluggish blood flow through the narrow sinusoids facilitates interactions between the sinusoidal-lining cells and lymphocytes. Compared to KCs and SECs, HSCs have less contact with lymphocytes via their cytosolic extensions penetrating through the endothelial fenestrations (Figure 14.4). However, extravasation of lymphocytes through endothelial fenestrations can occur in physiology when HSCs can directly interact with them. The physical contact between HSCs and immune cells increases upon disruption of the endothelial barrier, which occurs in transplantation during ischemic cold preservation and reperfusion of the donor liver [51,52]. Through such physical interactions, as well as by releasing cytokines and chemokines, HSCs are in a position to modulate immune cells and impact the outcome of liver transplantation.

14.4 HSCs PRODUCE INFLAMMATORY AND IMMUNOREGULATORY CYTOKINES AND CHEMOKINES

An important stimulator of the synthesis of chemokines and cytokines by murine and human HSCs is LPS [50,53–56]. Table 14.1 lists cytokines and chemokines expressed by rat HSCs and increased robustly within 1–3 h of stimulation with LPS [50]. LPS is delivered continuously to the liver via the portal vein and is effectively removed by KCs, endothelial cells, and hepatocytes [57,58]. Physiologically, up to 1 ng/mL LPS is

Table 14.1 HSC-expressed chemokines upregulated by LPS and their target cells [50]

Chemokine	Receptor	Cells recruited
CCL2	CCR2, CCR4	Monocytes, memory T cells, dendritic cells
CCL3	CCR1, CCR4, CCR5	Monocytes, T cells
CCL5	CCR1; CCR3; CCR5	Eosinophils, basophils, T cells
CCL7	CCR1; CCR2	Monocytes, macrophages
CCL9/CCL10	CCR1	$CD34^+$ immature myeloid cells, dendritic cells
CCL19	CCR7	Dendritic cells, B cells, memory T cells
CCL20	CCR6	T cells, neutrophils
CCL22	CCR4	Regulatory T cells
CXCL1	CXCR2	Neutrophils
CXCL2	CXCR2	Neutrophils, hematopoietic stem cells
CXCL3	CXCR2	Monocytes, macrophages
CXCL4	CXCR3	Neutrophils, and monocytes
CXCL5	CXCR2	Neutrophils
CXCL9	CXCR3	T cells
CXCL10	CXCR3; CCR3	T cells, macrophages, NK cells
CXCL11	CXCR3	T cells
CXCL13	CXCR5	B cells
CXCL16	CXCR6	T cells, NKT cells

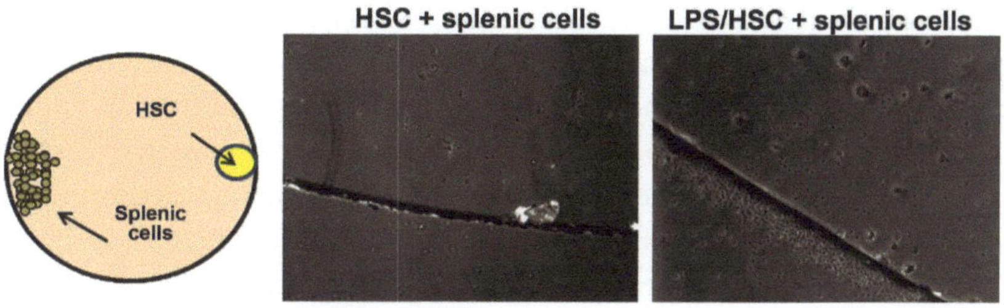

Figure 14.5 ***HSCs induce migration of immune cells.*** HSCs cultured on a cover slip were placed at an edge of a large petri dish and were incubated with 10 ng/mL LPS for 12 h. The medium was aspirated and fresh medium was added after washing the cells twice. Splenic mixed lymphocytes were then placed at the other edge of the petri dish and their migration was followed for 6 h. C.R. Gandhi, unpublished observation.

present in the normal portal venous blood and is not detected in the systemic circulation [59]. Circulating LPS increases in liver disease [59,60] and remains elevated after liver transplantation until restoration of graft function [61,62]. Thus, the effects of LPS on hepatic cells including HSCs become important in transplantation. The potential of HSCs in recruiting immune cells is illustrated by the spontaneous rapid migration of spleen-derived cells toward cultured HSCs, which increases strongly when HSCs are pretreated with LPS (Figure 14.5).

Initial evidence for a role of HSCs in hepatic inflammation and immune regulation was provided in 1993 (and confirmed later) by the observation that culture-activated human and murine HSCs produce monocyte- and lymphocyte-recruiting monocyte chemoattractant protein-1 (MCP-1 or CCL2), and that its synthesis was strongly increased by stimulation with TNF-α, IL1-α, and IFN-γ [40,41,63,64]. The pathophysiological relevance of this finding was provided by increased MCP-1 expression in portal tracts and active fibrous septa surrounding regenerative nodules in murine and human cirrhosis [65,66], and elevated expression of MCP-1 by HSCs of patients with chronic viral hepatitis [40,41]. Activated HSCs also recruit neutrophils by producing IL-8 (human) or its rat homolog, cytokine-induced neutrophil chemoattractant (CINC) and promote inflammation [67–70]. There is a positive correlation of serum and hepatic levels of IL-8 with severity of chronic viral hepatitis and cirrhosis [67], and marked neutrophil infiltration is associated with a rapid progression of alcohol-induced hepatitis to cirrhosis [71]. Thus recruitment of the inflammatory and immune cells (especially $CD8^+$ T cells) by HSCs expressing IL-8, MCP-1, and other chemokines, and bilateral interactions between the two cell types constitute a major mechanism of hepatic fibrogenesis [72–74].

In addition to the chemokines and inflammatory cytokines (TNF-α, IFN-β, IL-1, and IL-6) [50,53–56], activated HSCs produce anti-inflammatory and immunosuppressive IL-10, expression of which is increased by LPS stimulation [56,75]. LPS also induces expression of inducible nitric oxide (NO) synthase in both quiescent and activated rat HSCs, and LPS-stimulated HSCs induce iNOS in hepatocytes [53,54]. Therefore, high levels of NO produced through such mechanisms can additionally be a major mechanism of T-cell suppression [76,77] in the hepatic microenvironment. The following sections describe the current understanding of the mechanisms of HSC-induced modulation of DCs, regulatory T cells (immunosuppression), and conventional $CD4^+$/$CD8^+$ T cells that creates an environment favoring acceptance of the liver allograft.

14.5 HSC–DC INTERACTIONS

First described in 1973 [78], DCs are the professional, bone marrow-derived APCs that induce T-cell activation. Both conventional myeloid (m) DCs and nonconventional (Type-1 IFN-producing) plasmacytoid (p) DCs are present in the liver, mDCs being predominant. Despite being a minor population (<1%) of the NPCs, DCs play an important role in ischemia-reperfusion injury [79–83], hepatic fibrosis [84,85], and can regulate liver allograft tolerance [5,86]. Interestingly, compared to the peripheral DCs, liver DCs express low levels of MHC class II, adhesion molecules and co-stimulatory molecules, are resistant to maturation, and can downregulate effector T cell responses [5].

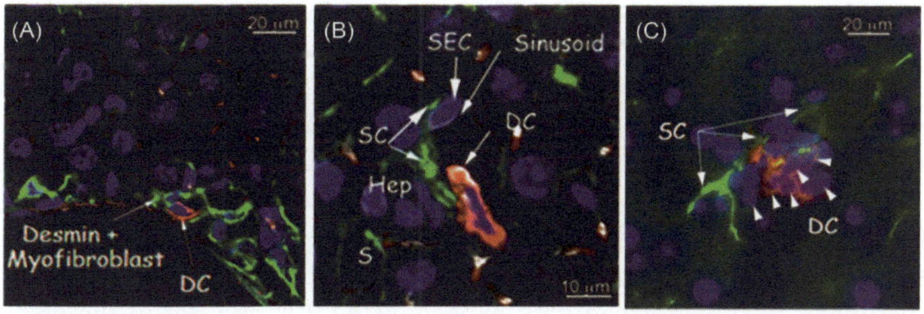

Figure 14.6 ***Co-localization of mouse HSC/DC in situ.*** Confocal micrographs of untreated (A and B) or Flt3L-treated (C) mouse liver cryostat sections stained for DCs (CD11c; red), HSCs (SC) or portal myofibroblasts (desmin; green), F-actin (phalloidin; gray, pseudo-colored from Alexa 647), and nucleus (Hoechst; blue). *Reproduced with permission from Sumpter et al. [94].*

KCs recruit circulating DCs by secreting MIP-1α [62], capture them using an *N*-acetylgalactosamine-specific C-type lectin-like receptor [87,88], and can render them tolerogenic via an IL-10-dependent mechanism [89]. However, murine KCs have been shown to induce activation of T cells [90], and, as described earlier, their role in liver transplantation tolerance has been questionable. HSCs, which are similar in number to KCs, also exhibit DC-modulating ability, and their direct interactions with DCs and other immune cells become prominent upon disruption of the endothelial barrier due to cold/ischemic preservation- or reperfusion-induced injury [52].

DCs are localized predominantly in portal areas (close to portal myofibroblasts) [91,92]. Physical association between DCs and HSCs is evident in steady-state mouse liver; following Flt3L administration-induced DC mobilization in the liver [93], significantly greater number of DCs accumulate close to HSCs [94] (Figure 14.6). This association is likely due to the potent DC chemoattractants MCP-1 and MIP-1α produced by HSCs, and can be recapitulated in co-culture. Mouse HSCs promote survival of syngeneic liver-derived myeloid DCs (mDCs) and impair the ability of DCs to induce proliferation of both $CD4^+$ and $CD8^+$ allogeneic T cells [94]; HSCs exert the same effects on allogeneic hepatic and splenic mDCs (unpublished observation). HSC-conditioned DCs express increased levels of indoleamine dioxygenase (IDO) that degrades tryptophan, an amino acid essential for T-cell activation/proliferation [95]. A primary mechanism of increased IDO expression involves HSC-derived IL-6 and subsequent IL-6-induced STAT3 activation in DCs. Increased IDO expression by HSC-conditioned DCs appears to be critical for their ability to downregulate T-cell responses in the presence of contemporaneously increased expression of cell surface co-stimulatory molecules.

Interestingly, exogenous IL-6 or IL-10 that are produced by HSCs, as well as culture supernatants from HSCs failed to increase IDO expression by liver mDCs indicating the requirement for cell–cell contact in this process [94]. Of note is the finding that HSCs downregulate suppressor of cytokine signaling 3 (SOCS3) expression in DCs [94], possibly via retinoic acid (RA) [96], and SOCS3 downregulation has been shown to increase IDO [97,98]. The role of RA released by activating HSCs appears important in DC regulation as all-trans RA (ATRA) has been found to promote T-cell suppressor phenotype in bone marrow-derived DCs by inducing arginase-1 and iNOS expression [99]. The interpretation of these data was that by releasing ATRA, HSCs contribute to DC-induced immunological tolerance, with a caveat that the experiments were performed with exogenous ATRA, and not in the presence of HSCs, without or with RA receptor (RAR) antagonist. However, HSCs have been shown to promote DC-induced Treg expansion that is blocked by RAR antagonist [100]. The molecules involved in HSC-DC contact required for IL-6-induced suppressor DC phenotype as well as the mechanisms of DC survival by HSCs remain undetermined. Nevertheless, HSC-induced increased IDO in DCs not only suppresses effector T cells by depleting tryptophan, but also through expansion of Tregs [56]. It has been shown that LPS-matured IDO^+ human DCs induce and expand Tregs [101–103], and activation of Tregs by IDO^+ DCs correlates strongly with their enhanced potential to suppress T cells [104].

Whereas the HSC-induced tolerogenic state of DCs is related to IDO expression, constitutive exposure to LPS under physiologic condition has been postulated to maintain liver DCs in an immature state. A role for LPS-induced increased expression of DNAX-activating protein of 12 kDa (DAP12), which is associated with IDO induction, is pertinent in this regard as DAP12-deficient DCs exhibit enhanced maturation upon TLR ligation [105,106]. This mechanism may be an important part of the phenomenon known as "endotoxin tolerance" [107]. It will be of interest to determine whether HSCs also contribute to DAP12-mediated mechanism of maintaining DC immaturity.

While HSCs modulate mDCs and render them tolerogenic, DCs in turn can alter the function of HSCs. Such bilateral HSC-DC interaction is illustrated by increased production of IFN-γ-induced protein 10 (IP-10), CXCL9, and IL-6 in co-culture than by either cell type cultured alone. However, the synthesis of MCP-1, CXCL1, MIP-1α (CCL3), and IL-10 by HSCs was not altered in the HSC-DC co-culture [94]. The mechanisms of these interactions that potentially alter the hepatic immunological environment are not elucidated.

14.6 HSCs AND CONVENTIONAL EFFECTOR T CELLS

Vinas et al. [108] observed that highly activated human HSCs exhibit phagocytic activity, and express antigen-presenting molecules (HLA-II DR, CD80, and CD86) that are upregulated by stimulation with IL-1β + IFN-γ. Because HSCs induced proliferation of allogeneic lymphocytes in an HLA-II-dependent manner, the authors suggested that they are nonprofessional APCs [108]. CD11c/MHC class II-expressing mouse HSCs were also reported to present antigen and modulate T-cell activation [74]. Accumulation of $CD8^+$ T cells in the proximity of HSCs in periportal area and along fibrous septa of CCl_4-treated rats led to the postulate that interactions between the two cell types is critical to fibrosis development. On the other hand, an earlier study reported higher rate of fibrosis progression with low incidence of $CD4^+$ T cells in HIV and HCV co-infected patients [72]. However, evidence that HSCs may induce immunological tolerance came from the accumulation and apoptosis of both $CD4^+$ and $CD8^+$ T cells in the fibrous area [109]. Because T-cell migration was induced by injecting concanavalin A in this study [109], the data suggested that HSCs induce chemotaxis and apoptosis of T cells. This aspect of HSC biology becomes important in transplantation as locally and systemically activated T cells must be rendered inactive to prevent destruction and rejection of the liver allograft.

Early work showed retention and apoptosis of adoptively transferred activated $CD8^+$ T cells in the liver; apoptosis of these $CD8^+$ T cells was suggested to be mediated by KCs via Fas/FasL mechanism [110,111]. Following liver allograft transplantation in mouse also, increased accumulation of both $CD4^+$ and $CD8^+$ T cells occurred, but their number decreased gradually due to high levels of apoptosis [112,113]. This study, however, did not determine which hepatic cell types induced T-cell apoptosis and the underlying mechanisms.

Experiments to discover immunoregulatory function of HSCs in context of liver transplantation led to the finding that the low-level expression of MHC class I and class II, CD80, CD86, and CD54 in mouse quiescent HSCs increased significantly upon their culture-induced activation [114]. The activated mouse HSCs suppressed alloantigen as well as anti-CD3-induced $CD4^+$ and $CD8^+$ T-cell proliferation (but not activation) and caused their apoptosis that was partly blocked by anti-B7-H1 Ab [114]. In contrast, IFN-γ-stimulated culture-activated mouse HSCs stimulated $CD4^+$ T-cell proliferation [115]. It was later observed that highly purified cultured mouse HSCs [116] as well as those stimulated with IFN-γ [100] were unable to stimulate proliferation of naïve $CD4^+$ T cells. Furthermore, consistent with an insufficient APC function, the highly purified HSCs failed to stimulate naïve OT-II TCR transgenic $CD4^+$ T cells [100,116].

A direct evidence for the tolerogenic potential of HSCs in allograft transplantation was provided by Chen et al. [117] when they co-transplanted allogeneic islet cells and activated HSCs under the renal capsule of a chemically induced diabetic mouse. The graft survived and was functional for more than 100 days. The role of IFN-γ-induced increased expression of B7-H1 by HSCs and increased incidence of Tregs in the kidney that contained co-transplanted HSCs and islet cells were suggested to contribute to

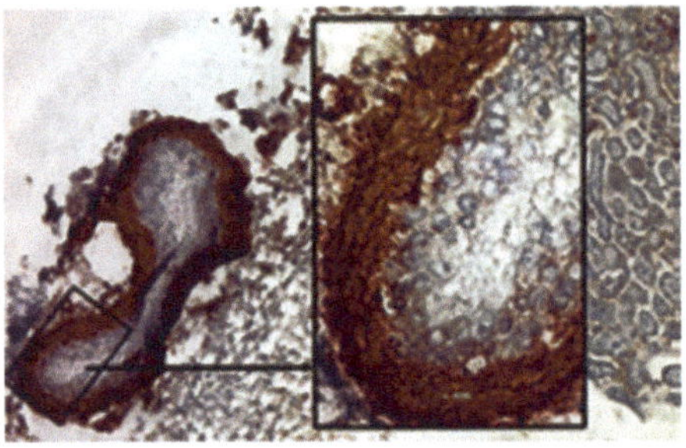

Figure 14.7 ***Formation of α-sma-positive capsule surrounding islets after co-transplantation with islet (BALB/c) + HSCs (B6) in chemically induced diabetic B6 recipients on day 122 after transplantation under the renal capsule.*** The cryostat sections of the kidney stained with α-sma mAb show a complete multi-layered α-sma-positive capsule around functional islets. *Reproduced with permission from Chen et al. [117].*

the allograft survival (Figure 14.7) [117,118]. Interestingly, these co-transplanted HSCs formed several layers of concentric circles around the islet cells, and were fully transformed as shown by intense α-sma staining. Because activated HSCs, *in vivo*, undergo apoptosis or revert to quiescent phenotype upon termination of the activation stimulus [119], how they survived and retained their fully activated phenotype in another organ [117] is an interesting question. Furthermore, whether these HSCs also produce increased amounts of ECM and how this impacts the graft survival remain undetermined. Nevertheless, these investigations affirmed that HSCs possess an impressive potential in transplantation tolerance.

Schlidberg et al. [120] reported that the tolerogenic function of murine and human HSCs becomes apparent upon activation, and was CD54-dependent and, unlike other studies [114,118] B7-H1-independent. They found that progressive activation of mouse HSCs in culture increases their CD54 expression and inhibits the DC-induced expression of IL-2 and IL-2 receptor in $CD8^+$ T cells, thereby preventing activation of the T cells. Because HSC–T-cell interactions have been reported to be a major mechanism of liver fibrosis [72–74], this observation indicates that HSCs also have an ability to limit fibrosis by inhibiting T-cell proliferation. This counter-regulatory effect of HSCs on T cells in limiting fibrogenesis is supported by the ability of activated HSCs to inhibit the synthesis of IFN-γ and IL-2 by T cells as well as their activation/proliferation via another B7 family member (B7-H4) in a contact-dependent manner [121].

An important consideration to understand the function of HSCs in liver allograft transplantation is to use the cell at an appropriate state of activation. Most of the work in the literature has been performed using fully activated cultured HSCs or cells obtained from the fibrotic liver. Furthermore, T cells or DCs can affect the activation level of HSCs. Because HSCs are activated rapidly to a transitional phenotype shortly following transplantation (Figure 14.3), it is critical to use very early-stage activated

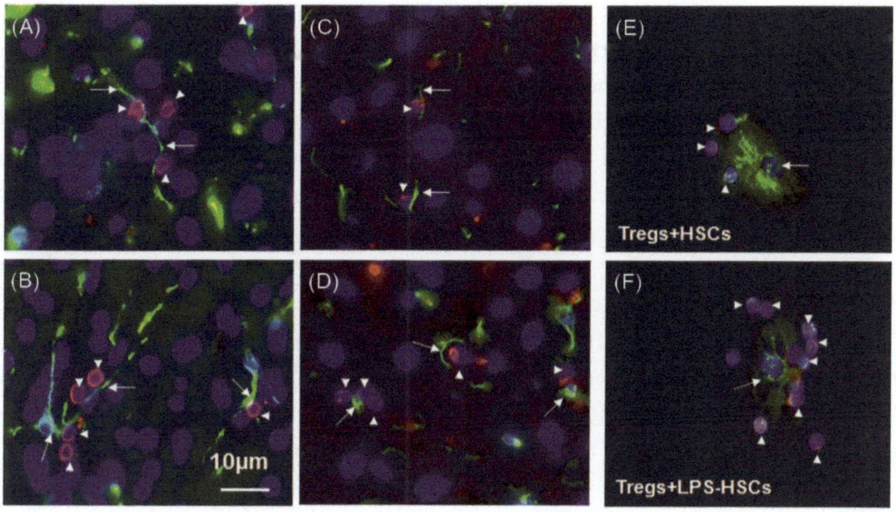

Figure 14.8 ***Close association of CD4+ T cells and Tregs with HSCs.*** (A and B) Sections of cold preserved livers (untreated and LPS-pretreated), reperfused with allogeneic $CD4^+$ T cells, showing the close interaction between HSCs (desmin; green stain-arrows) and $CD4^+$ T cells (CD4; red stain-arrowheads). (C and D) Sections of the cold preserved livers (untreated and LPS-pretreated), reperfused with allogeneic $CD4^+$ T cells showing the close interaction between HSCs (desmin; green stain-arrows) and Tregs (FoxP3; red stain-arrowheads). Magnification, 100 ×. (E and F) Co-culture of HSCs (untreated or LPS-pretreated) and purified-Tregs (on coverslips) showing the close interaction between HSCs (desmin; green stain arrows) and Tregs (FoxP3; red stain-arrowheads). Magnification 100 ×. *Reproduced with permission from Dangi et al. [56].*

HSCs. Indeed, incubation of HSCs starting at day 2 of culture with $CD4^+CD25^-$ T cells was found to suppress proliferation and induce apoptosis of the latter via Fas/FasL pathway [56]. LPS treatment of HSCs, however, did not augment T-cell apoptosis, but LPS treatment of mice *in vivo* and of HSCs *in vitro* promoted T-cell recruitment proximal to HSCs [56] (Figure 14.8). Because LPS levels are increased in liver transplantation, these observations provide evidence for significant contribution by HSCs to recruit and kill T cells and promote allograft tolerance.

14.7 HSCs AND REGULATORY T CELLS

Treg cells maintain immune homeostasis by suppressing the function of effector conventional $CD4^+$ and $CD8^+$ T cells. Pathogenesis of several autoimmune diseases such as multiple sclerosis, rheumatoid arthritis, and type 1 diabetes is associated with altered number and function of Tregs [122–124]. Increased number of Tregs is also found in cancer tumor sites including lung, pancreas, and breast, and by preventing antitumor immune responses, may be responsible for increased mortality. Thus, it is not surprising that their hepatic number is increased following transplantation of liver allograft and in chronic HCV infection [102,125–127]. Furthermore, portal venous administration of allogeneic (B6) spenocytes into BALB/c mice increases the number of $CD4^+CD25^+FoxP3^+$ T cells; adoptive transfer of these Tregs into BALB/c mice followed by transplantation of B6 heart induced tolerance and prolonged the allograft survival [128]. These data emphasize the tolerogenic potential of Tregs in allograft transplantation.

Both $CD4^+$ and $CD8^+$ Tregs are present naturally, the former being the predominant subtype. $CD4^+$ Tregs are broadly classified into two subsets: (i) naturally occurring $CD4^+CD25^+Foxp3^+$ T cells (nTregs) and (ii) inducible Tregs generated from $CD4^+CD25^-Foxp3^-$ T cells (iTregs) that are relatively unstable. Induction of Tregs requires TGFβ1 and ATRA for priming of naïve $CD4^+$ T cells to express CD25 and FoxP3 [129–132]. HSCs, being the major storage site of retinoids and capable of producing ATRA, can generate iTregs in hepatic microenvironment in presence of TGFβ1 released by KCs [133] or SECs [134].

Culture activated mouse HSCs were found to induce differentiation of Tregs but required ovalbumin (OVA) peptide-primed DCs and TGFβ1 [100]. Another study also reported induction of Tregs from $CD4^+$ T cells by HSCs in a TGFβ1- and ATRA-mediated manner and required the presence of DCs [116]. In both studies, HSCs were purified via sorting and selected based on their vitamin A autofluorescence and CD45-negativity. Because these highly purified HSCs did not express MHC class II, it was interpreted that the effect was due to secreted RA (blocked by RA receptor antagonist LE540), and thus HSCs were considered as bystander regulators of iTreg differentiation. SECs and KCs were unable to produce iTregs from $CD4^+$ cells [116]. Such tolerogenic environment can only be induced when HSCs are activated because quiescent HSCs express low levels of retinaldehyde dehydrogenases.

Other investigations found that IFN-γ or LPS-stimulated HSCs can expand naturally occurring Tregs [56,115,118]. Treg expansion by IFN-γ-stimulated HSCs was almost completely dependent on IL-2 [115], questioning the role of RA in HSC-induced Treg expansion in the liver. From the total $CD4^+$ T-cell population, HSCs caused selective expansion of nTregs, but not iTregs, that express Helios, a member of Icarus transcription factor family; at the same time HSCs caused apoptosis of conventional $CD4^+$ T cells [56]. Retention of exogenously infused Tregs by HSCs following cold preservation of the LPS pretreated liver [56] (Figure 14.8) and association of more than 50% Tregs with HSCs in transplanted liver allograft (Figure 14.9) further emphasize the importance of HSCs in hepatic tolerogenicity. Increased incidence of Tregs in the kidney following co-transplantation of HSCs and islet cells [118] also supports the view that regardless of the organ, HSCs have the ability to promote tolerance by increasing the Treg number. Indeed, following their adoptive transfer in mouse, allogeneic HSCs lodged in lymph nodes and the spleen, where the number of Ki67-positive Tregs increased indicating their active proliferation [56].

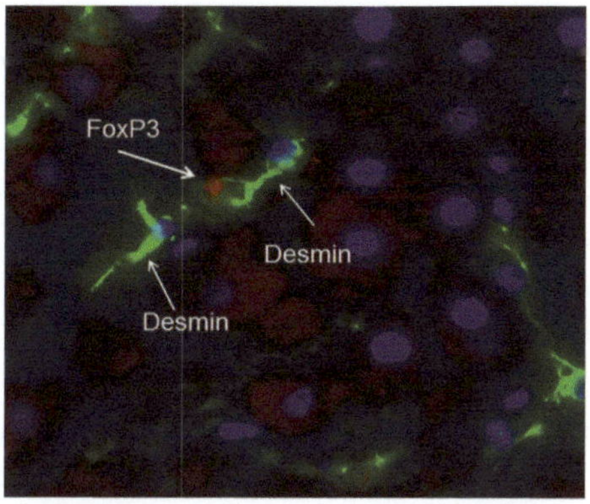

Figure 14.9 ***Direct association of HSCs with Tregs after transplantation.*** Following cold preservation (18h) and reperfusion (6h post transplant) of a mouse liver allograft, FoxP3-positive Tregs (red) are seen in close association with desmin-positive HSCs (green). Of the total, more than 50% of Tregs were associated with HSCs. The livers were kindly provided by Dr. Noriko Murase, and immunostaining and imaging were performed by Dr. Anil Dangi. Magnification 20×.

Perhaps a more important aspect of HSC-expanded Tregs was their enhanced ability to suppress effector T-cell activation/proliferation and superior *in vivo* stability [56]. This is due to the increased expression of FoxP3, GITR, CTLA4, and LAG-3 by HSC-expanded Tregs. It is worth mentioning that LPS pretreatment of HSCs increased their ability to expand Tregs, but did not affect their immunosuppressive function beyond that by unstimulated HSCs [56]. The impressive ability of HSCs to modulate Tregs may have significant potential for Treg therapy, which is being tested clinically to treat graft-versus-host disease, autoimmune diseases, and inflammatory diseases [135–137]. Even though KCs are also shown to expand Tregs, they are inferior to freshly isolated cells in their immunosuppressive potential and LPS treatment of KCs actually reverses the suppressive activity of KC-expanded Tregs [138]. These findings exemplify the importance of HSCs in promoting liver tolerogenicity. Current clinical trials use Tregs expanded *ex vivo* with anti-CD3/CD28 Ab, but their *in vivo* stability is poor [137,139]. While Dangi et al. [56] performed experiments using HSCs in their early stage of activation (d2–d5 of culture), fully activated HSCs were also found to expand Tregs in a similar way (A. Dangi and C.R. Gandhi, unpublished observation). This indicates that cryopreserved HSCs may be used to expand patient's Tregs *ex vivo* for autologous transfer.

HSCs produce IL-10, IL-6, and low levels of TNF-α constitutively and LPS increases the synthesis of all of these cytokines [50,53,54]. In co-culture with Tregs, the synthesis of IL-10 by both cell types increased significantly whereas that of TNF-α, IL-6, IL-1α, and IL-1β reduced [56]. The mechanisms of this interaction and selective increase in anti-inflammatory IL-10 and decrease in pro-inflammatory cytokines remain to be determined.

14.8 HSCs AND NKT CELLS

NKT cells express an invariant T-cell receptor and recognize endogenous as well as exogenous lipid and glycolipid ligands including those derived from bacteria [140–143]. The lipid antigens are displayed by DCs and KCs via CD1a–d (human) and CD1d (mouse), and presented to NKT cells that become activated and produce large amounts of IFN-γ and IL-4 and limit the infection [144]. Upon examining whether HSCs might perform similar function as DCs, culture- as well as *in vivo*-activated human HSCs were found to express CD1b and CD1c [108] indicating their ability to present lipid and glycolipid antigens to NKT cells. Winau et al. [145] later reported significant expression of CD1d in quiescent mouse HSCs and of CD1a–c and CD1d in fully activated human (passage 3–6) HSCs; both cell types also expressed CD1d, and induced activation of NKT cells by presenting lipid antigen and promoted their proliferation by producing IL-15. The ability of HSCs to perform this function was equivalent to that of DCs. In contrast, Ichikawa et al. [100] found that mouse HSCs were inferior to DCs in activating NKT cells, as determined by IFN-γ production, although they effectively presented α-galactosylceramide to CD1-restricted NKT cells [100]. However, because the number of HSCs is much greater than that of hepatic DCs, their function as NKT cell stimulators during infection should be considered to be of critical significance. These data thus provided an important aspect of HSC biology in contributing to protection against microbial infection along with macrophages including KCs and other cells of the innate immune system. Unfortunately, not much work is being done in this area but it will be critical to assess the ability of HSCs at different stages of activation to assess their role in NKT cell activation during progressive liver disease.

14.9 PERSPECTIVE

During the last decade, substantial evidence has emerged for a highly critical role of HSCs in immune regulation that impacts the fate of allograft transplantation. A decade prior to that, demonstration of HSC-expressed cytokines and chemokines triggered investigations to discover HSC-inflammatory cells and HSC-immune cells (especially T cells) interactions in the regulation of hepatic fibrosis. This work primarily addressed fully activated HSCs. In transplantation, however, HSCs are not fully activated (at least during early times following the surgery) and thus transitionally activated HSCs represent the correct phenotype that would interact with the immune cells after reperfusion-induced injury. This phenotype appears to be highly tolerogenic in its modulation of DCs and Tregs, and in inducing apoptosis of effector T cells. HSC-modulated Tregs may have therapeutic potential in immunological disorders where effector T cells should be suppressed. Although *ex vivo* expanded Tregs are used in clinical trials to mitigate immune responses, a major drawback of this practice is their short *in vivo* stability. HSC-expanded Tregs are superior immunosuppressors and more stable *in vivo* compared with freshly isolated Tregs or those expanded by other means. Thus, treatment with HSC-expanded Tregs in preventing liver allograft rejection may be an exciting possibility of replacing or reducing drug-induced immunosuppression. Although promising advances in understanding HSC-immune cell(s) interactions have been made, there are contradicting observations. This may be because several studies use co-culture of immune cells and HSCs in suspension. A possibility that the physiologically stationary HSCs might react differently in suspension, undergo apoptosis and their phagocytosis by DCs can alter the outcome of the experimental results should be considered. Furthermore, purification (selection) of HSCs based on their vitamin A autofluorescence will eliminate substantial number of vitamin A-deficient HSCs that may have distinct immunological properties/function, and thus the overall outcome may not represent an *in vivo* situation. Nevertheless, the overwhelming information to date clearly indicates that HSCs are critical in the regulation of hepatic immunological tolerance.

ACKNOWLEDGMENT

I thank Dr. Angus W. Thomson, University of Pittsburgh, for a critical review of this manuscript. This work was supported by a VA Merit Review Award (1IO1BX001174-01) and grants from NIH (PO1AIO81678 and R21AA020846).

REFERENCES

[1] Knolle PA, Schmitt E, Jin S, Germann T, Duchmann R, Hegenbarth S, et al. Induction of cytokine production in naïve CD4(+) T cells by antigen-presenting murine liver sinusoidal endothelial cells but failure to induce differentiation toward Th1 cells. Gastroenterology 1999;116:1428–40.

[2] Smedsrød B, Pertoft H, Gustafson S, Laurent TC. Scavenger functions of the liver endothelial cell. Biochem J 1990;266:313–27.

[3] Crispe IN, Dao T, Klugewitz K, Mehal WZ, Metz DP. The liver as a site of T-cell apoptosis: graveyard or killing field? Immunol Rev 2000;174:47–62.

[4] Crispe IN, Giannandrea M, Klein I, John B, Sampson B, Wuensch S. Cellular and molecular mechanisms of liver tolerance. Immunol Rev 2006;213:101–18.

[5] Thomson AW, Knolle PA. Antigen-presenting cell function in the tolerogenic liver environment. Nat Rev Immunol 2010;10:753–66.

[6] Limmer A, Ohl J, Kurts C, Ljunggren HG, Reiss Y, Groettrup M, et al. Efficient presentation of exogenous antigen by liver endothelial cells to $CD8^{+}$ T cells results in antigen-specific T-cell tolerance. Nat Med 2000;6:1348–54.

[7] Mehal WZ, Azzaroli F, Crispe IN. Immunology of the healthy liver: old questions and new insights. Gastroenterology 2001;120:250–60.

[8] Morelli AE, Thomson AW. Tolerogenic dendritic cells and the quest for transplant tolerance. Nat Rev Immunol 2007;7:610–21.

[9] Callery MP, Kamei T, Flye MW. The effect of portacaval shunt on delayed-hypersensitivity responses following antigen feeding. J Surg Res 1989;46:391–4.

[10] Ilan Y, Sauter B, Chowdhury NR, Reddy BV, Thummala NR, Droguett G, et al. Oral tolerization to adenoviral proteins permits repeated adenovirus-mediated gene therapy in rats with pre-existing immunity to adenoviruses. Gastroenterology 1998;27:1368–76.

[11] Yu S, Nakafusa Y, Flye MW. Portal vein administration of donor cells promotes peripheral allospecific hyporesponsiveness and graft tolerance. Surgery 1994;116:229–35.
[12] Starzl TE. History of clinical transplantation. World J Surg 2000;24:759–82.
[13] Sanchez-Fueyo A, Strom TB. Immunologic basis of graft rejection and tolerance following transplantation of liver or other solid organs. Gastroenterology 2011;140:51–64.
[14] Chomette G, Garnier H, Clot JP, Pinaudeau Y, Auriol M, Brocheriou C. Microscopic study on allogenic liver transplantation in the pig. Virchows Arch A Pathol Pathol Anat 1969;347:176–84.
[15] Calne RY, Sells RA, Pena JR, Davis DR, Millard PR, Herbertson BM, et al. Induction of immunological tolerance by porcine liver allografts. Nature 1969;223:472–6.
[16] Calne RY. Mechanisms in the acceptance of organ grafts. Br Med Bull 1976;32:107–12.
[17] Houssin D, Gigou M, Franco D, Bismuth H, Charpentier B, Lang P, et al. Specific transplantation tolerance induced by spontaneously tolerated liver allograft in inbred strains of rats. Transplantation 1980;29:418–9.
[18] Qian S, Demetris AJ, Murase N, Rao AS, Fung JJ, Starzl TE. Murine liver allograft transplantation: tolerance and donor cell chimerism. Hepatology 1994;19:916–24.
[19] Starzl TE, Kaupp Jr. HA, Brock DR, Butz Jr. GW, Linman JW. Homotransplantation of multiple visceral organs. Am J Surg 1962;103:219.
[20] Kamada N, Brons G, Davies HS. Fully allogeneic liver grafting in rats induces a state of systemic nonreactivity to donor transplantation antigens. Transplantation 1980;29:429–31.
[21] Zimmerman FA, Davies HS, Knoll PP, Gocke JM, Schmidt T. Orthotopic liver allografts in the rat. Transplantation 1984;37:406–10.
[22] Murase N, Demetris AJ, Kim DG, Todo S, Fung JJ, Starzl TE. Rejection of the multivisceral allografts in rats: a sequential analysis with comparison to isolated orthotopic small bowel and liver grafts. Surgery 1990;108:880–9.
[23] Murase N, Demetris AJ, Matsuzaki T, Yagihasi A, Todo S, Fung J, et al. Long survival in rats after multivisceral versus isolated small bowel allotransplantation under FK 506. Surgery 1991;110:87–98.
[24] Marsh JW, Gray E, Ness R, Starzl TE. Complications of right lobe living donor liver transplantation. J Hepatol 2009;51:715–24.
[25] Wiesner RH, Fung JJ. Present state of immunosuppressive therapy in liver transplant recipients. Liver Transpl 2011;17(Suppl. 3):S1–S9.
[26] Starzl TE, Koep LJ, Halgrimson CG, Hood J, Schroter GP, Porter KA, et al. Fifteen years of clinical liver transplantation. Gastroenterology 1979;77:375–88.
[27] Starzl TE, Demetris AJ, Trucco M, Murase N, Ricordi C, Ildstad S, et al. Cell migration and chimerism after whole-organ transplantation: the basis of graft acceptance. Hepatology 1993;17:1127–52.
[28] Gugenheim J, Samuel D, Reynes M, Bismuth H. Liver transplantation across ABO blood group barriers. Lancet 1990;336:519–23.
[29] Iwatsuki S, Iwaki Y, Kano T, Klintmalm G, Koep LJ, Weil R, et al. Successful liver transplantation from crossmatch-positive donors. Transplant Proc 1981;13:286–8.
[30] Mazariegos GV, Reyes J, Marino IR, Demetris AJ, Flynn B, Irish W, et al. Weaning of immunosuppression in liver transplant recipients. Transplantation 1997;63:243–9.
[31] Takatsuki M, Uemoto S, Inomata Y, Egawa H, Kiuchi T, Fujita S, et al. Weaning of immunosuppression in living donor liver transplant recipients. Transplantation 2001;72:449–54.
[32] Feng S, Ekong UD, Lobritto SJ, Demetris AJ, Roberts JP, Rosenthal P, et al. Complete immunosuppression withdrawal and subsequent allograft function among pediatric recipients of parental living donor liver transplants. JAMA 2012;307:283–93.
[33] Sato K, Yabuki K, Haba T, Maekawa T. Role of Kupffer cells in induction of tolerance after liver transplantation. J Surg Res 1996;63:433–8.
[34] Tian Y, Jochum W, Georgiev P, Moritz W, Graf R, Clavien PA. Kupffer cell-dependent TNF-alpha signaling mediates injury in the arterialized small-for-size liver transplantation in the mouse. Proc Natl Acad Sci USA 2006;103:4598–603.
[35] Akamatsu Y, Ohkohchi N, Doi H, Satomi S. Effect of elimination of donor Kupffer cells and/or recipient macrophages on acute rejection in liver transplantation. Hepatogastroenterology 2003;50:1105–10.

[36] Sun S, Wada T, Maemura K, Uchiura K, Hashino S, Diehl AM, et al. Hepatic allograft-derived Kupffer cells regulate T cell response in rats. Liver Transpl 2003;9:489–97.
[37] Knolle PA, Germann T, Treichel U, Uhrig A, Schmitt E, Hegenbarth S, et al. Endotoxin down-regulates T cell activation by antigen-presenting liver sinusoidal endothelial cells. J Immunol 1999;162:1401–7.
[38] Diehl L, Schurich A, Grochtmann R, Hegenbarth S, Chen L, Knolle PA. Tolerogenic maturation of liver sinusoidal endothelial cells promotes B7-homolog 1-dependent $CD8^+$ T cell tolerance. Hepatology 2008;47:296–305.
[39] Onoe T, Ohdan H, Tokita D, Hara H, Tanaka Y, Ishiyama K, et al. Liver sinusoidal endothelial cells have a capacity for inducing nonresponsiveness of T cells across major histocompatibility complex barriers. Transpl Int 2005;18:206–14.
[40] Marra F, Valente AJ, Pinzani M, Abboud HE. Cultured human liver fat-storing cells produce monocyte chemotactic protein-1. Regulation by proinflammatory cytokines. J Clin Invest 1993;92:1674–80.
[41] Czaja MJ, Geerts A, Xu J, Schmiedeberg P, Ju Y. Monocyte chemoattractant protein 1 (MCP-1) expression occurs in toxic rat liver injury and human liver disease. J Leukoc Biol 1994;55:120–6.
[42] Costa AM, Tuchweber B, Rubbia-Brandt L, et al. Early activation of hepatic stellate cells and perisinusoidal extracellular matrix changes during *ex vivo* pig liver perfusion. J Submicrosc Cytol Pathol 2001;33:231–40.
[43] Petrovic LM, Arkadopoulos N, Demetriou AA. Activation of hepatic stellate cells in liver tissue of patients with fulminant liver failure after treatment with bioartificial liver. Hum Pathol 2001;32:1371–5.
[44] Tanaka Y, Mak KM, Lieber CS. Immunohistochemical detection of proliferating lipocytes in regenerating rat liver. J Pathol 1990;160:129–34.
[45] Mabuchi A, Mullaney I, Sheard PW, Hessian PA, Mallard BL, Tawadrous MN, et al. Role of hepatic stellate cell/hepatocyte interaction and activation of hepatic stellate cells in the early phase of liver regeneration in the rat. J Hepatol 2004;40:910–6.
[46] Johnson SJ, Hines JE, Burt AD. Phenotypic modulation of perisinusoidal cells following acute liver injury: a quantitative analysis. Int J Exp Pathol 1992;73:765–72.
[47] Jaeschke H. Inflammation in response to hepatocellular apoptosis. Hepatology 2002;35:964–6.
[48] Canbay A, Friedman S, Gores GJ. Apoptosis: the nexus of liver injury and fibrosis. Hepatology 2004;39:273–8.
[49] Watanabe A, Hashmi A, Gomes DA, et al. Apoptotic hepatocyte DNA inhibits hepatic stellate cell chemotaxis via Toll-like receptor 9. Hepatology 2007;46:1509–18.
[50] Harvey SAK, Dangi A, Gandhi CR. The transcriptomic response of rat hepatic stellate cells to endotoxin: implications for hepatic inflammation and immune regulation. PLoS One 2013;8:e82159.
[51] Ohkohchi N. Mechanisms of preservation and ischemic/reperfusion injury in liver transplantation. Transplant Proc 2002;34:2670–3.
[52] Stolz DB, Ross MA, Ikeda A, Tomiyama K, Kaizu T, Geller DA, et al. Sinusoidal endothelial cell repopulation following ischemia/reperfusion injury in rat liver transplantation. Hepatology 2007;46:1464–75.
[53] Uemura T, Gandhi CR. Inhibition of DNA synthesis in cultured hepatocytes by endotoxin-conditioned medium of activated stellate cells is transforming growth factor-beta- and nitric oxide-independent. Br J Pharmacol 2001;133:1125–33.
[54] Thirunavukkarasu C, Uemura T, Wang LF, Watkins S, Gandhi CR. Normal rat hepatic stellate cells respond to endotoxin in LBP-independent manner to produce inhibitor(s) of DNA synthesis in hepatocytes. J Cell Physiol 2005;204:654–65.
[55] Seki E, De Minicis S, Osterreicher CH, et al. TLR4 enhances TGF-beta signaling and hepatic fibrosis. Nat Med 2007;13:1324–32.
[56] Dangi A, Sumpter TL, Kimura S, Stolz DB, Murase N, Raimondi G, et al. Selective expansion of allogeneic regulatory T cells by hepatic stellate cells: role of endotoxin and implications for allograft tolerance. J Immunol 2012;188:3667–77.
[57] Mimura Y, Sakisaka S, Harada M, Sata M, Tanikawa K. Role of hepatocytes in direct clearance of lipopolysaccharide in rats. Gastroenterology 1995;109:1969–76.
[58] Bikhazi AB, Jurjus AR, Kamal MT, Al-Housseini AM, Saab RN, Jaroudi WA, et al. Kinetics of lipopolysaccharide clearance by Kupffer and parenchyma cells in perfused rat liver. Comp Biochem Physiol C Toxicol Pharmacol 2001;129:339–48.

[59] Lumsden AB, Henderson JM, Kutner MH. Endotoxin levels measured by a chromogenic assay in portal, hepatic and peripheral venous blood in patients with cirrhosis. Hepatology 1988;8:232–6.
[60] Riordan SM, Skinner N, Nagree A, McCallum H, McIver CJ, Kurtovic J, et al. Peripheral blood mononuclear cell expression of toll-like receptors and relation to cytokine levels in cirrhosis. Hepatology 2003;37:1154–64.
[61] Yokoyama I, Gavaler JS, Todo S, Miyata T, Van Thiel DH, Starzl TE. Endotoxemia is associated with renal dysfunction in liver transplantation recipients during the first postoperative week. Hepatogastroenterology 1995;42:205–8.
[62] Yoneyama H, Ichida T. Recruitment of dendritic cells to pathological niches in inflamed liver. Med Mol Morphol 2005;38:136–41.
[63] Sprenger H, Kaufmann A, Garn H, Lahme B, Gemsa D, Gressner AM. Differential expression of monocyte chemotactic protein-1 (MCP-1) in transforming rat hepatic stellate cells. J Hepatol 1999;30:88–894.
[64] Mühlbauer M, Bosserhoff AK, Hartmann A, Thasler WE, Weiss TS, Herfarth H, et al. A novel MCP-1 gene polymorphism is associated with hepatic MCP-1 expression and severity of HCV-related liver disease. Gastroenterology 2003;125:1085–93.
[65] Marra F, Grandaliano G, Valente AJ, Abboud HE. Thrombin stimulates proliferation of liver fat-storing cells and expression of monocyte chemotactic protein-1: potential role in liver injury. Hepatology 1995;22:780–7.
[66] Marra F, DeFranco R, Grappone C, Milani S, Pastacaldi S, Pinzani M, et al. Increased expression of monocyte chemotactic protein-1 during active hepatic fibrogenesis: correlation with monocyte infiltration. Am J Pathol 1998;152:423–30.
[67] Masumoto T, Ohkubo K, Yamamoto K, Ninomiya T, Abe M, Akbar SM, et al. Serum IL-8 levels and localization of IL-8 in liver from patients with chronic viral hepatitis. Hepatogastroenterology 1998;45:1630–4.
[68] Maher JJ, Lozier JS, Scott MK. Rat hepatic stellate cells produce cytokine-induced neutrophil chemoattractant in culture and *in vivo*. Am J Physiol 1998;275:G847–53.
[69] Schwabe RF, Schnabl B, Kweon YO, Brenner DA. CD40 activates NF-kappa B and c-Jun N-terminal kinase and enhances chemokine secretion on activated human hepatic stellate cells. J Immunol 2001;166:6812–9.
[70] Paik YH, Schwabe RF, Bataller R, Russo MP, Jobin C, Brenner DA. Toll-like receptor 4 mediates inflammatory signaling by bacterial lipopolysaccharide in human hepatic stellate cells. Hepatology 2003;37:1043–55.
[71] Reeves HL, Burt AD, Wood S, Day CP. Hepatic stellate cell activation occurs in the absence of hepatitis in alcoholic liver disease and correlates with the severity of steatosis. J Hepatol 1996;25:677–83.
[72] Benhamou Y, Bochet M, Di Martino V, Charlotte F, Azria F, Coutellier A, et al. Liver fibrosis progression in human immunodeficiency virus and hepatitis C virus coinfected patients. The multivirc group. Hepatology 1999;30:1054–8.
[73] Schwabe RF, Bataller R, Brenner DA. Human hepatic stellate cells express CCR5 and RANTES to induce proliferation and migration. Am J Physiol Gastrointest Liver Physiol 2003;285:G949–58.
[74] Muhanna N, Horani A, Doron S, Safadi R. Lymphocyte-hepatic stellate cell proximity suggests a direct interaction. Clin Exp Immunol 2007;148:338–47.
[75] Wang SC, Ohata M, Schrum L, Rippe RA, Tsukamoto H. Expression of interleukin-10 by *in vitro* and *in vivo* activated hepatic stellate cells. J Biol Chem 1998;273:302–8.
[76] Roland CR, Mangino MJ, Duffy BF, Flye MW. Lymphocyte suppression by Kupffer cells prevents portal venous tolerance induction: a study of macrophage function after intravenous gadolinium. Transplantation 1993;55:1151–8.
[77] Roland CR, Walp L, Stack RM, Flye MW. Outcome of Kupffer cell antigen presentation to a cloned murine Th1 lymphocyte depends on the inducibility of nitric oxide synthase by IFN-gamma. J Immunol 1994;153:5453–64.
[78] Steinman RM, Cohn ZA. Identification of a novel cell type in peripheral lymphoid organs of mice. I. Morphology, quantitation, tissue distribution. J Exp Med 1973;137:1142–62.
[79] Bamboat ZM, Ocuin LM, Balachandran VP, Obaid H, Plitas G, DeMatteo RP. Conventional DCs reduce liver ischemia/reperfusion injury in mice via IL-10 secretion. J Clin Invest 2010;120:559–69.

[80] Tsung A, Zheng N, Jeyabalan G, Izuishi K, Klune JR, Geller DA, et al. Increasing numbers of hepatic dendritic cells promote HMGB1-mediated ischemia-reperfusion injury. J Leukoc Biol 2007;81:119–28.

[81] Zhang M, Ueki S, Kimura S, Yoshida O, Castellaneta A, Ozaki KS, et al. Roles of dendritic cells in murine hepatic warm and liver transplantation-induced cold ischemia/reperfusion injury. Hepatology 2013;57:1585–96.

[82] Yoshida O, Kimura S, Jackson EK, Robson SC, Geller DA, Murase N, et al. CD39 expression by hepatic myeloid dendritic cells attenuates inflammation in liver transplant ischemia-reperfusion injury in mice. Hepatology 2013;58:2163–75.

[83] Castellaneta A, Yoshida O, Kimura S, Yokota S, Geller DA, Murase N, et al. Plasmacytoid dendritic cell-derived IFN-α promotes murine liver ischemia/reperfusion injury by induction of hepatocyte IRF-1. Hepatology 2014;60:267–77.

[84] Connolly MK, Bedrosian AS, Mallen-St Clair J, Mitchell AP, Ibrahim J, Stroud A, et al. In liver fibrosis, dendritic cells govern hepatic inflammation in mice via TNF-alpha. J Clin Invest 2009;119:3213–25.

[85] Jiao J, Sastre D, Fiel MI, Lee UE, Ghiassi-Nejad Z, Ginhoux F, et al. Dendritic cell regulation of carbon tetrachloride-induced murine liver fibrosis regression. Hepatology 2012;55:244–55.

[86] Sumpter TL, Abe M, Tokita D, Thomson AW. Dendritic cells, the liver, and transplantation. Hepatology 2007;46:2021–31.

[87] Kudo S, Matsuno K, Ezaki T, Ogawa M. A novel migration pathway for rat dendritic cells from the blood: hepatic sinusoids-lymph translocation. J Exp Med 1997;185:777–84.

[88] Uwatoku R, Suematsu M, Ezaki T, Saiki T, Tsuiji M, Irimura T, et al. Kupffer cell-mediated recruitment of rat dendritic cells to the liver: roles of *N*-acetylgalactosamine specific sugar receptors. Gastroenterology 2001;121:1460–72.

[89] Ellett JD, Atkinson C, Evans ZP, Amani Z, Balish E, Schmidt MG, et al. Murine Kupffer cells are protective in total hepatic ischemia/reperfusion injury with bowel congestion through IL-10. J Immunol 2010;184:5849–58.

[90] You Q, Cheng L, Kedl RM, Ju C. Mechanism ofT cell tolerance induction by murine hepatic Kupffer cells. Hepatology 2008;48:978–90.

[91] Woo J, Lu L, Rao AS, Li Y, Subbotin V, Starzl TE, et al. Isolation, phenotype, and allostimulatory activity of mouse liver dendritic cells. Transplantation 1994;58:484–91.

[92] Matsuno K, Ezaki T, Kudo S, Uehara Y. A life stage of particleladen rat dendritic cells *in vivo*: their terminal division, active phagocytosis, and translocation from the liver to the draining lymph. J Exp Med 1996;183:1865–78.

[93] Steptoe RJ, Fu F, Li W, Drakes ML, Lu L, Demetris AJ, et al. Augmentation of dendritic cells in murine organ donors by Flt3 ligand alters the balance between transplant tolerance and immunity. J Immunol 1997;159:5483–91.

[94] Sumpter TL, Dangi A, Matta BM, Huang C, Stolz DB, Vodovotz Y, et al. Hepatic stellate cells undermine the allostimulatory function of liver myeloid dendritic cells via STAT3-dependent induction of IDO. J Immunol 2012;189:3848–58.

[95] Mellor AL, Munn DH. IDO expression by dendritic cells: tolerance and tryptophan catabolism. Nat Rev Immunol 2004;4:762–74.

[96] Feng T, Cong Y, Qin H, Benveniste EN, Elson CO. Generation of mucosal dendritic cells from bone marrow reveals a critical role of retinoic acid. J Immunol 2010;185:5915–25.

[97] Orabona C, Belladonna ML, Vacca C, Bianchi R, Fallarino F, Volpi C, et al. Cutting edge: silencing suppressor of cytokine signaling 3 expression in dendritic cells turns CD28-Ig from immune adjuvant to suppressant. J Immunol 2005;174:6582–6.

[98] Orabona C, Pallotta MT, Volpi C, Fallarino F, Vacca C, Bianchi R, et al. SOCS3 drives proteasomal degradation of indoleamine 2,3-dioxygenase (IDO) and antagonizes IDO-dependent tolerogenesis. Proc Natl Acad Sci USA 2008;105:20828–33.

[99] Bhatt S, Qin J, Bennett C, Qian S, Fung JJ, Hamilton TA, et al. All-trans retinoic acid induces arginase-1 and inducible nitric oxide synthase-producing dendritic cells with T cell inhibitory function. Immunology 2014;192:5098–108.

[100] Ichikawa S, Mucida D, Tyznik AJ, Kronenberg M, Cheroutre H. Hepatic stellate cells function as regulatory bystanders. J Immunol 2011;186:5549–55.

[101] Hill M, Tanguy-Royer S, Royer P, Chauveau C, Asghar K, Tesson L, et al. IDO expands human $CD4^+CD25^{high}$ regulatory T cells by promoting maturation of LPS-treated dendritic cells. Eur J Immunol 2007;37:3054–62.

[102] Li W, Kuhr CS, Zheng XX, Carper K, Thomson AW, Reyes JD, et al. New insights into mechanisms of spontaneous liver transplant tolerance: the role of Foxp3-expressing $CD25^+CD4^+$ regulatory T cells. Am J Transplant 2008;8:1639–51.

[103] Park MJ, Park KS, Park HS, Cho ML, Hwang SY, Min SY, et al. A distinct tolerogenic subset of splenic IDO(+)CD11b(+) dendritic cells from orally tolerized mice is responsible for induction of systemic immune tolerance and suppression of collagen-induced arthritis. Cell Immunol 2012;278:45–54.

[104] Sharma MD, Baban B, Chandler P, Hou DY, Singh N, Yagita H, et al. Plasmacytoid dendritic cells from mouse tumor-draining lymph nodes directly activate mature Tregs via indoleamine 2,3-dioxygenase. J Clin Invest 2007;117:2570–82.

[105] Sumpter TL, Packiam V, Turnquist HR, Castellaneta A, Yoshida O, Thomson AW. DAP12 promotes IRAK-M expression and IL-10 production by liver myeloid dendritic cells and restrains their T cell allostimulatory ability. J Immunol 2011;186:1970–80.

[106] Yoshida O, Kimura S, Dou L, Matta BM, Yokota S, Ross MA, et al. DAP12 deficiency in liver allografts results in enhanced donor DC migration, augmented effector T cell responses and abrogation of transplant tolerance. Am J Transplant 2014;14:1791–805.

[107] Biswas SK, Lopez-Collazo E. Endotoxin tolerance: new mechanisms, molecules and clinical significance. Trends Immunol 2009;30:475–87.

[108] Viñas O, Bataller R, Sancho-Bru P, Ginès P, Berenguer C, Enrich C, et al. Human hepatic stellate cells show features of antigen-presenting cells and stimulate lymphocyte proliferation. Hepatology 2003;38:919–29.

[109] Kobayashi S, Seki S, Kawada N, Morikawa H, Nakatani K, Uyama N, et al. Apoptosis of T cells in the hepatic fibrotic tissue of the rat: a possible inducing role of hepatic myofibroblast-like cells. Cell Tissue Res 2003;311:353–64.

[110] Mehal WZ, Juedes AE, Crispe IN. Selective retention of activated $CD8^+$ T cells by the normal liver. J Immunol 1999;163:3202–10.

[111] Crispe IN. Hepatic T cells and liver tolerance. Nat Rev Immunol 2003;3:51–62.

[112] Qian S, Lu L, Fu F, Li Y, Li W, Starzl TE, et al. Apoptosis within spontaneously accepted mouse liver allografts: evidence for deletion of cytotoxic T cells and implications for tolerance induction. J Immunol 1997;158:4654–61.

[113] Thai NL, Fu F, Qian S, Sun H, Gao L, Wang SC, et al. Cytokine mRNA profiles in mouse orthotopic liver transplantation. Graft rejection is associated with augmented TH1 function. Transplantation 1995;59:274–81.

[114] Yu MC, Chen CH, Liang X, Wang L, Gandhi CR, Fung JJ, et al. Inhibition of T-cell responses by hepatic stellate cells via B7-H1-mediated T-cell apoptosis in mice. Hepatology 2004;40:1312–21.

[115] Jiang G, Yang HR, Wang L, Wildey GM, Fung J, Qian S, et al. Hepatic stellate cells preferentially expand allogeneic $CD4^+CD25^+FoxP3^+$ regulatory T cells in an IL-2-dependent manner. Transplantation 2008;86:1492–502.

[116] Dunham RM, Thapa M, Velazquez VM, Elrod EJ, Denning TL, Pulendran B, et al. Hepatic stellate cells preferentially induce $Foxp3^+$ regulatory T cells by production of retinoic acid. J Immunol 2013;190:2009–16.

[117] Chen CH, Kuo LM, Chang Y, Wu W, Goldbach C, Ross MA, et al. *In vivo* immune modulatory activity of hepatic stellate cells in mice. Hepatology 2006;44:1171–81.

[118] Yang HR, Chou HS, Gu X, Wang L, Brown KE, Fung JJ, et al. Mechanistic insights into immunomodulation by hepatic stellate cells in mice: a critical role of interferon-gamma signaling. Hepatology 2009;50:1981–91.

[119] Iredale JP. Hepatic stellate cell behavior during resolution of liver injury. Semin Liver Dis 2001;21:427–36.

[120] Schildberg FA, Wojtalla A, Siegmund SV, Endl E, Diehl L, Abdullah Z, et al. Murine hepatic stellate cells veto CD8 T cell activation by a CD54-dependent mechanism. Hepatology 2011;54:262–72.

[121] Chinnadurai R, Grakoui A. B7-H4 mediates inhibition of T cell responses by activated murine hepatic stellate cells. Hepatology 2010;52:2177–85.

[122] Marek-Trzonkowska N, Myśliwec M, Siebert J, Trzonkowski P. Clinical application of regulatory T cells in type 1 diabetes. Pediatr Diabetes 2013;14:322–32.

[123] Michels-van Amelsfort JM, Walter GJ, Taams LS. $CD4^{+}CD25^{+}$ regulatory T cells in systemic sclerosis and other rheumatic diseases. Expert Rev Clin Immunol 2011;7:499–514.

[124] Venken K, Hellings N, Liblau R, Stinissen P. Disturbed regulatory T cell homeostasis in multiple sclerosis. Trends Mol Med 2010;16:58–68.

[125] Cabrera R, Tu Z, Xu Y, Firpi RJ, Rosen HR, Liu C, et al. An immunomodulatory role for CD4(+) CD25(+) regulatory T lymphocytes in hepatitis C virus infection. Hepatology 2004;40:1062–71.

[126] Ward SM, Fox BC, Brown PJ, Worthington J, Fox SB, Chapman RW, et al. Quantification and localisation of $FOXP3^{+}$ T lymphocytes and relation to hepatic inflammation during chronic HCV infection. J Hepatol 2007;47:316–24.

[127] Li W, Carper K, Zheng XX, Kuhr CS, Reyes JD, Liang Y, et al. The role of $Foxp3^{+}$ regulatory T cells in liver transplant tolerance. Transplant Proc 2006;38:3205–6.

[128] He F, Chen Z, Xu S, Cai M, Wu M, Li H, et al. Increased $CD4^{+}CD25^{+}Foxp3^{+}$ regulatory T cells in tolerance induced by portal venous injection. Surgery 2009;145:663–74.

[129] Sun CM, Hall J, Blank RB, Bouladoux N, Oukka M, Mora JR, et al. Small intestine lamina propria dendritic cells promote *de novo* generation of Foxp3 T reg cells via retinoic acid. J Exp Med 2007;204:1775–85.

[130] Denning TL, Wang YC, Patel SR, Williams IR, Pulendran B. Lamina propria macrophages and dendritic cells differentially induce regulatory and interleukin 17-producing T cell responses. Nat Immunol 2007;8:1086–94.

[131] Coombes JL, Siddiqui KRR, Arancibia-Carcamo CV, Hall J, Sun CM, Belkaid Y, et al. A functionally specialized population of mucosal $CD103^{+}$ DCs induce $Foxp3^{+}$ regulatory T cells via aTGF-b and retinoic acid-dependent mechanism. J Exp Med 2007;204:1757–64.

[132] Wang J, Han WG, Foks AC, Huizinga TW, Toes RE. Neutralization of IL-4 reverses the nonresponsiveness of $CD4^{+}$ T cells to regulatory T-cell induction in non-responder mouse strains. Mol Immunol 2010;48:137–46.

[133] Meyer DH, Bachem MG, Gressner AM. Modulation of hepatic lipocyte proteoglycan synthesis and proliferation by Kupffer cell-derived transforming growth factors type beta 1 and type alpha. Biochem Biophys Res Commun 1990;171:1122–9.

[134] Karrar A, Broome U, Uzunel M, Qureshi AR, Sumitran-Holgersson S. Human liver sinusoidal endothelial cells induce apoptosis in activated T cells: a role in tolerance induction. Gut 2007;56:243–52.

[135] Tang Q, Bluestone JA. Regulatory T-cell therapy in transplantation: moving to the clinic. Cold Spring Harb Perspect Med 2013;3(11).

[136] Trzonkowski P, Dukat-Mazurek A, Bieniaszewska M, Marek-Trzonkowska N, Dobyszuk A, Juścińska J, et al. Treatment of graft-versus-host disease with naturally occurring T regulatory cells. BioDrugs 2013;27:605–14.

[137] Singer BD, King LS, D'Alessio FR. Regulatory T cells as immunotherapy. Front Immunol 2014;5:46.

[138] Wiegard C, Frenzel C, Herkel J, Kallen KJ, Schmitt E, Lohse AW. Murine liver antigen presenting cells control suppressor activity of $CD4^{+}CD25^{+}$ regulatory T cells. Hepatology 2005;42:193–9.

[139] MacDonald KG, Orban PC, Levings MK. T regulatory cell therapy in transplantation: stability, localization and functional specialization. Curr Opin Organ Transplant 2012;17:343–8.

[140] Kronenberg M, Gapin L. The unconventional lifestyle of NKT cells. Nat Rev Immunol 2002;2:557–68.

[141] Zhou D, Mattner J, Cantu III C, Schrantz N, Yin N, Gao Y, et al. Lysosomal glycosphingolipid recognition by NKT cells. Science 2004;306:1786–9.
[142] Mattner J, Debord KL, Ismail N, Goff RD, Cantu C, Zhou DP, et al. Exogenous and endogenous glycolipid antigens activate NKT cells during microbial infections. Nature 2005;434:525–9.
[143] Kinjo Y, Wu D, Kim GS, Xing GW, Poles MA, Ho DD, et al. Recognition of bacterial glycosphingolipids by natural killer T cells. Nature 2005;434:520–5.
[144] Brigl M, Brenner MB. CD1: antigen presentation and T cell function. Annu Rev Immunol 2004;22:817–90.
[145] Winau F, Hegasy G, Weiskirchen R, Weber S, Cassan C, Sieling PA, et al. Ito cells are liver-resident antigen-presenting cells for activating T cell responses. Immunity 2007;26:117–29.

CHAPTER 15

Stellate Cell Depletion Models

Fiona Oakley and Derek A. Mann
Fibrosis Research Group, Institute of Cellular Medicine, Newcastle University, Newcastle upon Tyne, UK

15.1 INTRODUCTION

Quiescent hepatic stellate cells (qHSCs), also known as liver pericytes, Ito cells, or fat storing cells, reside in the space of Disse between the blood vessel and the hepatocytes. HSCs form part of the non-parenchymal cell compartment and comprise approximately 8–12% of the liver cell population [1]. The qHSCs are phenotypically characterized by storage of vitamin A, expression of lipogenic genes, desmin and the cell surface markers vimentin and glial fibrilary acidic protein (GFAP) [2,3].

In acute liver injury the HSCs become activated to hepatic myofibroblasts (HMs) via a process known as transdifferentiation. In their activated state HMs are highly proliferative and contractile cells that synthesize a scar matrix comprised of collagens, fibronectin, and laminin, secrete numerous fibrogenic molecules such as transforming growth factor beta (TGFβ) and connective tissue growth factor (CTGF) as well as inflammatory stimuli [2,4]. The physiological consequence of HM activation is to promote normal wound healing and tissue repair. However, repetitive epithelial injury and liver damage provide survival signals for HMs, causing these cells to persist and continue to deposit collagenous scar tissue. Over time this excess ECM deposition causes fibrosis and can progress to cirrhosis, which can cause organ failure or an increased susceptibility to liver cancer [2,4]. For many years it was presumed that fibrosis was essentially a "one-way" progressive disease and that the "best prognosis" of a successful therapy would be to halt the fibrotic response. However, studies in rodent models and later in clinical trials revealed that fibrosis is a dynamic process and that cessation of injury or effective treatment of the underlying cause of disease was associated with active remodeling of the scar matrix, indicative of resolution of fibrosis [5–7].

The first mechanistic insights into the cellular control of the fibrolytic response were described in 1998 when HM apoptosis was described as a key event contributing to the resolution of fibrosis [8]. Using the reversible carbon tetrachloride (CCl_4) model to induce central-lobular fibrosis in rats, it was shown that after cessation of liver injury, HM apoptosis or programmed cell death preceded the spontaneous resolution of the scar tissue. Loss of the scar-forming cell population removed the primary collagen-producing cell and the cellular source of proteins that prevent matrix

Stellate Cells in Health and Disease
DOI: http://dx.doi.org/10.1016/B978-0-12-800134-9.00015-4

degradation; tissue inhibitors of matrix metalloproteases (TIMP). Stimulation of HM apoptosis therefore provides a mechanism to promote fibrolysis and potentially the reversion of liver disease. However, it is important to only target HMs and not qHSCs for depletion because they play an important role in liver homeostasis, normal wound healing, and are the major source of vitamin A storage in the body.

15.2 IMMUNE CELL MEDIATED CLEARANCE OF HM

The development of liver fibrosis is complex and involves multiple interactions between liver cells and the immune system; therefore it is possible that inflammatory cells can influence fibrogenic processes. Kupffer cells (KCs), the resident liver macrophage, was reported to directly kill HMs in a Caspase 9 (an effecter Caspase in the apoptotic pathway) and receptor interacting protein (RIP) dependent manner [9]. However, it was independent of CD95L (FAS ligand) or TNFα. KC mediated HM cell death required direct cell contact and stimulation of KC with lipopolysaccharide (LPS). In an *in vitro* KC:HM co-culture system LPS treatment induced up to 90% apoptosis of HMs by KCs over a 48 h co-culture period. However, this could be blocked by dexamethasone, prostaglandin E2 (PGE2), and tumor necrosis factor-related apoptosis-inducing ligand (TRAIL) antagonists or by suppression of RIP. The authors reported that LPS primed KC released soluble factors, which stimulated KC dependent HM killing. LPS was ineffective at directly inducing HM apoptosis in monocultures, which led to the conclusion that sensitization of HM to apoptosis by LPS was not required for KC-induced HM cell death, rather LPS primes KC to release a soluble factor that stimulates HM apoptosis.

Natural killer (NK) cells comprise approximately 10–20% of hepatic lymphocytes [10] and therefore could potentially influence the fibrogenic process. NK cells were identified as a potential population of cells to eliminate HMs and exert anti-fibrotic effects. Administration of polyinosinis-polycytidylic acid (poly IC) to mice, which mimics viral induced innate immune activation, caused activation of NK cells, and release of interferon gamma (IFNγ) [11]. Specific depletion of NK cells with an anti-asialo GM-1 (ASGM1) antibody caused a significant increase in liver fibrosis in the 3,5-diethoxycarbonyl-1,4-dihydrocollidine (DDC) dietary model. Administration of poly IC to DDC fed mice ameliorated fibrosis and reduced levels of α-smooth muscle actin (α-sma) positive HMs. However, this protective effect was lost when NK cells were depleted with anti-ASGM1. The importance of IFNγ expression was highlighted by the observation that IFNγ null mice were more susceptible to liver fibrosis. Administration of poly IC increased the number of dual α-sma/TUNEL positive HM and reduced fibrosis in wild type but not IFNγ deficient mice, suggesting that a reduction in HM apoptosis contributed to the enhanced fibrosis observed in IFNγ null mice. Importantly, NK cells were shown to selectively target HMs and not hepatocytes *in vitro*. Moreover, NK cells did not target qHSCs and the efficiency of HM killing could be increased by stimulation with poly IC. Further investigation revealed that NK cells killed HMs via a specific interaction between NK group 2, member D (NKG2D) and its ligand retinoic acid early inducible gene 1 (RAE1). RAE1 is highly expressed by HMs but absent in qHSCs, suggesting that only HMs are susceptible to the cytotoxic actions of NK cells. In HMs retinoic acid signaling regulates RAE1 expression and blocking retinoic acid production or signaling via the retinoic acid receptor inhibited RAE1 expression. Furthermore, NK cells express high levels of TRAIL and HM killing by NK cells was suggested to require TRAIL [12]. It is important to note that NKG2D is also expressed on $CD8^+$ and γδT cells and as such may provide an additional level of control [13].

A major NK activating receptors is NKp46 and it has been shown to play a role in limiting liver fibrosis [14]. Murine HMs highly express currently unknown ligands of the NCR1 receptor (the murine orthologue of the NKp46 receptor) and deletion of NCR1 in mice was associated with increased numbers of α-sma+ HMs and Sirius Red staining in response to CCl_4 injury. To prove that NK cell killing of HMs was via NKp46, the authors generated an NKp46 blocking antibody. Pre-incubation of NK with the NKp46-Ig significantly reduced NK cell mediated HM killing. Human HMs express the ligand for NKp46 and their NK mediated apoptosis is NKp46 dependent.

Chronic alcohol can suppress NK cell mediated HM apoptosis [15]. Mice were fed an ethanol diet or a control diet for 8 weeks and then injured with CCl_4 for 2 weeks while continuing on the same diet. Increased fibrosis and reduced HM apoptosis was observed in ethanol-fed mice compared to the control diet group. The authors reported that ethanol feeding reduced expression of NKG2D and TRAIL on NK cells and that this inhibited NK cell mediated killing of HMs.

15.3 HM INTRINSIC SIGNALING EVENTS THAT LIMIT SCAR CELL SURVIVAL

The regulatory mechanisms controlling HM apoptosis and clearance after cessation of liver injury are still unclear, but nerve growth receptor (NGF) signaling has been identified as one potential mechanism [16]. Upon activation HSC derived HMs express the low affinity NGF receptor p75. In acute CCl_4 injury the peak in hepatic p75 levels correlate with the peak in HM apoptosis. Treatment of HMs with NGF dose dependently induced Caspase-dependent cell death. This response was associated with a diminution of NF-κB activity, which is known to regulate HM survival. Hence, NGF released during fibrotic liver injury binds to the p75 receptor and suppresses NF-κB activation, which in turn stimulates HM apoptosis and resolution of fibrosis. NGF may offer an interesting therapeutic approach if these effects can be replicated in human chronic liver disease.

As liver injury progresses, senescent HMs accumulate in the fibrotic liver and can limit fibrosis. Compared to highly proliferative activated HMs, senescent HMs exhibit reduced production of extracellular matrix proteins and TGFβ [17]. HMs in mice lacking the senescence regulators p53 and INK4/ARF fail to senesce and these mice develop more fibrosis upon CCl_4 challenge compared to control mice. NK cells were found to kill senescent human HMs more effectively than non-senescent cells *in vitro*. In the chronic CCl_4 model deletion of NK cells using the ASGM1 depleting antibody resulted in accumulation of senescent HMs and increased fibrosis compared to IgG treated mice. Conversely, stimulation of the immune system with Poly IC and NK cell activation via IFNγ stimulated removal of HMs and resolution of fibrosis.

15.4 CHEMICAL SYSTEMS

In 2001 a seminal paper was published showing that fibrolysis could be dramatically enhanced by chemically inducing HM apoptosis [18]. Inducing HM apoptosis with gliotoxin (GT) dramatically accelerated remodeling of scar matrix in CCl_4 injured rodents. The authors reported that the fungal metabolite GT efficiently promoted apoptosis of both rodent and human HMs *in vitro*. However, the effect of GT on hepatocytes was much less potent [19]. A single dose of GT was sufficient to induce HM apoptosis *in vivo* within 24 h, this being associated with a reduction in liver fibrosis as quantified by Sirius Red and α-sma staining. Since this initial report, numerous compounds and drugs targeting a plethora of receptors, enzymes, and signaling molecules have been used to induce HM apoptosis and shown to exert anti-fibrotic effects in rodent models of chronic liver disease [20].

Cannabinoid 1 (CB1) receptors are expressed on endothelial cells and HMs in the liver and signaling via the CB1 receptor was shown to be profibrogenic [21]. Conversely, they reported that signaling via Cannabinoid 2 (CB2) receptors on HMs and KCs is anti-inflammatory and limits fibrosis. Antagonism of CB1 receptors with SR141716A or deletion of CB1 in mice reduced fibrosis in the bile duct ligation (BDL), thioacetamide (TAA) and CCl_4 models of liver injury. CB1 antagonism also has beneficial effects on lipid metabolism and rimonabant, a CB1 receptor antagonist, was approved for clinical use in overweight patients and those with cardio-metabolic risks [22]. However, because rimonabant crosses the blood–brain barrier it was withdrawn due to the high rate of adverse effects of depression and mood [22]. Nonbrain penetrable CB1 antagonists and CB2 agonists are likely to be the focus of future drug discovery programs in this area.

Blockade of the renin–angiotensin system (RAS) using either angiotensin converting enzyme inhibitors (ACE) or AT1R blockers (losartan) can induce apoptosis of rodent and human HMs and is anti-fibrotic in rodents [23–27]. Two promising clinical trials have shown some promising anti-fibrotic effects of angiotensin blockers. The first was a small pilot study where a decrease in fibrosis stage was observed in 7/14 patients versus 1/9 control patients with hepatitis C infection that failed to respond to antiviral therapy over 6 months of losartan treatment [28]. The second study discovered that ~50% patients with hepatitis C infection that failed to respond to antiviral therapy regressed one fibrosis grade over 18 months of losartan therapy [29]. Larger-scale losartan trials in liver diseases of different etiologies in multiple centers are currently ongoing. However, it is important to consider that the RAS has effects outside of the liver, including the homeostatic maintenance of blood pressure as well as effecting cardiac and renal function. Drugs targeting the RAS are well tolerated and are safe for long periods in patients with heart and kidney disease and while the RAS system is an extremely promising anti-fibrotic target for liver disease it remains to be proven that losartan will be an effective long-term treatment for liver fibrosis.

One pathway regulating HM survival is signaling via the transcription factor NF-κB that is persistently upregulated in HMs and is critical for their survival [30]. Directly inhibiting NF-κB signaling using an IκBα super repressor or sulphasalazine, an IKK inhibitor, induced apoptosis of rodent and human HMs [31,32]. *In vivo*, administration of sulphasalazine to rats with established fibrosis selectively promoted clearance of HMs and accelerated fibrolysis. Intriguingly, in HMs the inhibitory checkpoint of the NF-κB signaling pathway is lost during HM activation as IκBα gene expression is switched off, which results in high levels of constitutively active RelA:p50 subunit dimers [33,34]. This was a puzzling discovery because IKK inhibitors including sulphasalazine induce HM apoptosis and accelerate fibrolysis. One explanation was that in addition to phosphorylating IκBα, active IKK's phosphorylate other proteins in the NF-κB signaling pathway. Of particular interest was IKKβ dependent phosphorylation of RelA at serine 536 (P-Ser536-RelA), a modification required for efficient RelA nuclear import. P-Ser536-RelA is a feature of rodent and human HMs but not other liver cells and treatment of HMs with a cell permeable P-Ser536-RelA competing peptide induced their apoptosis [24,35]. Administration of the peptide *in vivo* in the chronic CCl_4 and methionine choline deficient diet models was anti-fibrotic [35]. Unlike classic IKK inhibitors, the P-Ser536-RelA cell permeable peptide did not affect the hepatic inflammatory response after LPS challenge. Targeting protein modifications unique to HMs but only transient in other liver cells and organs may provide an improved drug targeting strategy.

15.5 DEVELOPMENT OF CARRIERS TO SELECTIVELY TARGET THERAPEUTICS TO HMs

Currently there are no effective anti-fibrotic drugs approved to treat liver fibrosis. This is likely to be because liver disease is complex, it develops over many years, and involves multiple cell:cell interactions, therefore the "perfect therapeutic" would selectively target HMs and efficiently promote their apoptosis but have negligible effects on other liver cells such as hepatocytes, biliary cells, and endothelial cells, other organs or inflammation and innate immunity. Achieving sufficient drug efficacy to sustain an anti-fibrotic response in situations of ongoing injury but minimizing adverse effects has been difficult to accomplish and has limited the ability of researchers and industry to successfully translate new anti-fibrotic drugs to the clinic. In an attempt to overcome some of these hurdles a number of research groups have developed novel strategies to selectively target drugs or siRNA to HMs to either modulate their function or induce their apoptosis. Another advantage of developing systems to selectively target HSC or develop genetic models to manipulate the expression of genes in HMs is to advance our mechanistic understanding of the different signaling pathways and proteins that control HM activation and ask how they contribute to disease pathogenesis. This molecular interrogation will identify new therapeutic targets and unravel the complexities of fibrotic disease. HMs are phenotypically characterized by their loss of retinol stores, downregulation of lipogenic genes but induction of α-sma and cell surface markers such as platelet derived growth factor receptor beta (PDGFRβ), insulin-like growth factor II receptor (IGFR II) and synaptophysin [3]. When designing cell-targeting strategies it is these phenotypic changes that occur during HSC to HM transdifferentiation that many researchers are exploiting to selectively target and deplete the scar-forming myofibroblasts to treat liver fibrosis.

15.6 ADVANTAGES OF HM TARGETED DELIVERY VEHICLES FOR THERAPY

1. Improve drug efficacy and lower the dose of compound required to achieve a therapeutic effect.
2. Extend compound half-life and reduce dose frequency.
3. Reduce adverse effects.

Administration of Gliotoxin or Sulphasalazine in the rat CCl_4 model accelerates resolution of fibrosis [18,32]. However, the long-term use of pro-apoptotic drugs is likely to be associated with undesirable side effects due to off target effects. To overcome this problem and achieve cell specific targeting of this pro-apoptotic compound novel technologies have been developed to selectively deliver drugs to HMs. The first approach was to exploit the fact that HMs express significantly higher levels of the protein synaptophysin on their cell membrane compared to qHSCs [36]. Phage display was used to generate a human recombinant single-chain antibody (scAb), C1–3 that binds to synaptophysin on the surface of HMs [37]. Fluorescent labeled C1–3 was shown to co-localize with α-sma+ HMs but not ED1+ macrophages or hepatocytes. Synaptophysin is also expressed on neuronal cells and is involved in synaptic vesicle exo-endocytosis. C1–3 was chemically conjugated to GT to produce C1–3-GT. *In vitro* experiments revealed that C1–3-GT was taken up by HM and efficiently induced their apoptosis. Bio-distribution using fluorescent labeled C1–3 revealed that C1–3 was not retained in normal livers but could be detected in fibrotic livers suggesting it could be used to selectively deliver GT to HMs. GT was chemically conjugated to the C1–3 scAb to generate C1–3-GT [38]. Administration of C1–3-GT or GT alone in the presence of established fibrosis and ongoing liver injury resulted in a reduction in α-sma+ HMs and fibrosis suggesting that C1–3 can be used to selectively target HMs *in vivo*. When compared directly, injection of GT depleted liver myofibroblasts by 30% and monocytes/macrophages by 50% but did not affect fibrosis in the sustained injury model. By contrast, administration of C1–3-GT decreased HM numbers by 60%, significantly reduced fibrosis severity but did not affect monocytes/macrophage numbers. One caveat to note is that the C1–3 technology is severely limited by drug-antibody conjugation chemistry.

The second system used to target GT to HMs utilized a mannose-6-phosphate modified human serum albumin (M6P-HSA) carrier [39], which binds the IGFR II, a protein highly upregulated on HSC during their transdifferentiation to activated HMs. M6P-HSA had previously been shown to be internalized by HMs via endocytosis and to rapidly accumulate in fibrotic rat liver when given by intravenous injection. M6P-HSA coupled the pro-apoptotic compound GT to M6P-HSA to generate GTX-M6P-HSA [40]. High performance liquid chromatography (HPLC) confirmed that the carrier to drug ratio was 0.4:1 and *in vitro* studies revealed that GTX-M6P-HSA selectively bound and reduced the viability of human HMs. Furthermore, in precision cut slices generated from rats with established liver fibrosis, GTX-M6P-HSA reduced HM numbers but did not affect the survival of other liver cells [40]. Finally, administration of GTX-M6P-HSA reduced α-sma+ HMs in the BDL model, suggesting that this targeting system achieves therapeutic effects in both the CCl_4 induced central-lobular fibrosis as well as the BDL model of periportal fibrosis.

M6P-HSA carriers could also be used to deliver drugs to HMs that target three key fibrogenic molecules/pathways: TGFβ/RAS [41–43]. Pharmacological blockade of the TGFβ type-1 receptor (ALK5) using small molecule inhibitors such as LY-364947 had previously been reported to suppress TGFβ1 and expression of fibrogenic genes in HMs [44]. Inhibiting the TGFβ1 pathway offers a strong rationale as an anti-fibrotic drug strategy, but it also plays key roles in immune cell activation and cellular differentiation; long-term blockade may have side effects [45]. LY-364947 coupled to M6P-HSA (LY-M6PHSA) was compared to the free drug form in its effectiveness to block TGFβ1 signaling and to act as an anti-fibrotic [41]. LY-M6PHSA reduced TGFβ1 induced activation of SMAD2 phosphorylation in human HMs as well as α-sma gene expression. Dual staining for HSA and α-sma+ HMs showed co-localization whereas $CD68^+$ KCs and $CD31^+$ endothelial cells were negative for HSA. In an acute CCl_4 injury model, LY-M6PHSA did not affect liver injury shown by similar elevation of liver transaminases but did significantly reduce collagen expression and numbers of α-sma+ HMs. These data suggest that LY-M6PHSA has potent anti-fibrotic activities in CCl_4 induced liver injury and is likely to be more selective in its therapeutic actions than free LY-364947.

The AT1 receptor blocker losartan was also attached to M6PHSA via a platinum linker [43]. Losartan-M6PHSA efficiently targeted HMs in BDL rats and IHC staining showed co-localization of HSA and losartan-M6PHSA. Daily administration of losartan-M6PHSA in the final 3 days of the model in BDL or CCl_4 injured rats with established fibrosis reduced α-sma+ HMs, Sirius red staining, and fibrotic gene expression. Anti-fibrotic effects were not observed with M6PHSA alone or 125 μg/kg of oral losartan.

Rho kinase signaling regulates many cellular processes including cytoskeletal organization, migration and HM activation [46]. The Rho-kinase small molecule inhibitor Y29632 reduced the fibrogenic potential of HMs and administration of the compound in a rat liver fibrosis model was fibro-protective [47]. However, Rho kinases are ubiquitously expressed and are key regulators of vascular homeostasis and blood pressure. Y29632 was coupled to the M6P-HSA carrier to generate Y29632-M6PHSA [42]. In rodent studies, doses of 30 mg/kg free Y29632 were required to achieve therapeutic effects; however, the effective dose was significantly reduced using M6PHSA carriers. Administration of Y29632:M6PHSA at 1.5 mg/kg was sufficient to reduce the number of αSMA+ HMs. Because no difference in desmin positive cells was observed, it was hypothesized that the inhibitor was preventing activation of HSC to HM; however, no data were provided to show that Rho kinases are inhibited in HSC *in vivo*.

A second system that targets HMs is the cyclic peptide (C*GRGDSPC*, where C* denoted the cyclizing cysteine), which binds to collagen IV. It was conjugated to HSA, yielding pCVI-HAS [48]. This complex targeted HMs *in vitro* and was internalized. When tested *in vivo* it was shown to bind HMs in fibrotic tissue but mainly associate with endothelial cells in normal rat liver.

IFNγ is a potent antiviral cytokine that has immunosuppressive, anti-proliferative, and anti-fibrotic activities [49]. However, because of its short plasma half-life, limited efficacy, and side effects IFNγ has limited therapeutic application in chronic liver disease. Many groups have tried to chemically attach IFNγ to nanoparticles or microspheres to extend its half-life or entrap IFNγ in liposomes to allow slow release, but these approaches have not been successful [50,51]. To reduce its adverse side effects and selectively target IFNγ to HM, IFNγ was chemically linked to a cyclic peptide (*CSRNLIDC*) that recognizes the PDGFRβ, either directly or via a peg linker [52]. These molecules were named PPB-IFNγ and PPB-Peg-IFNγ, respectively. PDGFRβ is highly expressed on HMs in culture and within areas of active fibrosis in diseased mouse and human tissue. Free IFNγ does not bind HMs in culture whereas PPB-modified IFNγ carriers display high affinity for HMs. When administered in an acute CCl_4 injury model in mice, both PPB-IFNγ and PPB-Peg-IFNγ accumulated in the injured liver within 10 min and histological staining of the *ex vivo* liver confirmed co-localization with desmin+ HMs. However, only the PPB-Peg-IFNγ moiety exerted anti-fibrotic activities, promoting a significant reduction in collagen production, TIMP1 levels, and α-sma+ HMs and an increase in the matrix degrading enzyme MMP-13. To ascertain whether PPB-Peg-IFNγ could reduce established fibrosis in chronic liver injury, mice were administered six doses of PBS vehicle, free IFNγ, or PPB-Peg-IFNγ in the final 2 weeks of an 8-week CCl_4 injury model. The vehicle group developed extensive bridging fibrosis, which was reduced by 70% in the PPB-Peg-IFNγ treated group. The fibroprotective effects of PPB-Peg-IFNγ were associated with reduced liver injury (transaminases) and α-sma+ HMs. HM proliferation was impaired by PPB-Peg-IFNγ, suggesting that this is the mechanistic basis of the fibroprotective activities of PPB-Peg-IFNγ. To determine if modification of IFNγ moieties reduced the pro-inflammatory side effects of IFNγ, body temperature, endothelial cell activation (eNOS), and serum cytokines IL-6 and TNFα were quantified. All markers were elevated in the Peg-IFNγ treated mice but levels were comparable between the control and PPB-Peg-IFNγ groups. In conclusion, HM-specific cell targeting of IFNγ was anti-fibrotic and minimized the side effects associated with IFNγ therapy.

Additional methods to modify PPB-Peg-IFNγ were then developed to improve delivery. The first strategy was to conjugate PPB-Peg-IFNγ to HSA to generate PPB-HSA-Peg-IFNγ [53]. Addition of HSA to PPB-Peg-IFNγ did not affect its binding to HMs *in vitro* and retained IFNγ signaling (STAT1 phosphorylation) and the ability to induce IFNγ dependent genes. The bioavailability of IFNγ and the anti-fibrotic actions of PPB-HSA-Peg-IFNγ were preserved in an acute model of CCl_4 injury in mice. A second approach was to synthesize a bicyclic peptide (CSRNLIDC-GGDGG-CSRNLIDC) called BiPPB [54]. The BiPPB molecule contains a C"C disulfide bond and GGDGG linker region, which acts as a spacer permitting the binding of BiPPB to dimeric PDGFRβ receptor complexes with greater affinity. This is because one peptide interacts with one half of the PDGFRβ dimer and the other peptide binds the second receptor in the complex. BiPPB was chemically conjugated to Peg-IFNγ to form mimγ-BiPPB and then tested in both acute and chronic CCl_4 models in mice. Administration of mimγ-BiPPB reduced α-sma+ HMs without inducing inflammation in the liver or other organs. Conversely, administration of the mim-peg-IFNγ induced significant infiltration of neutrophils, macrophages, and dendritic cells to the chronically injured liver, as well as other organs including the kidney and lung. In a more recent study deoxyribonucleic acid (DNA) constructs were generated that encode either IFNγ, mimIFNγ, and mimγ-BiPPB fusion proteins, which can be expressed in *E. coli* [55]. The mimγ-BiPPB fusion protein also bound HMs and exhibited anti-fibrotic activities in acute and chronic CCl_4 models in mice. Stimulation of murine macrophages with IFNγ induced STAT1 phosphorylation and nitric oxide production; however, these effects were ablated by conjugation of IFNγ to a carrier.

An alternative approach was to entrap IFNγ inside sterically stable liposomes (SSL) coated with the cyclic RGD peptides that target collagen IV [48] to selectively deliver IFNγ to HMs *in vivo* and *in vitro* [56]. The cRGD-SSL coated liposomes had ten fold higher affinity *in vitro* than liposomes alone (SSL). Bio-distribution studies proved that cRGD-SSL accumulated in HMs in BDL rats and that this was associated with a reduction in fibrosis, hydroxyl-proline, and collagen gene expression. Taken together these elegant studies demonstrate that HM-specific cell targeting carriers can reduce the unwanted side effects associated with IFNγ but retain, and potentially improve, its anti-fibrotic efficacy.

15.7 NANOPARTICLES AND VIRAL DELIVERY SYSTEMS

Delivering siRNA to HMs *in vivo* would allow researchers to modify HM gene expression and potentially influence the fibrogenic process. Polymers with *N*-acetylglucosamine (GlcNAc) have a strong affinity for vimentin and desmin [57]. The modified bio-reducible polymer polyethylenimine (PEI) with GlcNAc to develop PEI-D-GlcNAc, a low molecular weight carrier, was then conjugated to an siRNA directed against TGFβ1 [57,58]. The PEI-D-GlcNAc-SiRNA complex efficiently targeted HMs in culture and was subsequently internalized. Western blot of PEI-D-GlcNAc-siRNA treated cells confirmed knockdown of TGFβ and an associated reduction of α-sma levels compared to HMs treated with PEI-D-SiRNA only. Optical imaging was used to track an indocyanine green modified complex (PEI-D-GlcNAc-ICG-SiRNA) *in vivo*. The complex was retained in CCl_4 injured liver compared to normal liver and desmin staining revealed that up to 79% of desmin+ cells were also positive for the PEI-D-GlcNAc-ICG-SiRNA. This technology efficiently targets HMs and has been shown to modulate their gene expression and function; future work will determine if it will be a useful tool to help dissect the complex molecular events controlling HM phenotype *in vivo*.

The alkaloid oxymatrine (OM) has anti-fibrotic properties, but has a short half-life and undesirable adverse effects to HMs [59]. To selectively target OM, a polymer approach was employed. Polymersomes are low molecular weight, self-assembling polymeric vesicles based on poly(ethylene glycol)-b-ploy(ε-caprolactone) or PEG-b-PCL (PM). OM was entrapped in polymersomes coated with the previously described cyclic RGD peptides, which target collagen IV to yield RGD–PM–OM [48]. The RGD–PM–OM complex suppressed HM proliferation and reduced α-sma and collagen gene expression *in vitro*. When tested *in vivo* this formulation reduced BDL-induced fibrosis in rats when compared to OM or PM–OM only. In a follow-up study, RGD–PM–OM liposomes were shown to exhibit anti-fibrotic effects in the CCl_4 model in mice [60].

Adenoviral vectors can be used to efficiently deliver genes to cells via coxsackie and adenovirus receptor (CAR) receptors, but their lack of cell specificity makes them poor targeting systems. Specifically, in the liver adenoviral vectors have displayed considerable tropism for hepatocytes and KCs. To overcome these short falls, two unique strategies were developed to selectively direct adenoviral vectors to HMs. The first approach requires two adenoviral constructs, namely AxAw and AxCAL. The AxAw adenovirus expresses Cre under the control of a HM cell specific promoter (17COL (COL 1A2), desmin or GFAP) or CAG, a synthetic promoter that drives high-level gene expression in all mammalian cells as a positive control [61]. The second construct, AxCAL contained a LoxP flanked "snuffer" region upstream of an expression unit, which encodes for the gene of interest such as GFP, LacZ, or Smad7. Upon co-infection of both adenoviruses in a cell driving 17COL, desmin, GFAP, or CAG promoters, Cre is expressed and recombination removes the snuffer region allowing expression of the chosen gene. When rat qHSCs and activated HMs were co-infected with either desmin-Cre or GFAP-Cre AxAw adenovirus and the AxCAL GFP adenoviral reporter, only a small number of cells were positive for GFP. However, co-infection with AxAw 17COL-Cre induced modest GFP expression in quiescent cells but high expression in HMs and no expression in the hepatocellular carcinoma (HCC) cell lines HuH7 and HepG2. As expected, the AxAw CAG-Cre induced high-level GFP expression in all of the cell lines. Co-administration of AxAw 17COL-Cre and AxCAL GFP adenovirus to BDL rats caused up to 50% of HMs to be GFP positive, suggesting that the adenoviral system was able to target HMs *in vivo*. Next, the AxAw 17COL-Cre adenovirus was used to induce expression of AxCAL Smad7 adenovirus, an inhibitor of Smad signaling in Lx2 cells. Western blot confirmed that Smad7 expression was upregulated and that this correlated with a reduction in Collagen 1A2 gene expression. When tested in BDL or TAA rats, adenoviral induced Smad7 overexpression was detected in HMs and was associated with a decrease in Sirius Red staining and collagen 1A2 gene expression.

Finally, a system where the AxCAL expressed thymidine kinase (LNL-Tk), which renders cells susceptible to Ganciclovir (GCV) treatment, was developed. GCV induced cell death in HMs co-infected with AxAw 17COL-Cre or CAG-Cre with LNL-Tk adenovirus. To demonstrate *in vivo* efficacy, rats were injured with TAA for 4 weeks and then given one injection of 17COL-Cre and LNL-Tk adenovirus; this was followed by GCV daily concurrently with TAA for 2 additional weeks. Livers of GCV treated animals had less fibrosis, shown by reduced α-sma and Sirius Red area analysis compared to control rats.

The second strategy was to generate a fusion protein that had high affinity for both the adenovirus and the PDGFRβ to retarget the virus to HMs [62]. The PDGFRβ peptide CSRNLIDC was cloned upstream of a single chain antibody fragment (S11) raised against the adenoviral knob. The PDGF-S11 fusion protein increased adenoviral gene transfer of luciferase or GFP up to 60-fold into primary rat HMs but was unable to infect hepatocytes. Conversely, a scrambled peptide that doesn't bind PDGFRβ fused to S11 had no effect on gene transfer in HMs or hepatocytes. While this reagent showed promise *in vitro* it was not tested *in vivo*. One note of caution when using adenoviral systems: it is important to recognize that it is currently not possible to give repeated doses of adenovirus because the virus triggers an immune response.

Herpesvirus saimari (HVS) based vectors can stably transduce dividing cells for long periods of time both *in vitro* and *in vivo*. Rat HMs can be transduced with high efficiency (~45%) by an HVS using an HVS-GFP vector [63]. When administered *in vivo*, bioluminescent tracking of a HVS vector expressing luciferase showed long-term maintenance of up to 10 weeks in the liver. Immuno-histochemical staining for luciferase or open reading frame 73 (ORF73) of the HVS genome was located in both hepatocytes and HMs. Luciferase activity was also detected in the spleen, lung, and heart albeit at levels around 10–20% of the liver activity.

15.8 GENETIC SYSTEMS

Advances in genetic mouse model technology using the Cre-Lox systems have made cell-specific deletion of gene sequences possible. Such systems are needed to be able to interrogate the molecular mechanisms driving the fibrogenic activities of HMs. For many years these systems have been available for most liver cell types including hepatocytes, cholangiocytes, and macrophages. However, genetic models or "floxed" mice that permitted HM specific depletion or manipulate the HM gene expression profiles have been unavailable until recently.

In 2013, a transgenic mouse model that permitted selective depletion of HMs was published. The mouse expresses the herpes simplex-Thymidine kinase virus (*HSV-Tk*) under the control of the GFAP promoter that renders proliferating HMs vulnerable to killing by the antiviral drug GCV [64]. HMs were isolated from wild type (WT) and *GFAP HSV-Tk* (TG) mice and treated with 5–500 μM GCV. Apoptosis, shown by increased PARP cleavage, was induced in a dose-dependent manner in TG but not WT HM or hepatocytes upon GSV treatment *in vitro*. In this transgenic mouse model HMs must be proliferating to be susceptible to GSV mediated killing. WT and TG mice were given repeated doses of CCl_4 and allyl alcohol (AA) to induce centrilobular and periportal injury and HM activation and proliferation and then treated ± GSV. There was a 65% reduction in dual desmin/GFAP expressing HMs in CCl_4-AA-GSV treated TG mice compared to WT. This was accompanied by a decrease in the number of α-sma+ HMs and expression of PDGFRβ and collagen mRNA suggesting that HM depletion ameliorated the fibrotic response. These findings were then validated in the BDL model, where upon GSV treatment in TG mice desmin staining was reduced and TUNEL+ apoptotic cells were increased by ~50%. These data suggest that desmin+ HMs are proliferating in biliary injury and contribute to periportal fibrosis. Importantly, GSV administration and HM depletion did not affect mouse behavior or alter hepatic inflammation and CYP2E1 activity. However, the reduction in liver damage was shown biochemically by serum transaminases and HMGB1 release and histologically by 4HNE staining and pathological scoring of hepatic necrosis. The authors concluded that HMs were not only the professional scar-forming cells of the liver but also played a key role in the amplification of acute liver injury. In a follow-up publication the *GFAP HSV-Tk* (TG) mouse was used to investigate the role of HMs in ischemic reperfusion (IR) and bacterial sepsis (endotoxin exposure) [65]. Acute liver injury was induced by administering CCl_4 daily over 3 consecutive days to promote activation and proliferation of HMs then subjected mice to IR or endotoxin treatment. Administration of GSV in acute CCl_4 injury caused a 64–72% depletion in HMs in TG mice. When exposed to acute CCl_4 followed by warm IR injury WT mice developed profound parenchymal damage and elevated liver transaminase, alanine transaminase (ALT); however, this was markedly reduced in TG mice. Neutrophil infiltration was modestly reduced in the HM depleted TG mice compared to GSV treated WT. In the CCl_4 and endotoxin treated mice, HM depletion lead to a significant decrease in TUNEL+ apoptotic hepatocytes, suggesting that HMs contribute to cellular damage and intensify the injury process. The authors concluded that HMs contribute to the pathogenesis of IR and endotoxin-induced hepatic injury and that TNFα and endothelin-1 are important mediators of this effect. One caveat that is important to note about the use of GFAP systems is that cholangiocytes have previously been reported to express GFAP; therefore, this promoter has the potential to affect biliary cell numbers [66–68].

The PDGFRβ is expressed on HMs and is an early feature of their activation [69]. Therefore, the possibility that the PDGFRβ-Cre mouse, where Cre is expressed under the control of a fragment of the PDGFRβ gene could be used to selectively target liver pericytes and HMs was explored [70,71]. To test the cell specificity, the PDGFRβ-Cre mouse was crossed with the Ail4 reporter mouse, where tdTomato is expressed in cells after Cre mediated recombination. The progeny displayed a highly efficient recombination in the livers upon expression of PDGFRβ-Cre. tdTomato expression co-localized with desmin staining in both the normal uninjured liver and in the chronic CCl_4 injured liver, was suggestive that recombination occurred in qHSCs and HMs. It was hypothesized that selectively depleting the α_v integrin subunit, which is required for activation of the highly potent profibrogenic cytokine TGFβ1 from its latent form in HMs would be fibro-protective. The *Itgav*$^{flox/flox}$ mouse (encoding for the α_v integrin subunit) was crossed with the PDGFRβ-Cre mouse, generating *Itgav*$^{flox/flox}$; PDGFRβ-Cre mice and HM specific deletion of the α_v integrin subunit was confirmed in isolated HMs by western blot. In the chronic CCl_4 liver injury model *Itgav*$^{flox/flox}$; PDGFRβ-Cre mice were significantly protected from developing fibrosis, shown by reduced Sirius Red staining, hydroxyproline (collagen) content, and α-sma+ HMs. The mechanistic basis of the anti-fibrotic effect was reported to be due to reduced TGFβ1 signaling. HMs isolated from *Itgav*$^{flox/flox}$; PDGFRβ-Cre mice were less efficient at inducing TGFβ1 dependent luciferase activity in the MINK lung TGFβ-luc reporter than control mice and in the CCl_4 model, levels of P-Smad3, a downstream target of TGFβ1 signaling, were reduced in the *Itgav*$^{flox/flox}$; PDGFRβ-Cre injured mice compared to control. The α_v integrin subunit has five potential dimer partners (β_1, β_3, β_5 β_6, and β_8) that can bind or activate TGFβ and all but β_6 are expressed on HMs. To address the potential contribution of the β subunit-binding partners in liver fibrosis either global knockouts of β_3, β_5 or β_6 and floxed β_8 (global KO is embryonic lethal) crossed with the PDGFRβ-Cre were injured with CCl_4. No fibro-protective effects were observed with HM specific deletion of the β_8 integrin or in the global knockouts for the β_3, β_5, and β_6 subunits suggesting that redundancy occurs within the system. This work proved that the PDGFRβ-Cre mouse could be used to efficiently manipulate gene expression in HMs and help unravel the mechanistic events controlling their function. It is important to recognize that efficient Cre mediated recombination occurs in pericytes in other organs, including lung and kidney. In the same study Henderson et al. showed that *Itgav*$^{flox/flox}$; PDGFRβ-Cre mice were protected from pulmonary fibrosis induced by bleomycin and renal fibrosis in the unilateral ureteric obstruction model.

Fate tracing studies were used to very elegantly demonstrate that the lecithin-retinol acyltransferase Cre (Lrat-cre) would provide an effective system to selectively target HMs. This was proved using a mouse containing a bacterial artificial chromosome (BAC) where the LRAT promoter was driving Cre expression [66]. When crossed with either the ZsGreen or tdTomato reporter mouse labeling of Cre positive cells was restricted to HMs, showing ~99% recombination efficiency in HMs; however, no recombination was detected in hepatocytes, KCs, cholangiocytes, or endothelial cells. Upon injury with either CCl_4 or TAA the levels of ZsGreen positive cells increased and co-localized to α-sma+ HMs. However, when LRAT-Cre tdTomato mouse was crossed with the collagen-GFP reporter mouse, in the BDL model subpopulations of collagen-expressing (GFP positive) cells were negative for tomato, suggesting that portal fibroblasts (PFs) were negative for LRAT and arise from a different cell population in the liver. The LRAT promoter faithfully labels HMs and could be used in the future to drive Cre expression in HMs. We must also consider that recombination was observed in cells containing lipid droplets suggesting this promoter may target qHSCs as well as HMs.

Recent data generated from CCl_4 and BDL models of liver fibrosis performed in the collagen-GFP reporter mouse suggest that both HMs and PFs comprise the liver myofibroblast population [72]. While HMs were the major cell type in CCl_4 induced liver injury, forming up to 87% of the fibrogenic cells, in the BDL model, PFs rapidly activate and at day 5 postsurgery comprise ~70% of the scar cell population. However, as this cholestatic disease progresses, HMs increase in number to ~50% of the fibrogenic cells as defined by the number of GFP/vitamin A dual positive cells. These studies certainly highlight the importance of considering which is the right genetic system to use for the different liver fibrosis models when performing detailed molecular studies.

Recently, in another fate tracing study, tamoxifen inducible Vimentin-CreER mice were crossed with mTomato/mGFP mice to generate a reporter mouse where all cells are red (expressing tomato) but switch to green (GFP) upon tamoxifen induced Cre mediated recombination [73]. All cells were red in the control-uninjured livers of mice, which had not received tamoxifen suggesting that spontaneous recombination did not occur. Vimentin-CreER+ mTomato/mGFP+ mice were given four injections of CCl_4 and six injections of tamoxifen over a 4-week period to induce Cre expression during ongoing liver injury. In tamoxifen treated mice there was a clear switch from red to GFP expression in cells localized to the fibrotic bands within the liver. GFP+ cells were isolated from injured livers by fluorescence-activated cell sorting (FACS) and were shown to contain vitamin A and retinoid and RT-PCR analysis confirmed that they expressed collagen and α-sma, suggesting that the cells had HSC/HM properties. Interestingly, up to 45% of cells in the liver remained GFP+ as long as 30–45 days after cessation of liver injury. These cells were localized to peri-sinusoidal areas of the liver but their fibrogenic gene expression profile was higher than that of naïve or unchallenged HSC, suggesting that they were being held in a "reverted state." It was also suggested that the GFP+ deactivated or reverted cells were being held in "primed" state and responded more efficiently to challenge with fibrogenic stimuli [73]. This work suggests that both HM deactivation and HM apoptosis or clearance provides mechanisms to promote fibrolysis and remodeling of scar matrix. However, it is important to note that recombination was not restricted to the liver and also occurred in the kidney and GI tract, including the stomach, colon, and small intestine.

A desperately needed advance of this technology is to develop conditional systems where an inducible promoter, such as tamoxifen or doxycycline regulated promoters, is driving HM specific Cre expression. This will allow researchers to silence or induce specific genes in a temporal manner. The ability to manipulate gene expression in HMs at precise time-points during the establishment or resolution of fibrosis will allow researchers to ask complex biological questions about how specific signaling events in HMs alter their phenotype and contribute to liver fibrosis and cancer.

15.9 THE ROLE OF HM DEPLETION IN LIVER REGENERATION AND CANCER

The fibrotic actions of the HMs have been well characterized for many years; however, the role of HMs in tissue regeneration was less clear. A recent study revealed that HMs are negative regulators of hepatocyte proliferation. Serotonin (5-HT) signaling via the 5HT-2B receptor (5HT-2BR) expressed on HM activated expression of TGFβ, a powerful suppressor of hepatocyte proliferation, via an ERK/JunD mediated pathway [74]. Administration of a 5HT-2B receptor antagonist in the chronic CCl_4 model was anti-fibrotic and pro-regenerative. Furthermore, selective depletion of HMs using C1–3 conjugated to gliotoxin (C1–3-GT) in the acute CCl_4 and the BDL models significantly increased hepatocyte proliferation compared to C1–3 alone. Finally, pharmacological blockade of 5HT-2B receptors or C1–3-GT mediated HM depletion promoted increased liver regeneration in the partial hepatectomy model compared to drug vehicle or C1–3 alone. 5HT-2B receptors are expressed on multiple cells types in multiple organs and not restricted to the HMs; therefore it was important to ask if the pro-regenerative effect of blocking 5HT-2B receptor signaling was independent of HMs. No additive effect of combining 5HT-2B receptor blockade and C1–3-GT treatment was observed in liver regeneration, suggesting that the effects of 5-HT are mediated by 5HT-2B receptors expressed on HMs and not through 5HT-2B receptor signaling on other cell types. Serotonin has pro- and anti-regenerative activities in the liver; 5-HT binding 5HT2A receptors on hepatocytes promote hepatocyte proliferation [75] but also stimulate 5HT-2B receptors expressed on HMs to increase TGFβ production, which in turn inhibits hepatocyte proliferation. HM persistence likely contributes to the pathogenesis of liver disease by tipping the balance from liver regeneration to replacement of lost parenchymal tissue with scar matrix.

The fibrotic liver is a damaged and highly inflamed organ that has been considered a pro-cancerous environment. In fact, majority of primary liver tumors develop on the background of inflammation and fibrosis and advanced fibrosis/cirrhosis is a risk factor for HCC [76]. Inflammation and diet have been linked with the development of HCC in mice; however, the contribution of HMs to the establishment of liver cancer is less well understood [77,78]. HM senescence has provided us with some clues. A recent study has provided data showing that ablation of senescent HMs leaves highly activated and proliferating HMs that promotes liver fibrosis and enhance epithelial cell transformation into liver cancer [79]. Furthermore, these proliferating p53 deficient HMs secrete factors that induce a tumor-promoting M2 macrophage phenotype and enhance proliferation of pre-malignant cells. Conversely, p53 positive senescent HM release factors that skew macrophage polarization towards a pro-inflammatory M1 tumor killing phenotype and limit cancer formation. These studies highlight that when targeting HMs for anti-fibrotic therapy, there is a delicate balance between promoting liver regeneration and creating a pro-tumorigenic niche.

15.10 FUTURE PERSPECTIVES

Liver fibrosis is the only one of the top five causes of death that is increasing and therapeutic options are currently limited because of a lack of effective anti-fibrotic drugs if the underlying cause of disease cannot be treated [80]. Over the past 20 years there have been significant advances in our understanding of the complex cell:cell interactions and molecular events that control fibrogenic and fibrolytic processes. Collectively this information has been used to design new therapeutic approaches to start to treat liver fibrosis. The next step is to translate one of the targeting strategies or therapeutics to the clinic and prove that it can be used to selectively deliver an effective anti-fibrotic compound to HMs in patients to reverse fibrosis where injury is ongoing.

Fibrosis is a multi-organ disease and is estimated to be a causal factor in ~45% of deaths in the western world with the lungs, kidneys, skin, and heart among the most common organs affected [81]. Recent studies have provided evidence suggesting that some of the signaling pathways that control HM phenotype and fibrotic properties may be common to fibroblasts/myofibroblasts in other organs. It is possible that some of the proteins described to achieve HM cell specific targeting or HM specific transgenic mouse models are shared with myofibroblasts in other organs. This raises the possibility that discoveries made in the liver could be translated to either deliver therapeutics or cell specifically knockout genes in scar forming myofibroblasts in the lung, kidney, skin, or heart.

REFERENCES

[1] Geerts A. History, heterogeneity, developmental biology, and functions of quiescent hepatic stellate cells. Semin Liver Dis 2001;21:311–35.

[2] Bataller R, Brenner DA. Liver fibrosis. J Clin Invest 2005;115:209–18.

[3] Friedman SL. Hepatic stellate cells: protean, multifunctional, and enigmatic cells of the liver. Physiol Rev 2008;88:125–72.

[4] Henderson NC, Iredale JP. Liver fibrosis: cellular mechanisms of progression and resolution. Clin Sci (Lond) 2007;112:265–80.

[5] Friedman SL, Bansal MB. Reversal of hepatic fibrosis—fact or fantasy? Hepatology 2006;43:S82–88.

[6] Poynard T, et al. Impact of pegylated interferon alfa-2b and ribavirin on liver fibrosis in patients with chronic hepatitis C. Gastroenterology 2002;122:1303–13.

[7] Honkoop P, de Man RA, Zondervan PE, Schalm SW. Histological improvement in patients with chronic hepatitis B virus infection treated with lamivudine. Liver 1997;17:103–6.

[8] Iredale JP, et al. Mechanisms of spontaneous resolution of rat liver fibrosis. Hepatic stellate cell apoptosis and reduced hepatic expression of metalloproteinase inhibitors. J Clin Invest 1998;102:538–49.

[9] Fischer R, Cariers A, Reinehr R, Haussinger D. Caspase 9-dependent killing of hepatic stellate cells by activated Kupffer cells. Gastroenterology 2002;123:845–61.

[10] Gao B, Jeong WI, Tian Z. Liver: an organ with predominant innate immunity. Hepatology 2008;47:729–36.

[11] Radaeva S, et al. Natural killer cells ameliorate liver fibrosis by killing activated stellate cells in NKG2D-dependent and tumor necrosis factor-related apoptosis-inducing ligand-dependent manners. Gastroenterology 2006;130:435–52.

[12] Radaeva S, et al. Retinoic acid signaling sensitizes hepatic stellate cells to NK cell killing via upregulation of NK cell activating ligand RAE1. Am J Physiol Gastrointest Liver Physiol 2007;293:G809–816.

[13] Stern-Ginossar N, Mandelboim O. An integrated view of the regulation of NKG2D ligands. Immunology 2009;128:1–6.

[14] Gur C, et al. NKp46-mediated killing of human and mouse hepatic stellate cells attenuates liver fibrosis. Gut 2012;61:885–93.

[15] Jeong WI, Park O, Gao B. Abrogation of the antifibrotic effects of natural killer cells/interferon-gamma contributes to alcohol acceleration of liver fibrosis. Gastroenterology 2008;134:248–58.

[16] Oakley F, et al. Hepatocytes express nerve growth factor during liver injury: evidence for paracrine regulation of hepatic stellate cell apoptosis. Am J Pathol 2003;163:1849–58.
[17] Krizhanovsky V, et al. Senescence of activated stellate cells limits liver fibrosis. Cell 2008;134:657–67.
[18] Wright MC, et al. Gliotoxin stimulates the apoptosis of human and rat hepatic stellate cells and enhances the resolution of liver fibrosis in rats. Gastroenterology 2001;121:685–98.
[19] Orr JG, et al. Mechanism of action of the antifibrogenic compound gliotoxin in rat liver cells. Hepatology 2004;40:232–42.
[20] Elsharkawy AM, Oakley F, Mann DA. The role and regulation of hepatic stellate cell apoptosis in reversal of liver fibrosis. Apoptosis 2005;10:927–39.
[21] Teixeira-Clerc F, et al. CB1 cannabinoid receptor antagonism: a new strategy for the treatment of liver fibrosis. Nat Med 2006;12:671–6.
[22] Mallat A, Teixeira-Clerc F, Lotersztajn S. Cannabinoid signaling and liver therapeutics. J Hepatol 2013;59:891–6.
[23] Jonsson JR, et al. Angiotensin-converting enzyme inhibition attenuates the progression of rat hepatic fibrosis. Gastroenterology 2001;121:148–55.
[24] Oakley F, et al. Angiotensin II activates I kappaB kinase phosphorylation of RelA at Ser 536 to promote myofibroblast survival and liver fibrosis. Gastroenterology 2009;136:2334–44. e2331.
[25] Bataller R, et al. Systemic infusion of angiotensin II exacerbates liver fibrosis in bile duct-ligated rats. Hepatology 2005;41:1046–55.
[26] Bataller R, et al. Angiotensin II induces contraction and proliferation of human hepatic stellate cells. Gastroenterology 2000;118:1149–56.
[27] Yang L, et al. Attenuated hepatic inflammation and fibrosis in angiotensin type 1a receptor deficient mice. J Hepatol 2005;43:317–23.
[28] Sookoian S, Fernandez MA, Castano G. Effects of six months losartan administration on liver fibrosis in chronic hepatitis C patients: a pilot study. World J Gastroenterol 2005;11:7560–3.
[29] Colmenero J, et al. Effects of losartan on hepatic expression of nonphagocytic NADPH oxidase and fibrogenic genes in patients with chronic hepatitis C. Am J Physiol Gastrointest Liver Physiol 2009;297:G726–734.
[30] Elsharkawy AM, et al. Persistent activation of nuclear factor-kappaB in cultured rat hepatic stellate cells involves the induction of potentially novel Rel-like factors and prolonged changes in the expression of IkappaB family proteins. Hepatology 1999;30:761–9.
[31] Watson MR, et al. NF-kappaB is a critical regulator of the survival of rodent and human hepatic myofibroblasts. J Hepatol 2008;48:589–97.
[32] Oakley F, et al. Inhibition of inhibitor of kappaB kinases stimulates hepatic stellate cell apoptosis and accelerated recovery from rat liver fibrosis. Gastroenterology 2005;128:108–20.
[33] Mann J, et al. MeCP2 controls an epigenetic pathway that promotes myofibroblast transdifferentiation and fibrosis. Gastroenterology 2010;138:705–14. 714 e701–4.
[34] Oakley F, et al. Basal expression of IkappaBalpha is controlled by the mammalian transcriptional repressor RBP-J (CBF1) and its activator Notch1. J Biol Chem 2003;278:24359–70.
[35] Moles A, et al. Inhibition of RelA-Ser536 phosphorylation by a competing peptide reduces mouse liver fibrosis without blocking the innate immune response. Hepatology 2013;57:817–28.
[36] Cassiman D, et al. Synaptophysin: a novel marker for human and rat hepatic stellate cells. Am J Pathol 1999;155:1831–9.
[37] Elrick LJ, et al. Generation of a monoclonal human single chain antibody fragment to hepatic stellate cells—a potential mechanism for targeting liver anti-fibrotic therapeutics. J Hepatol 2005;42:888–96.
[38] Douglass A, et al. Antibody-targeted myofibroblast apoptosis reduces fibrosis during sustained liver injury. J Hepatol 2008;49:88–98.
[39] Beljaars L, et al. Albumin modified with mannose 6-phosphate: a potential carrier for selective delivery of antifibrotic drugs to rat and human hepatic stellate cells. Hepatology 1999;29:1486–93.
[40] Hagens WI, et al. Cellular targeting of the apoptosis-inducing compound gliotoxin to fibrotic rat livers. J Pharmacol Exp Ther 2008;324:902–10.
[41] van Beuge MM, et al. Enhanced effectivity of an ALK5-inhibitor after cell-specific delivery to hepatic stellate cells in mice with liver injury. PLoS One 2013;8:e56442.

[42] van Beuge MM, et al. Increased liver uptake and reduced hepatic stellate cell activation with a cell-specific conjugate of the Rho-kinase inhibitor Y27632. Pharm Res 2011;28:2045–54.
[43] Moreno M, et al. Reduction of advanced liver fibrosis by short-term targeted delivery of an angiotensin receptor blocker to hepatic stellate cells in rats. Hepatology 2010;51:942–52.
[44] de Gouville AC, et al. Inhibition of TGF-beta signaling by an ALK5 inhibitor protects rats from dimethylnitrosamine-induced liver fibrosis. Br J Pharmacol 2005;145:166–77.
[45] de Gouville AC, Huet S. Inhibition of ALK5 as a new approach to treat liver fibrotic diseases. Drug News Perspect 2006;19:85–90.
[46] Tada S, et al. A selective ROCK inhibitor, Y27632, prevents dimethylnitrosamine-induced hepatic fibrosis in rats. J Hepatol 2001;34:529–36.
[47] Yee Jr. HF. Rho directs activation-associated changes in rat hepatic stellate cell morphology via regulation of the actin cytoskeleton. Hepatology 1998;28:843–50.
[48] Beljaars L, et al. Successful targeting to rat hepatic stellate cells using albumin modified with cyclic peptides that recognize the collagen type VI receptor. J Biol Chem 2000;275:12743–51.
[49] Farrar MA, Schreiber RD. The molecular cell biology of interferon-gamma and its receptor. Annu Rev Immunol 1993;11:571–611.
[50] Cleland JL, Jones AJ. Stable formulations of recombinant human growth hormone and interferon-gamma for microencapsulation in biodegradable microspheres. Pharm Res 1996;13:1464–75.
[51] Gu F, Younes HM, El-Kadi AO, Neufeld RJ, Amsden BG. Sustained interferon-gamma delivery from a photocrosslinked biodegradable elastomer. J Control Release 2005;102:607–17.
[52] Bansal R, et al. Novel engineered targeted interferon-gamma blocks hepatic fibrogenesis in mice. Hepatology 2011;54:586–96.
[53] Bansal R, Prakash J, de Ruijter M, Beljaars L, Poelstra K. Peptide-modified albumin carrier explored as a novel strategy for a cell-specific delivery of interferon gamma to treat liver fibrosis. Mol Pharm 2011;8:1899–909.
[54] Bansal R, Prakash J, De Ruiter M, Poelstra K. Targeted recombinant fusion proteins of IFNgamma and mimetic IFNgamma with PDGFbetaR bicyclic peptide inhibits liver fibrogenesis *in vivo*. PLoS One 2014;9:e89878.
[55] Bansal R, Prakash J, De Ruiter M, Poelstra K. Interferon gamma peptidomimetic targeted to hepatic stellate cells ameliorates acute and chronic liver fibrosis *in vivo*. J Control Release 2014;179: 18–24.
[56] Du SL, et al. Cyclic Arg–Gly–Asp peptide-labeled liposomes for targeting drug therapy of hepatic fibrosis in rats. J Pharmacol Exp Ther 2007;322:560–8.
[57] Kim SJ, Ise H, Goto M, Akaike T. Interactions of vimentin- or desmin-expressing liver cells with *N*-acetylglucosamine-bearing polymers. Biomaterials 2012;33:2154–64.
[58] Kim SJ, et al. Imaging and therapy of liver fibrosis using bioreducible polyethylenimine/siRNA complexes conjugated with *N*-acetylglucosamine as a targeting moiety. Biomaterials 2013;34:6504–14.
[59] Yang J, et al. Targeted delivery of the RGD-labeled biodegradable polymersomes loaded with the hydrophilic drug oxymatrine on cultured hepatic stellate cells and liver fibrosis in rats. Eur J Pharm Sci 2014;52:180–90.
[60] Chai NL, et al. Oxymatrine liposome attenuates hepatic fibrosis via targeting hepatic stellate cells. World J Gastroenterol 2012;18:4199–206.
[61] Kinoshita K, et al. Targeted and regulable expression of transgenes in hepatic stellate cells and myofibroblasts in culture and *in vivo* using an adenoviral Cre/loxP system to antagonise hepatic fibrosis. Gut 2007;56:396–404.
[62] Schoemaker MH, et al. PDGF-receptor beta-targeted adenovirus redirects gene transfer from hepatocytes to activated stellate cells. Mol Pharm 2008;5:399–406.
[63] Smith PG, et al. Herpesvirus saimiri-based vector biodistribution using noninvasive optical imaging. Gene Ther 2005;12:1465–76.
[64] Puche JE, et al. A novel murine model to deplete hepatic stellate cells uncovers their role in amplifying liver damage in mice. Hepatology 2013;57:339–50.
[65] Stewart RK, et al. A novel mouse model of depletion of stellate cells clarifies their role in ischemia/reperfusion- and endotoxin-induced acute liver injury. J Hepatol 2014;60:298–305.

[66] Mederacke I, et al. Fate tracing reveals hepatic stellate cells as dominant contributors to liver fibrosis independent of its aetiology. Nat Commun 2013;4:2823.
[67] Yang L, et al. Fate-mapping evidence that hepatic stellate cells are epithelial progenitors in adult mouse livers. Stem Cells 2008;26:2104–13.
[68] Michelotti GA, et al. Smoothened is a master regulator of adult liver repair. J Clin Invest 2013;123:2380–94.
[69] Pinzani M, Gesualdo L, Sabbah GM, Abboud HE. Effects of platelet-derived growth factor and other polypeptide mitogens on DNA synthesis and growth of cultured rat liver fat-storing cells. J Clin Invest 1989;84:1786–93.
[70] Foo SS, et al. Ephrin-B2 controls cell motility and adhesion during blood-vessel-wall assembly. Cell 2006;124:161–73.
[71] Henderson NC, et al. Targeting of alphav integrin identifies a core molecular pathway that regulates fibrosis in several organs. Nat Med 2013;19:1617–24.
[72] Iwaisako K, et al. Origin of myofibroblasts in the fibrotic liver in mice. Proc Natl Acad Sci USA 2014;111(32):E3297–305.
[73] Troeger JS, et al. Deactivation of hepatic stellate cells during liver fibrosis resolution in mice. Gastroenterology 2012;143:1073–83. e1022.
[74] Ebrahimkhani MR, et al. Stimulating healthy tissue regeneration by targeting the 5-HT(2)B receptor in chronic liver disease. Nat Med 2011;17:1668–73.
[75] Lesurtel M, et al. Platelet-derived serotonin mediates liver regeneration. Science 2006;312:104–7.
[76] Wallace MC, Friedman SL. Hepatic fibrosis and the microenvironment: fertile soil for hepatocellular carcinoma development. Gene Expr 2014;16:77–84.
[77] Park EJ, et al. Dietary and genetic obesity promote liver inflammation and tumorigenesis by enhancing IL-6 and TNF expression. Cell 2010;140:197–208.
[78] Sun B, Karin M. Inflammation and liver tumorigenesis. Front Med 2013;7:242–54.
[79] Lujambio A, et al. Non-cell-autonomous tumor suppression by p53. Cell 2013;153:449–60.
[80] Friedman SL, Sheppard D, Duffield JS, Violette S. Therapy for fibrotic diseases: nearing the starting line. Sci Transl Med 2013;5:167sr161.
[81] Wynn TA. Cellular and molecular mechanisms of fibrosis. J Pathol 2008;214:199–210.

CHAPTER 16

Pancreatic Stellate Cells

Minoti V. Apte, Romano C. Pirola and Jeremy S. Wilson
Pancreatic Research Group, South Western Sydney Clinical School, University of New South Wales, Ingham Institute for Applied Medical Research, Sydney, Australia

16.1 INTRODUCTION

Prominent scarring (fibrosis) of the pancreas is a characteristic histological feature of two major diseases of the gland—chronic pancreatitis and pancreatic cancer [1,2] (Figure 16.1). Histologically, the large areas of dense fibrosis as evident on trichrome staining (which stains connective tissue) or Sirius Red staining (which stains collagen specifically) have been long documented by pathologists examining pancreatic sections from patients with these conditions. Yet, up until just over 15 years ago, this fibrosis was usually dismissed as an epiphenomenon or an end result of chronic disease, with most of the research into the pathophysiology of chronic pancreatitis and pancreatic cancer, being focused on the primary sites of injury (acinar cells in the case of chronic pancreatitis) and ductal and acinar cells (in the case of pancreatic cancer).

In contrast to the pancreas, fibrosis in the liver has been a highly active area of research for many decades. Indeed, the first description of the hepatic stellate cell (HSC) was recorded by the famous pathologist Karl von Kupffer as early as in 1876. This cell is now well established to be the key cell responsible for fibrosis in the liver [3] (described in Chapters 4 and 6). The characteristic features of HSCs are detailed elsewhere in this book (see Chapters 1 and 2), but the ability of these cells to store vitamin A (retinoids) as lipid droplets in their cytoplasm is a specific identifier of a cell type that is part of "a stellate cell system" in the body (i.e., retinoid storing cells that are present not only in the liver but also in extrahepatic organs including the lungs, kidney, intestine, adrenal glands, spleen) where the cells are distributed in a perivascular location similar to pericytes [4]. The pancreatic counterpart of the HSC was described over a century after Kupffer's report, in 1982, by Watari from Japan, after he administered vitamin A to mice and examined the pancreas by fluorescence as well as electron microscopy [5]. Watari reported the presence of cells exhibiting a rapidly fading blue-green fluorescence that is characteristic of vitamin A stores. A subsequent study by Ikejiri [6] in 1990 described vitamin A autofluorescence in both rat and human pancreas, supporting Watari's findings. However, it was to be another 8 years before a major step forward was achieved in the field, with the development of methods to

Stellate Cells in Health and Disease
DOI: http://dx.doi.org/10.1016/B978-0-12-800134-9.00016-6

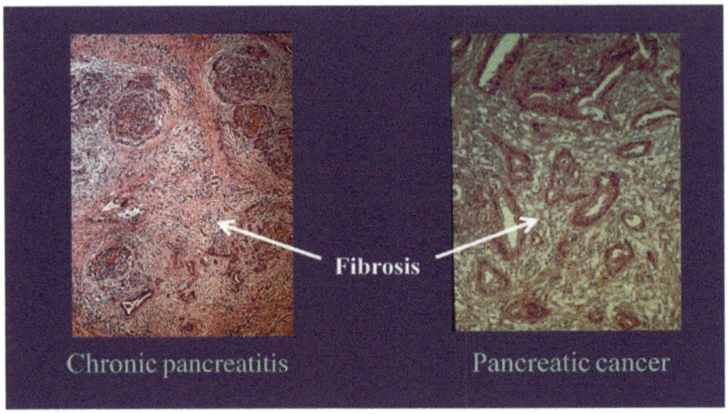

Figure 16.1 Hematoxylin and eosin (H&E) stained photomicrographs of pancreatic sections from a patient with chronic pancreatitis (A) and pancreatic cancer (B). Both sections show abundant fibrosis (stromal reaction) surrounding (i) remnants of acini and distorted ducts in chronic pancreatitis and (ii) tumor elements in pancreatic cancer.

isolate pancreatic stellate cells (PSCs) from rat and human pancreas, providing a much-needed tool for in-depth studies of their biology [7,8]. This chapter details the "state of the art" with respect to current knowledge about the function of PSCs. As is becoming increasingly evident, these cells are not only critical to pancreatic fibrogenesis, but also perform several major functions that influence pancreatic physiology in health and disease.

16.2 PANCREATIC STELLATE CELLS—ISOLATION AND CHARACTERIZATION

Taking their cues from the HSC field, researchers interested in pancreatic fibrogenesis first set about confirming the reports by Watari and Ikejiri, using immunostaining for stellate cell selective markers (desmin, glial fibrillary acidic protein, nestin, vimentin, and neuroectodermal markers such as nerve growth factor (NGF) and neural cell adhesion molecule) to identify PSCs in normal rat pancreas [7,8]. Importantly, these PSC selective markers differentiate the cells from fibroblasts. These studies (reviewed in Ref. [9]) indicated that PSCs in healthy pancreas are located in close proximity to the basolateral aspects of acinar cells (Figure 16.2A) and are also observed around pancreatic ducts and blood vessels. Morphometric analysis of desmin-stained PSCs suggests that in the normal pancreas, PSCs comprise 4–7% of total parenchymal cells. The cell cytoplasm exhibits the presence of abundant vitamin A containing lipid droplets and as such, the cells are in their quiescent (non-activated state). An abundant rough endoplasmic reticulum (RER), vacuoles (vitamin A-containing lipid droplets) surrounding a central nucleus and collagen fibrils are evident upon electron microscopy.

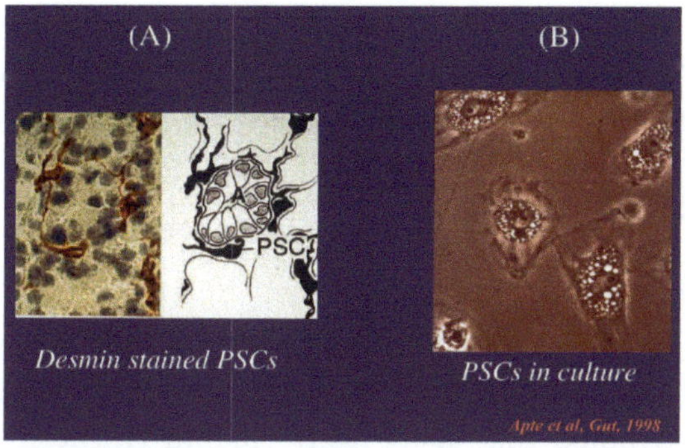

Figure 16.2 (A) Desmin staining of rat PSCs *in situ*. A rat pancreatic section immunostained for the PSC selective marker desmin shows desmin positive (PSCs) on the left and a corresponding line diagram on the right. Desmin positive PSCs are located close to the basolateral aspect of acinar cells. The PSCs exhibit a central cell body and long cytoplasmic processes encircling the base of acinar cells. (B) Freshly isolated rat PSCs in culture showing a flattened polygonal shape with abundant lipid droplets surrounding the central nucleus.

16.2.1 PSC isolation

In 1998, Apte et al. [7] reported a method of isolation of PSCs from normal rat pancreas, based upon the fact that in their healthy, quiescent state PSCs contain abundant vitamin A-containing lipid droplets, which gives the cells a certain buoyancy and allows them to be separated from other pancreatic cells by differential centrifugation through a Nycodenz gradient. The cells obtained by this method are uncontaminated by other cells such as acinar, ductal, or endothelial cells. When plated on culture plates, the cells assume a flattened, polygonal shape showing the characteristic opalescent lipid vacuoles surrounding a central nucleus (Figure 16.2B). When these cells are exposed to ultraviolet (UV) light at 338 nm they emit a transient blue-green fluorescence typical of vitamin A. In health, PSCs are thought to play a central role in maintaining normal extracellular matrix (ECM) turnover in the gland, because they produce both ECM proteins as well as matrix degrading enzymes (matrix metalloproteinases, MMPs).

When cultured on plastic for 48 h or so, the quiescent vitamin A-containing cells transform into a myofibroblast-like phenotype. The cells become elongated/spindle-shaped with several cytoplasmic projections, and they invariably lose their stored vitamin A and acquire positivity for alpha smooth muscle actin (α-sma), a cytoskeletal protein. More importantly, they synthesize and secrete excessive amounts of ECM proteins associated with alteration of their secreted MMP profile. Bachem et al. [8] described a method of isolation of pre-activated PSCs using the outgrowth method from the pancreas of patients with chronic pancreatitis or from a rat model of chronic pancreatitis. For this method of PSC isolation, pancreatic tissue is first finely minced and then placed on culture dishes containing an optimized culture medium.

Pre-activated PSCs proliferate from the outer margins of the minced tissue pieces, while the central parts of the tissue are lost through necrosis. This is a useful method for the isolation of activated PSCs from diseased pancreas, but is limited by the fact that it does not allow assessment of the process of transformation of the cells from their quiescent to activated state.

16.2.2 PSCs in health

As noted earlier, in the normal pancreas, PSCs are in their quiescent state, characterized by the presence of vitamin A-containing lipid droplets in their cytoplasm. Studies with HSCs have determined that apart from vitamin A, the lipid droplets comprise cholesterol, phospholipids, cholesterol ester, triglyceride, and free fatty acids [10]. It is likely that the lipid droplets in PSCs are of a similar composition, although this has not yet been assessed. The exact mechanism(s) responsible for the accumulation of these lipid droplets in both HSCs and PSCs remains a mystery. Based on their findings that albumin is endogenously expressed in PSCs and co-localized with vitamin A in lipid droplets in PSCs, Kim et al. [11] have suggested that albumin is an essential protein in the formation of the droplets. Transfection of activated PSCs with expression plasmids for albumin led to a re-accumulation of lipid droplets in PSCs, accompanied by a reversal to quiescence as assessed by morphology and a resistance to the activating effects of the known fibrogenic growth factor transforming growth factor beta (TGFβ). Interestingly, the authors also subsequently demonstrated that albumin was a downstream effector of peroxisome proliferator activated receptor gamma (PPARγ), a nuclear receptor known to inhibit PSC activation after binding to its ligand glitazone [12]. While these reports shed some light on lipid accumulation within PSCs, the mechanisms mediating their loss during PSC activation remain un-elucidated.

Because the earliest studies on PSC biology were related to the possible role of these cells in fibrogenesis, the capacity of PSCs for ECM protein production has been well studied. PSCs actively synthesize and secrete ECM proteins such as collagen (particularly types I and IV), fibronectin, and laminin as well as a range of matrix degrading enzymes and their inhibitors (MMPs and tissue inhibitors of MMPs (TIMPs), respectively) [9]. Thus, in the healthy pancreas, PSCs are thought to maintain normal ECM turnover by balancing ECM production with ECM degradation. As detailed later in the chapter, this balance is severely deranged upon activation of PSCs, whereby excessive ECM production associated with decreased ECM degradation leads to overt fibrosis. Recent studies have taken advantage of the advances in proteomic analysis techniques to compare the proteomes of quiescent versus activated rat and human PSCs [13,14]. As may be expected, several proteins were found to be differentially expressed in the two phenotypic states of the cell, with activated PSCs exhibiting upregulation of proteins relevant to cell metabolism, growth, motility, invasion, and the cytoskeleton.

Before moving on to a detailed description of the fibrogenic role of activated PSCs in disease, it may be pertinent to discuss the more recently discovered nonfibrogenic functions of non-activated PSCs, as they may be very relevant to the role of these cells in the normal pancreas. These include the capacity of PSCs to act as (i) progenitor cells, (ii) immune cells, or (iii) intermediary cells in cholecystokinin (CCK)-stimulated exocrine pancreatic secretion.

16.2.2.1 Progenitor function

The first study to provide some evidence in support of a possible progenitor role for PSCs was reported by Mato et al. [15], who isolated and expanded a population of mitoxantrone-resistant pancreatic cells from lactating rats. Mitoxantrone is a drug that acts via multidrug transporter systems. The mitoxantrone-resistant cell population expressed a stem cell marker—ABCG2 transporter (ATP binding cassette G2 transporter) and was able to secrete insulin when incubated in an appropriate "differentiation" medium. However, whether such "engineered" cells are truly reflective of normal PSCs needs to be questioned. More intriguingly, a recent study by Kordes et al. [16] has reported that freshly isolated and clonally expanded rat PSCs expressed the stem cell markers nestin, CD133, SOX9, and GDF3 and more importantly, when injected into rats that had undergone partial hepatectomy, were able to home to the liver and differentiate into hepatocytes, thereby contributing to liver regeneration. The authors concluded that the expression of stem cell markers, the transplantability of the cells, their long survival in circulation, and their ability to differentiate into functional hepatocytes strongly supported a progenitor role for PSCs.

16.2.2.2 Immune function

A role for PSCs in innate immunity was first proposed by Shimizu et al. [17], based on their observations both *in vitro* using cultured PSCs incubated with damaged acinar cells or neutrophils and *in vivo* using a mouse model of acute pancreatitis induced by bile duct ligation and a rat model of spontaneous chronic pancreatitis. The authors reported that PSCs had the ability to internalize necrotic acinar cells and apoptotic neutrophils via both endocytic and phagocytic pathways, suggesting that the cells may exercise a locally protective innate immune function to abrogate pancreatic damage during early necroinflammation. This notion is supported by the finding that PSCs express toll-like receptors (TLRs) 3, 4, 5, and 9, all of which recognize pathogen associated molecular patterns (PAMPs) [18,19]. As for a role in acquired immunity by acting as antigen presenting cells, it was found that rat PSCs failed to express any markers suggestive of this cell type. Thus, PSCs did not express major histocompatibility complex (MHC) class II molecules or human leukocyte antigen-D antigen related (HLA-DR) molecules, making it unlikely that the cells were part of an acquired immune function [20]. This contrasts with the findings in HSCs, which have been reported to process lipid and protein antigens and to present them to natural killer (NK) cells via CD1d and to CD4/CD8 T cells, respectively [21,22]. The observed differences between PSCs and HSCs with respect to a role in acquired immunity likely reflect the microenvironment in which these cells reside. HSCs are consistently exposed to foreign antigens from the gastrointestinal tract via the portal circulation, thereby inducing the development of an antigen-presenting capacity in these cells. In contrast, PSCs are located in a relatively protected location and are therefore not exposed to the same load of exogenous antigens as HSCs, making it less essential for the cells to acquire an antigen-presenting function.

16.2.2.3 Role in CCK-induced pancreatic exocrine secretion

It has long been acknowledged that pancreatic digestive enzyme secretion in response to a meal is predominantly mediated by the secretagogue CCK [23]. There has been some debate in the recent literature about the presence of functional CCK receptors on human acinar cells (as against the well-established presence of CCK receptors on rodent acinar cells) [24,25]. This consideration and the observed close proximity of PSCs to acinar cells *in vivo* raised the possibility of a role for PSCs in CCK-induced exocrine secretion. Human PSCs have now been shown to express both CCK receptors 1 and 2 [26,27]; in addition, PSCs have been shown to produce the neurotransmitter acetylcholine in response to physiological concentrations of CCK [27]. Moreover, co-culture of human PSCs with acinar cells led to an increase in amylase production by acinar cells, an effect that was inhibited by the muscarinic receptor blocker atropine. It is possible therefore that at least in humans, PSCs play a significant intermediary role in CCK-regulated exocrine secretion.

16.2.2.4 Origin of PSCs

As noted earlier, it is now generally agreed that PSCs form part of a larger stellate cell system in the human body. This system comprises retinoid-storing cells in the parenchyma and/or around blood vessels in the liver as well as other extrahepatic organs. Buccholz et al. [13] performed transcriptomic analyses of HSCs, PSCs, and skin fibroblasts to show that the two stellate cell types were more similar to each other and both were significantly different from fibroblasts. HSCs and PSCs were homologous with regard to genes for retinoid metabolism, motility, ECM proteins, and growth factors. Given that HSCs and PSCs express both mesenchymal as well as neurotrophic proteins, there has been some debate as to whether these cells are of mesodermal or neuroectodermal origin. Initial studies with HSCs appeared to support the neuroectodermal theory [28,29], however, recent definitive cell lineage tracing studies using a mesoderm-specific MesP1Cre have clearly shown that HSCs arise from the mesoderm [30]. In view of the similarities between HSCs and PSCs, it is likely that the latter also originate from the mesoderm, but lineage tracing studies have not yet been performed with these cells.

16.3 ROLE OF PSCs IN PANCREATIC DISEASE

Over the past 15 years (since the first reports of methods to isolate PSCs), significant progress has been made with regard to our understanding of the key influence that these cells have in pancreatic inflammatory diseases (pancreatitis) and in pancreatic cancer.

16.3.1 Activation of PSCs

The earliest studies of PSC biology were focused on the transformation of PSCs from a quiescent to activated phenotype, the latter being responsible for the synthesis of excessive ECM proteins that comprised fibrous tissue [7]. On the basis of knowledge about the variety of factors that are known to be upregulated during pancreatic injury/necroinflammation, researchers have assessed the response of cultured PSCs to a wide range of potential activators including (Table 16.1): (i) proinflammatory cytokines, chemokines, and growth factors, e.g., interleukins, TNFα, CXCR12, TGFβ, connective tissue growth factor (CTGF), and platelet derived growth factor (PDGF); (ii) alcohol and its metabolites acetaldehyde and fatty acid ethyl ester (FAEE) (given the known role of alcohol in pancreatitis) [31]; (iii) endotoxin (lipopolysaccharide, LPS) [32]; in view of the known association between alcohol abuse and endotoxemia [33] and the positive correlation between circulating LPS levels and severity of pancreatitis [34]; (iv) oxidant stress (which occurs during both acute and chronic pancreatitis) [35,36]; (v) increased pancreatic pressure (secondary to the compartment syndrome in chronic pancreatitis) [37]; hyperglycemia (since diabetes is a known association of chronic pancreatitis) [38]; (vi) angiotensin (given the upregulation of the renin–angiotensin system during pancreatitis) [39]. More factors are added to this list everyday with some of the more recent factors including endothelin 1 (a vasoconstrictor secreted by endothelial cells) [40]; cyclooxygenase 2 (COX-2, which catalyzes conversion of arachidonic acid to prostaglandin) [41]; pigment epithelium derived factor, a non-inhibitory member of the serine protease inhibitor gene family [42]; galectin 1 (a beta galactoside-binding lectin) [43]; hypoxia, usually a consequence of vasoconstriction and inadequate oxygenation in dense fibrosis [44]; proteases (prematurely activated within the gland during pancreatitis) [45]; and the hemostatic protein fibrinogen [46].

Table 16.1 Pancreatic stellate cell activating factors

Ethanol and its metabolites (acetaldehyde, FAEEs)	Oxidant stress
Inflammatory mediators (cytokines, growth factors)	Angiotensin
Pigment epithelium-derived factor	Proteases
Galectin 1	Hypoxia
Endothelin-1	Cyclooxygenase 2 (COX-2)
Fibrinogen	Hyperglycemia

Upon exposure to the previous factors *in vitro*, PSC activation has usually been assessed by various parameters, the most common being loss of vitamin A, α-sma expression, proliferation, synthesis of ECM proteins and MMPs, contractility, migration, and cytokine release. It is well established that PSC activating factors may alter one or more of the previously mentioned functions, but may not necessarily affect all the functions at the same time. Thus, TGFβ and its downstream effector molecule CTGF exert potent fibrogenic effects on PSCs by inducing ECM synthesis [47–49], but TGFβ has a minimal mitogenic effect on the cells [47]. In contrast, PDGF is a potent proliferative and chemotactic factor [47,50], with little effect on ECM synthesis by PSCs [47]. Interleukins IL1, IL6, and IL13 and the cytokines TNFα and monocyte chemotactic protein 1 (MCP1) induce collagen synthesis and α-sma expression by PSCs [48,51]. Interestingly, while PSCs respond to paracrine stimulation by exogenous cytokines, the cells can also produce their own cytokines including TGFβ, CTGF, MCP1, IL1, 8, and 15 and RANTES (regulated on activation normal T cell expressed and secreted) [52,53]. Synthesis of endogenous cytokines by PSCs has been shown to be stimulated by other cytokines such as TGFβ and CTGF and also by alcohol and its oxidative metabolite acetaldehyde [48,54]. Furthermore, an autocrine loop between specific cytokines has also been recently reported to exist in PSCs. In this regard, Aoki et al. [55] have shown that IL1 and IL6 synthesized by PSCs can each stimulate autocrine secretion of TGFβ by the cells, and vice versa. On the other hand, the Th2 lymphokine IL13 suppresses TGFβ secretion by rat PSCs, although it causes PSC proliferation [56]. Whether the IL13 findings are relevant to human pancreatitis is currently unclear, because human PSCs do not express the IL13 receptor. Nevertheless, the ability of PSCs to produce factors that can induce activation in an autocrine manner indicates that PSC may be able to achieve a perpetually activated state, even in the absence of the original triggers. This would allow progression of injury, for example in chronic pancreatitis even in the absence of any intermittent acute necroinflammatory episodes.

In view of the established role of alcohol in acute and chronic pancreatitis, researchers have examined the effects of alcohol (ethanol) and its metabolites on PSC activation *in vitro*. Apte et al. [57] have shown that alcohol, at levels seen with social drinking (10 mM) as well as heavy drinking (50 mM), directly activates PSCs in culture. This effect is most likely due to the oxidative metabolism of ethanol via alcohol dehydrogenase (the isoform ADH1, is known to be expressed and active in PSCs) and the subsequent generation of oxidant stress within the cells. The role of reactive oxygen species in alcohol-related PSC activation is further supported by the fact that alcohol induces PDGF-mediated NADPH oxidase activity (an enzyme that catalyzes free radical formation) within the cells [58]. Most importantly, Apte et al. [57] have reported that ethanol can activate PSCs from their quiescent phenotype without requiring preactivation of the cells. These findings suggest that *in vivo*, PSC activation may occur at the very earliest stages of alcohol intake, even without concurrent inflammation. Further perpetuation of this activation may occur during episodes of necroinflammation, eventually leading to overt fibrosis. In addition to PSC activation, ethanol has been shown to inhibit apoptosis of PSCs, thereby increasing the survival of these cells [59]. Of particular interest is the finding that alcohol enhances the anti-apoptotic effects of endotoxin on PSCs [59], indicating a synergistic effect of the two compounds on the cells leading to increased survival combined with increased activation, thereby facilitating pancreatic fibrosis. While oxidative metabolites of alcohol have been shown to activate PSCs, the same is not the case for the non-oxidative metabolites of alcohol, FAEEs. However, a recent report indicates that palmitic acid ethyl ester (one of the abundant FAEEs found in the pancreas) induces specific signaling pathways in PSCs (detailed later in the chapter) [60].

Given the previously described role of oxidant stress in alcohol-mediated PSC activation, considerable attention has been paid to the response of PSCs to pro-oxidant complexes *in vitro*. Thus iron sulfate/ascorbic acid or hydrogen peroxide have been shown to activate PSCs and this activation is prevented by the antioxidant vitamin E [57,61,62]. The capacity of PSCs to generate reactive oxygen species within the cells has also been well demonstrated because the cells express NADPH oxidase (an enzyme found in phagocytic cells such as neutrophils) [63]. Activation of PSCs upon exposure to high pressure conditions (as produced by placing cultured cells in a sealed chamber loaded with helium gas) has been shown to be mediated by oxidant stress, because the activating effects of increased pressure are prevented in the presence of antioxidants such as epigallocatechin gallate (a green tea polyphenol) or *N*-acetyl cysteine [64]. It is possible that oxidant stress may be the final common pathway that mediates activation of PSCs exposed to several different activating factors ranging from proinflammatory mediators to environmental toxins.

16.3.2 Signaling pathways in PSCs

In order to understand the mechanisms mediating the functional responses of PSCs to exogenous and endogenous stimuli, intracellular signaling pathways have been examined, many of which are well known to regulate specific functions in other cell types (Table 16.2). A major aim of these studies was to identify molecules or pathways that could be therapeutically targeted to modulate PSC behavior, thereby inhibiting/retarding the fibrogenic process. The main findings in this area are summarized next.

The proliferative effect of PDGF on PSCs is mediated by extracellular signal regulated kinase (ERK1/2, a major component of the mitogen activated protein kinase (MAPK) pathway), and JAK/STAT (Janus activated/signal induced activation of transcription), while the chemotactic effect of PDGF is mainly mediated by phosphatidylinositol 3 kinase (PI3K) [65–67]. The existence of crosstalk between PI3K and ERK in PSCs has also been described, such that modulation of one pathway changes

Table 16.2 Pancreatic stellate cells—signaling pathways

Signaling pathways	PSC functions
Mitogen activated protein kinase (MAPK)	α-sma expression, proliferation, migration ECM protein synthesis
Phosphatidylinositol 3 kinase (PI3K)	Migration, proliferation, ECM protein synthesis
Protein kinase C (PKC)	ECM protein synthesis
Hedgehog	Migration
JAK-STAT	Proliferation
Smads	ECM protein synthesis
Rho, Rho kinase	Actin cytoskeleton, stress fiber formation
Transcription factors (AP-1, NF-κB, Gli-1)	Activation, migration, proliferation, ECM protein synthesis
Peroxisome proliferator activated receptor gamma (PPARγ)	α-sma expression, proliferation, phagocytosis

the function of the other. Cell migration is also influenced by Indian Hedgehog (IHH) [68]; this peptide is part of the Hedgehog protein family that plays an important role in the developing pancreas because it controls patterning and differentiation processes. Binding of IHH to its receptor system (Smoothened and patched-1), which is expressed by PSCs, leads to the movement of the transcription factor Gli-1 from the cytosol to the nucleus where it induces genes controlling PSC migration. This process is associated with relocation of matrix type metalloproteinases (MT1-MMP) to the cell surface. The subsequent degradation of basement membrane around the cells facilitates local invasion. With regard to the potent profibrogenic factor, TGFβ, it has been shown that binding of this growth factor to its receptor on PSCs induces SMAD 2 and 3 function [69]. Interestingly, TGFβ also induces the expression of its own mRNA in PSCs and this process is mediated by ERK1/2 [69].

PSC activation by alcohol, acetaldehyde, FAEE, or oxidant stress appears to be mediated by the MAPK pathway with all three major components—ERK1/2, p38 kinase (p38K), and c-jun amino terminal kinase (JNK)—playing a role [70,71]. In addition, alcohol and acetaldehyde have been reported to activate protein kinase C (PKC) and PI3K, two pathways upstream of the MAPK cascade [71]. The synergistic effects of alcohol and endotoxin noted earlier are likely mediated via an LPS-induced upregulation of its receptor TLR4 on PSCs [19]. As noted previously, PSCs also express other TLR receptors (2, 3, and 5), and binding of relevant ligands results in activation of NF-κB in the cells [18]. These findings are relevant to the observed LPS-induced decrease in apoptosis of PSCs because NF-κB can induce anti-apoptotic proteins such as IAPs (inhibitors of apoptosis proteins) [72] and may explain the LPS-induced inhibition of PSC apoptosis.

Regulation of the PSC actin cytoskeleton has recently been examined because increased α-sma expression is one of the consistent features of PSC activation.

Masamune et al. [73] have reported that the actin cytoskeleton and stress fiber formation in PSCs as well as the alteration of cell shape are regulated by the small GTP protein Rho and its downstream effector Rho kinase.

For most of the previously mentioned signaling pathways stimulated via binding of ligands such as growth factors and cytokines to relevant receptors, modulation of intracellular calcium signaling is a common downstream event. Such modulation has now been demonstrated in PSCs, not only in response to growth factors and cytokines, but also most recently to extracellular nucleotides (purines and pyrimidines) that act via P2X and P2Y receptors on PSCs to stimulate fibrogenesis [74].

MicroRNAs (small non-coding RNAs) have been attracting increasing attention in cell biology in general, and are now implicated in the regulation of a range of cellular functions such as differentiation, proliferation, apoptosis, protein synthesis, and so on. Shen et al. [75] have reported that miR-15b and miR-16 influence apoptosis in rat PSCs by targeting the anti-apoptotic factor Bcl-2. A recent analysis of microRNA profiles of quiescent and activated rat PSCs has revealed upregulation of 42 microRNAs and downregulation of 42 others in activated PSCs [76]. Using Integrated Pathway Analysis, the authors of this study report that the differentially expressed microRNAs are implicated in cell development, cell growth and proliferation, cell movement, cell death, and cell survival. While these are important contributions to the field, further studies are required to clarify the roles of specific microRNAs in the different phenotypic states of PSCs.

As would be expected, inhibition of the pathways mediating PSC activation noted earlier can facilitate PSC quiescence. Thus, the quiescence of PSCs upon exposure to retinol or its metabolites (ATRA and 9 cisRA) has been shown to be mediated via inhibition of MAPK [77]. Similarly, curcumin (a polyphenol in turmeric) reduces PDGF-induced PSC proliferation, by inhibiting the ERK pathway and cytokine-induced PSC activation by inhibiting activation of the transcription factor AP-1 [78]. It is noteworthy that the only transcription factor that directly inhibits PSC activation upon being stimulated by its ligand troglitazone is the PPARγ [79].

16.4 PSCs IN ACUTE PANCREATITIS

Although the initial focus of research into PSC biology was their possible role in the fibrosis of chronic pancreatitis (detailed later), it has now become evident that PSCs are also important players in acute pancreatitis, particularly during the repair/regeneration phase of the disease (Figures 16.3 and 16.4). The primary site of injury in acute pancreatitis is thought to be the acinar cell [80]. Regardless of the causative factor (commonly gallstone disease or alcohol abuse), the prevailing view, based on extensive studies with a variety of experimental animal models, is that the initial insult causes subcellular changes in the acinar cell that predispose the cell to autodigestion via

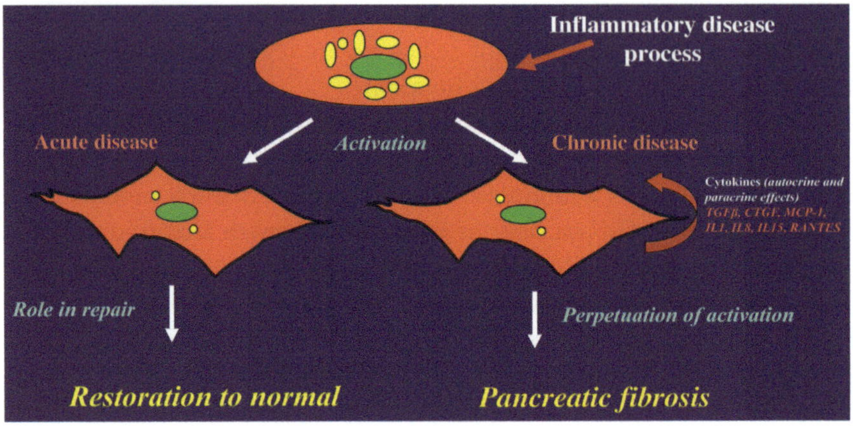

Figure 16.3 ***Role of pancreatic stellate cells in acute and chronic pancreatitis.*** PSCs are activated early during both acute and chronic pancreatic injury. In acute pancreatitis, activated PSCs play a role in pancreatic repair/regeneration via synthesis and secretion of ECM proteins that act as a scaffold and provide essential stimuli for acinar and ductal cell regeneration. In chronic pancreatitis, the initial activation of PSCs is perpetuated by endogenous production of cytokines by the PSCs, which act on the cells via autocrine pathways. The resultant excessive synthesis of ECM proteins and imbalance between ECM production and degradation leads to pathological fibrosis.

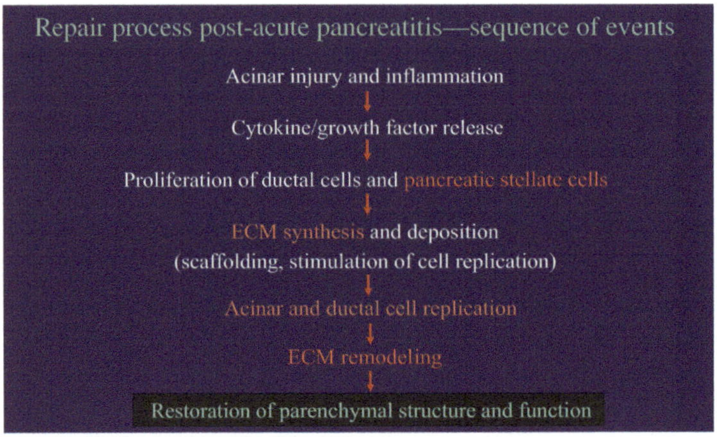

Figure 16.4 ***Repair/regeneration post-acute pancreatitis.*** The figure depicts a flow diagram denoting the sequence of events from injury to repair during acute pancreatitis.

premature intracellular activation of pancreatic digestive enzymes. Early in the course of injury, cytokine/growth factor levels in the pancreas are increased secondary to release from acinar cells themselves, as well as from inflammatory cells that home to the injured pancreas [32,81,82]. These proinflammatory mediators stimulate proliferation of ductal cells as well as interstitial cells (particularly PSCs) [47,81].

While local proliferation of resident PSCs accounts for the majority of the increased PSC numbers seen during acute necroinflammation of the pancreas, a recent study has shown that circulating bone marrow cells may also contribute to the PSC population during injury. Using a chimeric approach whereby bone marrow from enhanced green fluorescent protein (GFP) mice was transplanted into irradiated recipient rats injected with DBTC to produce pancreatitis, Sparmann et al. [83] reported that α-sma + ve/GFP + ve bone marrow derived PSCs contribute to approximately 18% (3 times that in control rats) of the activated PSC population in the injured pancreas. Activated PSCs in areas of injury produce ECM proteins that act as scaffolding and support for the regenerating acinar and ductal cells during the repair/recovery phase of acute pancreatitis. In addition, the ECM mediates intracellular signaling through its interaction with cell surface receptors such as integrins; these interactions are crucial for cell differentiation and growth [84]. Indeed, using conditional β1 integrin knockout mice, Riopel et al. [85] have shown that loss of β1 integrin expression in PSCs was associated with decreased ECM deposition, loss of acinar to acinar contact, increased acinar apoptosis, and decreased acinar proliferation.

PSCs are also thought to play a critical role in recovery from severe necrotizing pancreatitis where loss of exocrine parenchyma is quite significant. Zimmerman et al. [86] examined pancreatic sections from patients with severe necrotizing pancreatitis and described severe loss of acinar cells associated with areas termed as hypercellular regenerative spheres—islands of vascular granulation tissue, ductular cells, stellate cells, and residual lobular elements. The authors described the formation of pilot ductules that extended from the margins of the regenerative spheres into the surrounding tissue. These ductules were composed of flattened metaplastic cells (that were positive for ductal as well as acinar cell markers), surrounded by a mantle of α-sma positive activated PSCs (Figure 16.5). The authors speculated that this mantle of PSCs provided

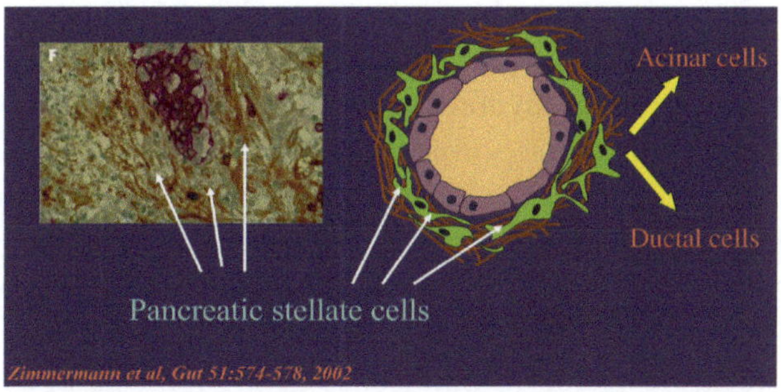

Figure 16.5 ***Role of pancreatic stellate cells in acute pancreatitis repair.*** Activated PSCs form a mantle around the pilot ductule lined by flattened, cytokeratin positive epithelial cells. The ECM produced by PSCs, plays a role in providing essential stimuli for re-differentiation of the cells lining the ductile, into mature acinar and duct cells.

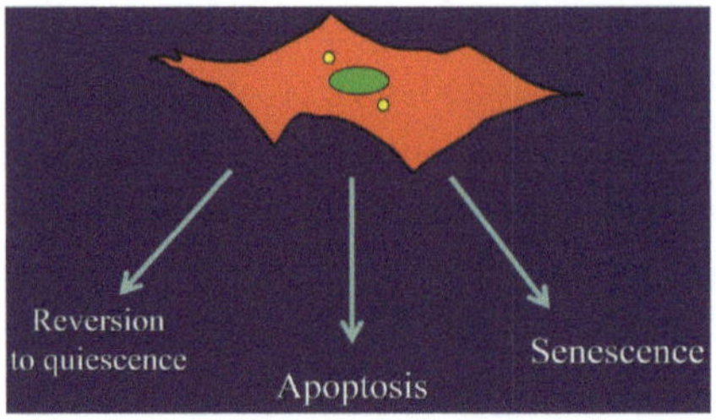

Figure 16.6 ***Fate of activated pancreatic stellate cells.*** Diagram depicting the three possibilities for the fate of activated PSCs—(i) reversion to quiescence; (ii) apoptosis; and (iii) senescence.

the necessary stimulus for the replication and differentiation of the epithelial cells in the pilot ductule, eventually facilitating regeneration of pancreatic parenchyma.

The final step in the recovery from acute pancreatitis involves removal of excess ECM and restitution of the cell population to its normal proportions, with the latter implying a reversal of the increased numbers of PSCs to their normal proportion in the parenchyma. This could be achieved in three possible ways (Figure 16.6): (i) reversion to quiescence; (ii) apoptosis; and (iii) senescence of activated PSCs. Partial reversion to quiescence has been reported *in vitro* upon exposure to retinol and its metabolites, albumin, or upon culture on Matrigel (a basement membrane like matrix) [11,77,87] and these observations are supported by *in vivo* findings of decreased PSC activation upon ATRA treatment, albeit in a model of pancreatic cancer (discussed later) [88]. Apoptosis of PSCs has been better studied and has been reported both *in vitro* and *in vivo* upon withdrawal of alcohol in the alcohol-fed endotoxin challenged model of chronic pancreatitis [59]. Evidence from a study by Li et al. [89] suggests that pancreatic stone protein, known to be upregulated in both acute and chronic pancreatitis, induces loss of PSCs by apoptosis and also causes increased effective MMP activity (by significantly decreasing the synthesis of TIMPs by PSCs), which could then facilitate fibrolysis.

PSC senescence has only attracted attention recently, with Fitzner et al. [90] reporting that after long-term culture (6 weeks) PSCs express the senescence marker senescence associated β-galactosidase (SA-βGal) and are highly susceptible to immune cell-mediated cytotoxicity. The authors further reported that in the rat model of dibutyltin chloride (DBTC)-induced pancreatitis, PSCs express activation markers as well as senescence markers leading them to the speculation that inflammation, PSC activation, and senescence are timely coupled processes in the injured pancreas. However, PSC apoptosis was not assessed in this study. The relative contributions of the processes of apoptosis versus senescence versus reversion to quiescence in the removal of activated PSCs after pancreatic injury remain to be clarified.

16.5 PSCs IN CHRONIC PANCREATITIS

To fully elucidate the role of PSCs in chronic pancreatitis, it was essential that the cells were examined not only *in vitro*, as described earlier, but also in their natural *in vivo* microenvironment. To that end, researchers turned to examination of sections from patients with chronic pancreatitis and to experimental animal models of the disease.

16.5.1 Human studies

Using standard staining techniques such as H&E, Masson's trichrome for connective tissue, Sirius Red for collagen, and immunohistochemical staining for PSC selective markers, it is now well established that areas of fibrosis in chronic pancreatitis sections, exhibit the presence of α-sma positive activated PSCs. Furthermore, dual staining techniques (i.e., immunohistochemistry for α-sma and *in situ* hybridization for procollagen mRNA) have provided unequivocal evidence that the α-sma positive PSCs are the predominant source of collagen I in the fibrotic pancreas [91].

It is also noteworthy that the expression of growth factors known to activate PSCs and relevant growth factor receptors are upregulated in chronic pancreatitis tissue. Strong staining for TGFβ is observed in spindle shaped cells in the fibrotic areas, and in acinar cells adjacent to areas of fibrosis, while acinar cells away from fibrotic bands are negative for TGFβ [91]. These findings indicate that TGFβ secreted by acinar cells may exert paracrine effects on PSCs resulting in increased collagen production by the cells and support the activation of PSCs observed upon exposure to TGFβ *in vitro*. The expression of the receptor for the mitogenic and chemotactic factor PDGF is also increased in fibrotic areas of chronic pancreatitis [91]. This may explain the heightened response of PSCs to PDGF during pancreatic injury and may be one of the mechanisms mediating PSC proliferation in the injured pancreas. Another selective PSC marker that is upregulated in chronic pancreatitis is NGF [92], a neurotrophic factor expressed by neuronal cells as well as PSCs. Because NGF is postulated to contribute to the pain of chronic pancreatitis via stimulation of neurite growth, PSCs (as a source of NGF) could be implicated in pancreatic pain.

Fibrotic areas of human chronic pancreatitis sections also exhibit increased levels of oxidant stress markers such as 4-hydroxynonenal (4HNE, a lipid peroxidation product) [35]. This is a relevant finding given that PSCs are known to be activated upon exposure to oxidant stress *in vitro*, as discussed earlier.

As noted earlier, diabetes is a well recognized complication of chronic pancreatitis [93], and is most likely secondary to beta cell destruction due to progressive fibrosis. Recent evidence from animal studies, described later, suggests a role for PSCs in beta cell dysfunction as well as in the production of progressive islet fibrosis.

16.5.2 Animal models

Observations with human tissue described in previous sections were essential confirmatory evidence of the association of PSC activation and pancreatic fibrosis in chronic pancreatitis. However, studies on pancreatic sections are inherently limited by their "point-in-time" nature, because chronological changes cannot be assessed by this approach. In order to determine the initiation and progression of chronic pancreatitis, researchers had to turn to experimental animal models (mostly rodent models) of the disease. Many different models of chronic pancreatitis have been described in the literature to date. Detailed descriptions of each of these models are beyond the scope of this chapter, but the approaches used to produce chronic pancreatitis in rats have been summarized: (i) intravenous injection of an organotin compound DBTC [94]; (ii) severe hyperstimulation obstructive pancreatitis (SHOP), involving intraperitoneal (IP) injections of supramaximal doses of caerulein (a synthetic analog of CCK, a major pancreatic secretagogue) + bile-pancreatic duct ligation [95]; (iii) trinitrobenzene sulfonic acid (TNBS) injection into the pancreatic duct [91]; (iv) spontaneous chronic pancreatitis in WBN/Kob rats [96]; (v) repeated IP injections of a superoxide dismutase inhibitor [97]; (vi) intragastric high-dose alcohol administration + repeated caerulein injections [98,99]; (vii) chronic alcohol administration (liquid diet) with repeated cyclosporin and caerulein injections [100], and (viii) chronic alcohol administration with repeated endotoxin (LPS) injections [19]. Mouse models of pancreatic fibrosis include: (i) repetitive pancreatic injury induced by repeated injections of supramaximal caerulein [101]; (ii) transgenic mice overexpressing IL-1β [102]; and (iii) transgenic mice overexpressing TGFβ or the EGF receptor ligand heparin binding epidermal growth factor-like growth factor (HB-EGF) [103].

Overall, these models have demonstrated that the expression of factors known to activate PSCs *in vitro*, are upregulated in pancreatic tissue. These included oxidant stress, PDGF and its receptor, TGFβ and two TGFβ regulated genes SM22a and Cygd/STAP. The studies have also confirmed that (i) PSCs are activated early during the course of injury, with the most likely activating factors being toxins such as alcohol or cytokines, growth factors, and oxidant stress (acting via paracrine or autocrine pathways); and (ii) activated PSCs were the predominant sources of collagen in the pancreas (as observed in human sections). It is important to add a cautionary note here with regard to the immediate clinical relevance of the models described above. All but one of the mentioned models were produced using nonphysiological methods (e.g., administration of supraphysiological doses of caerulein or injections of toxins into the pancreatic duct). The approach that would be closest to simulating the human situation is the model based on chronic alcohol intake and repeated endotoxin challenge [19], because raised serum endotoxin levels (secondary to alcohol-induced increases in gut permeability) have been well demonstrated in heavy drinkers [104]. This model showed that alcohol

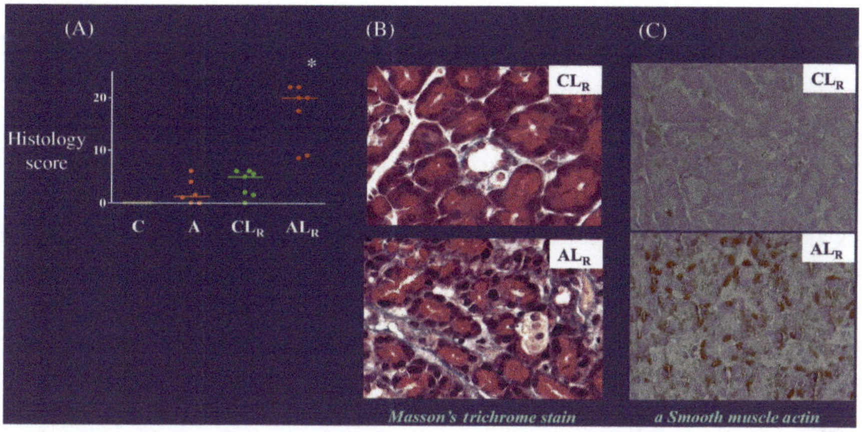

Figure 16.7 ***Rat model of alcoholic chronic pancreatitis involving chronic ethanol administration and repeated (one injection per week for 3 weeks) endotoxin (LPS) challenge.*** (A) Graphical representation of histological injury in 4 groups of rats as assessed by scoring for vacuolization, necrosis, inflammatory infiltrate, hemorrhage, and edema: (i) Control diet fed rat, no alcohol (C); (ii) Alcohol fed rat (A); (iii) Control diet fed rat challenged with repeated LPS injections (CL_R); (iv) Alcohol-fed rat challenged with repeated LPS injections (AL_R). Alcohol and LPS alone caused minimal histological damage to the pancreas. ALr animals exhibited the highest histological injury scores compared to the other 3 groups. 25 HPF/section, 3 sections per rat were examined (*$P < 0.001$ AL_R *vs* CL_R, *A and C; n = 7 rats/group*). (B) Masson's trichrome staining for pancreatic connective tissue: representative micrographs from A and AL_R animals showing increased fibrosis (blue Masson's staining) in the latter group. (C) α-sma staining for activated PSCs: representative micrographs from A and AL_R animals showing significantly increased α-sma in the latter group.

and endotoxin exerted synergistic effects on the pancreas, causing acinar atrophy associated with increased PSC activation and decreased PSC apoptosis, thereby enhancing the survival of PSCs and facilitating increased fibrosis (Figure 16.7).

As was the case with acute pancreatitis, it is now evident that a small proportion of the increased numbers of PSCs observed in models of chronic pancreatitis are derived from the bone marrow. Two recent studies have used a gender mismatch and chimeric approach (GFP labeled bone marrow derived cells (BMDC) from donor male mice, transplanted into irradiated wild type female mice) with subsequent induction of chronic pancreatitis in the recipient mice by repeated caerulein injections or by injection of DBTC into the pancreatic duct [105,106]. The authors reported that between 5% and 18% of the proliferating PSCs expressed GFP and were therefore bone marrow derived. However, the specific contribution of these BMDCs to progression of chronic pancreatitis remains unclear. Additional studies with different models of chronic pancreatitis would be useful to fully delineate the role of BMDCs in the pathophysiologic process of the disease.

16.5.2.1 PSCs in pancreatic diabetes

Saito et al. [107] used PSC selective markers such as GFAP and the activation marker α-sma to demonstrate the presence of increasing numbers of PSCs within fibrotic areas around and within the islets of Goto-Kakizaki rats (a model of type 2 diabetes). *In vitro* studies by Kikuta et al. [108] have demonstrated a direct interaction between PSCs and beta cells. The authors report that in the presence of PSCs, beta cells exhibit increased apoptosis and produce significantly less insulin. In turn, beta cells (specifically the beta cell line INS-1) have been shown to increase PSC proliferation [109]. Surprisingly, however, these authors reported that PSC migration, ECM production, and cytokine production were decreased upon exposure to conditioned medium from INS-1 cells. They speculate that this may represent a capacity of beta cells to protect themselves from PSC-induced damage. Obviously, further work is needed to fully elucidate the possible role of PSCs in pancreatic diabetes.

16.6 REVERSAL OF PANCREATIC FIBROSIS IN CHRONIC PANCREATITIS

Our improved understanding of processes of PSC activation has helped the development of potentially useful anti-fibrotic therapies *in vivo* (albeit tested only in experimental models so far). Approaches/therapeutic strategies that have been reported to prevent/retard fibrosis in animal models, include: (i) cytokine suppression—(a) TGFβ inhibition—using TGFβ neutralizing antibodies [110], a herbal medicine Saiko-keishi-to [111] or a plant alkaloid halofuginone, which inhibits downstream Smad3 phosphorylation [112]; (b) TNFα inhibition—using a TNFα antibody [113], soluble TNFα receptors or an inhibitor of TNFα production pentoxifylline [114]; (ii) antioxidants—vitamin E (the subclass tocotrienol has been shown to induce PSC death via apoptosis and autophagy) [115,116], oxypurinol and allopurinol, both xanthine oxidase inhibitors [114,117], ellagic acid, a plant derived polyphenol with antioxidant, anti-inflammatory, and anti-fibrosis activities [118], and salvianolic acid, a herbal medicine with free radical scavenging properties [119]; (iii) antiinflammatory agents—protease inhibitors such as camostat mesilate, which inhibit proinflammatory cytokine production by monocytes [120] and the synthetic carboxamide derivative IS-741, which suppresses macrophage infiltration into the pancreas, with a consequent decrease in *in vivo* PSC activation [121]; (iv) modulation of signaling molecules using the PPARγ ligand troglitazone [79]; (v) angiotensin inhibition using the ACE inhibitor captopril or the angiotensin II receptor inhibitor losartan [122]; (vi) inhibition of collagen synthesis—administering siRNA for a collagen-specific chaperone in vitamin A-coupled liposomes that target PSCs [123].

In regards to alcoholic chronic pancreatitis, clinicians usually advocate abstinence in a bid to prevent disease progression in their patients. However, epidemiological evidence regarding the benefits of abstinence has been contradictory. Vonlaufen et al. [59] have recently provided the first experimental evidence to support the advice of abstinence to patients. The authors have reported that withdrawal of alcohol from the diet after established early pancreatitis in a rat model results in complete reversal of pancreatic fibrosis [59], an effect attributed to the fact that the alcohol-induced inhibition of PSC apoptosis (described earlier) is removed in the absence of alcohol, which enables the loss of activated PSCs through cell death and interrupts the fibrogenic process.

16.7 PSCs IN PANCREATIC CANCER

As evidence of the key role of PSCs in the fibrosis of chronic pancreatitis was being consolidated, it became apparent that there was a large gap in knowledge with regard to the fibrogenic process in another major disease of the pancreas with well documented and abundant stromal tissue, namely pancreatic ductal adenocarcinoma (PDAC). Indeed, a typical histological section of PDAC shows islands of tumor cells embedded in a profuse desmoplastic/fibrotic reaction (Figure 16.1). Interestingly, Binkley et al. [124] have described a significant overlap in gene expression of the stromal compartments of chronic pancreatitis and pancreatic cancer. In view of this and the known increased risk of pancreatic cancer in patients with chronic pancreatitis [125,126], it was logical to question whether PSCs were responsible for the fibrosis of pancreatic cancer and if so, whether they interacted with cancer cells to influence disease progression.

Human pancreatic cancer sections stained for PSC selective markers have unequivocally shown that activated PSCs are present in stromal areas [127] (Figure 16.8A). Moreover, using dual staining methods for α-sma and procollagen mRNA (as described earlier for chronic pancreatitis) it has been confirmed that PSCs are the main source of collagen in the stromal reaction [127]. Importantly, recent studies indicate that α-sma positive PSCs can be detected as early as at the premalignant stage of PDAC, that is, surrounding PanIN (pancreatic intraepithelial neoplasia) lesions both in human pancreatic tissue [128] (Figure 16.8B) as well as in pancreatic sections from a genetically modified mouse model of the disease (Figure 16.8C). Similarly, periostin (a cell adhesion protein solely expressed by PSCs) has been observed in intraductal papillary mucinous neoplasms (IPMNs) of the human pancreas [129], further supporting the idea that PSCs are activated early in the neoplastic process and that they may be influential in cancer progression.

Evidence to support the previously mentioned role of PSCs in cancer has been sought via both *in vitro* (cultured cells) and *in vivo* (subcutaneous, orthotopic, and transgenic models) approaches. As summarized next, these studies indicate a close interaction of PSCs not only with cancer cells themselves, but also with other cell types in the stroma such as endothelial cells and immune cells.

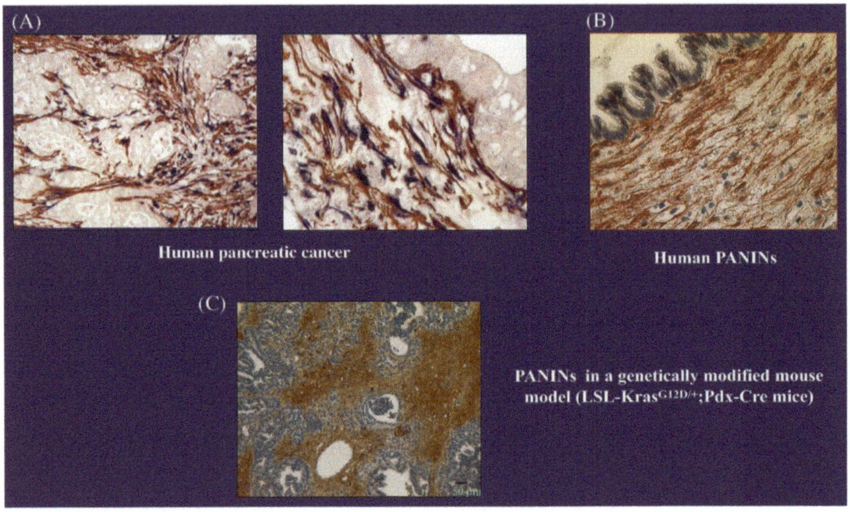

Figure 16.8 ***Role of pancreatic stellate cells in pancreatic cancer.*** (A) Representative photomicrographs (low and high power) of dual stained pancreatic cancer sections (immunostaining for α-sma (brown) and *in situ* hybridization for procollagen mRNA (blue)), showing that cells staining positive for α-sma also exhibit the presence of procollagen mRNA, indicating that activated PSCs are the predominant source of collagen in the desmoplasia of pancreatic cancer. (B) Representative photomicrograph of a human PanIN showing the presence of α-sma positive, activated PSCs. (C) Representative photomicrograph of PanINs in a genetically modified mouse model (LSL-Kras$^{G12D/+}$; Pdx-Cre mice) showing the presence of α-sma positive, activated PSCs.

16.7.1 *In vitro* studies

These have usually involved co-culture of PSCs with the cell of interest or exposure of one cell type to conditioned medium from the other (and vice versa).

16.7.1.1 PSC–cancer cell interactions

It has now been consistently demonstrated that PSCs are activated (as evidenced by increased proliferation, migration, and ECM synthesis) upon exposure to cancer cells (either directly or via conditioned media) [130]. On the other hand, cancer cells exposed to PSCs exhibit increased proliferation and decreased apoptosis (which effectively improves survival) and increased migration. The latter effect is associated with enhanced epithelial–mesenchymal transition (EMT) in cancer cells as indicated by increased expression of mesenchymal markers such as vimentin and Snail and a concomitant decrease in expression of epithelial markers such as E-cadherin [131]. It is possible that this PSC-induced EMT in cancer cells facilitates their migration. In view of the observed presence of activated PSCs around the early neoplastic lesions (PanINs) [128], it is noteworthy that interactions between PSCs and PanIN cells have also now been confirmed *in vitro* [132]. Exposure of PSCs to PanIN cells *in vitro* resulted in

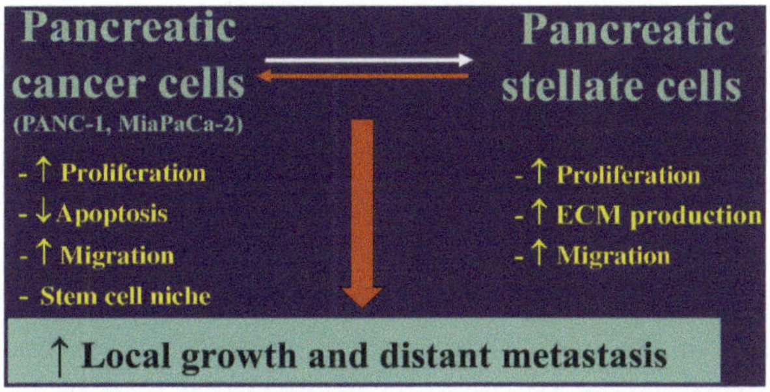

Figure 16.9 ***Interactions between pancreatic cancer cells and pancreatic stellate cells.*** Diagram depicts the effects of pancreatic cancer cells on PSCs and vice versa. Pancreatic cancer cells induce proliferation and migration of PSCs and stimulate ECM production by the cells. In turn, PSCs induce cancer cell proliferation but inhibit cancer cell apoptosis (thus increasing their overall survival). PSCs also stimulate cancer cell migration and stimulate "stemness" as assessed by increased expression of stem cell markers. This interaction between pancreatic cancer cells and PSCs sets up a facilitatory environment for pancreatic tumor growth and metastasis.

activation and increased proliferation, fibronectin production, and MMP expression in PSCs. These findings indicate that preneoplastic cells have the capacity to influence PSC functions even at very early stages of the disease. Reciprocal effects of PSCs on PanIN cells may also be postulated, albeit through indirect evidence such as the report by Funahashi et al. [133] showing that inhibition of COX-2 (the inducible form of cyclooxygenases expressed by PSCs) using nimesulide, slows the progression of precursor lesions in a genetically engineered mouse model (GEMM) of pancreatic cancer.

Keeping in mind the known heterogeneity of pancreatic cancer, particularly with respect to disease progression, researchers have asked whether all PSCs exert uniformly similar effects on cancer cells. In this regard, a recent study by Ikenaga et al. [134] has reported that within PSC isolates from pancreatic cancer patients, there is a subset of cells that overexpress CD10 (a cell membrane associated MMP). The authors found that CD10 + ve PSCs induced significantly more invasion and proliferation of cancer cells than CD10-ve cells, suggesting that functional heterogeneity of PSCs may influence their effects on tumor progression.

Because chemoresistance and recurrence are common features of pancreatic cancer, there has been some interest in the possible role of PSCs in these processes. Hamada et al. [135] have recently reported that co-culture of PSCs with the cancer cell lines AsPC-1 or SUIT-2 results in an increase in stem cell marker expression (nestin, ABCG2, and LIN28) in some cancer cells, prompting the speculation that PSCs facilitate the selection and survival of a cancer stem cell niche, which may then be responsible for recurrence of tumor.

These findings provide strong evidence that pancreatic cancer cells recruit host PSCs to their immediate vicinity from the earliest stages of the carcinogenic process and that PSCs reciprocate by facilitating cancer cell growth, survival, and invasion (Figure 16.9).

Mediators of this interaction are now being sought in an effort to identify therapeutically targetable pathways. There is evidence that increased ECM synthesis by PSCs in response to cancer cells is likely mediated by TGFβ1 and fibroblast growth factor 2 (FGF2) [136]. Factors implicated in cancer cell-induced PSC proliferation include PDGF, COX-2 [137], and trefoil factor 1 (TFF1) [138] (a stable secretory protein that is upregulated in pancreatic cancer but not expressed in the normal pancreas), with ERK1/2 identified as the responsible signaling pathway [138]. Factors mediating the effects of PSCs on cancer cells have not yet been characterized, although there is evidence that PSC-induced proliferation of cancer cells may be mediated, at least in part, by PDGF [139]. Additional candidate factors in PSC secretions worthy of study as possible mediators include hepatocyte growth factor (HGF), TGFβ, insulin-like growth factor (IGF), EGF and proinflammatory cytokines. Interestingly, ERK1/2 and Akt have been shown to be the signaling pathways that regulate cancer cell migration, invasion, and colony formation in response to PSCs [139,140].

16.7.1.2 PSC–endothelial cell interactions

Neo-angiogenesis is a major feature of all cancers and while the interaction between cancer cells and endothelial cells has been well documented, that between PSCs and endothelial cells has only recently received some attention. PSCs have now been shown to induce proliferation and tube formation (a measure of angiogenesis) of human microvascular endothelial cells (HMEC-1) *in vitro* [141] and these effects are thought to be mediated via vascular endothelial growth factor (VEGF). Most recently, Patel et al. [142] have implicated the HGF (known to be principally secreted by PSCs)—cMET (receptor for HGF expressed on endothelial cells) pathway in the inductive effects of PSCs on endothelial cells. They have shown that targeting this pathway using a monoclonal antibody to HGF significantly decreases PSC-induced HMEC1 proliferation and tube formation. Further, they have demonstrated that p38 kinase and PKC are the downstream signaling pathways mediating the effects of PSCs on HMEC1 cells.

16.7.1.3 PSC–immune cell interactions

The desmoplastic reaction in PDAC is known to contain a prominent immune cell infiltrate [128,143–145] but these cells appear to be unable to exert their immunogenic effects to destroy cancer cells. There is some evidence to suggest that PSCs may aid the process of immune evasion by cancer cells. PSCs exert a chemotactic effect on $CD8^+$ T cells and this effect is mediated by the PSC-derived chemokine CXCL12 [143]. It is thought that this effect of PSCs explains the observed sequestration of $CD8^+$ T cells in stromal areas *in vivo*, reducing the number of T cells from infiltrating peritumoral areas to exert their anti-tumor effects. PSCs have also been shown to interact with mast cells, inducing their activation and production of tryptase and

IL13, both of which increase cancer cell proliferation, while at the same time activating PSCs [146]. More recently, PSCs have been implicated in inducing the migration of myeloid derived suppressor cells (MDSC) into pancreatic tumors, via the synthesis of IL16 [147]. As the name suggests, MDSCs suppress immune cell function, thereby aiding the lack of immune surveillance in pancreatic cancer.

16.7.1.4 PSC–neural cell interactions

Significant neural remodeling (neural hypertrophy and increased neural density) is known to occur in pancreatic cancer [148]. However, there have not yet been any studies assessing the interactions of PSCs and neural cells. Because PSCs express several neurotrophic factors (as noted earlier), it would not be unreasonable to postulate that PSCs may also act as neural elements in cancer stroma, in addition to influencing the growth and proliferation of nerves via secretion of NGF and ECM components such as collagen and fibronectin. Indeed, the recent observation of a close correlation between extent of desmoplasia and degree of neural invasion in pancreatic cancer [149] lends support to this concept.

16.7.2 *In vivo* studies

The *in vitro* studies described earlier provide strong evidence for an active role of PSCs in pancreatic cancer pathobiology. However, because confirmatory findings in *in vivo* settings are key to ascribing clinical relevance to these studies, researchers have assessed stromal–tumor interactions in subcutaneous, orthotopic, and transgenic models of pancreatic cancer. One of the earliest such studies by Bachem et al. [136] used an immunocompromised mouse model where pancreatic cancer was produced by subcutaneous injections of either a mixture of cancer cells and PSCs or of cancer cells alone. The authors reported a significantly increased growth rate in tumors in the presence of PSCs compared to controls (without PSCs). These larger tumors were due to both the fibrosis produced by the PSCs and also the increased proliferation of cancer cells themselves, supporting the *in vitro* observation of a proliferative effect of PSCs on cancer cells.

Orthotopic models of pancreatic cancer have a distinct advantage over subcutaneous tumor models because the tumors are produced within the organ of interest and the cancer and stromal cells are in their natural microenvironment. In addition, the metastatic process (a hallmark of pancreatic cancer [150]) can be readily studied in such models in order to elucidate mechanisms of cancer progression. Lohr et al. [151] have reported that orthotopic injections of pancreatic cancer cells (PANC-1 cell line) transfected with TGFβ1 cDNA into immunodeficient mice resulted in tumors that exhibited a prominent and extensive stromal reaction; although they did not study the mechanism of fibrosis in this study, it is likely that TGFβ from the injected cancer cells induced activation of fibroblasts/PSCs in the host (mouse) pancreas.

Other, more recent orthotopic models have involved the injection of human cancer cells (MiaPaCa-2, AsPC-1, BxPC-1) with or without human pancreatic stellate cells (hPSCs) into the mouse pancreas [139–141,152]. Local tumor growth, and more importantly, regional and distant metastasis were significantly increased in the presence of PSCs. The histological appearance of tumors produced by cancer cells alone was of a mass of tumor cells, completely devoid of any stromal reaction. In contrast, tumors formed by cancer cells + PSCs were akin to human pancreatic cancer with bands of desmoplasia surrounding cancer cells, positive staining for α-sma in fibrotic areas, increased cancer cell proliferation, and decreased apoptosis (as assessed by cytokeratin and TUNEL staining, respectively). These findings support the *in vitro* observations of interactions between PSCs and cancer cells that may facilitate cancer progression (increased local growth and distant metastasis).

Traditionally, metastasis has been thought to be limited to cancer cells alone that move from the primary tumor to distant sites. Intriguingly, however, Xu et al. [141] have provided strong evidence to suggest that PSCs may also accompany cancer cells to metastatic sites. Using a gender mismatched approach whereby orthotopic tumors were produced by injection of a mixture of female pancreatic cancer cells (AsPC-1, derived from a female patient) + male PSCs into the pancreas of female nude mice, the authors showed that metastatic nodules in these mice exhibited the presence of y-chromosome positive PSCs. This implies an ability of PSCs to intravasate/extravasate to and from blood vessels, which has been confirmed *in vitro* using transendothelial migration assays [141]. Notably this transendothelial migration of PSCs was further stimulated in the presence of cancer cells, an effect thought to be mediated by PDGF in cancer cell secretions [141]. Thus, PSCs appear to aid cancer cells not only within the pancreas but also at distant metastatic sites by possibly facilitating their seeding, survival, and proliferation. A similar finding was subsequently reported in a model of lung cancer [153], raising the possibility that metastasis can no longer be considered as a phenomenon limited to cancer cells alone.

Pancreatic cancer is notorious for its resistance to chemotherapy and radiotherapy [150]. The reasons for this are not fully known, but recent studies with orthotopic models suggest that PSCs may play a role by producing the dense desmoplastic stroma that may sequester chemotherapeutic agents thereby reducing effective amounts of the drug from reaching the cancer cells. Mantoni et al. [154] have shown that PSCs may be responsible for radioresistance because they can protect cancer cells from radiation via a β1 integrin-dependent pathway.

While the xenograft studies (using human cells) described earlier have been very useful in delineating the role of stromal-tumor interactions in pancreatic cancer, they lack an important component, namely the immune system, because the models necessarily involve immunocompromised animals. Genetically engineered mouse models provide the opportunity to study pancreatic cancer as it develops spontaneously in an immunocompetent setting. Many different mouse models have now been described in the literature [155], including: (i) KPC mice ($Kras^{LSL\text{-}G12D/+}$; $Trp53^{LSL\text{-}R172H/+}$; $Pdx^{cre/+}$), (ii) KPGC mice ($Kras^{LSL\text{-}G12D/+}$; $Trp53^{LSL\text{-}R172H/+}$; $R26^{LSL\text{-}GFP/+}$; $Pdx^{cre/+}$), and (iii) TGFβ type II receptor organ specific knockout in the mouse pancreas ($Kras^{LSL\text{-}G12D/+}$; $TGF\beta r2^{floxflox}$; $Ptf1a^{cre/+}$). Progressive lesions are observed in these animals starting with preinvasive changes (pancreatic intraepithelial neoplasms, PanINs) to obvious carcinoma and metastasis. Interestingly, as noted earlier, activated α-sma positive PSCs have been identified in the earliest PanIN lesions [128].

As discussed earlier, human pancreatic cancer tissue exhibits significant infiltration by immune cells such as T cells, B cells, NK cells, neutrophils, and macrophages as well as myeloid-derived suppressor cells (as the name suggests, MDSCs have a largely immunosuppressive function) [128,143–145]. Yet, pancreatic cancer cells appear to be able to successfully evade any attacks from these immune cells. One of the mechanisms mediating this evasive ability is the production of granulocyte-macrophage colony-stimulating factor by cancer cells that suppresses anti-tumor T-cell immunity [156]. Recent work suggests that PSCs may also help with immune evasion. PSCs in the stroma of PanIN lesions and around cancer cells produce galectin-1, a β-galactoside-binding protein, that binds to *N*-acetyllactosamine on membrane glycoproteins and induces T-cell apoptosis, thus suppressing the immune response [157]. Activated PSCs reduce the migration of CD8 positive T cells towards cancer cells in both human PDAC and the KPC mouse model of pancreatic cancer [143]. Fibroblast activation protein-α (FAP-α), expressed by stromal cells, can also disrupt anti-tumor immunity. Depletion of the cells expressing FAP-α enabled immune response-associated tumor regression, supporting the notion that FAP-α might act as an immune suppressor in pancreatic cancer [158]. A recent study has suggested the involvement of another type of immune cell, the mast cell, in pancreatic cancer progression [159]. Orthotopic cancer growth was found to be significantly hampered in mast cell deficient Kit mice, while the reconstitution of mast cells in these mice from the bone marrow of wild type restored tumor growth. Interestingly, as described earlier, PSCs have been shown to activate mast cells *in vitro* [158], suggesting crosstalk between these two cell types in the stroma. When considered together, these findings indicate that PSCs may negatively modulate immune responses in pancreatic cancer.

16.7.3 Therapeutic targeting of stroma in pancreatic cancer

Therapeutic modalities mainly targeting cancer cells alone (surgery, radiotherapy, and chemotherapy) have unfortunately been unable to significantly improve pancreatic cancer outcome over many decades. Given the accumulating evidence of the importance of the stroma in pancreatic cancer progression, strategies are now being developed to target not only cancer cells but also the desmoplastic reaction, with a particular focus on inhibiting PSC activation (via inhibition of signaling pathways mediating activation) and reducing the stroma. The majority of studies are restricted to pre-clinical models, although a few have now been tested in clinical trials. Treatments that have been assessed to date in both pre-clinical and clinical settings are summarized next.

16.7.3.1 Inhibition of the Hedgehog pathway

PSCs express the receptors Patched and Smoothened that bind Hedgehog ligands (Sonic, Indian, and Desert Hedgehog). In transgenic models of pancreatic cancer, inhibition of Smoothened by cyclopamine marginally increased median survival by 6 days [160]. Inhibition of the receptor by IPI-926 (a cyclopamine derivative) in combination with gemcitabine, increased the delivery of the chemotherapeutic agent to cancer cells, but had only a transient effect on improving blood vessel density and extension of median survival [161]. Subsequently, Hwang et al. [152] used another smoothened inhibitor AZD8542 in an orthotopic model of pancreatic cancer. AZD8542 reduced tumor volume, metastasis, and Hedgehog downstream signaling activity. Based on these encouraging pre-clinical reports, clinical trials using Hedgehog inhibitors were commenced. Unfortunately, the phase II trial with IPI-926 had to be abandoned prematurely due to decreased survival of patients in the treatment arm. The lack of translation of the pre-clinical findings to the clinical setting may reflect the fact that pre-clinical models do not fully capture the heterogeneity of human pancreatic cancer, or that the pre-clinical findings need to be better confirmed using a range of experimental settings.

Nanoparticle albumin complexed paclitaxel (nab-paclitaxel) ± *gemcitabine* depleted the stroma in a patient-tumor-derived subcutaneous xenograft model and increased perfusion via an increase in blood vessel diameter with consequent improved delivery of gemcitabine to tumor cells [162]. In a recent Phase 3 trial, this combination was found to significantly improve overall survival as well as progression-free survival compared to gemcitabine alone (8.5 months vs. 6.7 months and 5.5 months vs. 3.7 months, respectively) [163]. Although the improvements may be regarded as modest, the results support the concept that targeting the stroma in addition to cancer cells may be a potentially beneficial approach.

16.7.3.2 Activation of CD40 using a CD40 agonist monoclonal antibody

In the KPC mouse model of pancreatic cancer, activation of CD40, a member of the TNF receptor superfamily, activated macrophages in the stroma and resulted in apoptosis of cancer cells as well as a reduction in stromal collagen [164]. In a Phase I study involving a small number of chemotherapy-naïve advanced pancreatic cancer patients, the antibody in combination with gemcitabine exhibited evidence of anti-tumor activity, but with heterogeneous responses particularly with regard to metastatic lesions [165]. Thus, larger randomized controlled trials will be needed before the role of a CD40 agonist monoclonal antibody in pancreatic cancer treatment can be clearly determined.

16.7.3.3 PEGylated human recombinant PH20 hyaluronidase (PEGPH20)

This compound enzymatically degrades one of the predominant components of the ECM, hyaluronan [166]. PEGPH20 treatment of KPC mice caused stromal depletion, decompression of tumor vessels, and increased permeability of the endothelium to macromolecules. When combined with gemcitabine, PEGPH20 treatment improved the delivery of gemcitabine to tumor cells inhibiting tumor growth and extending the median survival of the mice [166,167]. However, a Phase 1b trial with PEGPH20 has resulted only in a partial response in patients with advanced pancreatic cancer. A Phase 2 multicenter, randomized clinical trial is currently underway to evaluate PEGPH20 as a first-line therapy for patients with stage 4 metastatic pancreatic cancer (Study 202 (Halo 109–202)).

16.7.3.4 Angiotensin II receptor antagonists

As noted earlier, angiotensin II has been shown to induce PSC activation. Angiotensin II receptor blockade, using olmasartan decreased tumor growth in a subcutaneous xenograft model of pancreatic cancer [168], while losartan decreased α-sma positive cells, and reduced collagen and hyaluronan production in the stroma of pancreatic cancer in an orthotopic mouse model [169]. A Phase II trial (NCT01821729) is currently recruiting patients for treatment with a combination of losartan, Folfirinox, and proton beam radiation.

Other approaches, albeit currently limited to pre-clinical studies include (i) the anti-fibrotic agent Pirfenidone reported to decrease the growth of tumors produced by the injection of a mixture of pancreatic cancer cells and PSCs, but not that of tumors produced by cancer cells alone [170]; (ii) the phytonutrients ellagic acid and embelin, which have been reported to decrease proliferation and increase apoptosis of cancer cells as well as stellate cells resulting in significantly reduced tumor volumes in a xenograft model of pancreatic cancer [171]; and (iii) all trans retinoic acid (ATRA) reported to decrease tumor growth in a transgenic model of pancreatic cancer [88].

While the findings of these studies support the concept that targeting the stroma would represent a potentially useful addition to the armamentarium of anti-pancreatic cancer strategies, the picture is somewhat clouded because clinical translation of the observed pre-clinical benefits remains a challenge. Two recent pre-clinical studies have brought to the fore the reality that stromal targeting in pancreatic cancer will require careful thought and further study before clinically effective treatments can be identified. Rhim et al. [172] reported that deletion of Sonic Hedgehog in a genetically engineered mouse model decreased stroma but increased vascularity of tumors, resulting in an overall increase in aggressiveness of the cancer, while Ozdemir et al. [173] used a mouse model of pancreatic cancer where myofibroblasts were conditionally depleted and reported decreased stromal collagen and decreased vascularity, but increased EMT in cancer cells and decreased survival. A discussion of the limitations and strengths of these recent studies is beyond the scope of this chapter, but it is clear that stromal–tumor interactions in pancreatic cancer represent complex and rapidly changing events (possibly depending on disease stage) and warrant further detailed study via next generation *in vitro* approaches such as three-dimensional/organotypic cultures and *in vivo* experiments involving a variety of clinically relevant animal models.

16.8 SUMMARY AND CONCLUSIONS

Compared to the HSC field, studies elucidating PSC biology are relatively recent. Nonetheless, over the past 16 years there has been a steep improvement in our understanding of pancreatic fibrogenesis and the central role of PSCs in this process. It is also becoming increasingly evident that PSCs have functions over and beyond the regulation of pathologic fibrosis in the pancreas, with the cells likely playing important roles in health as immune and/or progenitor cells and as intermediary cells in digestive enzyme secretion (at least in humans). The concerted research efforts currently underway to better clarify the role of PSCs in chronic pancreatitis and pancreatic cancer (the latter summarized in Figure 16.10) provide strong hope for the development of effective novel therapies in the near future for these difficult to treat diseases of the pancreas.

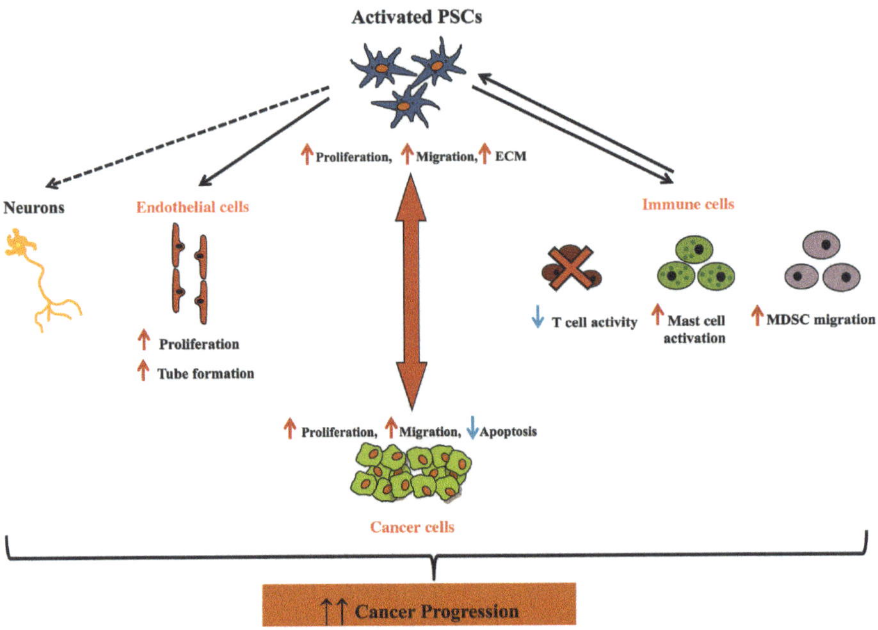

Figure 16.10 Diagram depicting the interactions between pancreatic stellate cells and pancreatic cancer cells, as well as between PSCs and other stromal cells including immune cells, endothelial cells, and neurons. Solid arrows indicate the availability of strong experimental evidence in support of the interactions between PSCs and relevant cells. Broken arrow indicates that the interaction between PSCs and neurons while postulated has not yet been experimentally examined.

ACKNOWLEDGMENTS

Authors were supported by grants from the National Health and Medical Research Council of Australia, Australian Research Council, and the Cancer Council of New South Wales.

REFERENCES

[1] Apte M, Pirola R, Wilson J. New insights into alcoholic pancreatitis and pancreatic cancer. J Gastroenterol Hepatol 2009;24(Suppl. 3):S51–6.
[2] Kloppel G. Chronic pancreatitis, pseudotumors and other tumor-like lesions. Mod Pathol 2007;20(Suppl. 1):S113–31.
[3] Kocabayoglu P, Friedman SL. Cellular basis of hepatic fibrosis and its role in inflammation and cancer. Front Biosci (Schol Ed) 2012;5:217–30.
[4] Wake K. Perisinusoidal stellate cells (fat-storing cells, interstitial cells, lipocytes), their related structure in and around the liver sinusoids, and vitamin A-storing cells in extrahepatic organs. Int Rev Cytol 1980;66:303–53.
[5] Watari N, Hotta Y, Mabuchi Y. Morphological studies on a vitamin A-storing cell and its complex with macrophage observed in mouse pancreatic tissues following excess vitamin A administration. Okajimas Folia Anat Jpn 1982;58:837–58.
[6] Ikejiri N. The vitamin A-storing cells in the human and rat pancreas. Kurume Med J 1990;37:67–81.
[7] Apte MV, Haber PS, Applegate TL, et al. Periacinar stellate shaped cells in rat pancreas—identification, isolation, and culture. Gut 1998;43:128–33.
[8] Bachem MG, Schneider E, Gross H, et al. Identification, culture, and characterization of pancreatic stellate cells in rats and humans. Gastroenterology 1998;115:421–32.
[9] Apte MV, Pirola RC, Wilson JS. Pancreatic stellate cells: a starring role in normal and diseased pancreas. Front Physiol 2012;3:344.
[10] Yamada M, Blaner WS, Soprano DR, Dixon JL, Kjeldbye HM, Goodman DS. Biochemical characteristics of isolated rat liver stellate cells. Hepatology 1987;7:1224–9.
[11] Kim N, Yoo W, Lee J, et al. Formation of vitamin A lipid droplets in pancreatic stellate cells requires albumin. Gut 2009;58:1382–90.
[12] Kim N, Choi S, Lim C, Lee H, Oh J. Albumin mediates PPAR-gamma or C/EBP-alpha-induced phenotypic changes in pancreatic stellate cells. Biochem Biophys Res Commun 2010;391:640–4.
[13] Buchholz M, Kestler HA, Holzmann K, et al. Transcriptome analysis of human hepatic and pancreatic stellate cells: organ-specific variations of a common transcriptional phenotype. J Mol Med 2005;83:795–805.
[14] Paulo JA, Urrutia R, Banks PA, Conwell DL, Steen H. Proteomic analysis of an immortalized mouse pancreatic stellate cell line identifies differentially-expressed proteins in activated vs nonproliferating cell states. J Proteome Res 2011;10:4835–44.
[15] Mato E, Lucas M, Petriz J, Gomis R, Novials A. Identification of a pancreatic stellate cell population with properties of progenitor cells: new role for stellate cells in the pancreas. Biochem J 2009;421:181–91.
[16] Kordes C, Sawitza I, Haussinger D. Hepatic and pancreatic stellate cells in focus. Biol Chem 2009;390:1003–12.
[17] Shimizu K, Kobayashi M, Tahara J, Shiratori K. Cytokines and peroxisome proliferator-activated receptor gamma ligand regulate phagocytosis by pancreatic stellate cells. Gastroenterology 2005; 128:2105–18.
[18] Masamune A, Kikuta K, Watanabe T, Satoh K, Satoh A, Shimosegawa T. Pancreatic stellate cells express Toll-like receptors. J Gastroenterol 2008;43:352–62.
[19] Vonlaufen A, Xu ZH, Joshi S, et al. Bacterial endotoxin—a trigger factor for alcoholic pancreatitis? Findings of a novel physiologically relevant model. Gastroenterology 2007;133:1293–303.
[20] Shimizu K, Hashimoto K, Tahara J, Imaeda H, Andoh A, Shiratori K. Pancreatic stellate cells do not exhibit features of antigen-presenting cells. Pancreas 2012;41:422–7.

[21] Unanue ER. Ito cells, stellate cells, and myofibroblasts: new actors in antigen presentation. Immunity 2007;26:9–10.
[22] Winau F, Hegasy G, Weiskirchen R, et al. Ito cells are liver-resident antigen-presenting cells for activating T cell responses. Immunity 2007;26:117–29.
[23] Owyang C, Logsdon CD. New insights into neurohormonal regulation of pancreatic secretion. Gastroenterology 2004;127:957–69.
[24] Ji B, Bi Y, Simeone D, Mortensen RM, Logsdon CD. Human pancreatic acinar cells lack functional responses to cholecystokinin and gastrin. Gastroenterology 2001;121:1380–90.
[25] Murphy JA, Criddle DN, Sherwood M, et al. Direct activation of cytosolic Ca^{2+} signaling and enzyme secretion by cholecystokinin in human pancreatic acinar cells. Gastroenterology 2008;135:632–41.
[26] Berna MJ, Seiz O, Nast JF, et al. CCK1 and CCK2 receptors are expressed on pancreatic stellate cells and induce collagen production. J Biol Chem 2010;285:38905–14.
[27] Phillips PA, Yang L, Shulkes A, et al. Pancreatic stellate cells produce acetylcholine and may play a role in pancreatic exocrine secretion. Proc Natl Acad Sci USA 2010;107:17397–402.
[28] Cassiman D, Roskams T. Beauty is in the eye of the beholder: emerging concepts and pitfalls in hepatic stellate cell research. J Hepatol 2002;37:527–35.
[29] Friedman SL. Molecular regulation of hepatic fibrosis, an integrated cellular response to tissue injury. J Biol Chem 2000;275:2247–50.
[30] Cassiman D, Barlow A, Vander Borght S, Libbrecht L, Pachnis V. Hepatic stellate cells do not derive from the neural crest. J Hepatol 2006;44:1098–104.
[31] Apte MV, Wilson JS. Alcohol-induced pancreatic injury. Best Pract Res Clin Gastroenterol 2003;17:593–612.
[32] Vonlaufen A, Apte MV, Imhof BA, Frossard JL. The role of inflammatory and parenchymal cells in acute pancreatitis. J Pathol 2007;213:239–48.
[33] Parlesak A. Alcohol, altered gut permeability and endotoxins. Comprehensive handbook of alcohol related pathology, vol. 2.; 2005. p. 965–75.
[34] Ammori BJ, Leeder PC, King RF, et al. Early increase in intestinal permeability in patients with severe acute pancreatitis: correlation with endotoxemia, organ failure, and mortality. J Gastrointest Surg 1999;3:252–62.
[35] Casini A, Galli A, Pignalosa P, et al. Collagen type I synthesized by pancreatic periacinar stellate cells (PSC) co-localizes with lipid peroxidation-derived aldehydes in chronic alcoholic pancreatitis. J Pathol 2000;192:81–9.
[36] Uden S, Bilton D, Nathan L, Hunt LP, Main C, Braganza JM. Antioxidant therapy for recurrent pancreatitis: placebo-controlled trial. Aliment Pharmacol Ther 1990;4:357–71.
[37] Jalleh RP, Aslam M, Williamson RC. Pancreatic tissue and ductal pressures in chronic pancreatitis. Br J Surg 1991;78:1235–7.
[38] Nomiyama Y, Tashiro M, Yamaguchi T, et al. High glucose activates rat pancreatic stellate cells through protein kinase C and p38 mitogen-activated protein kinase pathway. Pancreas 2007;34:364–72.
[39] Skipworth JR, Szabadkai G, Olde Damink SW, Leung PS, Humphries SE, Montgomery HE. Review article: pancreatic renin–angiotensin systems in health and disease. Aliment Pharmacol Ther 2011;34:840–52.
[40] Jonitz A, Fitzner B, Jaster R. Molecular determinants of the profibrogenic effects of endothelin-1 in pancreatic stellate cells. World J Gastroenterol 2009;15:4143–9.
[41] Aoki H, Ohnishi H, Hama K, et al. Cyclooxygenase-2 is required for activated pancreatic stellate cells to respond to proinflammatory cytokines. Am J Physiol Cell Physiol 2007;292:C259–68.
[42] Samkharadze T, Erkan M, Reiser-Erkan C, et al. Pigment epithelium-derived factor associates with neuropathy and fibrosis in pancreatic cancer. Am J Gastroenterol 2011;106:968–80.
[43] Masamune A, Satoh M, Hirabayashi J, Kasai K, Satoh K, Shimosegawa T. Galectin-1 induces chemokine production and proliferation in pancreatic stellate cells. Am J Physiol Gastrointest Liver Physiol 2006;290:G729–36.
[44] Rebours V, Albuquerque M, Sauvanet A, et al. Hypoxia pathways and cellular stress activate pancreatic stellate cells: development of an organotypic culture model of thick slices of normal human pancreas. PLoS One 2013;8:e76229.

[45] Masamune A, Kikuta K, Satoh M, Suzuki N, Shimosegawa T. Protease-activated receptor-2-mediated proliferation and collagen production of rat pancreatic stellate cells. J Pharmacol Exp Ther 2004;14:14.
[46] Masamune A, Kikuta K, Watanabe T, et al. Fibrinogen induces cytokine and collagen production in pancreatic stellate cells. Gut 2009;58:550–9.
[47] Apte MV, Haber PS, Darby SJ, et al. Pancreatic stellate cells are activated by proinflammatory cytokines: implications for pancreatic fibrogenesis. Gut 1999;44:534–41.
[48] Mews P, Phillips P, Fahmy R, et al. Pancreatic stellate cells respond to inflammatory cytokines: potential role in chronic pancreatitis. Gut 2002;50:535–41.
[49] Schneider E, Schmid-Kotsas A, Zhao J, et al. Identification of mediators stimulating proliferation and matrix synthesis of rat pancreatic stellate cells. Am J Physiol 2001;281:C532–43.
[50] Phillips PA, Wu MJ, Kumar RK, et al. Cell migration: a novel aspect of pancreatic stellate cell biology. Gut 2003;52:677–82.
[51] Michalski CW, Gorbachevski A, Erkan M, et al. Mononuclear cells modulate the activity of pancreatic stellate cells which in turn promote fibrosis and inflammation in chronic pancreatitis. J Transl Med 2007;5:63.
[52] Andoh A, Takaya H, Saotome T, et al. Cytokine regulation of chemokine (IL-8, MCP-1, and RANTES) gene expression in human pancreatic periacinar myofibroblasts. Gastroenterology 2000;119:211–9.
[53] Shek FW, Benyon RC, Walker FM, et al. Expression of transforming growth factor-b1 by pancreatic stellate cells and its implications for matrix secretion and turnover in chronic pancreatitis. Am J Pathol 2002;160:1787–98.
[54] Karger A, Fitzner B, Brock P, et al. Molecular insights into connective tissue growth factor action in rat pancreatic stellate cells. Cell Signal 2008;20:1865–72.
[55] Aoki H, Ohnishi H, Hama K, et al. Autocrine loop between TGF-beta1 and IL-1beta through Smad3- and ERK-dependent pathways in rat pancreatic stellate cells. Am J Physiol Cell Physiol 2006;290:C1100–8.
[56] Shinozaki S, Mashima H, Ohnishi H, Sugano K. IL-13 promotes the proliferation of rat pancreatic stellate cells through the suppression of NF-kappaB/TGF-beta1 pathway. Biochem Biophys Res Commun 2010;393:61–5.
[57] Apte MV, Phillips PA, Fahmy RG, et al. Does alcohol directly stimulate pancreatic fibrogenesis? Studies with rat pancreatic stellate cells. Gastroenterology 2000;118:780–94.
[58] Hu R, Wang YL, Edderkaoui M, Lugea A, Apte MV, Pandol SJ. Ethanol augments PDGF-induced NADPH oxidase activity and proliferation in rat pancreatic stellate cells. Pancreatology 2007;7:332–40.
[59] Vonlaufen A, Phillips P, Xu ZH, et al. Alcohol withdrawal promotes regression of pancreatic fibrosis via induction of pancreatic stellate cell (PSC apoptosis). Gut 2011;60:238–46.
[60] Masamune A, Kikuta K, Satoh M, Suzuki N, Shimosegawa T. Fatty acid ethyl esters activate activator protein-1 and mitogen-activated protein kinases in rat pancreatic stellate cells. Pancreatology 2004;4:311.
[61] Kikuta K, Masamune A, Satoh M, Suzuki N, Satoh K, Shimosegawa T. Hydrogen peroxide activates activator protein-1 and mitogen-activated protein kinases in pancreatic stellate cells. Mol Cell Biochem 2006;291:11–20.
[62] Kikuta K, Masamune A, Satoh M, Suzuki N, Shimosegawa T. 4-hydroxy-2, 3-nonenal activates activator protein-1 and mitogen-activated protein kinases in rat pancreatic stellate cells. World J Gastroenterol 2004;10:2344–51.
[63] Masamune A, Watanabe T, Kikuta K, Satoh K, Shimosegawa T. NADPH oxidase plays a crucial role in the activation of pancreatic stellate cells. Am J Physiol Gastrointest Liver Physiol 2008;294:G99–G108.
[64] Asaumi H, Watanabe S, Taguchi M, Tashiro M, Otsuki M. Externally applied pressure activates pancreatic stellate cells through the generation of intracellular reactive oxygen species. Am J Physiol Gastrointest Liver Physiol 2007;293:G972–8.
[65] Jaster R, Sparmann G, Emmrich J, Liebe S. Extracellular signal regulated kinases are key mediators of mitogenic signals in rat pancreatic stellate cells. Gut 2002;51:579–84.
[66] Masamune A, Shimosegawa T. Signal transduction in pancreatic stellate cells. J Gastroenterol 2009;44:249–60.

[67] McCarroll JA, Phillips PA, Kumar RK, et al. Pancreatic stellate cell migration: role of the phosphatidylinositol 3-kinase(PI3-kinase) pathway. Biochem Pharmacol 2004;67:1215–25.
[68] Shinozaki S, Ohnishi H, Hama K, et al. Indian hedgehog promotes the migration of rat activated pancreatic stellate cells by increasing membrane type-1 matrix metalloproteinase on the plasma membrane. J Cell Physiol 2008;216:38–46.
[69] Ohnishi H, Miyata T, Yasuda H, et al. Distinct roles of Smad2-, Smad3-, and ERK-dependent pathways in transforming growth factor-beta1 regulation of pancreatic stellate cellular functions. J Biol Chem 2004;279:8873–8. [Epub 18/12/03].
[70] Gukovskaya AS, Mouria M, Gukovsky I, et al. Ethanol metabolism and transcription factor activation in pancreatic acinar cells in rats. Gastroenterology 2002;122:106–18.
[71] McCarroll JA, Phillips PA, Park S, et al. Pancreatic stellate cell activation by ethanol and acetaldehyde: is it mediated by the mitogen-activated protein kinase signaling pathway? Pancreas 2003;27:150–60.
[72] Bhanot UK, Moller P. Mechanisms of parenchymal injury and signaling pathways in ectatic ducts of chronic pancreatitis: implications for pancreatic carcinogenesis. Lab Invest 2009;89:489–97.
[73] Masamune A, Kikuta K, Satoh M, Satoh K, Shimosegawa T. Rho kinase inhibitors block activation of pancreatic stellate cells. Br J Pharmacol 2003;140:1292–302.
[74] Hennigs JK, Seiz O, Spiro J, et al. Molecular basis of P2-receptor-mediated calcium signaling in activated pancreatic stellate cells. Pancreas 2011;40:740–6.
[75] Shen J, Wan R, Hu G, et al. miR-15b and miR-16 induce the apoptosis of rat activated pancreatic stellate cells by targeting Bcl-2 *in vitro*. Pancreatology 2012;12:91–9.
[76] Masamune A, Nakano E, Hamada S, Takikawa T, Yoshida N, Shimosegawa T. Alteration of the microRNA expression profile during the activation of pancreatic stellate cells. Scand J Gastroenterol 2014;49:323–31.
[77] McCarroll JA, Phillips PA, Santucci N, Pirola R, Wilson J, Apte M. Vitamin A induces quiescence in culture-activated pancreatic stellate cells—potential as an anti-fibrotic agent? Pancreas 2003;27:396.
[78] Masamune A, Suzuki N, Kikuta K, Satoh M, Satoh K, Shimosegawa T. Curcumin blocks activation of pancreatic stellate cells. J Cell Biochem 2006;97:1080–93.
[79] Shimizu K, Shiratori K, Kobayashi M, Kawamata H. Troglitazone inhibits the progression of chronic pancreatitis and the profibrogenic activity of pancreatic stellate cells via a PPARgamma-independent mechanism. Pancreas 2004;29:67–74.
[80] Mossner J. New advances in cell physiology and pathophysiology of the exocrine pancreas. Dig Dis 2010;28:722–8.
[81] Elsasser HP, Adler G, Kern HF. Time course and cellular source of pancreatic regeneration following acute pancreatitis in the rat. Pancreas 1986;1:421–9.
[82] Norman J. The role of cytokines in the pathogenesis of acute pancreatitis. Am J Surg 1998;175:76–83.
[83] Sparmann G, Kruse ML, Hofmeister-Mielke N, et al. Bone marrow-derived pancreatic stellate cells in rats. Cell Res 2010;20:288–98.
[84] Brizzi MF, Tarone G, Defilippi P. Extracellular matrix, integrins, and growth factors as tailors of the stem cell niche. Curr Opin Cell Biol 2012;24:645–51.
[85] Riopel MM, Li J, Liu S, Leask A, Wang R. beta1 integrin-extracellular matrix interactions are essential for maintaining exocrine pancreas architecture and function. Lab Invest 2013;93:31–40.
[86] Zimmermann A, Gloor B, Kappeler A, Uhl W, Friess H, Buchler MW. Pancreatic stellate cells contribute to regeneration early after acute necrotising pancreatitis in humans. Gut 2002;51:574–8.
[87] Apte MV, Yang L, Phillips PA, et al. Extracellular matrix composition significantly influences pancreatic stellate cell gene expression pattern: role of transgelin in PSC function. Am J Physiol Gastrointest Liver Physiol 2013;305:G408–17.
[88] Froeling FE, Feig C, Chelala C, et al. Retinoic acid-induced pancreatic stellate cell quiescence reduces paracrine Wnt-beta-catenin signaling to slow tumor progression. Gastroenterology 2011;141: 1486–97. 97 e1–14.
[89] Li L, Bachem MG, Zhou S, et al. Pancreatitis-associated protein inhibits human pancreatic stellate cell MMP-1 and -2, TIMP-1 and -2 secretion and RECK expression. Pancreatology 2009;9:99–110.
[90] Fitzner B, Muller S, Walther M, et al. Senescence determines the fate of activated rat pancreatic stellate cells. J Cell Mol Med 2012 http://dx.doi.org/10.1111/j.582-4934.2012.01573.x [Epub ahead of print].

[91] Haber P, Keogh G, Apte M, et al. Activation of pancreatic stellate cells in human and experimental pancreatic fibrosis. Am J Pathol 1999;155:1087–95.
[92] Friess H, Zhu ZW, di Mola FF, et al. Nerve growth factor and its high-affinity receptor in chronic pancreatitis. Ann Surg 1999;230:615–24.
[93] Forsmark CE. Management of chronic pancreatitis. Gastroenterology 2013;144:1282–91. e3.
[94] Emmrich J, Weber I, Sparmann GH, Liebe S. Activation of pancreatic stellate cells in experimental chronic pancreatitis in rats. Gastroenterology 2000;118:A166.
[95] Murayama KM, Barent BL, Gruber M, et al. Characterization of a novel model of pancreatic fibrosis and acinar atrophy. J Gastrointest Surg 1999;3:418–25.
[96] Ohashi K, Kim JH, Hara H, Aso R, Akimoto T, Nakama K. WBN/Kob rats. A new spontaneously occurring model of chronic pancreatitis. Int J Pancreatol 1990;6:231–47.
[97] Matsumura N, Ochi K, Ichimura M, Mizushima T, Harada H, Harada M. Study on free radicals and pancreatic fibrosis–pancreatic fibrosis induced by repeated injections of superoxide dismutase inhibitor. Pancreas 2001;22:53–7.
[98] Tsukamoto H, Towner SJ, Yu GS, French SW. Potentiation of ethanol-induced pancreatic injury by dietary fat. Induction of chronic pancreatitis by alcohol in rats. Am J Pathol 1988;131:246–57.
[99] Uesugi T, Froh M, Gabele E, et al. Contribution of angiotensin II to alcohol-induced pancreatic fibrosis in rats. J Pharmacol Exp Ther 2004;17:17.
[100] Gukovsky I, Lugea A, Shahsahebi M, et al. A rat model reproducing key pathological responses of alcoholic chronic pancreatitis. Am J Physiol Gastrointest Liver Physiol 2008;294:G68–79.
[101] Neuschwander-Tetri BA, Burton FR, Presti ME, et al. Repetitive self-limited acute pancreatitis induces pancreatic fibrogenesis in the mouse. Dig Dis Sci 2000;45:665–74.
[102] Marrache F, Tu SP, Bhagat G, et al. Overexpression of interleukin-1beta in the murine pancreas results in chronic pancreatitis. Gastroenterology 2008;135:1277–87.
[103] Blaine SA, Ray KC, Branch KM, Robinson PS, Whitehead RH, Means AL. Epidermal growth factor receptor regulates pancreatic fibrosis. Am J Physiol Gastrointest Liver Physiol 2009;297: G434–41.
[104] Bode C, Fukui H, Bode JC. Hidden endotoxin in plasma of patients with alcoholic liver disease. Eur J Gastroenterol Hepatol 1993;5:257–62.
[105] Marrache F, Pendyala S, Bhagat G, Betz KS, Song Z, Wang TC. Role of bone marrow-derived cells in experimental chronic pancreatitis. Gut 2008;57:1113–20.
[106] Watanabe T, Masamune A, Kikuta K, et al. Bone marrow contributes to the population of pancreatic stellate cells in mice. Am J Physiol Gastrointest Liver Physiol 2009;297:G1138–46.
[107] Saito R, Yamada S, Yamamoto Y, et al. Conophylline suppresses pancreatic stellate cells and improves islet fibrosis in Goto-Kakizaki rats. Endocrinology 2011;153:621–30.
[108] Kikuta K, Masamune A, Hamada S, Takikawa T, Nakano E, Shimosegawa T. Pancreatic stellate cells reduce insulin expression and induce apoptosis in pancreatic beta-cells. Biochem Biophys Res Commun 2013;433:292–7.
[109] Li F, Chen B, Li L, et al. INS-1 cells inhibit the production of extracellular matrix from pancreatic stellate cells. J Mol Histol 2013;45:321–7.
[110] Menke A, Yamaguchi H, Gress TM, Adler G. Extracellular matrix is reduced by inhibition of transforming growth factor beta1 in pancreatitis in the rat. Gastroenterology 1997;113:295–303.
[111] Su SB, Motoo Y, Xie MJ, Taga H, Sawabu N. Antifibrotic effect of the herbal medicine Saiko-keishi-to (TJ-10) on chronic pancreatitis in the WBN/Kob rat. Pancreas 2001;22:8–17.
[112] Zion O, Genin O, Kawada N, et al. Inhibition of transforming growth factor beta signaling by halofuginone as a modality for pancreas fibrosis prevention. Pancreas 2009;38:427–35.
[113] Hughes CB, Gaber LW, Mohey el-Din AB, et al. Inhibition of TNF alpha improves survival in an experimental model of acute pancreatitis. Am Surg 1996;62:8–13.
[114] Pereda J, Sabater L, Cassinello N, et al. Effect of simultaneous inhibition of TNF-alpha production and xanthine oxidase in experimental acute pancreatitis: the role of mitogen activated protein kinases. Ann Surg 2004;240:108–16.
[115] Gomez JA, Molero X, Vaquero E, Alonso A, Salas A, Malagelada JR. Vitamin E attenuates biochemical and morphological features associated with development of chronic pancreatitis. Am J Physiol Gastrointest Liver Physiol 2004;287:G162–9. [Epub 4/3/04].

[116] Vaquero EC, Rickmann M, Molero X. Tocotrienols: balancing the mitochondrial crosstalk between apoptosis and autophagy. Autophagy 2007;3:652–4.
[117] Tasci I, Deveci S, Isik AT, et al. Allopurinol in rat chronic pancreatitis: effects on pancreatic stellate cell activation. Pancreas 2007;35:366–71.
[118] Suzuki N, Masamune A, Kikuta K, Watanabe T, Satoh K, Shimosegawa T. Ellagic acid inhibits pancreatic fibrosis in male Wistar Bonn/Kobori rats. Dig Dis Sci 2009;54:802–10.
[119] Lu XL, Dong XY, Fu YB, et al. Protective effect of salvianolic acid B on chronic pancreatitis induced by trinitrobenzene sulfonic acid solution in rats. Pancreas 2009;38:71–7.
[120] Gibo J, Ito T, Kawabe K, et al. Camostat mesilate attenuates pancreatic fibrosis via inhibition of monocytes and pancreatic stellate cells activity. Lab Invest 2005;85:75–89.
[121] Kaku T, Oono T, Zhao H, et al. IS-741 attenuates local migration of monocytes and subsequent pancreatic fibrosis in experimental chronic pancreatitis induced by dibutyltin dichloride in rats. Pancreas 2007;34:299–309.
[122] Madro A, Korolczuk A, Czechowska G, et al. RAS inhibitors decrease apoptosis of acinar cells and increase elimination of pancreatic stellate cells after in the course of experimental chronic pancreatitis induced by dibutyltin dichloride. J Physiol Pharmacol 2008;59(Suppl. 2):239–49.
[123] Ishiwatari H, Sato Y, Murase K, et al. Treatment of pancreatic fibrosis with siRNA against a collagen-specific chaperone in vitamin A-coupled liposomes. Gut 2012;62:1328–39.
[124] Binkley CE, Zhang L, Greenson JK, et al. The molecular basis of pancreatic fibrosis: common stromal gene expression in chronic pancreatitis and pancreatic adenocarcinoma. Pancreas 2004;29:254–63.
[125] Duell EJ, Lucenteforte E, Olson SH, et al. Pancreatitis and pancreatic cancer risk: a pooled analysis in the International Pancreatic Cancer Case-Control Consortium (PanC4). Ann Oncol 2012;23:2964–70.
[126] Maisonneuve P, Lowenfels AB. Epidemiology of pancreatic cancer: an update. Dig Dis 2010;28:645–56.
[127] Apte MV, Park S, Phillips PA, et al. Desmoplastic reaction in pancreatic cancer: role of pancreatic stellate cells. Pancreas 2004;29:179–87.
[128] Apte MV, Wilson JS, Lugea A, Pandol SJ. A starring role for stellate cells in the pancreatic cancer microenvironment. Gastroenterology 2013;144:1210–9.
[129] Erkan M, Kleeff J, Gorbachevski A, et al. Periostin creates a tumor-supportive microenvironment in the pancreas by sustaining fibrogenic stellate cell activity. Gastroenterology 2007;132:1447–64.
[130] Apte MV, Wilson JS. Dangerous liaisons: pancreatic stellate cells and pancreatic cancer cells. J Gastroenterol Hepatol 2012;27(Suppl. 2):69–74.
[131] Kikuta K, Masamune A, Watanabe T, et al. Pancreatic stellate cells promote epithelial–mesenchymal transition in pancreatic cancer cells. Biochem Biophys Res Commun 2010;403:380–4.
[132] Pandol S, Gukovskaya A, Edderkaoui M, Dawson D, Eibl G, Lugea A. Epidemiology, risk factors, and the promotion of pancreatic cancer: role of the stellate cell. J Gastroenterol Hepatol 2012;27(Suppl. 2): 127–34.
[133] Funahashi H, Satake M, Dawson D, et al. Delayed progression of pancreatic intraepithelial neoplasia in a conditional Kras(G12D) mouse model by a selective cyclooxygenase-2 inhibitor. Cancer Res 2007;67:7068–71.
[134] Ikenaga N, Ohuchida K, Mizumoto K, et al. $CD10^+$ pancreatic stellate cells enhance the progression of pancreatic cancer. Gastroenterology 2010;139:1041–51. 51 e1–8.
[135] Hamada S, Masamune A, Takikawa T, et al. Pancreatic stellate cells enhance stem cell-like phenotypes in pancreatic cancer cells. Biochem Biophys Res Commun 2012;421:349–54.
[136] Bachem MG, Schunemann M, Ramadani M, et al. Pancreatic carcinoma cells induce fibrosis by stimulating proliferation and matrix synthesis of stellate cells. Gastroenterology 2005;128:907–21.
[137] Yoshida S, Ujiki M, Ding XZ, et al. Pancreatic stellate cells (PSCs) express cyclooxygenase-2 (COX-2) and pancreatic cancer stimulates COX-2 in PSCs. Mol Cancer 2005;4:27.
[138] Arumugam T, Brandt W, Ramachandran V, et al. Trefoil factor 1 stimulates both pancreatic cancer and stellate cells and increases metastasis. Pancreas 2011;40:815–22.
[139] Vonlaufen A, Joshi S, Qu C, et al. Pancreatic stellate cells: partners in crime with pancreatic cancer cells. Cancer Res 2008;68:2085–93.
[140] Hwang RF, Moore T, Arumugam T, et al. Cancer-associated stromal fibroblasts promote pancreatic tumor progression. Cancer Res 2008;68:918–26.

[141] Xu Z, Vonlaufen A, Phillips PA, et al. Role of pancreatic stellate cells in pancreatic cancer metastasis. Am J Pathol 2010;177:2585–96.
[142] Patel MB, Pothula SP, Xu Z, et al. The role of the hepatocyte growth factor/c-MET pathway in pancreatic stellate cell–endothelial cell interactions: antiangiogenic implications in pancreatic cancer. Carcinogenesis 2014;35(8):1891–900.
[143] Ene-Obong A, Clear AJ, Watt J, et al. Activated pancreatic stellate cells sequester $CD8^+$ T cells to reduce their infiltration of the juxtatumoral compartment of pancreatic ductal adenocarcinoma. Gastroenterology 2013.
[144] Hamada S, Masamune A, Shimosegawa T. Inflammation and pancreatic cancer: disease promoter and new therapeutic target. J Gastroenterol 2013.
[145] Ino Y, Yamazaki-Itoh R, Shimada K, et al. Immune cell infiltration as an indicator of the immune microenvironment of pancreatic cancer. Br J Cancer 2013;108:914–23.
[146] Ma Y, Hwang RF, Logsdon CD, Ullrich SE. Dynamic mast cell–stromal cell interactions promote growth of pancreatic cancer. Cancer Res 2013;73:3927–37.
[147] Mace TA, Bloomston M, Lesinski GB. Pancreatic cancer-associated stellate cells: a viable target for reducing immunosuppression in the tumor microenvironment. Oncoimmunology 2013;2:e24891.
[148] Ceyhan GO, Schafer KH, Kerscher AG, et al. Nerve growth factor and artemin are paracrine mediators of pancreatic neuropathy in pancreatic adenocarcinoma. Ann Surg 2010;251:923–31.
[149] Ceyhan GO, Bergmann F, Kadihasanoglu M, et al. Pancreatic neuropathy and neuropathic pain—a comprehensive pathomorphological study of 546 cases. Gastroenterology 2009;136:177–86. e1.
[150] Hanahan D, Weinberg RA. Hallmarks of cancer: the next generation. Cell 2011;144:646–74.
[151] Lohr M, Schmidt C, Ringel J, et al. Transforming growth factor-beta1 induces desmoplasia in an experimental model of human pancreatic carcinoma. Cancer Res 2001;61:550–5.
[152] Hwang RF, Moore TT, Hattersley MM, et al. Inhibition of the hedgehog pathway targets the tumor-associated stroma in pancreatic cancer. Mol Cancer Res 2012;10:1147–57.
[153] Duda DG, Duyverman AM, Kohno M, et al. Malignant cells facilitate lung metastasis by bringing their own soil. Proc Natl Acad Sci USA 2010;107:21677–82.
[154] Mantoni TS, Lunardi S, Al-Assar O, Masamune A, Brunner TB. Pancreatic stellate cells radioprotect pancreatic cancer cells through beta1-integrin signaling. Cancer Res 2011;71:3453–8.
[155] Guerra C, Barbacid M. Genetically engineered mouse models of pancreatic adenocarcinoma. Mol Oncol 2013;7:232–47.
[156] Bayne LJ, Beatty GL, Jhala N, et al. Tumor-derived granulocyte-macrophage colony-stimulating factor regulates myeloid inflammation and T cell immunity in pancreatic cancer. Cancer cell 2012;21:822–35.
[157] Tang D, Yuan Z, Xue X, et al. High expression of Galectin-1 in pancreatic stellate cells plays a role in the development and maintenance of an immunosuppressive microenvironment in pancreatic cancer. Int J Cancer 2012;130:2337–48.
[158] Kraman M, Bambrough PJ, Arnold JN, et al. Suppression of antitumor immunity by stromal cells expressing fibroblast activation protein-alpha. Science 2010;330:827–30.
[159] Chang DZ, Ma Y, Ji B, et al. Mast cells in tumor microenvironment promotes the *in vivo* growth of pancreatic ductal adenocarcinoma. Clin Cancer Res 2013;17:7015–23.
[160] Feldmann G, Dhara S, Fendrich V, et al. Blockade of hedgehog signaling inhibits pancreatic cancer invasion and metastases: a new paradigm for combination therapy in solid cancers. Cancer Res 2007;67:2187–96.
[161] Olive KP, Jacobetz MA, Davidson CJ, et al. Inhibition of Hedgehog signaling enhances delivery of chemotherapy in a mouse model of pancreatic cancer. Science 2009;324:1457–61.
[162] Von Hoff DD, Ramanathan RK, Borad MJ, et al. Gemcitabine plus nab-paclitaxel is an active regimen in patients with advanced pancreatic cancer: a phase I/II trial. J Clin Oncol 2011;29:4548–54.
[163] Von Hoff DD, Ervin T, Arena FP, et al. Increased survival in pancreatic cancer with nab-paclitaxel plus gemcitabine. N Engl J Med 2013;369:1691–703.
[164] Beatty GL, Chiorean EG, Fishman MP, et al. CD40 agonists alter tumor stroma and show efficacy against pancreatic carcinoma in mice and humans. Science 2011;331:1612–6.
[165] Beatty GL, Torigian DA, Chiorean EG, et al. A phase I study of an agonist CD40 monoclonal antibody (CP-870,893) in combination with gemcitabine in patients with advanced pancreatic ductal adenocarcinoma. Clin Cancer Res 2013;19:6286–95.

[166] Jacobetz MA, Chan DS, Neesse A, et al. Hyaluronan impairs vascular function and drug delivery in a mouse model of pancreatic cancer. Gut 2013;62:112–20.
[167] Provenzano PP, Cuevas C, Chang AE, Goel VK, Von Hoff DD, Hingorani SR. Enzymatic targeting of the stroma ablates physical barriers to treatment of pancreatic ductal adenocarcinoma. Cancer cell 2012;21:418–29.
[168] Masamune A, Hamada S, Kikuta K, et al. The angiotensin II type I receptor blocker olmesartan inhibits the growth of pancreatic cancer by targeting stellate cell activities in mice. Scand J Gastroenterol 2013;48:602–9.
[169] Chauhan VP, Martin JD, Liu H, et al. Angiotensin inhibition enhances drug delivery and potentiates chemotherapy by decompressing tumour blood vessels. Nat Commun 2013;4:2516.
[170] Kozono S, Ohuchida K, Eguchi D, et al. Pirfenidone inhibits pancreatic cancer desmoplasia by regulating stellate cells. Cancer Res 2013;73:2345–56.
[171] Edderkaoui M, Lugea A, Hui H, et al. Ellagic acid and embelin affect key cellular components of pancreatic adenocarcinoma, cancer, and stellate cells. Nutr Cancer 2013;65:1232–44.
[172] Rhim AD, Oberstein PE, Thomas DH, et al. Stromal elements act to restrain, rather than support, pancreatic ductal adenocarcinoma. Cancer cell 2014;25:735–47.
[173] Ozdemir BC, Pentcheva-Hoang T, Carstens JL, et al. Depletion of carcinoma-associated fibroblasts and fibrosis induces immunosuppression and accelerates pancreas cancer with reduced survival. Cancer cell 2014;25:719–34.

INDEX

Note: Page numbers followed by "*f*" and "*t*" refer to figures and tables, respectively.

A

Absent, small, or homeotic disc 1 (ASH1), 52
Acetaldehyde, 166–167
Acetaldehyde dehydrogenase (ALDH), 166
Activating transcription factor-6 (ATF6), 215–216
Activation of hepatic stellate cells, 7–9, 16–18, 164–165, 231*f*
  in vitro versus *in vivo*, 18–19
Activin, 70
Acute pancreatitis, role of pancreatic stellate cells in, 281–284, 282*f*, 283*f*, 284*f*
ADAM, 44
Adaptive immune cells
  HSC fate and matrix degradation, control of, 181
  interaction with hepatic stellate cells, 169–170
  role in modulating HSC activity and fibrogenesis, 194–195
Adenosine, 167–168
Adenosine monophosphate-activated protein kinase (AMPK), 76–77
Adenosine triphosphate (ATP), 46
Adipokines, 74–77
Adiponectin
  on HSC biology, opposing actions of, 76, 76*f*
  metabolic actions of, 75*t*
Alanine aminotransferase (ALT), 32
Alcohol dehydrogenase (ADH), 166
  in retinoid homeostasis, 167
Alcohol dehydrogenase III (ADH3), 54–55
Alcoholic liver disease (ALD), 163
  fibrosis, pathophysiology of, 163–165
Alcoholic steatohepatitis (ASH), 32, 35–36, 163
  adaptive immunity–hepatic stellate cells interaction in, 169–170
  hepatic stellate cells activation, therapeutic activities for reducing, 170
  hepatic stellate cells and, 165–166
  innate immune system–hepatic stellate cells interaction in, 168–169
Alcohol-induced liver fibrosis (ALD), 31–32
Alcohol metabolism, 167–168
Alkaloid oxymatrine (OM), 260
Angiogenesis, 64, 67–68
Angiogenic cytokines, 67–68
Angiotensin II receptor antagonists, 133, 297–298
Anti-desmin antibody, 9–10
Antigen presenting cells (APC), 93–94
Apelin, 67–68, 76–77
Apoptosis, 50
  hepatic stellate cells, 178
Arg–Gly–Asp (RGD) peptide, 19
Aspartate aminotransferase (AST), 32
Autoimmune hepatitis (AIH), 33–34
Autophagy, 47
  LPS/HSC-induced, in hepatocytes, 215–216, 217*f*

B

BAMBI, 55, 68–69
Basic fibroblast growth factor (BFGF), 112
Bile duct ligation (BDL), 16
Biliary fibrosis, 36, 37*f*
Bone morphogenetic protein, and activin membrane-bound inhibitor (BAMBI), 47
Budd–Chiari syndrome, 36

C

Cancer-associated fibroblasts (CAFs), 146–147
Cancer cell–PSC interactions, 290–292, 291*f*
Cannabinoid receptors, 254
Carbon monoxide (CO), 135
Carbon tetrachloride (CCl_4), 16, 50, 65–66, 89–90, 115–118, 148, 165–166, 187–188, 251–252, 256–257
Carriers development, 255–256
CD40 activation, using CD40 agonist monoclonal antibody, 296
Celecoxib, 65–66
Cell adhesion, 19
Cell culture-based studies, 127–128
Cellular senescence, 50–51, 92–93, 178

Chemical systems, 254–255
Chemokines, 70–72, 71*t*
 inflammatory/immunoregulatory, 232–234, 233*t*
Chemotaxis, of activated hepatic stellate cells, 48
Cholangiocytes
 interaction with hepatic stellate cells, 199
 interaction with mesenchymal cells, 200–201
Cholangiopathies
 mesenchymal cell–cholangiocytes interaction in, 200–201
Chronic heart failure, 36
Chronic liver diseases (CLDs), 35–38, 107, 114–121
 fibrosis resolution, cellular origins of, 118–119, 120*f*
 hepatic stellate cells, proangiogenic role of, 197–199
 matrix degradation pattern, changes in, 114–116, 116*f*
 MMP-mediated matrix remodeling, 117–118
Chronic pancreatitis
 pancreatic fibrosis, reversal of, 288–289
 role of pancreatic stellate cells in, 285–288
  animal models, 286–288, 287*f*
  human studies, 285
Chronic viral hepatitis, 30–32, 34
 HBV infection and, 30–31
 HCV infection and, 30–31
 HDV infection and, 30–31
 in HIV-infected patients, 31
Cirrhosis, 29
c-Jun N-terminal kinase pathways, 34–35
Collagenase, 15
Collagen proportionate area (CPA) method, 36–37
Complement activation, 169
Connective tissue growth factor (CTGF), 44, 70, 164–165
Constitutive androstane receptor (CAR), 98
Contractility, 49
Contraction, 127–129, 133
Conventional effector T cells, 236–239, 238*f*, 239*f*
CPI-17, 128
Cre-lox-system, 90–91
Culture models, of hepatic stellate cells, 6–7, 8*f*, 15
 co-culture, 20–21
 in vitro 3D culture systems, 21–22
 in vitro -versus *in vivo*-activated hepatic stellate cells, 18–19
 isolation, 15–16, 16*f*, 17*f*
 single cell culture, 17–18, 18*f*
 two-dimensional single cell culture, 19–20
Cyclic adenosine monophosphate (cAMP), 51–52, 65–66
Cytochrome P450 2E1 (CYP2E1), 45
Cytokines
 HSC-secreted, role of, 155–156
 inflammatory/immunoregulatory, 232–234
 production and signaling, in stellate cells, 63
  angiogenic cytokines, 67–68
  chemokines, 70–72, 71*t*
  interferons, 74
  interleukins, 74
  osteopontin, 77
  platelet-derived growth factor, 63–67, 64*f*
  transforming growth factor-beta, 68–70
  tumor necrosis factor superfamily, 72–73

D

Dendritic cells (DCs)
 HSC fate and matrix degradation, control of, 180
 interactions with hepatic stellate cells, 234–236, 235*f*
Deoxyribonuclease (DNase), 15
Desmin, 15–16, 89
Diethylynitrosamine (DEN), 147–148
Disability-adjusted life years (DALYs), 163
Discoidin domain receptor 2 (DDR2), 167

E

Endoplasmic reticulum (ER), 45, 165
Endothelial cell–PSC interactions, 292
Endothelial-to-mesenchymal transition (EndMT), 87–88, 99
Endothelin, 129–133
 biosynthetic pathway, 131*f*
 and smooth muscle cell signaling, 130*f*
 synthesis in normal and injured liver, 132*f*
Endothelin-1 (ET-1), 49, 67–68, 212–213
Endotoxemia, 212–217
 hepatic vasoconstriction during, 212–213
 hepatocyte injury during, 213–215
Engelbreth-Holm-Swarm murine tumor, 9
Epidermal growth factor (EGF), 46, 48, 89, 112
Epidermal growth factor receptor (EGFR), 44
Epithelial-to-mesenchymal transition (EMT), 44, 87–88, 99–100
 involvement in biliary fibrosis, 201–202
ERK5, 66–67
Erlotinib, 148
Ethanol metabolism

with hepatic stellate cells, interaction of, 166–168
pathways of, 166
Etiology-driven liver fibrosis, 34–38, 35*f*, 37*f*
Extracellular matrix (ECM), 19–21, 29, 34
alterations between normal and fibrotic liver, 42, 43*f*
cellular sources of, 44
changes in liver fibrosis, 108–109, 109*f*
composition, 42–44
deposition and degradation, 150–152
homeostasis, 42–44
matrix turnover, 47
Extracellular signal-regulated kinase (ERK), 65–68

F

Farnesoid X receptor (FXR), 54, 98
Fat-storing cells, 1–3
Fibrogenesis, 47
Fibrosis
in alcoholic liver disease, pathophysiology of, 163–165
definition of, 41
hepatic. *See* Hepatic fibrosis
and portal hypertension, 138
regression
macrophages, role of, 189–190
reversibility, 176–177
Fibrosis resolution
HSC fate and matrix degradation, control of, 179–182
by immune cells, 179–182
liver, sinusoidal endothelial cells, 182
HSC clearance during, mechanisms of, 176*f*, 177–179
Fluorescence-activated cell sorting (FACS) analysis, 15
Follistatin, 70
Free fatty acids (FFAs), 36, 45, 53–54

G

Gelatinase A, 34–35
Genetic systems, 262–265
Ghrelin, 77
Girdin. *See* Gα-Interacting Vesicle (GIV)
Glial fibrillary acidic protein (GFAP), 9–10, 15–16, 89–90, 210
Granulocyte/macrophage-colony stimulating factor (GM-CSF), 187
Gα-Interacting Vesicle (GIV), 51–52

H

Health, pancreatic stellate cells in, 274–277
CCK-induced pancreatic exocrine secretion, role in, 276
immune function, 275–276
origin of PSCs, 276
progenitor function, 275
Hedgehog pathway, inhibition of, 296
Hematoxylin and eosin (H&E), 272*f*
Hepatic fibrosis, 29, 41
activation
new and emerging pathways of, 51–55
alcohol-induced, 31–32
autoimmune hepatitis, 33
chronic viral hepatitis, 30–32
ECM changes in, 108–109, 109*f*
etiology-driven, 34–38, 35*f*, 37*f*
hepatic stellate cells and, 6, 45–51, 89–97
initiation, 45–46
perpetuation, 46–49
regression, 49–51
and hepatocellular carcinoma, 145
nonalcoholic fatty liver disease, 32–33
nonalcoholic steatohepatitis, 32–33
pathogenesis of, 42–45
ECM cellular sources, 44
ECM composition and matrix homeostasis, 42–44
pathological and clinical associations, 42, 43*f*
primary biliary cirrhosis, 29, 33
primary sclerosing cholangitis, 33
reversibility of, 90–91
Hepatic immunological tolerance, stellate cells in, 227
conventional effector T cells, 236–239, 238*f*, 239*f*
DCs–HSCs interaction, 234–236, 235*f*
hepatic immune system, 227–228
inflammatory/immunoregulatory chemokines, 232–234, 233*t*
inflammatory/immunoregulatory cytokines, 232–234
liver tolerogenicity, 228–230
NKT cells, 242
regulatory T cells, 239–241, 241*f*
transplantation, 230–232
Hepatic myofibroblasts (HMs)
depletion in liver regeneration and cancer, role of, 265–266
immune cell mediated clearance of, 252–253

Hepatic myofibroblasts (HMs) (*Continued*)
intrinsic signaling events limiting scar cell survival, 253–254
targeted delivery vehicles, advantages of, 254–255
Hepatic stellate cells (HSCs), 2*f*, 185–186
activation of, 7–9, 16–18, 164–165, 231*f*
and alcoholic steatohepatitis, 165–166
apoptosis, 178
and cancer promotion, 147–150
culture models, 6–7, 8*f*, 15
cytokine production and signaling in, 63
depletion model, 217–219
discovery of, 1–3
fibrogenesis, vascular mediators and, 138–139
fibrotic response, 107–108
and hepatic fibrosis, 45–51
initiation, 45–46
perpetuation, 46–49
regression, 49–51
in hepatocyte injury during endotoxemia, 212–217
immune surveillance decreased by, 152
inactivation of, 91–92
interaction with ethanol metabolism, 166–168
interaction with innate immune system, 168–169
interaction with non-parenchymal cells, 185
isolation of, 6–7, 8*f*, 15–16, 16*f*, 17*f*
and liver fibrosis, 6, 89–97
and liver regeneration, 211–212
markers for, 9–10, 10*f*
morphological characteristics of, 4–5, 4*f*, 5*f*
perspective of, 11
phenotypes and liver injury, 210
phenotype to an inactivated state, reversion of, 178–179
and portal hypertension, 136–138, 137*f*
as primary target of anti-fibrotic therapy, 89–90
proangiogenic role in chronic liver disease, 197–199
in promoting HCC development, functional role of, 153*t*–154*t*
quiescent, 89
as regulators of hepatocellular damage, 94–95
as regulatory bystanders, 93–94
role in fibrosis progression during ALD pathogenesis, 164–165
role in inflammation and immune regulation, 232–234
role in liver regeneration, 96
-secreted cytokines, role of, 155–156
senescence, 178
senescent HSCs, distinct functions of, 92–93
as target for fibrosis/cirrhosis reversibility, 175
transdifferentiation of, 7–9, 17–19
tumor angiogenesis promoted by, 155
as tumor microenvironment component, 149–150
tumor-promoting role in premaligant microenvironment, 147–148
vitamin A homeostasis by, 3–4
Hepatic vasoconstriction, 212–213
Hepatitis B virus (HBV) infection, 29–30
and chronic viral hepatitis, 30–31
Hepatitis C virus (HCV) infection, 29–30
and chronic viral hepatitis, 30–31
and etiology-driven liver fibrosis, 34–35
Hepatitis D virus (HDV) infection
and chronic viral hepatitis, 30–31
Hepatocellular carcinoma (HCC), 29, 42
ECM deposition and degradation, 150–152
fibrosis and, 145
hepatic stellate cells co-cultures with, 20–21
premaligant microenvironment, 145–146
tumor microenvironment, 146–147
Hepatocyte growth factor (HGF), 46
Hepatocytes
hepatic stellate cells co-cultures with, 20–21
injury, during endotoxemia, 213–215
LPS/HSC-induced autophagy in, 215–216, 217*f*
survival, 209
Herpes simplex virus thymidine kinase gene (HSV-Tk), 89–90
Herpesvirus saimari (HVS), 262
HIV-infected patients, chronic viral hepatitis in, 31
HIV-1 infection, and etiology-driven liver fibrosis, 34–35
Human antigen R (HuR), 65
Human leukocyte antigen (HLA), 33
4-Hydroxynonenal (4-HNE), 35–36, 67, 166–167
5-Hydroxytryptamine (5-HT), 96
5-Hydroxytryptamine 2B receptor (5-HT2B), 96
Hypoxia, ethanol-induced, 167–168
Hypoxia-inducible factor 1α (HIF-1α), 48, 75–76, 167

I

Immune cell-mediated allograft rejection, 228*f*
Immune cell–PSC interactions, 292–293
Immune cells
HSC fate and matrix degradation, control of, 179–182
role in modulating HSC activity and fibrogenesis, 190–196, 191*f*
Immune surveillance, decreased by hepatic stellate cells, 152
Immune system, 227–228
Immunoregulation, 49
Inactivated phenotype, reversion of activated HSCs to, 51
Inactivation of hepatic stellate cells (iHSCs), 91–92
Inflammasomes, 169
Inflammatory bowel disease (IBD), 33
Innate immune system
HSC fate and matrix degradation, control of, 179–181
interaction with hepatic stellate cells, 168–169
Inositol-requiring enzyme 1 (IRE1), 215–216
Inositol requiring enzyme 1 alpha (IRE1α), 53–54
Insulin-like growth factor-1 (IGF-1), 66
Integrins, 44
Intercellular cell adhesion molecule type 1 (ICAM-1), 34–35, 72–73
Interferons, 74
Interferon-γ (IFN-γ), 74
Interleukin-1 (IL-1), 74, 112
Interleukin-1β (IL-1β), 89, 178
Interleukin-6 (IL-6), 74, 89, 93, 107–108, 165
and etiology-driven liver fibrosis, 34–35
Interleukin-8 (IL-8), 165
Interleukin-10 (IL-10), 74
Interleukin-13 (IL-13), 98
Interleukin-17 (IL-17), 74
Interleukin 17A (IL-17A), 47, 89
Interleukin 22 (IL-22), 50–51
Interleukin-25 (IL-25), 98
Interleukins, 74
Intracellular retinol-binding protein, 3–4
In vitro 3D culture systems, 21–22
In vitro-versus *in vivo*-activated hepatic stellate cells, 18–19
In vivo studies, 129
IQ motif containing GTPase activating protein 1 (IQGAP1), 68–69
Isolation of hepatic stellate cells, 6–7, 8*f*, 15–16, 16*f*, 17*f*
Ito cells, 1–3

K

Kupffer cells (KC), 1–7, 4*f*, 5*f*, 9–10
in hepatic inflammation and fibrogenesis, 187–188
hepatic stellate cells co-cultures with, 20–21
interaction with hepatic stellate cells, 186–187, 188*f*

L

Lecithin-retinol acyltransferase (LRAT), 46–47, 94–95
Leptin
on HSC biology, opposing actions of, 75–76, 76*f*
metabolic actions of, 74–75, 75*t*
Lipopolysaccharide (LPS), 186
-induced gene transcripts in HSCs, alterations in, 215*t*
role in HSCs–innate immune system interaction, 168
stimulated HSCs, 215–216
Liver fibrosis. *See* Hepatic fibrosis
Liver macrophages–hepatic stellate cells interaction, 186–187, 188*f*
Liver regeneration (LR)
hepatic stellate cells and, 211–212
HSCs role in, 96
Liver sinusoidal endothelial cells (LSECs), 15
hepatic stellate cells co-cultures with, 20–21
HSC fate and matrix degradation, control of, 182
interaction with hepatic stellate cells, 196–197
Liver tolerogenicity, 228–230
Liver X receptor (LXR), 54

M

Macrophages, 111–112, 118–119, 120*f*
heterogeneity, 188–189
liver macrophages–hepatic stellate cells interaction, 186–187, 188*f*
monocyte-derived
HSC fate and matrix degradation, control of, 179–180
restorative, 180, 182
role in fibrosis regression, 189–190

Malondialdehyde (MDA), 166–167
Malondialdehyde–acetaldehyde (MAA), 166–167
Matrix metallopeptidase-2 (MMP-2), 34–35
Matrix metalloproteases (MMPs), 42–44, 47, 107
  classification of, 110*t*
  fibrosis resolution, cellular origins of, 118–119, 120*f*
  in liver, cellular sources of, 111–112
  matrix degradation pattern, changes in, 114–116, 116*f*
  -mediated matrix remodeling, 117–118
  pro-enzyme activation, 112–113
  structure of, 111, 111*f*
  substrate profile, 110
  transcriptional regulation of, 112
Metalloproteinases (MMPs), 177
Methyl-CpG binding protein 2 (MeCP2), 52
Microcirculation, 129
Micro RNAs (miRNAs), 52, 66, 68
Microsomal ethanol oxidizing system (MEOS), 166
MIP-2, 165
miR-133a, 68–69
miR-29b, 66
Mitogen-activated protein kinase (MAPK), 151
MLC phosphatase targeting subunit 1 (MYPT1), 128
Monocyte chemotactic protein 1 (MCP-1), 48, 165
  and etiology-driven liver fibrosis, 34–35
Morphological characteristics, of hepatic stellate cells, 4–5, 4*f*, 5*f*
Myocardin-related transcription factor A (MRTF-A), 52
Myofibroblasts, 44
  as primary target of anti-fibrotic therapy, 89–90
  sources of, 88*f*
Myosin light chain kinase (MLCK), 128
Myristoylated alanine rich protein kinase C substrate (MARCKS), 66–67

N

Nanoparticles, 260–262
Natural killer (NK) cells
  HSC fate and matrix degradation, control of, 181
  role in modulating HSC activity and fibrogenesis, 192–194
Natural killer T (NKT) cells
  in hepatic immunological tolerance, 242
  HSC fate and matrix degradation, control of, 181
  role in modulating HSC activity and fibrogenesis, 192–194
Nerve growth factor (NGF) receptor p75, 15–16
Nestin, 15–16
Neural cell–PSC interactions, 293
NF-κB, 34–35, 47, 50, 72–73, 178, 255
Nicotinamide adenine dinucleotide phosphate (NADPH), 48
Nitric oxide (NO), 134
  signaling, in sinusoidal endothelial cells, 135*f*
Nitric oxide synthase (NOS)
  constitutive, 134
  inducible, 134
Nonalcoholic fatty liver disease (NAFLD), 29–30, 32–33, 35–36, 45, 54–55
Nonalcoholic steatohepatitis (NASH), 32–33, 35–38, 54–55
Non-parenchymal cells–hepatic stellate cells interaction, 185
  adaptive immune cells, 194–195
  cholangiocytes, 199
  immune cells, 190–196, 191*f*
  macrophages, 186–190
    heterogeneity, 188–189
    liver macrophages, 186–187, 188*f*
    role in fibrosis regression, 189–190
  NK cells, 192–194
  NKT cells, 192–194
  pericytes, 196–197
  portal fibroblasts, 200
  proangiogenic role in chronic liver disease, 197–199
Nonstructural protein 3 (NS3) protein, 34–35

O

Osteopontin (OPN), 77
Oxidative stress, 166–167

P

p53 upregulated modulator of apoptosis (PUMA), 167
p75, 73, 89
Pancreatic cancer, role of pancreatic stellate cells in, 289–298, 290*f*
  in vitro studies, 290–293, 291*f*
  in vivo studies, 293–295
  stroma, therapeutic targeting of, 295–298
Pancreatic diabetes, role of pancreatic stellate cells in, 285–288

Pancreatic disease, role of pancreatic stellate cells in, 277–281
activation, 277–279, 277*t*
signaling pathways, 279–281, 280*t*
Pancreatic fibrosis, reversal of, 288–289
Pancreatic stellate cells (PSCs), 271
in acute pancreatitis, 281–284, 282*f*, 283*f*, 284*f*
characterization of, 272–277
in chronic pancreatitis, 285–288
animal models, 286–288, 287*f*
human studies, 285
pancreatic fibrosis, reversal of, 288–289
in health, 274–277
CCK-induced pancreatic exocrine secretion, role in, 276
immune function, 275–276
origin of PSCs, 276
progenitor function, 275
isolation of, 273–274
in pancreatic cancer, 289–298, 290*f*
in vitro studies, 290–293, 291*f*
in vivo studies, 293–295
stoma, therapeutic targeting of, 295–298
in pancreatic diabetes, 285–288
in pancreatic disease, 277–281
activation, 277–279, 277*t*
signaling pathways, 279–281, 280*t*
Patatin-like phosphatase domain-containing 3 (PNPLA3), 54–55
PEGylated human recombinant PH20 hyaluronidase (PEGPH20), 297
Pentoxifylline, 65–66
Pericytes, 125–126, 126*f*
role in modulating HSC activity and fibrogenesis, 196–197
Peroxisome proliferator-activated receptor gamma (PPARγ), 47, 54–55, 73, 89
Phosphatidic acid, 65
Phosphoinositide 3 kinase (PI3K), 66, 151
Placental growth factor (PIGF), 67–68
Plasminogen activator inhibitor (PAI), 68
Platelet-derived growth factor (PDGF), 9, 46, 48–49, 52–53, 63–67, 89, 112, 148, 164–165, 210
isoform–receptor interaction, 64*f*
Porcine liver, stellate cells in, 2*f*
Portal fibroblasts (PFs), 87–90, 97–99
characterization of, 97–98
fibrosis and, 138
interaction with hepatic stellate cells, 200
stellate cells and, 136–138, 137*f*
therapeutic implications, 139
Precision-cut liver slices (PCLS), 21–22
Pregnane X receptor (PXR), 98
Premaligant microenvironment (PME), 145–146
ECM deposition and degradation in, 150–152
hepatic stellate cells, tumor-promoting role of, 147–148
Primary biliary cirrhosis (PBC), 29, 33, 36, 37*f*
Primary sclerosing cholangitis (PSC), 33, 36, 37*f*
Pro-collagen 1α protein, 167
Pro-enzyme activation, 112–113
Pronase, 15
Prostaglandin E2 (PGE2), 65–66
Prostaglandins, 135–136
Protein kinase-like ER kinase (PERK), 215–216

Q

Quiescent hepatic stellate cells (qHSCs), 89, 251

R

Rab11, 68–69
Ras-MAPK and phosphoinositide 3-kinase (PI3K) pathway, 48–49
Reactive oxygen species (ROS), 34–35, 45, 48
Regulatory T cells, 239–241, 241*f*
Renin–angiotensin system (RAS), 254–255
Resistin, 76–77
Retinoic acid early inducible 1 (RAE1), 181
Retinoic acid receptor (RAR), 4–5
Retinoids, 3–4, 6–11
homeostasis, alcohol dehydrogenase in, 167
loss, 46–47
Retinoid X receptor (RXR), 4–5
Rolipram, 65–66
Rough endoplasmic reticulum (rER), 87–88

S

Scar-in-a-Jar model, 22
Senescence, 50–51, 178
Senescence-associated secretory phenotype (SASP), 93
Senescent hepatic stellate cells, distinct functions of, 92–93
Serotonin. *See* 5-Hydroxytryptamine (5-HT)
Signal transducer and activator of transcription 3 (STAT3), 51–52
Single cell culture, 17–18, 18*f*
two-dimensional, 19–20

Sinusoidal anatomy, 126*f*
Sinusoidal capillarization, 108–109
Sinusoidal endothelial cells, 4–5, 45–46
  nitric oxide singnaling in, 135*f*
Smad2, 69
Smad3, 69
Small heterodimer partner (SHP), 98
α-Smooth muscle actin (α-SMA), 9–10, 15–16, 89, 145–146
STAT-1, 65
Stellate cell contractile signaling, 128*f*
Stellate cell depletion models, 251
  carriers development, 255–256
  chemical systems, 254–255
  genetic systems, 262–265
  hepatic myofibroblasts
    depletion in liver regeneration and cancer, role of, 265–266
    immune cell mediated clearance of, 252–253
    intrinsic signaling events limiting scar cell survival, 253–254
    targeted delivery vehicles, advantages of, 254–255
  nanoparticles and viral delivery systems, 260–262
Stellate endothelial cells, 1–3
Stroma in pancreatic cancer, therapeutic targeting of, 295–298
  angiotensin II receptor antagonists, 297–298
  CD40 activation, using CD40 agonist monoclonal antibody, 296
  Hedgehog pathway, inhibition of, 296
  PEGylated human recombinant PH20 hyaluronidase, 297
Suppressor of cytokine signaling 1 (SOCS 1), 167
Synaptophysin, 15–16, 89
Synemin, 89

T

Thioacetamide (TAA), 94–95
Tissue inhibitor of matrix metalloproteinase-1 (TIMP-1), 89, 111, 113–118, 120*f*, 177–178
Tissue inhibitors of metalloproteinase-2 (TIMP-2), 113, 178
Tissue inhibitors of metalloproteinases (TIMPs), 42–44, 47–49, 68, 111–114
  activation of, 113–114
  classification of, 113
  regulation of, 113–114
  target definition, 113
Toll-like receptor 4 (TLR4), 55, 187
Toll-like receptors (TLRs)
  role in HSCs–innate immune system interaction, 168
Transdifferentiation of hepatic stellate cells, 7–9, 17–19
Transforming growth factor-beta (TGF-β), 42–47, 51–52, 55, 68–70, 164–165
Transforming growth factor-β1 (TGF-β1), 9, 87–89, 93–94, 96, 98
Transplantation, importance of HSCs in, 230–232
Tumor-associated macrophages (TAMs), 146
Tumor microenvironment (TME), 146–147
  ECM deposition and degradation in, 150–152
  hepatic stellate cells as component of, 149–150
  tumor angiogenesis promoted by, 155
Tumor necrosis factor (TNF)
  superfamily, 72–73
Tumor necrosis factor-related apoptosis inducing ligand (TRAIL), 45, 73, 180
Tumor necrosis factor-α (TNF-α), 48–49, 112, 165, 178
Two-dimensional single cell culture, 19–20

U

Unfolded protein response (UPR), 47, 53–54

V

Vascular endothelial growth factor (VEGF), 45–46, 48–49, 67, 107–108, 182
Vascular mediators, and stellate cell fibrogenesis, 138–139
Vasodilator-stimulated phosphoprotein (VASP), 68–69
Vimentin, 15–16
Viral delivery systems, 260–262
Vitamin A homeostasis
  by hepatic stellate cells, 3–4
Vitamin D receptor (VDR), 98
von Kupffer, Karl Wilhelm, 1–3, 2*f*

W

Wound healing, 107

X

X-box transcription factor 1 (XBP1s), 53–54

Kolibri 360 GmbH
Ferdinand-Jühlke-Straße 7
99095 Erfurt, Deutschland
produktsicherheit@kolibri360.de